OT 48
Operator Theory: Advances and Applications
Vol. 48

Birkhäuser Verlag
Basel · Boston · Berlin

Topics in Operator Theory

Ernst D. Hellinger Memorial Volume

Volume Editors

L. de Branges
I. Gohberg
J. Rovnyak

1990

Birkhäuser Verlag
Basel · Boston · Berlin

Volume Editorial Office:

Raymond and Beverly Sackler Faculty of Exact Sciences
School of Mathematical Sciences
Tel-Aviv University
69978 Tel-Aviv
Israel

Library of Congress Cataloging-in-Publication Data

Topics in operator theory: Ernst D. Hellinger memorial volume /
volume editors, L. de Branges, I. Gohberg, J. Rovnyak.
p. cm. – (Operator theory, advances and applications ; vol. 48)
Includes bibliographical references.
ISBN 3-7643-2532-1 – ISBN 0-8176-2532-1 (U.S.)
1. Operator theory. 2. Hellinger, Ernst, b. 1883. I. Hellinger,
Ernst, b. 1883. II. De Branges, Louis. III. Gohberg, I. (Israel),
1928– IV. Rovnyak, James. V. Series: Operator theory, advances
and applications ; v. 48.
QA329.T643 1990
515'. 724 – dc20

Deutsche Bibliothek Cataloging-in-Publication Data

Topics in operator theory: Ernst D. Hellinger memorial
volume / vol. ed. L. de Branges ... – Basel ; Boston ; Berlin :
Birkhäuser, 1990
(Operator theory ; Vol. 48)
ISBN 3-7643-2532-1
NE: DeBranges, Louis [Hrsg.]; GT

Printed in Germany on acid-free paper
ISBN 3-7643-2532-1
ISBN 0-8176-2532-1

DEDICATED TO THE MEMORY OF ERNST DAVID HELLINGER

Ernst Hellinger 1883-1950
Portrait made in Evanston, Illinois, Spring 1949

TABLE OF CONTENTS

Operator Theory:
Advances and Applications, Vol. 48

Ernst David Hellinger 1883–1950: Göttingen, Frankfurt Idyll, and the New World

JAMES ROVNYAK

Ernst Hellinger was a student of Hilbert and a pioneer in operator theory and modern analysis. His dissertation is an early landmark of spectral theory, and together with Toeplitz he produced some of the first results of operator theory. His work also influenced the theory of Jacobi matrices, continued fractions, moment problems, and integral equations, as well as integration theory and its applications. He is remembered for his wit, strength of personality, and uncommon ability to make friends.

Breslau 1883–1904

The world in which Ernst Hellinger was born and reared no longer exists. This was German Silesia, which is today part of Poland. Hellinger was born September 30, 1883, in Striegau, the son of Emil and Julie Hellinger. His only sibling was a twelve-year younger sister, Hanna. Emil Hellinger was a businessman who dealt in materials used to manufacture paper. The family moved to Breslau, today Wrocław, when Ernst was a child.

Hellinger's mathematical talent was identified by a gifted teacher, Prof. Dr. Th. Maschke, at the König-Wilhelms-Gymnasium in Breslau. Maschke's pupils, beside Hellinger, included Max Born (1882–1970), Richard Courant (1888–1972), Heinz Hopf (1894–1971), Otto Stern (1888–1969), and Wolfgang Sternberg (1887–1953). Born described Maschke in his autobiography[1]: he was "a small, sturdy man with a kindly round face and a big moustache. His way of teaching mathematics was different from that of all my previous teachers and was quite modern. He did not state theorems and give demonstrations, but he told us something about the meaning of the whole theory to which a theorem belonged, about its history and its practical applications." Constance Reid[2] writes that Maschke "was an exponent of the 'Socratic method' of leading students to discovery on their own ... he loved

Emil and Julichen Hellinger

science and had the ability to inspire scientific excitement in his pupils." It was Maschke who inspired Hellinger to choose mathematics as a career. Emil Hellinger did not immediately take to the idea, but Maschke spoke with him and convinced him that Ernst should study mathematics at the university.

Hellinger graduated from the König-Wilhelms-Gymnasium in the spring of 1902. He attended the university in Heidelberg in the summer of 1902, hearing lectures by K. Fischer, Koenigsberger, Landsberg, and Valentiner. He returned to Breslau in the same year, and at the university in Breslau he attended lectures by Abegg, Ebbinghaus, Franz, London, O. E. Meyer, E. R. Neumann, Rosanes, Cl. Schäfer, and Sturm. Of these early teachers, Koenigsberger and Rosanes made the deepest impression on Hellinger.

Among the outstanding students living in Breslau at the turn of the century was Otto Toeplitz (1881–1940)[3]. Toeplitz was born in Breslau, but he attended a different secondary school, the Johannes-Gymnasium, where his father and grandfather were mathematics teachers. Hellinger and Toeplitz were best friends. The two-year older Toeplitz had a missionary's zeal for mathematics, and he communicated his enthusiasm to Hellinger. They studied together, and Hellinger's mother, seeing the pair absorbed in mathematics, was reminded of her grandfather, who was a Talmud scholar. Hellinger also knew Born and Courant in Breslau, though not as well. The four friends—Hellinger, Toeplitz, Born, and Courant—went to Göttingen for their "serious" studies.

Göttingen and Marburg 1904–1914

The atmosphere of Göttingen at the beginning of the twentieth century is one of the legends of mathematics. The stars of the faculty were David Hilbert (1862–1943), Felix Klein (1849–1925), and Hermann Minkowski (1864–1909). They attracted a remarkable group of students and young assistants, including Born, Courant, Hellinger, and Toeplitz, and many others such as Alfred Haar (1885–1933), Eduard Helly (1884–1943), Erhard Schmidt (1876–1959), and Hermann Weyl (1885–1955), to name a few. The academic and social climate of Göttingen at this time was immensely exciting to those who participated[4].

Hellinger, Toeplitz, Born, and Courant were known as the "Breslau group" in Göttingen, and all were accepted into Hilbert's circle. Born came first in 1904 on the advice of Toeplitz. Born's stepmother gave him a letter of introduction to Minkowski, and he met Hilbert through Minkowski. Born was Hilbert's assistant in 1904–1905. Hellinger arrived in Göttingen later in 1904. He was Hilbert's assistant in 1905–1907. Toeplitz received his doctorate in Breslau in 1905, but he had been aiming for Göttingen all along and evidently arrived in 1905. Toeplitz took his *Habilitation* in Göttingen in 1907 and was *Privatdozent* there. The younger Courant came to Göttingen last, in 1907. Hilbert was by then impressed by the students from Breslau and took Courant as his assistant in 1908–1910, replacing Haar, who had the job in 1907–1908.

As a student, Hellinger participated in the Hilbert–Minkowski seminar on mathematical physics. He heard lectures by Carathéodory, Herglotz, Hilbert, Husserl, Klein, Minkowski, Prandtl, C. Runge, Voigt, Wiechert, and Zermelo. Beginning already in his first semester in Göttingen and continuing through his term as Hilbert's assistant, Hellinger prepared *Lesezimmer* notes of Hilbert's lectures. In later years, Hellinger's notes of Hilbert's lectures were evidently much copied and circulated widely. They cover a wide range of topics and include:

Wintersemester 1904/1905
Variationsrechnung

Sommersemester 1905
Einführung in die Theorie der Integralgleichungen

Wintersemester 1905/1906
Mechanik
Theorie der quadratischen Formen von unendlich vielen Variablen

Sommersemester 1906
Integralgleichungen

Wintersemester 1906/1907
Mechanik der Kontinua

Hellinger gave up his assistantship with Hilbert in 1907 in order to concentrate on his dissertation, which he completed that year under Hilbert's direction. His doctoral examination took place on July 17, 1907.

Although Hellinger was a Hilbert protégé, he also had an important association with Felix Klein. Klein's most vigorous research years were behind him, but he remained the pre-eminent authority on nineteenth-century mathematics, and he was a powerful figure in the German academic world. Hellinger was Klein's assistant in 1907–1909. Max Dehn (1878–1952), another student of Hilbert, provides us with a picture of these two mentors, in remarks that he made on the occasion of Hellinger's retirement from Northwestern University. An outline of Dehn's remarks survives[5], and his main points are easily reconstructed:

> Hilbert was a person of ruthless clarity, incomparable, with indefatigable strength, together with a deep feeling for the moral obligation in regard to scientific work. The mathematician has not the right to choose the problems. Perhaps they are hard to be worked out. Science—like a goddess—sets them, and the mathematician has to labor long years to try to master them, trying one way or another to overcome the difficulties, but optimistic that the problem given by the goddess will be solved, perhaps not by oneself, but what does that matter! In contrast, Klein was like a man on a high mountain top with far-seeing eyes. Klein saw the entire mathematical landscape as a unit and sought to discover its main principle of order. Through Hilbert and Klein, Hellinger was connected with the great achievements in mathematics at the turn of the century. He was very lucky to find such teachers, but he possessed unique talents that made these contacts bear fruit.

Hellinger left Göttingen in 1909 for Marburg, where he remained until 1914. He took his *Habilitation* in Marburg in 1909 and became *Privatdozent.* Of Hellinger's Marburg years, Dehn said that they were happy years, which he shared with friends having interests in theology, archeology, philosophy, and geography. Dehn added that these friends had a great importance for Hellinger's ideas about value in human life.

Hellinger's Mathematical Work in Göttingen and Marburg

Hellinger was among the first to demonstrate the potential of Hilbert's program in analysis, and he championed Hilbert's point of view throughout his career. He was co-author with Toeplitz of some of the first papers of operator theory. In his dissertation, he introduced what has come to be called the Hellinger integral, and he created multiplicity theory and used this to exhibit a complete set of unitary invariants for any selfadjoint operator. Along with others such as Hermann Weyl, Hellinger was an early pioneer in the applications of operator theory in analysis, particularly in the area of the problem of moments, continued fractions, and related topics. The rich interaction between abstract and classical analysis, which today we call modern analysis, had its origins in this period. This was the essence of Hilbert's program in analysis, and Hellinger played an important role in setting it on course.

To understand Hellinger's contributions, it is therefore necessary to begin with Hilbert. Hilbert described his program in analysis in a paper which was intended for a lecture at the International Congress in Rome in 1908[6]. He said that, like Siegfried before the magic fire, he sought the beautiful prize of a unified method in algebra and analysis (*einer methodisch-einheitlichen Gestaltung von Algebra und Analysis*) for treating problems in which a function is the unknown quantity in an integral, differential, or functional relationship. Hilbert had begun with a study of integral equations, but since a function could be replaced by its sequence of Fourier coefficients, the work soon evolved into a theory of quadratic forms and linear equations in infinitely many unknowns. In this way, Hilbert set up a powerful analogy between analysis and linear algebra. The goal of his program was to treat problems of analysis: integral and differential equations, continued fractions, moment problems, complex analysis, and applications in mathematical physics. Hellinger told this anecdote about Hilbert's lectures[7]. After proving that the eigenvalues of a selfadjoint operator are real, Hilbert said: *Damit, meine Herren, werden wir die Riemannsche Vermutung beweisen* ("With that, gentlemen, we shall prove the Riemann hypothesis"). There is more to the story. Israel Gohberg remembers being told, by someone who knew Erhard Schmidt, that Hilbert suggested the Riemann hypothesis to Schmidt as a dissertation topic. He was to prove it by constructing a selfadjoint operator H such that the determinant of $\frac{1}{2} + iH$ is equal to the zeta function. Schmidt, of course, did not succeed but wrote a dissertation with Hilbert in 1905 on a different topic that grew out of the effort. Hilbert published the main part of his work in the area in a series of papers beginning in the same year. The papers were reprinted in book form under the title *Grundzüge einer allgemeinen Theorie der linearen Integralgleichungen*[8].

The applications envisioned by Hilbert were not necessarily to be found on the surface, however. Hilbert said in his 1908 Rome paper that the underlying theory of equations in infinitely many unknowns had first to be worked out. Of central importance was the case of linear equations, which was treated in depth in Hilbert's *Grundzüge.* Hellinger and Toeplitz also took up the challenge in joint work [**1,4**]. The idea was to construct a calculus of bounded infinite matrices parallel to the matrix calculus of linear algebra. The paper [**1**], published in 1906, is a tentative version of the theory. The full treatment appears four years later in the well-known Hellinger and Toeplitz paper [**4**] in the *Mathematische Annalen.* This paper is an early embodiment of the spirit of operator theory. In it, Hellinger and Toeplitz introduce the concepts of equivalence, congruence, similarity, and unitary equivalence for bounded infinite matrices. Köthe[9] has written a critical analysis of this paper and discussed its importance, and we shall not cover the same ground. It is worth noting one result from [**4**], which is still cited today. It is a forerunner of the closed graph theorem, which states, in modern language[10]:

HELLINGER–TOEPLITZ THEOREM. *An everywhere defined symmetric operator on a Hilbert space is continuous.*

To Hilbert, the theory of infinitely many equations in infinitely many unknowns meant, above all, the spectral theory of quadratic forms. Hellinger's most penetrating work as a young mathematician was in the area of spectral theory. Again, we must first understand Hilbert's view in order to appreciate Hellinger's contributions. Hilbert proved the first version of the spectral theorem in his *Grundzüge.* Consider a bounded symmetric quadratic form

$$K(x) = \sum_{p,q} a_{pq} x_p x_q$$

on square summable sequences $x = (x_1, x_2, x_3, \ldots)$ of real numbers. Hilbert proved the existence of an associated *Spektralform,*

$$\sigma(\mu; x) = \sum_{p,q} \sigma_{pq}(\mu) x_p x_q,$$

which in modern terms is the quadratic form associated with a resolution of the identity. The spectral theorem takes the form of a Stieltjes integral representation

$$K(x) = \int \mu \, d\sigma(\mu; x)$$

of the quadratic form in terms of the *Spektralform.* Integration is over an interval containing the spectrum of the form. Hilbert used the term *Spektrum* thinking of applications in the theory of oscillations ([**11**], p. 136).

In the finite-dimensional case, the spectral theorem is equivalent to a representation of the quadratic form as a sum of squares

$$K(x) = \sum_n \lambda_n \xi_n^2,$$

where the new variables

$$\xi_n = \sum_j \rho_{nj} x_j$$

are obtained from the original variables by an orthogonal transformation. In geometric language, the surface $K(x) = 1$ is reduced to principal axes by an orthogonal transformation. This form of the spectral theorem had great appeal to Hilbert, who sought a generalization to infinitely many variables. He succeeded in the case of completely continuous forms, but not in the general case because of the possibility of *Streckenspektrum*, that is, continuous spectrum.

Hilbert studied special cases of *Streckenspektrum*. He constructed examples using Jacobi forms

$$J(x) = 2c_1x_1x_2 + 2c_2x_2x_3 + 2c_3x_3x_4 + \cdots$$

and showed their relationship to several classical orthogonal systems. For example (*Grundzüge*, p. 155), the choice

$$c_1 = c_2 = \cdots = \frac{1}{\sqrt{2}}$$

leads to the trigonometric functions, and the choice

$$c_n = \frac{n}{\sqrt{4n^2 - 1}}, \qquad n = 1, 2, 3, \ldots,$$

leads to Legendre polynomials.

In his dissertation [**2**], Hellinger completed Hilbert's theory by writing a given quadratic form as an integral of squares of eigen-differentials. He applied this to a multiplicity theory and to a solution of the problem of orthogonal equivalence. Hellinger's starting point was Hilbert's form of the spectral theorem, but he needed a new kind of integral to express the result. He invented the Hellinger integral for this purpose. If $f(t)$ and $g(t)$ are functions of bounded variation, integrals of the form

$$\int_0^x (df)^2/dg \qquad \text{and} \qquad \int_0^x \sqrt{df}\sqrt{dg}$$

are defined as limits of finite sums

$$\sum_j (\Delta f_j)^2/\Delta g_j \qquad \text{and} \qquad \sum_j \sqrt{\Delta f_j}\sqrt{\Delta g_j}$$

corresponding to partitions of the interval from 0 to x. Hellinger conceived of the process as a way to compare the relative rates of growth of nondecreasing functions[11].

Hellinger showed that if the bounded symmetric quadratic form

$$K(x) = \sum_{p,q} a_{pq} x_p x_q$$

has simple spectrum, then it admits a representation in the form

$$K(x) = \int \mu \frac{\left(d \sum_j \rho_j(\mu) x_j\right)^2}{d\rho(\mu)}$$

for a system of eigen-differentials

$$d \sum_j \rho_j(\mu) x_j$$

which are normalized with respect to a basis function $\rho(\mu)$. The properties of eigen-differentials upon integration are those of a resolution of the identity. The basis function is a nondecreasing function whose points of increase coincide with the spectrum of the quadratic form. In general, the representation takes the form of a direct sum of such expressions,

$$K(x) = \sum_\alpha \int \mu \frac{\left(d \sum_j \rho_j^{(\alpha)}(\mu) x_j\right)^2}{d\rho^{(\alpha)}(\mu)},$$

with at most countably many summands. Hellinger measures multiplicity by the integer-valued function that counts points of increase of basis functions. His condition for orthogonal equivalence is, roughly, that basis functions have identical null sets and multiplicity functions. The elegant exposition is characteristic of Hellinger's writing throughout his career.

Hellinger's dissertation was viewed as groundbreaking work by his contemporaries. Literature from the period sometimes refers to the spectral theory of quadratic forms as the Hilbert and Hellinger theory[12]. Hermann Weyl's 1908 dissertation with Hilbert is a related work. It explores different aspects of Hilbert's examples of *Streckenspektrum*, namely, the connections with classical analysis. As a historical point, we note that Hilbert conceived the notion of an eigen-differential: it appears plainly in special cases in Hilbert's *Grundzüge*. Hellinger derived the general theory, and Weyl worked out key examples.

In [**3**], Hellinger gave extensions and new derivations of the main results of his dissertation. This paper was reprinted as Hellinger's *Habilitationsschrift* in Marburg. Here Hellinger generalized the notion of eigenvalue to the continuous spectrum. Eigenvectors are not always square summable, but certain integrals of them are square summable and admit normalization. Following a method of E. Hilb[13], Hellinger constructed a resolution of the identity using Cauchy's integral formula applied to resolvents in the complex domain.

A significant improvement to Hellinger's theory was added by Hahn in a paper published in 1912[14]. Hahn showed that Hellinger's basis functions can be ordered with respect to inclusion of their points of increase. Hahn also showed how to reduce the Hellinger integral to that of Lebesgue. In its modern formulation, the Hellinger–Hahn theory gives a complete set of unitary invariants for any selfadjoint operator. A standard treatment is that of Stone[15], which adapts the theory to unbounded operators.

Hilbert had great admiration for Stieltjes' theory of continued fractions, and he hoped that there would be a future application of the theory of quadratic forms to Stieltjes' theory. In the case of finitely many unknowns, the connections between continued fractions, Jacobi matrices, and the reduction of a quadratic form to principal axes are achievements of nineteenth-century mathematicians such as Jacobi, Kronecker, and E. Heine (original citations are given in [**8**], p. 1586). In the case of infinitely many unknowns, Hellinger and Toeplitz worked out the general theory in a series of papers. The bounded case was treated by Toeplitz in a 1910 paper[16] and by Hellinger and Toeplitz in their 1914 paper [**5**]. Consider a Jacobi form

$$J(x) = \sum_p (a_p x_p^2 - 2b_p x_p x_{p+1})$$

with

$$b_p \neq 0, \qquad m \leq a_p \leq M, \qquad m \leq b_p \leq M,$$

for all $p \geq 1$. Hellinger and Toeplitz [**5**] showed that the associated continued fraction

$$\cfrac{1}{a_1 - \lambda - \cfrac{b_1^2}{a_2 - \lambda - \cfrac{b_2^2}{a_3 - \lambda - \cdots}}}$$

converges for complex values of λ not in the real interval $[m, M]$, and the limit has a Stieltjes representation

$$\int_m^M \frac{d\rho(\mu)}{\mu - \lambda},$$

where $\rho(\mu)$ is a nondecreasing function on $[m, M]$. The Darboux polynomials, that is, the denominators of the approximants of the fraction, form a system of orthogonal polynomials with respect to $\rho(\mu)$:

$$\int_m^M \pi_p(\mu)\pi_q(\mu)\, d\rho(\mu) = \begin{cases} 0, & p \neq q, \\ 1, & p = q. \end{cases}$$

The quadratic form has simple spectrum. The basis function for the form is $\rho(\mu)$, and the spectral decomposition

$$J(x) = \int \lambda \, \frac{(d\rho_j(\lambda)x_j)^2}{d\rho(\lambda)}$$

can be computed from the continued fraction. Conversely, every bounded quadratic form with simple spectrum can be realized as a Jacobi form in infinitely many ways. The paper [**5**] is thus an early work of concrete spectral theory. In it, Hellinger and Toeplitz explicitly compute the unitary invariants of the bounded selfadjoint operator corresponding to a Jacobi matrix.

Contributions to Klein's Efforts in Mathematics Education and to the *Encyklopädie der Mathematischen Wissenschaften*

It was not unusual for young assistants, as part of their apprenticeship, to help in a major way in the writing of books for professors. The German verb *ausarbeiten* covers this kind of writing (the word means to "elaborate" or "work out"). As Klein's assistant, Hellinger performed such work in the first two parts of Klein's *Elementarmathematik vom höheren Standpunkte aus* [**14**]. On the title pages of each appear the credit, "*ausgearbeitet von E. Hellinger.*"

Since 1890, Klein had given lectures in Göttingen relating the development of mathematics to secondary school teaching. Klein credits a team of assistants for the production of *Elementarmathematik*, but Hellinger played a leading role. From notes of Klein's lectures in the winter semester of 1907/1908 and the summer semester of 1908, Hellinger literally wrote Volume I: *Arithmetik*, *Algebra*, *Analysis*, and Volume II: *Geometrie*. He prepared lithograph copies which became the first edition of *Elementarmathematik*; a second lithograph edition was prepared in 1911. The two volumes were distributed by B. G. Teubner, Leipzig. After they sold out, Julius Springer took over publication with third editions of Volumes I and II in 1924/1925.

Klein's purpose in *Elementarmathematik* was to influence the education of secondary school teachers of mathematics. Topics range from practical advice, such as how to teach logarithms, to subtle and difficult concepts, such as proofs of the transcendency of e and π. Klein believed that the best ideas of mathematics were not properly assimilated in the school curriculum. *Elementarmathematik* has timeless qualities that make it still a useful work.

Hellinger's survey article [**6**] on the mechanics of continua was written for the *Encyclopädie der Mathematischen Wissenschaften*. The *Encyclopädie* was a high-level effort to catalog the development of the mathematical sciences and their applications in authoritative survey articles. Felix Klein edited the fourth volume of the *Encyclopädie* and he must have solicited the article [**6**] from Hellinger specifically for the *Encyclopädie*. Hellinger wrote the *Lesezimmer* notes for Hilbert's lectures on this topic in 1906/1907, and he had closely followed Born's work on theoretical mechanics. The article [**6**] has been cited as a source for the ideas of Hilbert and Born on the minimal principle of elasticity theory[17]. Born's work appeared in a prize-winning paper of 1906.

Frankfurt 1914–1939

Hellinger left Marburg to accept a professorship in a new university, the *Königliche Universität Frankfurt am Main*, which opened in October 1914. The private founding of the university on the initiative of local citizens is apparently unique in the history of European universities. The starting capital was provided in large degree through the "extraordinary generosity" of the Jewish community in Frankfurt[18]. Inflation after the first world war forced the university to give up its private status, and today it is the *Johann Wolfgang Goethe-Universität Frankfurt am Main*. Hellinger was one of the original professors, an *Extraordinarius*. Besides Hellinger, the original mathematics professors in Frankfurt were Arthur Schoenflies (1853–1928) and Ludwig Bieberbach (1886–1982).

Hellinger volunteered for military service in the First World War. According to Hanna Meissner, he tested radio equipment for airplanes and later became a technical instructor. Later, in the 1930's, Hellinger was classified in official documents as a *Frontkämpfer*, which was strictly defined and would at the very least imply that he was in a combat operational area. Hellinger's personal file in Frankfurt records that he fought at the "Somme, Vogesen, Arras, Flandern, Avre, Noyon, Oise." A letter of July 12, 1915, from the Dean of the faculty in Frankfurt to military authorities states that Hellinger had received an order to report to the Territorial Reserves E/2 Limburg Battalion (*Landsturm E/2 Batl., Limburg*) on July 20. The Dean asked that the date be extended to July 31 to permit Hellinger to complete his duties to students that semester. The Dean's appeal was supported by a letter signed by twelve faculty members, including Schoenflies and Bieberbach. In a letter of September 9, 1915, Hellinger informed the faculty in Frankfurt that his military assignment was *Flieger-Funker, Funker-Lehr-Kommando, Fliegertruppe,*

Döberitz. A letter of August 1, 1916, from the Dean in Frankfurt to military authorities states that the faculty had received information that Hellinger had arrived on the western front. The university appealed for leave, but there is no record that leave was given. Nothing further is known from documentary evidence.

Hanna Meissner denied emphatically that Hellinger ever was, in any sense, a front fighter. She reported that his political views shifted sharply from the right to the left during and after the war. Hellinger himself spoke little of his past in later years. His Ph.D. student, Richard Beth, reports that when conversation came to personal matters, Hellinger would turn it away saying something like, "but that is not important." Maria Dehn Peters says that her memory is that Hellinger became involved in reading enemy communications. André Weil has the impression that Hellinger was awarded an Iron Cross. The first-class Iron Cross was a high distinction — about 200,000 were awarded; the second-class Iron Cross was given in large numbers — about 5,200,000 were awarded. It is unlikely that more will ever be known[19]. In any case, the official status of *Frontkämpfer* was crucial to Hellinger in the 1930's. It insulated him from the first round of persecutions of Jews by the Nazis.

The post-war era was a golden period for mathematics at the university. Hellinger was promoted to *Ordinarius* in 1920. Schoenflies was at this time *Rektor* of the university. By 1922 Hellinger was the only remaining mathematician on the faculty in Frankfurt of the original three. An excellent mathematics faculty was formed at this time: Max Dehn, Paul Epstein (1870–1939), Carl Ludwig Siegel (1896–1981), and Otto Szász (1884–1952) all came to Frankfurt in 1921 and 1922. The atmosphere was unique. In fact, the faculty of Hellinger, Dehn, Epstein, Siegel, and Szász created an environment that might truly be called a mathematical idyll.

Wilhelm Magnus was a student in Frankfurt and wrote a dissertation with Max Dehn in 1929. Describing the period, years later[20], Magnus quoted Goethe:

> Proverbially it was a time when, in a particular locality, all human endeavors interacted in such a fortunate way that the recurrence of a similar period could be expected only after many years and in very different places under exceptionally favorable circumstances.

Of Hellinger, Magnus says:

> Hellinger was probably the most widely appreciated teacher among the mathematicians. He ... was very well prepared. His lectures were highly polished, but he never forgot to mention the motivation for a theorem and he always pointed out connections between different parts of mathematics. His presentation was less austere than Siegel's. He liked to make entertaining remarks and to suspend the need for concentration for a moment, giving the audience a brief respite. He was an outstanding psychologist in the best sense: He always knew exactly how far a student could go, and his advice was given in a tactful manner which, at the same time, left no room for doubt.

Siegel described their studies in the history of mathematics[21]:

> I already mentioned that at Dehn's instigation the seminar on the history of mathematics was held each semester from 1922 to 1935. In addition to Dehn and Epstein, Hellinger and I also played important roles in these seminars, but due to his outstanding and universal learning, Dehn was in a sense our spiritual leader, and we always followed his advice when choosing topics for each semester. As I look back now, those communal hours in the seminar are some of the happiest memories of my life. Even then I enjoyed the activity which brought us together each Thursday afternoon from four to six. And later, when we had been scattered over the globe, I learned through disillusioning experiences elsewhere what rare good fortune it is to have academic colleagues working unselfishly together without thought to personal ambition, instead of just issuing directives from their lofty positions.
>
> It was the rule in those seminars to study the more important mathematical discoveries from all epochs in the original: everyone involved was expected to have studied the text at hand in advance and to be able to lead the discussion after a group reading. In this manner we studied ancient mathematicians and for many semesters devoted ourselves to a detailed study of Euclid and Archimedes. Another time we spent several semesters on the development of algebra and geometry from the Middle Ages to the mid 17th Century, in the course of which we became thoroughly familiar with the works of Leonardo Pisano, Vieta, Cardano, Descartes and Desargues. Our joint study of the ideas from which infinitesimal calculus developed in the 17th century was also rewarding. Here we dealt with the discoveries of Kepler, Huygens, Stevin, Fermat, Gregory and Barrow, among others.

Otto Toeplitz was a frequent visitor to the history seminar. Along with Neugebauer and Stenzel, Toeplitz was a founder of the journal, *Quellen und Studien zur Geschichte der Mathematik, Astronomie und Physik.* Although Hellinger and Toeplitz both published in the history of mathematics, their collaboration did not extend to this area. Another visitor in Frankfurt was André Weil, who was drawn by the dual attractions of Siegel and the history seminar. Weil has written on his own impressions of the Frankfurt history seminar, which he said was perhaps Dehn's most original creation[22].

Hellinger had both a talent and appetite for administrative responsibilities. He served as *Dekan der Naturwissenschaftlichen Fakultät* in Frankfurt from October 1927 until October 1928, and he was well known for his many years of work in student aid. Hellinger's administrative proclivity was described by Max Dehn in his remarks at Hellinger's retirement from Northwestern University:

> In Frankfurt, Hellinger found also a very fruitful field of activity, combined with his scientific activities. It was his responsibility to help the most promising students through administering stipends, scholarships and similar things. The love for administrative activity is rather deeply rooted in Hellinger's soul, all kinds of

such activities. He was very much interested, for example, in the organization of the German post office, and a highlight of his life was the moment when a girl in a post office with whom he spoke over the telephone to send a telegram asked him 'You are, I assume, the Post Office No. 13?'

Hellinger never married, but he had a special gift for making friends, and throughout his life, his friends sometimes became as close as family. He had a particular way with children, whom he treated as grown-ups. Eva Toeplitz Wohl, daughter of Otto Toeplitz, wrote: "Hellinger was not only a close family friend but also a kind of 'favorite uncle' of us children. Whenever he visited it was a feast for us. ... He was full of fun ... I believe that both of us remaining 'Toeplitz children' have always tried to imitate his special ways with children."

Study in Frankfurt, in his apartment on the Marbachweg

In Frankfurt, Hellinger was a *Hausfreund* with Max Dehn and his wife Toni[23]. He became also a special uncle to the Dehn children, Helmut, Maria, and Eva. Hellinger enjoyed good food, and he invited the children to lunch both at *Pfeil's Weinstube* and at his apartment at Marbachweg 339. They were invited individually in order to make it personal. At his apartment they were impressed by the view of the *Taunus* mountains from the many windows of his sunny study, his big desk with intriguing mathematical ornaments including sparkling glass polyhedra in gold, white (clear), and ruby, and a picture on the wall of a man doing a pole vault with Hellinger's own caption, *Der Körper schwinget sich wohl in die Höh-o-öh, Der Geist allein bleibt auf den Kanapee* ("The body flies high o-ho, the mind alone rests on the couch" — the caption is a paraphrase of a religious saying of the day which substituted "soul" for "body"; with Hellinger's typical ironic humor, it expresses his low opinion of competetive sport and physical activity, perhaps part of a bitter reaction to his war experiences, while affirming the supremacy of the mind). Hellinger hated the radio as it reminded him of the war. The Dehns kept no radio in their house out of respect for his feelings.

Daily life in Frankfurt had a rhythm. The Thursday history seminar, described by Siegel, was accompanied by a *Historische Kaffee* at the Dehns' home. The Dehn children were barely aware of the scholarly studies that were the point of these meetings, but they remember the essential gorgeous pastries that were brought by Hellinger, as well as the geometrically precise cuttings made by their mother so that all could taste them. Holidays such as Christmas and New Year were celebrated with the Dehns but also included other friends such as the Epsteins. This included decorating a tree, singing carols, and going to hear the bells of the ancient churches. In the spring, C. L. Siegel presided at a Great Asparagus Feast in the garden of the *Faust* restaurant.

Another view of the study, with the many windows looking to the *Taunus* mountains

Summer holidays were taken in Kampen on the Isle of Sylt (North Sea). There would be long discussions with the Dehns whether he should go, but in the end Hellinger always went and built his elaborate sand castle as was the custom at this resort area. At all times of the year, there were Institute Walks by faculty and students in the *Taunus* mountains. The faculty and students played together as hard as they worked together. An album of photographs from such outings, assembled by students for Hellinger, has been preserved and was loaned to the author by Maria Dehn Peters. A group picture in the album is playfully labeled, *unser Gesangverein*. Another photograph shows Hellinger "conducting" the group from atop a chair; in the next, Otto Szász is *Dirigent.*

"*unser Gesangverein*" — July 30, 1927, at the *Gasthof* in Fuchstanz, near Feldberg, in the *Taunus* mountains. Present are HELLINGER (far left), PAUL EPSTEIN (far right), WILHELM MAIER (behind Epstein), HENRI JORDAN (wearing bow tie, next to Hellinger), RUTH MOUFANG (woman on right, wearing a dark belt), OTT-HEINRICH KELLER (behind Moufang and next to Maier), OTTO SZÁSZ (sitting on ground on the left, with moustache), and CORNELIUS LANCZOS (sitting on ground on the right).

There was much good natured teasing among friends — for example, Siegel was teased about his dramatization of events, Hellinger for his Silesian accent. Temperamentally, Hellinger and Dehn were quite different. Hellinger's mind was organized like a "Bohr atom," while Dehn's mind dashed all over the place like an "electron cloud," in the words of Maria Dehn Peters. Mrs. Peters also said that Hellinger was angered by stupidity and insensitivity. Although this anger never involved his mathematical associates, on occasion they drew it out of him as a release of his smoldering feelings.

With his friend Max Born, Hellinger shared an interest in physics. Born spent a brief period in Frankfurt, from 1919 to 1921. Born and his wife rented rooms to Hellinger and his sister Hanna, in what he later described (Born, *My Life*) as "the nicest house we ever had, with a big garden full of fruit trees, pears and peaches, which brought us a rich harvest." Born said that of all his friends in Göttingen, Hellinger "was the one most interested in mathematical physics." Born's relations with mathematicians were evidently not always smooth. As a student, he had a difficult experience with Felix Klein. Born said that, in Frankfurt, it "was very convenient for me to have a 'tame mathematician' whom I could consult whenever I had a problem beyond my own resources."

After Born moved to Göttingen in 1921, he continued to consult Hellinger as the development of quantum mechanics unfolded. He wrote to Hellinger on October 1, 1925, with questions about *Streckenspektrum*. Born said in his letter that *Streckenspektrum* had important significance in quantum mechanics. He asked Hellinger to look over proofs from his paper with Heisenberg and Jordan[24] and point out any difficulties. Born was particularly concerned about the adaptation of Hellinger's *Habilitationsschrift* [**3**] to unbounded quadratic forms. Born, Heisenberg, and Jordan state in their paper that Hellinger's methods correspond completely to the physical sense of the problems. The paper includes a careful account of Hellinger's work adapted to unbounded forms; what hand Hellinger had in the mathematical details is unknown.

Born wrote to Hellinger again on September 10, 1926, asking for a meeting to discuss various mathematical points in quantum theory. He wished to hear Hellinger discourse on various notions. Hellinger's papers include fragments of notes from their meeting, in Hellinger's hand. They discussed quadratic forms, orthogonality, normalization of eigenfunctions, and Schrödinger's equation. Born mentioned the possibility of a joint work, but this never came about. The two Born letters are translated in an addendum at the end of this article.

Hellinger's papers also include seminar notes on quantum theory from about the same time. Later, in 1936, after he was dismissed from his university position by the Nazis, Hellinger participated in a "private seminar" on quantum theory. The 1936 seminar notes cover the experimental background of quantum mechanics, mathematical foundations, and the classical theory as a limiting case. Another participant of the 1936 private seminar was the experimental physicist K. W. Meissner, who later married Hellinger's sister Hanna. Meissner was in trouble with the Nazis because his first wife, also a physicist, was Jewish. The Nazis pressured Meissner to separate from his wife, but instead he and his wife emigrated to the United States[25]. They settled at Purdue University. Meissner's marriage to Hanna Hellinger came after his first wife died.

The mathematical idyll in Frankfurt was ended by the Nazis. It did not happen suddenly, but the clouds grew increasingly ominous. It was not uncommon in the Nazi period that academic colleagues adopted the "du" and first-name form of address to show solidarity. In the Dehn household, prior to the Nazi days the names "Hellinger," "Dehn," and "Frau Dehn," were always used, according to Dr. Helmut M. Dehn. Max Dehn and his wife switched to first-naming not only with Hellinger, but also with other colleagues, including Epstein and Siegel (Szász had always been called "Shasha"). In the end, nothing could save them.

Siegel, Dehn, and Hellinger on September 26, 1938

Siegel left a moving account of the fates of his colleagues in his article in the *The Mathematical Intelligencer*. Briefly, Szàsz was dismissed in 1933. Dehn, Epstein, and Hellinger were protected at first because of their military service but lost their positions in 1935. Epstein left voluntarily, before the authorities could fire him; he committed suicide in 1939 after receiving a summons from the Gestapo[26]. Szász, Dehn, and Hellinger survived and emigrated to the United States. Toeplitz was in Kiel 1913–1928 and then Bonn 1928–1933. Dismissed by the Nazis in 1933, he remained in Germany until 1938. He emigrated to Jerusalem and died there in 1940. Other Jewish mathematicians throughout Germany suffered similar fates[27].

Hellinger's formal dismissal from his professorship in Frankfurt resulted from racial laws adopted at the Nazi party rally in Nürnberg on September 15, 1935. Hellinger was notified in a letter of October 15, 1935, that he would be retired in accordance with these laws (RGBl.I S.1146). The racial laws were later implemented in the first decree of the Reich Citizenship Law of November 14, 1935 (RGBl.I S.13). The university trustees cited the latter statute in the formal dismissal of Hellinger, which came in a letter of December 23, 1935. Retirement was effective December 31, 1935. A letter of December 17, 1935, recognizes Hellinger's service as *Frontkämpfer* in the First World War in setting his retirement stipend[28].

Hellinger believed that he could remain in Germany, but his situation became critical after *Kristallnacht*, November 9–10, 1938, when the Nazis rioted against the Jews. In the following days, Jews were rounded up in such numbers that the jails could not hold them for transportation to concentration camps. On November 11, 1938, Wilhelm Magnus visited Hellinger and invited him in his parents' name to stay with them in their apartment on the Kettenhofweg. Magnus' father was a physical chemist and was himself in political difficulty at the time. Hellinger declined out of consideration for the Magnus family[29]. Dehn fled from Frankfurt, but Hellinger decided to stay. Siegel reported that Hellinger told him, after *Kristallnacht*, that he would "see just how far beyond the traditional standards of justice and ethics the authorities would go in his case." In the days after *Kristallnacht*, Hellinger was in constant telephone communication with Otto Toeplitz. Eva Toeplitz Wohl says that the uncertainty eventually became too great for Hellinger, and he gave himself up to the Gestapo. André Weil reports that he was told by someone that Hellinger, who never wore his war decoration in ordinary life, "put it on on the day, after *Kristallnacht*, when he was expecting the Gestapo, or the S. S., to come for him and to take him away."

Hellinger was arrested on November 13, 1938. He was held temporarily in the Festhalle and then taken to Dachau. Eva Toeplitz Wohl said they heard that in the camp, Hellinger acted heroically, protecting weaker inmates when guards beat them. He was released from Dachau after about six weeks on the condition that he leave the country. André Weil was told that Siegel, with characteristic bravery, went personally to negotiate the release.

Hellinger's Mathematical Work in Frankfurt

In Frankfurt, as in Göttingen and Marburg, Hellinger was occupied with the spectral theory of quadratic forms, continued fractions, and equations in infinitely many unknowns. He published four works in this period: three were major projects and contributions to Hilbert's program; the fourth was in the history of mathematics.

In [**7**], Hellinger returned to the subject of his paper with Toeplitz [**5**], namely, the spectral theory of Jacobi forms and their associated continued fractions. In [**5**] it is assumed that the Jacobi forms are bounded. The general case treated in [**7**] is a highly regarded work. Hellinger's analysis was inspired by Hamburger's theory of the moment problem on the real line.

Consider a Jacobi form

$$J(x) = \sum_p (a_p x_p^2 - 2b_p x_p x_{p+1})$$

and associated continued fraction

$$\cfrac{1}{a_1 - \lambda - \cfrac{b_1^2}{a_2 - \lambda - \cfrac{b_2^2}{a_3 - \lambda - \cdots}}}$$

as before, but without the boundedness assumptions on the coefficients. Let $\pi_1(\lambda), \pi_2(\lambda), \ldots$ be the corresponding Darboux polynomials. Hellinger showed that if the series $\sum_p |\pi_p(\lambda)|^2$ diverges for one nonreal number λ, it diverges for all nonreal numbers λ (theorem of invariability). In this case the continued fraction converges as a limit of generalized approximants

$$\cfrac{1}{a_1 - \lambda - \cfrac{b_1^2}{a_2 - \lambda - \ddots - \cfrac{b_{n-1}^2}{a_n - \lambda - b_n h}}}.$$

with uniform convergence in the parameter h. The limit is independent of h and has a Stieltjes representation

$$\frac{1}{\pi} \int_{-\infty}^{\infty} \frac{d\sigma(\mu)}{\mu - \lambda}.$$

Hellinger gave a sufficient condition for convergence, namely, that the numbers $b_1, b_2, \ldots$ have a finite lower bound. If the series $\sum_p |\pi_p(\lambda)|^2$ converges for any nonreal value of λ, the continued fraction diverges. Hellinger's convergence theory for the continued fraction follows a limit-point and limit-circle analysis after Weyl.

Similar results were obtained about the same time by several other authors. Ahiezer[30] gives this historical perspective:

> The first solution and discussion of the extended moment problem
>
> $$s_k = \int_{-\infty}^{\infty} u^k \, d\sigma(u), \qquad (k = 0, 1, 2, \ldots)$$
>
> is due to H. Hamburger, after whom that problem is usually named. Like his famous predecessors, Hamburger makes wide use of continued fractions. Almost simultaneously with Hamburger this problem was solved and advanced in certain directions by use of other methods, by R. Nevanlinna, M. Riesz, E. Hellinger and T. Carleman. This work revealed important connections between the moment problem and many branches of analysis. Of particular importance are the connections with functional analysis, the theory of functions and the spectral theory of operators.

The Hellinger–Toeplitz encyclopedia article [**8**] on integral equations and equations in infinitely many unknowns is a monumental survey of literature up to 1923. There is a clue to the origin of the work in the 1912 preface to Hilbert's *Grundzüge*. Hilbert apologizes for relatively few literature citations, but he says that Toeplitz is working on a comprehensive text that will remedy the deficiency. Perhaps Toeplitz' project evolved into [**8**]. The article [**8**] served as a handbook for researchers at the time of its publication. It combines elements of research, scholarship, history, and survey writing, and as such met the ideals that Felix Klein held for the *Encyclopädie*. J.D. Tamarkin wrote a long review of [**8**] for the *Bulletin of the American Mathematical Society*. Tamarkin praised the article's usefulness as a reference work and guide to future developement. He also took the authors to task for inadequate coverage of Carleman's work and the E. H. Moore school of General Analysis. In view of the audacious nature of the undertaking in [**8**], these omissions are relatively minor. The article [**8**] has outlived its usefulness as a handbook for researchers, but it will probably never be replaced as the most complete and authoritative source for the historical development of integral equations and the algebraic foundations of functional analysis and operator theory.

Hellinger's survey article [**9**] on Hilbert's work on integral equations and equations in infinitely many unknowns, which appears in Hilbert's collected works, may be viewed as a companion to [**8**]. It is a definitive statement of Hilbert's work in this area.

In the history of mathematics, Hellinger and Dehn participated in an important revaluation of the Scottish mathematician James Gregory (1638–1675).

An International Congress was held in Edinburgh and St. Andrews on July 4 and 5, 1938, to commemorate the 300-year anniversary of Gregory's birth. E.T. Whittaker invited Dehn and Hellinger to participate. A result of the Congress was a new appreciation of Gregory as a genius of the first rank. Hellinger and Dehn contributed [**10**] to the proceedings. Hellinger wrote a detailed review of the proceedings of the Congress for *Mathematical Reviews* (Vol. 1, 1940, pp. 129–130).

Evanston 1939–1950

Hellinger became part of the migration of European mathematicians to the United States in the 1930's, which so influenced the post-war development of mathematics. Hellinger's story is a case study of both the problems the émigrés encountered and the friendly reception they received from most colleagues and students[31].

As Hellinger's plight became known, friends in Europe and the United States worked to collect money to allow him to leave Germany, and to find a position for him. According to Hanna Meissner, many people were active in these efforts. One group was in Great Britain: Adolf Prag said that he and others attempted to raise money in Britain for security for a transit permit for Hellinger. In the United States, there seemed to be no possibility of finding a position on the East Coast. Hermann Weyl, who was in Princeton, sent an invitation on behalf of the German Mathematicians Relief Fund, but the fund's balance was down to $43.75, and there was no money to offer (Constance Reid, *Courant*, p. 213). A temporary position, involving no university funds, was created at Northwestern University. The money to pay for the position was put at the disposal of Northwestern University by "certain individuals in the city of Chicago." The Dean at Northwestern, Addison Hibbard, wrote on January 18, 1939, offering Hellinger a lectureship with a salary of $1,400. The Dean also informed Hellinger that Northwestern University had cabled the American Consul General in Stuttgart to assist in facilitating his departure. The letter is among Hellinger's private papers.

Hellinger reported the Northwestern offer to the German authorities in a letter of February 18, 1939, and he requested permission to live permanently outside of Germany. He asked that his German retirement salary be paid into an account in Breslau for his mother, and this was done[32]. Hellinger arrived in New York on March 13, 1939. He spent a few days in New York and Princeton before going on to Chicago on March 19. He applied for U. S. citizenship immediately upon his arrival in Chicago, on March 24, 1939. He received his first papers July 7, 1939 and became a U. S. citizen on September 5, 1944.

Hellinger's position at Northwestern was extremely fragile in his early years there. Northwestern considered the position a purely temporary one. The Dean wrote to Hellinger on October 16, 1939 (the letter is among Hellinger's papers):

> I am sure you are aware of the fact that the money turned over to the College for your stipend will not be available after this semester. Consequently there can be no appointment for you on the College faculty. I am writing this note to you at this time so that you may fully understand the situation in advance.

The Mathematics Department at Northwestern and powerful friends—Hermann Weyl, Richard Courant, Oswald Veblen—rallied in new appeals to private committees to contribute further support. On October 23, 1939, Weyl wrote to Hellinger advising that Northwestern should apply to the Emergency Committee in Aid of Displaced Scholars which was based in New York. He added:

> In writing to the Emergency Committee it would be of great importance if Northwestern could make some kind of commitment for the future when the stipend, which is usually not granted for more than a year, will have expired. In fact, it has been the declared policy of the Committee not to make any grants unless such commitments are made; whether they still stick to it as strictly as in former years I do not know.

Weyl said that he and Veblen might go to New York "to discuss the matter with key men in the Committee." Northwestern remained inflexible in its position. Writing to the Chicago Jewish Welfare Committee on November 1, 1939, the Dean said:

> I have written Professor Hellinger to the effect that his services would be terminated this next February. However, since Dr. Hellinger has no direction in which to turn, the Department of Mathematics has urged me to take up with you the possibility of your helping Professor Hellinger for another semester or another year. The longer he is in this country, of course, the greater are his chances of ultimately finding a position where his services are actually needed and where he can be absorbed into some faculty.

The Chicago Jewish Welfare Committee offered $1,000 on the condition that it be matched. In New York, the Emergency Committee in Aid of Displaced Foreign Scholars balked over Northwestern's refusal to hold out any hope for future support but in the end gave $500. How the final numbers came out is unknown, but in some such way, Hellinger's position was continued for the spring of 1940. Matters improved after this. Northwestern took over the funding of Hellinger's salary in the fall of 1940. In 1940-1945, he was hired as Lecturer in Mathematics for five successive one-year terms with twelve month salaries of $2,000, $2,200, $2,200, $2,700, $3,200. These were entry-level salaries for the day. It was typical that the immigrant mathematicians were originally paid not on the basis of their experience but as entry-level faculty. He received additional sums for summer teaching. The figures are from original salary notices sent to Hellinger by Northwestern University.

Hellinger's position at Northwestern was not stable, however, due to wartime budget constraints. President Franklyn B. Snyder of Northwestern University wrote to Hellinger on February 12, 1942 (the letter is among Hellinger's papers):

> ...I say to you quite frankly, and with sincere regret, that the University cannot here and now assure you of employment after August 31, 1942. This is not a notification of termination of employment. It is, however, a definite statement that you should not at this time assume reappointment.

The position became secure, finally, at the end of the war. On September 1, 1945, Hellinger became Professor of Mathematics with a five-year term. His salary increased at this point but was never high.

Happy reunion of Dehn and Hellinger,
at Pocatello Train Station in Idaho

In the Mathematics Department at Northwestern University, Hellinger found a friendly and supportive environment and new friends. He worked with undergraduates, directed Master's theses, and produced Ph.D. students. Students loved him. It is reported that in his lectures, Hellinger spoke in an accented quiet tone and spiced the discussion with fascinating tales of the history of mathematics. He stressed the beauty of an elegant solution of a problem. Students signed petitions on his behalf when he had difficulties over his position. In February 1948, on no particular occasion, twenty-one students contributed to a fund to present Hellinger with a leather chair for his office[33]. One time when he was ill, students arranged to take meals to him in his apartment. Hellinger never spoke of his past to students, but they noted his sensitivity to freedom. Once a mouse was trapped in a wastebasket. Hellinger carefully picked it up and left the room. On returning, he said, "I took it outside and assisted her to escape." He complained against screen doors as devices to trap flies indoors and opened the doors to let the flies out[34].

Hellinger's closest friends in Evanston were W.T. Reid and his wife Idalia. Idalia Reid remembers Hellinger as probably the most remarkable man she ever knew. He made an overwhelming impression on her at their first meeting. She described him as a gentle man, perceptive, warm and understanding. He retained an old-world grace and wit. Three or four times a week, Hellinger would eat with the Reids, at home or in a restaurant. Hellinger would appear at their door, a cane in one hand, a gift behind his back, and a twinkle in his eye. They vacationed together in a lodge at Pike's Peak. When they were together, they did not talk about personal matters or problems. They talked about the meaning of life and being a human being. Mostly they laughed. Idalia Reid remembers Hellinger as a "fun man" and a beautiful person.

Hellinger retired from Northwestern University in 1949 at the mandatory retirement age of 65. The Mathematics Department pleaded to keep him, and a feisty exchange took place between Mathematics Department Chairman H. T. Davis and Northwestern President F. B. Snyder. Davis told Snyder that they were unable to replace Hellinger[35], and young analysts in their early thirties were being paid $9,000–$12,000. Hellinger had offers of $4,700 from the Illinois Institute of Technology and $5,400 from the University of Tennessee. Snyder responded that keeping Hellinger would undermine the retirement plan, and it was to Hellinger's interest that he leave Northwestern while he was still vigorous, "when such attractive offers are open to him as you outline in your letter."

On the occasion of Hellinger's retirement, a festive banquet was held in his honor, and there was a great outpouring of affection from 142 people in attendance. W.T. Reid was master of ceremony, and Max Dehn was the principal speaker. A second speaker was Willy Hartner, who had been a student in Frankfurt in the 1920's and later became *Professor für Geschichte der Naturwissenschaften* in Frankfurt. Letters were read by others who could not be present, including Siegel, Wall, Magnus, Emil Artin, L.R. Ford, M.H. Stone, Szász, and Weyl. A proposal was made to publish Hellinger's principal mathematical works in one volume. This never came about, although the effort continued up to 1959.

Hellinger, in the Rocky Mountains, Colorado, August, 1949

Hellinger's last year was difficult. Troubled by financial insecurity, he took a temporary position at the Illinois Institute of Technology in 1949. He became ill in November of that year and never recovered. The two women who were closest to him — his sister, Hanna Meissner, and Toni Dehn — never knew if he knew that he had cancer. He did know, and he talked about it with André Weil. Weil recalls that Hellinger "mentioned a dream he thought he had had during an operation, to the effect that someone had said that nothing could be done: it seems that, in spite of the anesthesia, he had actually heard something said by the surgeon. There was also a painful episode, when he fell from his bed, late at night, and burnt himself severely by contact with the radiator, but refrained from calling for help for fear of disturbing the nurses." In his last days, Toni Dehn came from North Carolina to be at his bedside. She stayed until the doctor sent her away. He died in Chicago on March 28, 1950[36].

Efforts to restore Hellinger and Dehn to their positions in Frankfurt after the Second World War were frustrated by bureaucratic obstacles. An offer was made to Hellinger, but it came too late for him to have seriously considered it. He had no desire to return to Germany[37]. Dehn's last position was at Black Mountain College in North Carolina, where he died in 1952[38].

Hellinger's death was sadly noted in Europe and the United States. His scientific obituary was written by Wilhelm Magnus[39]. Aurel Wintner, writing to W.T. Reid shortly after Hellinger's death[40], paid this tribute:

> He undoubtedly was one of the leaders in his generation, and I believe that this country has lost much by not affording him such a position to which he would have been entitled and in which he could have exerted his influence on American mathematics to a fuller extent.

Hellinger's Mathematical Work in Evanston

Two papers from this period deal with quadratic forms and continued fractions. The paper **[11]** is a survey of the spectral theory of quadratic forms with an emphasis on examples. The paper **[12]** with Wall concerns nonselfadjoint generalizations of Hellinger's earlier work **[7]**. Hellinger and Wall study the continued fraction

$$\cfrac{1}{b_1 + z - \cfrac{a_1^2}{b_2 + z - \cfrac{a_2^2}{b_3 + z - \cdots}}}$$

under the assumption that

$$a_p > 0 \qquad \text{and} \qquad \operatorname{Im} b_p \geq 0$$

for all $p \geq 1$. This makes the associated Jacobi matrix J dissipative. They show that under these assumptions, the resolvents of J exist typically in only one half-plane. The theory of the symmetric case carries over otherwise with respect to the principle of invariability, limit-point and limit-circle analysis, Stieltjes and asymptotic representation, and moment problems. When the continued fraction converges, it represents the entry in the first row and first column of the resolvent matrix

$$(J + zI)^{-1}$$

in the upper half-plane. This example seems not to be subsumed in the modern theories of nonselfadjoint operators. An account of Hellinger's work on continued fractions is given in the book of H. S. Wall: *Continued Fractions*, Van Nostrand, New York, 1948.

Hellinger's last published work is the historical paper with Max Dehn [**13**]. This paper gives a detailed appraisal of the mathematical work of James Gregory, taking into account the scholarship of the tercentenary celebration of Gregory's birth for which Hellinger and Dehn wrote [**10**].

Addenda and Notes

Max Born and the Sum of Squares Representation in Quantum Mechanics

Hellinger lost most of his papers and books when he left Germany. There is a story that some of his personal books turned up in a mathematics library in Oklahoma after the Second World War, but this is unconfirmed. Two letters from Max Born seem to have been special to Hellinger, and he managed to save them. They detail the role of the Hilbert and Hellinger view of spectral theory in the early formulation of quantum mechanics. The Hilbert and Hellinger theory of quadratic forms was constructed in order to extend the "sum of squares" representation of symmetric quadratic forms to infinitely many variables. The Born letters show that this representation is not only desirable on mathematical grounds, but it also has physical significance. In particular, continuous spectrum is essential from the physical point of view.

Göttingen, October 21, 1925

Dear Hellinger,

In some weeks you will receive proofs from a work, for which Heisenberg, Jordan, and I are responsible. The reason for this communication is that we make a well-intentioned but frivolous and completely unauthorized use of your *Habilitationsschrift*. I would be very grateful to you if you would glance at it and call our attention to the crudest stupidities. I have indeed already once made use of continuous spectrum, in crystal vibrations; but there the infinite forms served only for "approximation" of finite ones, and therefore the normalization of the differential solutions had to be obtained differently from what is admitted by the orthodox interpretation of your article. Now, however, your catechism is strictly followed. Overall these new matters lie much deeper. We believe to have the root (better: a principal root) of quantum theory, namely a generalization of mechanics which automatically produces the discontinuities of quantum theory. The fundamental idea is this: In the classical theory, let a coordinate x be developed in a Fourier series: $x(t) = \sum a_n e^{i\omega n t}$. Here the a_n depend in a continuous way on certain constants of integration I, for example, the total energy. In place of the series, one can take the Fourier coefficients $a_n(I)$ to describe the state. Quantum theory was previously written so that one took also the constants I to be discontinuous: $I = mh$; then the $a_n(mh) = a_{nm}$ are dependent on two discrete indices n, m, but in a hideously unsymmetric way. In addition one cannot understand the fundamental empirical law of spectrum (Ritz combination principle), whereby the frequencies $\omega_n(mh)$ have the form $W_n - W_m$; for they have indeed the form $\omega_n(mh) = n \cdot \omega(mh)$, that is, they are equidistant (harmonic), but they do not have that simple symmetric dependence on n and m. Our new quantum mechanics goes in this way: We describe the state in an atom by means of a representation of harmonics, as in the classical theory, but we assume that in

$$a_{nm}\, e^{i\omega(nm)t}$$

n, m occur symmetrically (respectively, in ω skew symmetric); we furthermore postulate that there exists a numerical sequence $W_1, W_2, \ldots$ from which $\omega(n,m) = W_n - W_m$ is derived through formation of differences. Then we have for the description of the state an infinite matrix (Hermitian), concisely,

$$\underline{a} = \left(a_{nm}\, e^{i\omega(nm)t}\right).$$

The beauty is now this: in place of the multiplication law of classical Fourier coefficients, which one needs, for example, for the calculation of energy

$$x(t)^2 = \left[\sum a_n e^{i\omega n t}\right]^2 = \sum b_n e^{i\omega n t}, \qquad b_n = \sum_k a_k a_{n-k},$$

there appears simply matrix multiplication; then one has, for example,

$$\underline{a} \cdot \underline{b} = \Big[\sum_k a_{nk} b_{km}\, e^{i\left(\omega(nk)+\omega(km)\right)}\Big],$$

and because $\omega(nm) = W_n - W_m$ this is simply

$$= \Big[\sum_k a_{nk} b_{km} e^{i\omega(nm)t}\Big].$$

The Ritz principle is therefore directly the sufficient condition in order that the product of two coordinate matrices with periodic elements of the form $e^{i\omega t}$ is again such a matrix. This idea is due to Heisenberg. Jordan and I then developed the matrix–analysis and –mechanics and showed that one progresses rather far with it. The work, of which we send you the proofs, is already the second, but I think you will understand the sense of the matter. An important discovery is also this: in physics one calls the W_n "terms" and their totality the "term spectrum". The detection of these comes out of the orthogonal transformation of a quadratic form $\sum_{nm} \mathcal{H}_{nm}\, x_n x_m$ in a quadratic sum, and indeed this is directly $\sum_n W_n\, y_n^2$; that is, the physical term spectrum is identical with the mathematical eigenvalue spectrum of the form (the observerable frequency spectrum arises first through formation of differences $\omega_{mn} = W_m - W_n$). Now there naturally enters, as it physically must be, besides the point spectrum (the physicists say: term sequences or term series) also continuous spectrum, which provides the continuous frequency spectrum of optics. Here your work therefore becomes of value; only unfortunately our forms are not bounded. We have therefore only been able to conjecturally formulate the necessary theorems, and these moreover in a very rough form. Here we would be *very* thankful for your examination and help .

Whether the entirety is physically correct, it must first be shown. In any case, the matrix representation is the only one possible which corresponds to the established facts of quantum theory. The entire picture appears very happily, we believe, to come a considerable step closer to the quantum secrets.

Now I ask you to direct your answer on the proofs to Dr. W. Heisenberg, Göttingen, Physik. Inst., Bunsenstr., since my wife and I leave Göttingen on Monday the 26-th and on the 29-th to travel to America, from Hamburg to New York with the "Westfalia" [Hafag?]. From 16.11. until 31.1. I will give lectures in Boston, Mass. Inst. of Technology, then we want to travel around a bit and in March return. It would be very nice if you could let me know with a post card before Monday, if you are able to fulfill my request on the proofs. My wife sends hearty best wishes. Let all be well with you and help us a bit. With hearty greetings to Dehn, Madelung, etc.

Yours

Born

Göttingen, September 10, 1926

Dear Hellinger,

Tomorrow morning my wife and I travel south, in order to bring our oldest Irene to the school in Salem on the Bodensee, where she is to stay a year. On the return trip in about 8 days (between the 18-th and 20-th), we will be in Frankfurt perhaps one day, and I would very much like to speak with you. Please send a card to me in Salem, Baden, *Schlossschule*, if you will be in Frankfurt at this time and what your new address is (one finds these things in the mathematical register). I would like namely to squeeze something mathematically out of you. For some time I have been concerned with continuous spectrum, which is closely linked in the new quantum mechanics with aperiodic processes ("motions"). Thereby, your old work on continuous spectrum remains still the single tangible thing that I have found. But it does not entirely suffice, for as always I do not understand it adequately. The point which physics demands, is that your "basis function" remain arbitrary. Here a new physical principal enters that is connected with the axioms of statistical mechanics. As far as I can see, it can be brought into the following form:

Quantum theory leads (following Schrödinger) to differential equations in the form (for the simplest case of one independent variable x):

$$\frac{d^2\Psi}{dx^2} + \big[W - V(x)\big]\Psi = 0.$$

Here W is the "parameter" (otherwise denoted λ), which has the significance of energy; $V(x)$ is a given function, the potential energy, and $\Psi(x)$ is to be so determined that it is finite and continuous on the entire x-axis, inclusive of infinity. Then for judicious choice of $V(x)$ there are:

1) discrete eigenvalues $W_1, W_2, \ldots$, which are all < 0;

2) a continuous spectrum $W \geq 0$.

If one considers the latter, one can easily see that, always, if $V(x)$ tends to zero sufficiently fast for $x \to \pm\infty$, the solutions $\Psi(W, x)$ behave asymptotically like $\cos\left(\sqrt{W} + \alpha\right)$ for large absolute value of x. It now appears to me that the empirically determined results of statistical mechanics, which give the "relative weights" of the energy steps, work out correctly if one normalizes the eigenfunctions of the discrete spectrum as usual, namely,

$$\int_{-\infty}^{\infty} \Psi_n(x)^2\, dx = 1,$$

but in the continuous spectrum correspondingly sets:

$$\lim_{a \to \infty} \frac{1}{2a} \int_{-a}^{a} \Psi(W, x)^2\, dx = 1.$$

Thereby the basis function is obviously entirely determined, and one can attach to every piece of the W-axis a "weight" which is comparable with the weight 1 that occurs for discrete W_n-values. But all of that stands blurred in my mind, and I find here despite Herglotz and Courant, nobody who would be willing or able to help me. Alone I also cannot manage it. Therefore I would like to ask you if you have any interest in such things and would grant me a discussion (one-sided, naturally; I should say, "lesson"). Perhaps a joint work could arise out of this.

If you are able and willing, then write to me at the enclosed address. Then I would send you the proofs of a small work, from which you can see where I want to go physically. That would make the conversation easier.

I was in England in August, first in Cambridge, then with the British association in Oxford. It was quite gratifying. Then I was 10 days with my children in Langeoog, who were there 5 weeks with a young lady. Hedi was at home the entire time, where she could assure a better recovery.

What have you been up to, and what plans do you have? With best wishes, also from my wife,

Yours

Born

We pass through Frankfurt tomorrow afternoon. Stay there 4^{54}–5^{18}; train Hamburg-Stuttgart or Hamburg-Basel (not yet entirely determined).

Publications of Ernst Hellinger

Key to reviews: BAMS =*Bulletin of the American Mathematical Society;* FM =*Jahrbuch über die Fortschritte der Mathematik;* JDMV = *Jahresbericht der Deutschen Mathematiker Vereinigung;* MR = *Mathematical Reviews;* Zbl = *Zentralblatt für Mathematik und ihre Grenzgebiete.*

[**1**] "Grundlagen für eine Theorie der unendlichen Matrizen," together with O. TOEPLITZ, *Nachrichten der Gesellschaft der Wissenschaften*, Göttingen, Math.-Phys. Klasse (1906), 351–355. FM **37**, p. 191 (Jahnke).

[**2**] "Die Orthogonalinvarianten quadratischer Formen von unendlichvielen Variablen," Inaugural–Dissertation, Georg-August-Universität Göttingen, 1907, 86pp. FM **38**, pp. 153–156 (F. Meyer).

[**3**] "Neue Begründung der Theorie quadratischer Formen von unendlichvielen Veränderlichen," *Journal für die reine und angewandte Mathematik* **136** (1909), 210–271. *Habilitationsschrift*, Marburg. FM **40**, pp. 393–395 (Toeplitz).

[4] "Grundlagen für eine Theorie der unendlichen Matrizen," together with O. TOEPLITZ, *Mathematische Annalen* **69** (1910), 289–330. FM **41**, p. 381 (Toeplitz).

[5] "Zur Einordnung der Kettenbruchtheorie in die Theorie der quadratischen Formen von unendlichvielen Veränderlichen," together with O. TOEPLITZ, *Journal für die reine und angewandte Mathematik* **144** (1914), 212–238, 318. FM **45**, pp. 516–517 (Hellinger).

[6] "Die allgemeinen Ansätze der Mechanik der Kontinua," *Encyklopädie der Mathematischen Wissenschaften*, Artikel IV 30, Bd. IV, 4, pp. 601–694, 1913. FM **45**, p. 1012 (Lampe).

[7] "Zur Stieltjesschen Kettenbruchtheorie," *Mathematische Annalen* **86** (1922), 18–29. Reprinted in *Festschrift zum 60. Geburtstag von D. Hilbert*, Julius Springer, Berlin, 1922; reprinted 1982. FM **48**, pp. 535–536 (Hamburger).

[8] "Integralgleichungen und Gleichungen mit unendlichvielen Unbekannten," together with O. TOEPLITZ. *Encyklopädie der Mathematischen Wissenschaften* II C13, Bd.II,I,2, 1335–1601 (1927). Reprinted with a preface by E. HILB and subject index: Teubner, Leipzig, 1928. Reprinted by Chelsea, New York, N. Y., 1953. JDMV **37–38**, pp. 125–128 (Hammerstein); BAMS 36 15–18 (J. D. Tamarkin).

[9] "Hilberts Arbeiten über Integralgleichungen und unendliche Gleichungssysteme," in David Hilbert, *Gesammelte Abhandlungen*, Bd. I, 94–145, Julius Springer, Berlin, 1935. FM **61**, p. 22–23 (Feigl); Zbl **13**, pp. 56–57 (Tamarkin).

[**10**] "On James Gregory's Vera Quadratura," together with M. DEHN, *The James Gregory Tercentenary Memorial Volume*, 468–478 (1939). MR **1**,33 (D.J. Struik). MR **1**,129 (Hellinger). Communicated under the title "James Gregory's thoughts and constructions regarding transcendental numbers," at the celebration of Gregory's 300-th birthday, Edinburgh, July 1938. Zbl **61**, p. 4 (J. E. Hofmann).

[**11**] "Spectra of quadratic forms in infinitely many variables," *Northwestern University Studies in Mathematics and the Physical Sciences, No. 1: Mathematical Monographs*, Vol. 1, pp. 133–172. Northwestern University, Evanston, Illinois, 1941. MR **3**,210 (F. J. Murray).

[**12**] "Contributions to the analytic theory of continued fractions and infinite matrices," together with H. S. WALL, *Annals of Mathematics* (2) **44**, 103–127 (1943). MR **4**,244 (J. A. Shohat); Zbl **60**, p. 165 (E. Frank).

[**13**] "Certain mathematical achievements of James Gregory," together with M. DEHN, *American Mathematical Monthly* **50**, 149–163 (1943). Zbl **60**, p. 9 (J. E. Hofmann).

Work as assistant to Felix Klein, together with C. H. Müller

[**14**] FELIX KLEIN, *Elementarmathematik vom höheren Standpunkte aus*, Bände I–III, Die Grundlehren der mathematischen Wissenschaften, Bände 14,15,16, Springer, Berlin, 1968. JDMV **36**, 102–103 (Feigl).

Bd. I, *Arithmetik, Algebra, Analysis.* Vierte Auflage. Ausgearbeitetet von E. Hellinger. Für den Druck fertig gemacht und mit Zusätzen versehen von Fr. Seyfarth. (Reprinting of 1933 edition. English translation from the third German edition: Macmillan, New York, 1932; reprinting, Dover, New York, 1953.)

Bd. II, *Geometrie.* Dritte Auflage. Ausgearbeitet von E. Hellinger. Für den Druck fertig gemacht und mit Zusätzen versehen von Fr. Seyfarth. (Reprinting of 1925 edition. English translation from the third German edition: Macmillan, New York, 1939.)

Bd. III, *Präzisions– und Approximationsmathematik.* Dritte Auflage. Ausgearbeitet von C. H. Müller. Für den Druck fertig gemacht und mit Zusätzen versehen von Fr. Seyfarth. (Reprinting of 1928 edition.)

Dissertations Directed by Ernst Hellinger

Karl Siemon, Frankfurt a. M., 1922
Über die Fläche des absoluten Betrages analytischer Funktionen komplexen Arguments

Henri Jordan, Frankfurt a. M., 1930
Eine neue Methode zur Herleitung asymptotischen Ausdrücke in Anwendung auf die Bessel'schen Funktionen

Richard A. Beth, Frankfurt a. M., 1935
Untersuchungen über die Spektraldarstellung von J-Formen

Heinz Nagel, Frankfurt a. M., 1936
Über die aus quadrierbaren Hermite'schen Matrizen entstehenden Operatoren

Mason Eaton Wescott, Northwestern University, 1939–1940
A Set of Newton Polynomials Analogous to LaGuerre's Polynomials

Richard Harlan Stark, Northwestern University, 1942
Some Classes of Monotone Functions

Martinus Hendricus Mac Esser, Northwestern University, 1944
Spectral Expansion for Self-Adjoint Differential Transformations and for Arbitrary Self-Adjoint Transformations

Evelyn Frank, Northwestern University, 1945
C–Fractions

Acknowledgements

Special thanks are due to Mrs. Hanna Hellinger Meissner for critical help and information on her brother's life. Mrs. Meissner did not see this article before her death on February 8, 1989, but she approved the main biographical details in an earlier preliminary summary. Mrs. Meissner took a great interest in the project to produce the memorial volume dedicated to Ernst Hellinger, and she generously turned over to the author Hellinger's remaining mathematical and personal papers.

The author acknowledges assistance of archivists Patrick M. Quinn of the Northwestern University Library, Gerrit Walther of the Johann Wolfgang Goethe-Universität Frankfurt am Main, and Frederic F. Burchsted of the Archives of American Mathematics at the University of Texas.

The author thanks J. M. Anderson, Richard A. Beth, Ralph Boas, Louis and Tatiana de Branges, Diane de Branges, Robert B. Burckel, P. L. Butzer, Dr. Helmut M. Dehn, Michael A. Dritschel, George Gasper, Wilhelm and Gertrud Magnus, Wilhelm Maier, Karen V. H. Parshall, Maria Dehn Peters, Adolf Prag, Constance Reid, Mrs. W. T. Reid, Virginia Rovnyak, David Rowe, Norbert Schappacher, Hans A. Schmitt, Wolfgang Schwarz, Larry Smith, Helen Stange, Robert and Helen Sternberg, André Weil, Eva Toeplitz Wohl, and Daniel Zelinsky for information and valuable assistance.

Notes

1. Max Born, *My Life*, Taylor and Francis Ltd., London, 1978.
2. Constance Reid, *Courant*, Springer, New York, 1976.
3. Toeplitz has previously been honored as one of the founders of operator theory. Information on his life and work is given in separate articles by Max Born, Heinrich Behnke, and Gottfried Köthe in *Integral Equations and Operator Theory* 4 (1981), 278–297. The Toeplitz Memorial Conference in Operator Theory, held in Tel Aviv, May 11–15, 1981, resulted in the publication of additional memorial articles and research papers in a 588 page volume: *Toeplitz Centennial, Operator Theory: Advances and Applications*, Vol. 4, ed. I. Gohberg, Birkhäuser Verlag, Basel, 1982. Articles by Gottfried Köthe, Uri Toeplitz (son of Otto Toeplitz), and Israel Gohberg appear in the *Bonner Mathematische Schriften*, No. 143, Sonderband: Otto Toeplitz 1881–1940, Bonn, 1982. Toeplitz' scientific biography was written by Abraham Robinson: "Otto Toeplitz", *Dictionary of Scientific Biography*, Vol. 13, p. 428, Charles Scribner's Sons, New York, 1976.
4. Max Born, *My Life*; Constance Reid, *Hilbert–Courant*, Springer, New York, 1986.
5. A two-page typed summary of Dehn's talk is in the University Archives of Northwestern University Library. It was apparently prepared by someone other than Dehn, who worked from Dehn's notes. Dehn's sketches of Hilbert and Klein are tersely but clearly outlined. The version reproduced here is not represented as Dehn's spoken words, which are not recorded.
6. David Hilbert, "Wesen und Ziele einer Analysis der unendlichvielen unabhängigen Variablen," *Gesammelte Abhandlungen*, Bd. I, pp. 56–72, Julius Springer, Berlin, 1935.
7. Letter from André Weil of October 10, 1989. Weil knew Hellinger from his visits to Frankfurt from the late 1920's until the Nazis destroyed the faculty, and he met him repeatedly in Chicago from 1947 until his death in 1950.
8. D. Hilbert, *Grundzüge einer allgemeinen Theorie der linearen Integralgleichungen*, Leipzig, 1912; reprinting, Chelsea, New York, 1953. The original articles appeared in the *Nachrichten der Gesellschaft der Wissenschaften*, Göttingen, Math.-Phys. Klasse (1904), 49–91, 213–259; (1905), 307–338; (1906), 157–227, 439–480; (1910), 355–417.
9. Gottfried Köthe, "Scientific works of Otto Toeplitz," *Jahresbericht der Deutschen Mathematiker Vereinigung* **66** (1963), 8–16; English translation, *Integral Equations and Operator Theory* 4 (1981), 289–297.
10. We generally use the language of quadratic forms for historical accuracy, but sometimes we use modern language to better make a point. We simplify some formulas when nothing essential is lost. For example, we combine discrete and continuous spectra in integral formulas. In Hellinger's day, these were separated. The original works, or Hellinger's own survey articles [**8,9,11**], should be consulted for full statements.
11. The Hellinger integral continues to find applications today in some fields. An application in measure theory appears in S. Kakutani, "On equivalence of infinite product

measures," *Annals of Mathematics* **49** (1948), 214–224. The problem considered by Kakutani is to detect the equivalence or orthogonality of two product measures, $\mu = \Pi\,\mu_n$ and $\nu = \Pi\,\nu_n$, when the factor measures are pairwise equivalent. Kakutani's theorem is that the measures are equivalent or orthogonal according as a certain infinite product $\Pi\,\rho(\mu_n, \nu_n)$ is positive or zero. Kakutani attributes to von Neumann the recognition of the expression $\rho(\mu_n, \nu_n)$ as a Hellinger integral, namely,

$$\rho(\mu_n, \nu_n) = \int \sqrt{d\mu_n}\sqrt{d\nu_n}\,.$$

It was further noted by von Neumann that this expression, which is called a quasi-metric by Kakutani, can be converted to a true metric by introducing appropriate density functions. Dropping the subscript n, we can write the metric in the form

$$\begin{aligned} d(\mu, \nu) &= \Big(\int |\varphi - \psi|^2\, d\tau\Big)^{1/2} \\ &= 2\Big(1 - \int \sqrt{d\mu}\sqrt{d\nu}\Big)^{1/2}, \end{aligned}$$

where τ is a measure such that both μ and ν are absolutely continuous with respect to τ and φ and ψ are the corresponding density functions. This metric is widely used in modern statistical information theory, where either it or its square often bears the name *Hellinger distance* (*Encyclopedia of Statistical Sciences*, Vol. 3, Wiley, New York, 1983). The underlying statistical notions are associated with the name of A. Bhattacharyya; see I. J. Good, "C. 369. Invariant distance in statistics: some miscellaneous comments," J. Statist. Comput. and Simulation, to appear, and Thomas Kailath, "The divergence and Bhattacharyya distance measures in signal selection," *IEEE Trans. Commun. Tech.*, **Com–15** (1967), 52–60. Kakutani's theorem may be found in Daoxing Xia, *Measure and Integration Theory on Infinite-Dimensional Spaces*, Academic Press, New York, 1972. The Hellinger integral is used in different statistical applications, such as U. Grenander and G. Szegö, *Toeplitz Forms and their Applications*, University of California Press, Berkeley and Los Angeles, 1958, p. 207. Some applications use multivariate versions of the Hellinger integral. A study of matrix function theory based on an operator extension of the Hellinger integral appears in Yu. L. Shmul'yan, "A Hellinger operator integral," *Mat. Sbornik* (N.S.) **49 (91)** (1959), 381-430; *Amer. Math. Soc. Translations* (2) **22** (1962), 289–337. The author thanks Hamparsum Bozdogan, I. J. Good, Thomas Kailath, Marvin Rosenblum, and Daoxing Xia for helpful discussions on these uses of the Hellinger integral.

12. In a February 15, 1910, lecture to the Mathematische Gesellschaft in Göttingen, Otto Toeplitz reported on some results connected with the "Hilbert–Hellingerschen Theorie der beschränkten Formen" (*Jahresbericht der Deutschen Mathematiker Vereinigung*, Bd. 19 (1910), Nos. 3./4., 2. Abteilung, p. 77). See also Hermann Weyl, "Singuläre Integralgleichungen," *Mathematische Annalen* **66** (1909), p. 290, and "Über

gewöhnliche Differentialgleichungen mit Singularitäten und die zugehörigen Entwicklungen willkürlicher Funktionen," ibid. **68** (1910), p. 220.

13. E. Hilb, "Über die Auflösung linearer Gleichungen mit unendlich vielen Unbekannten," *Sitzungsber. d. Phys.-Med. Societ. in Erlangen* **40** (1908), 84–89.

14. H. Hahn, "Über die Integrale des Herrn Hellinger und die Orthogonalinvarianten der quadratischen Formen von unendlich vielen Veränderlichen," *Monatshefte Mathematik und Physik* **23** (1912), 161–224.

15. M. H. Stone, *Linear Transformations in Hilbert Space*, Amer. Math. Soc. Colloq. Publ., Vol. 15, Providence, 1932.

16. Otto Toeplitz, "Zur Theorie der quadratischen Formen von unendlichvielen Veränderlichen," *Nachrichten der Gesellschaft der Wissenschaften*, Göttingen, Math.-Phys. Klasse (1910), 489–506.

17. Georg Hamel, *Theoretische Mechanik*, Die Grundlehren der mathematischen Wissenschaften, Bd. 52, Springer, Berlin, 1967. See p. 375.

18. Von Paul Kluke, *Die Stiftungsuniversität Frankfurt am Main*, Waldemar Kramer, Frankfurt am Main, 1972, p. 54. Our account of this period also draws on the history of the mathematics department in Frankfurt by Wolfgang Schwarz and Jürgen Wolfart, "Zur Geschichte der Mathematischen Seminars der Universität Frankfurt am Main von 1914 bis 1945," Frankfurt, June 1988. There is also a new history of the university in Frankfurt by Notker Hammerstein, *Die Johann Wolfgang Goethe-Universität Frankfurt am Main*, Bd. I, Alfred Metzner Verlag, Neuwied/Frankfurt, 1989.

19. Definitive evidence on Hellinger's military service would have been in Prussian army records which were destroyed in a bombing raid in Potsdam in April 1945. The statement that Hellinger fought at the Somme, Vogesen, etc., appears in the Schwarz and Wolfart history of the Frankfurt mathematics seminar; Professor Schwarz confirmed that it is based on a document in Hellinger's personal file. The correspondence from the Dean in Frankfurt, as well as Hellinger's report on his *Döberitz* assignment, etc., are also from Hellinger's file in Frankfurt, with most of it duplicated in the files of the Prussian Minister of Education. The author is grateful to archivists in the *Bundesarchiv–Militärarchiv* in Freiburg (Federal Republic of Germany) and the *Zentrales Staatsarchiv* in Merseburg (German Democratic Republic) for assistance. The numbers of iron crosses awarded are from Hans-Peter Stein, *Symbole und Zeremoniell in deutschen Streitkräften*, Verlag E. S. Mittler & Sohn, Herford und Bonn, 1986, p. 59.

20. Wilhelm Magnus, "Vignette of a cultural episode," *Wilhelm Magnus, Collected Papers*, eds. G. Baumslag and B. Chandler, pp. 623–629, Springer, New York, 1984.

21. Carl Ludwig Siegel, "On the history of the Frankfurt mathematics seminar," *The Mathematical Intelligencer* **1** (1979), 223–230.

22. André Weil, Collected Papers, Vol. III, p. 460, Springer, New York, 1979.

23. The details of Hellinger's personal life in Frankfurt and his relationship with the Dehn family are from children of Max and Toni Dehn, Maria Dehn Peters (letters of September 6 and 27, 1989) and Dr. Helmut M. Dehn (letter of August 16, 1989). Both cite the rich friendships, the examples of serious work and exuberant play, and the breadth and depth of the minds of the mathematicians in Frankfurt as profound influences in their own lives. Most of the photographs used in this article were kindly given to the author by Maria Dehn Peters.

24. M. Born, W. Heisenberg, and P. Jordan, "Zur Quantenmechanik. II," *Zeitschrift für Physik* **35** (1926), 557–615.

25. Notker Hammerstein, *Die Johann Wolfgang Goethe-Universität Frankfurt am Main*, Bd. I, Alfred Metzner Verlag, Neuwied/Frankfurt, 1989, pp. 393–394.

26. Epstein corresponded with Toni Dehn after the Dehns emigrated. He told her of his intended *Freitod*. Helmut Dehn related this in his letter of August 16, 1989. Dr. Dehn characterized Epstein as a highly cultured person: "educated, just, sensitive, modest, but proud (in the best sense of the word), at home with the classics of German literature and music, himself a fine pianist."

27. The situation throughout Germany is detailed in a number of sources: (1) Max Pinl (in part with Auguste Dick), "Kollegen in einer dunklen Zeit I–IV, Nachtrag und Berichtigung," *Jahresbericht der Deutschen Mathematiker-Vereinigung* **71** (1969), 167–228; ibid. **72** (1971), 165–181; ibid. **73** (1972), 153–208; ibid. **75** (1974), 166–208; ibid. **77** (1976), 161–164. (2) Max Pinl and Lux Furtmüller, "Mathematicians under Hitler," *Leo Baeck Year Book* **13** (1973), pp. 129–182. (3) N. Schappacher unter Mitwirkung von M. Kneser, "Streiflichter auf das Verhältnis von Mathematik zu Gesellschaft und Politik in Deutschland seit 1890 unter besonderer Berücksichtigung der Zeit des Nationalsozialismus," *Festband zum 100–jährigen Bestehen der Deutschen Mathematiker-Vereinigung*, Vieweg-Verlag, Braunschweig, 1990.

28. The originals of the letters of December 17, 1935, and December 23, 1935, are among Hellinger's private papers. Copies of the two letters and the letter of October 15, 1935, are in Hellinger's personal file (*Rektorat*) in Frankfurt. These and other documents were made available to the author through the university archives of the Johann Wolfgang Goethe-Universität Frankfurt am Main.

29. Communication from Wilhelm and Gertrud Magnus, May, 1989. Magnus said that Hellinger played an important role in his life, both as a teacher and as an older friend. Gertrude Magnus was a student in Hellinger's last class in Analytic Geometry in 1934–35.

30. N. I. Ahiezer, *The Classical Moment Problem*, Moscow, 1961; English translation, Hafner, New York, 1965. The quotation is from the preface of the English translation and takes minor liberties such as omitting literature citations given by Ahiezer.

31. Lipman Bers, "The migration of European mathematicians to America," *History of Mathematics*, Vol. 1, pp. 231–243, Amer. Math. Soc., Providence, 1988. Ivan Niven, "The threadbare thirties," ibid. pp. 209–229. Nathan Reingold, "Refugee mathemati-

cians in the United States of America, 1933–1941: reception and reaction," ibid. pp. 175–200.

32. In Evanston, he continued to work the bureaucracy to aid his mother. In a letter to the German authorities from Evanston, dated September 17, 1940, he reported that his salary was raised to $2,000. In a reply dated November 9, 1940, receipt of the information was acknowledged, and the German authorities responded that only nine months of his retirement salary could therefore be paid to his mother that year. In July 1941 Hellinger and his sister Hanna succeeded in having their mother join them in Chicago. In this period, correspondence from the German authorities to Hellinger was addressed to "Ernst *Israel* Hellinger," because he was a Jew (his middle name was David). Hellinger acknowledged the form of address by initializing his letters "E. I. H."
33. The chair incident is reported in a letter of February 27, 1948, from Mathematics Department Chairman H. T. Davis to Simeon Leland, Dean of the College of Liberal Arts, in the University Archives of Northwestern University.
34. Letter of May 18, 1988, from Helen Stange, a former student and mathematics librarian at Northwestern University. A view of Hellinger as a teacher is also given in a letter by two former Northwestern University students, Robert and Helen Sternberg, in the *The Mathematical Intelligencer* **2** (1980), 62.
35. E. J. McShane of the University of Virginia told the author that he was approached by Northwestern University about this time, but he was not interested. The correspondence between Davis and Snyder is in the University Archives of Northwestern University.
36. The details of Hellinger's last days are from Dr. Helmut M. Dehn, Maria Dehn Peters, and André Weil. See also Siegel's account in the *The Mathematical Intelligencer* and Hanna Meissner's essay in this volume.
37. This is consistently reported by many sources. Only fragments of the correspondence between Hellinger and Frankfurt from this period seem to survive. A letter in his personal file (*Rektorat*) in Frankfurt suggests that Hellinger once actually did entertain the idea, if only fleetingly. A letter from the *Rektor* in Frankfurt, dated December 20, 1949, expresses profuse gratitude to Hellinger for writing that he might respond positively to their recall offer (the letter of Hellinger to which the *Rektor* refers is not preserved). The *Rektor* states that the faculty regarded him as their number one priority for restoration, and rules were changed to allow him to maintain his position until age 70, when he would be emerited. In a hand-written note dated February 4, 1950, Hellinger indicates that he was touched by the warmth of the *Rektor*'s December 20 letter. He says he is too weak to give a proper response and refers to a letter to the dean (not preserved) for his final decision (presumably negative). The correspondence is evidence of a moment of reconciliation, no doubt all that was possible under the circumstances. The tone of reconciliation did not last, and other correspondence shows that Hellinger clearly was not happy with the proposals being made to him. Among those working hard for the restoration of both Hellinger and Dehn was Willy Hartner. Additional details are in university archives in Frankfurt.

38. W. Magnus and R. Moufang, "Max Dehn zum Gedächtnis," *Math. Annalen* **127** (1954), 215–227; W. Magnus, "Max Dehn," *The Mathematical Intelligencer* **1** (1978), 132–143.

39. Wilhelm Magnus, "Ernst Hellinger," *Dictionary of Scientific Biography*, Vol. VI, pp. 235–236, Charles Scribner's Sons, New York, 1972.

40. William T. Reid papers, Archives of American Mathematics, University Archives, University of Texas at Austin.

Department of Mathematics
Mathematics–Astronomy Building
University of Virginia
Charlottesville, Virginia 22903–3199
U. S. A.

Operator Theory:
Advances and Applications, Vol. 48

Ernst Hellinger

HANNA MEISSNER†

How I, at age 92, remember my brother, 38 years after his death.

My parents had two children, each like an only child. My brother was twelve years older than I. I adored him—he was probably the best children's nurse my parents ever had. His love and understanding of children lasted all his life. When I was little he would interrupt his studies for school to fulfill my wishes, be it to make me a geometrical drawing or to repair a doll's broken arm or head.

When I started school, my brother started university. He had no doubts that he would study mathematics, but it needed several visits of Ernst's mathematics teacher in the *Gymnasium* to convince our father that these were the right and only studies for him. While he studied at the University of Breslau, he lived at home. He always had time for me, no matter how busy he was or how much time his students, friends, or his student organization required. He had many friends, mathematicians and others. There was a group of seven or eight young men who had been friends through the twelve years of the *Gymnasium* and stayed friends forever.

My brother was not a gregarious type, but all his life he had a talent for making friends and keeping them, no matter where he was. I was about ten years old when my brother left Breslau to accept an assistantship in the mathematical department in Göttingen, to Felix Klein and David Hilbert. He came always home during the vacations. He had a strong feeling for family—all his life he wrote home once a week, longer or shorter letters in his characteristic but barely readable handwriting which only my mother could decipher. It was always a feast for me when he was at home and he made time for me, no matter how busy he was. He took great interest in my development, answered my questions with great patience, and it was due to his influence that I left the private school and entered the girl's *Gymnasium*, the first and only one in Breslau at that time.

† *Hanna Meissner died on Wednesday, February 8, 1989. She was visited two days before by Tatiana de Branges. They had tea, and there was lively conversation about the project to produce the Hellinger memorial volume. Hanna Meissner expressed again that she was glad to have contributed to the project and that* **Integral Equations and Operator Theory** *made it possible to honor his work.*

Shortly after he wrote his doctoral thesis under Hilbert, he left Göttingen for the University of Marburg as *Privatdozent*. He made friends with other young faculty members. When he came home he would tell amusing stories, sometimes quite disrespectful of university bureaucracy, the older professors, and the life in a small university town. He had an odd sense of humor, sometimes biting wit, and while he would do everything for friends, he had strong dislikes for some people and would judge hard and not always justly.

The new University of Frankfurt am Main was opened in August 1914. My brother was offered a professorship, at that time quite an honor for a young Jew, thirty years of age. The opening took place with all the pomp of the old universities. One of many stories my brother told was the following: There were two separate entrances for students and faculty. Ernst looked always much younger than he was. As expected of him, he stood in the row of professors waiting for the opening of the doors. When the *Pedell* (*Pedells* in the German universities were very important people with a responsibility for the orderly procedure of official functions) rushed up to him, lifted him out of the row by his collar and said, "The entrance for students is over there!" (Since Ernst liked to say, "It is not important how a story happened, it is important how it could have happened," we never knew if the lifting by the collar was a nice exaggeration.)

My brother's time at the university was interrupted by the First World War. He was not drafted, but volunteered his services in the second year. He became a *Flieger-Funker*: that means he had to test the new radio equipment in the new war planes. While he was standing around in airports, he smoked his first cigarette. He got along well with all his comrades and kept his sense of humor about people and the army's red tape. Always very much interested in politics, he had been quite conservative before the war, but his political views got a sharp drift to the left during and after the war.

I joined my brother after the war in Frankfurt to finish my studies (not mathematics). We lived at that time in the house of Professor Max Born with his charming wife and two little daughters. Max Born had been a friend of my brother's since student days, and now Ernst became also a very good friend of his wife. The two little daughters adored him and used as much time as possible to be up in our apartment. While Ernst had never been attracted by young unmarried women and had been a determined bachelor, he was usually very popular with the wives of his friends and colleagues, and some of them joined him in fast and lasting friendships. He became one of the family, the uncle of the children with whom they could talk about their joys and worries and without whom no family feast was complete. While he liked small gatherings of friends, he was not of a gregarious nature. He abhorred large parties; he also liked to be alone, take long walks, spend his vacations alone in beautiful but very quiet places at the ocean or in the mountains, and could be very happy not to make any acquaintance or speak to anyone.

Despite the very confusing and unpleasant events after the war, life at the university was peaceful. Cooperation among the mathematicians was close, not marred by personal striving and ambition. Teaching was very important. My brother was an outstanding teacher. He spent very much time on careful preparation of his lectures and could arouse interest in lesser gifted students. He was very popular with students and famous for witty and amusing remarks. He had infinite patience in personal conferences with students who had questions. His interest in students went beyond mathematics. For many years he held an honorary position on the board of the student Aid Society, and like everything that he did, he took his position very seriously. Wilhelm Magnus, one of his students‡, now professor emeritus at New York University, and Ludwig Siegel, colleague and friend, give interesting descriptions of the relations with his students [Wilhelm Magnus, "Vignette of a cultural episode," Collected Papers, pp. 623–629, Springer-Verlag, New York, 1984; Carl Ludwig Siegel, "On the history of the Frankfurt mathematics seminar," The Mathematical Intelligencer, Vol. 1, no. 4 (1979), pp. 223–230].

I left Frankfurt in 1922, but returned after six years to a position which kept me away from the university. While I still saw my brother often and his friends sometimes, our relationship was less close than in former years. He was still the same understanding brother, but his humor and wit were more sarcastic and his judgment of people sharper and more critical.

My brother had always been extremely orderly and methodical in his private affairs, everything had its place and so had everyone. Change was uncomfortable. While he had some understanding of the political situation, the possibility of a radical change was far from his thoughts. A typical example: It was January 1933. I was in bed with a flu, listening to the radio, and heard to my greatest disgust that Hitler had become *Reichskanzler.* My brother visited me soon after and I told him what I had heard. He looked at me worried, felt my forehead and said: "How much fever do you have again?" It took all my strength to convince him that I was perfectly well and so was the radio.

The big changes at the University began in 1935. Ernst, Max Dehn, and others lost their positions. Ernst had his apartment, his books, his desk, paper, and pens and some friends who were either pure "Arians" or could not accept the seriousness of the situation. I had left Germany in 1933 for the USA, but there was no way to convince my brother to leave. When I met my mother and brother in the spring of 1938 in Italy, Ernst again refused definitely to undertake any steps to leave Germany. Then came November 1938 and the *Kristallnacht* when all Jewish men and men of Jewish descent were arrested and transported to concentration

---

‡ *Magnus took courses from Hellinger, but he was a doctoral student of Max Dehn.*

camps. I received a telegram from Siegel telling me in coded language that Ernst had been transported to the Dachau concentration camp. With the help of friends, a position as lecturer at Northwestern University was found for him. He spent six weeks in Dachau and was discharged because a university position was found for him and therefore the restricted visa policy of the USA did not apply. He had to leave Germany as soon as possible. I learned to know a quality of my brother which I had not known and because of his aversion to change had not expected. He adjusted quite well in the USA to completely strange surroundings, to different methods of teaching and learning; he made friends quickly at the University and became popular with students. He never spoke with anyone about his experience in Dachau. He told me about two decisions he had made, when he stood waiting with many others for the train which should bring them to their destiny. When he saw an SS man shooting down a poor old man who could not stand erect and upright, he decided 1) to leave Germany as soon as possible, and 2) to survive the camp with as little damage as possible to his fundamental personality, no matter how difficult. And so he did.

Despite the considerable distance between Evanston (Northwestern University) and the south of Chicago, where I lived, we saw each other frequently during his first three years. As in my youth, he was the one who had to remind me frequently not to forget the humor in difficulties arising in my job and not to take myself too seriously. But I felt that fundamentally he was not as well balanced as he liked to appear. Having loved Germany as he did, he could not get over the abolishment of justice and the cruelties.

He had always had a very strong bond with our mother. As conditions grew worse and worse in Germany, he insisted—despite my objections—that at age 84 she should come to us. She did so in 1941 under terrible difficulties and was happy to be here.

Ernst stayed at Northwestern as full professor until he reached retirement age in 1949. Since his pension would have been very small after only those few years, he was fortunate that the Illinois Institute of Technology offered him a visiting professorship for two years. He did not want to return to Germany although the German government had offered him return to his old position. During his first semester at IIT, he became very ill and he died of cancer in March 1950. Again I detected a contradiction in my brother's personality I never understood. He did not want to know what his illness was, the word cancer was never mentioned, he did not ask his doctor, nor us or his friends. If he knew that it was hopeless, he did not mention it. He died without having suffered too much.

Lafayette, Indiana, USA
February 1988

Operator Theory:
Advances and Applications, Vol. 48

ON THE REPRESENTATION OF CERTAIN HOLOMORPHIC FUNCTIONS DEFINED ON A POLYDISC

Jim Agler

We derive a concrete representation of the hereditary holomorphic functions h defined on $\mathbb{D}^d \times \mathbb{D}^d$ that are nonnegative on all d-tuples of commuting contractions. This gives rise to solutions on the polydisc $\mathbb{D}^d$ of the Herglotz representation problem and the realization problem for contractive functions, but in a norm that corresponds to the ordinary H^∞ norm only when $d = 1$ or 2.

1. INTRODUCTION

Let $\mathcal{F}_d$ denote the collection of all commuting d-tuples of contractive operators acting on a complex Hilbert space $\mathcal{H}$ of infinite dimension. $T = (T_1, ..., T_d)$ acting on a complex Hilbert space $\mathcal{H}$ of infinite dimension. If ϕ is a holomorphic function on the polydisc $\mathbb{D}^d$ in $\mathbb{C}^d$, set

$$\|\phi\|_{d,\infty} = \sup_{r<1, T\in\mathcal{F}_d} \|\phi(rT)\|. \tag{1.1}$$

H_d^∞ will denote the Banach space of holomorphic functions on $\mathbb{D}^d$ such that $\|\phi\|_{d,\infty} < \infty$. If $d = 1$ or $d = 2$ then it is known ([**N**], [**An**], [**Sz.-N,F**]) that $H_d^\infty = H^\infty(\mathbb{D}^d)$ and if $d > 2$ it is known that $H_d^\infty \neq H^\infty(\mathbb{D}^d)$ ([**P**] or [**V**] implies that there exists $\phi \in H^\infty(\mathbb{D}^d)$ with $\|\phi\|_\infty < \|\phi\|_{d,\infty}$). However in general, $H^\infty(\mathbb{D}^d) \subseteq H_d^\infty$ and $\|\phi\|_\infty \leq \|\phi\|_{d,\infty}$.

In [**A2**] the author generalized the classical Nevanlinna-Pick interpolation theorem ([**N**] and [**Pi**]) to the space H_d^∞ by establishing the following result.

Theorem 1.2. *Let* $\lambda_1, ..., \lambda_n \in \mathbb{D}^d$ *and let* $z_1, ..., z_n \in \mathbb{C}$. *There exists* $\phi \in H_d^\infty$ *with* $||\phi||_{d,\infty} \leq 1$ *and* $\phi(\lambda_i) = z_i$ *for each* i *if and only if there exist* d *positive semidefinite matrices* $(a_{ij}^1), ..., (a_{ij}^d)$ *such that*

$$1 - \overline{z}_i z_j = \sum_{r=1}^{d} (1 - \overline{\lambda}_{ir} \lambda_{jr}) a_{ij}^r.$$

for all i, j.

The work in this paper grew out of the desire to generalize another piece of classical function theory to H_d^∞, namely the Herglotz representation theorem.

Theorem 1.3. ([**Her**]). ψ *is a holomorphic function on* $\mathbb{D}$ *with* $Re\ f \geq 0$ *and* $f(0) = 1$ *if and only if there exists a probability measure supported on* $\partial\mathbb{D}$ *such that*

$$\psi(\lambda) = \int_{\partial\mathbb{D}} \frac{1 + e^{i\theta}\lambda}{1 - e^{i\theta}\lambda} d\mu(e^{i\theta})$$

for all $\lambda \in \mathbb{D}$.

Since the statement and the usual classical proofs of Theorem 1.3 are too concrete (from the point of view of functional analysis) to be easily generalized to several variables we reformulate Theorem 1.3 as the following theorem (see [**F**]). If $\mathcal{H}$ is a Hilbert space, let $\mathcal{L}(\mathcal{H})$ denote the space of bounded linear transformations of $\mathcal{H}$.

Let $\mathcal{C}$ be a Hilbert space.

Theorem 1.4. ψ *is a holomorphic* $\mathcal{L}(\mathcal{C})$*-valued function on* $\mathbb{D}$ *with* $Re\ \psi \geq 0$ *and* $\psi(0) = 1$ *if and only if there exists a Hilbert space* $\mathcal{M}$, *an isometry* $V : \mathcal{C} \to \mathcal{M}$, *and a unitary* *operator $U \in \mathcal{L}(\mathcal{M})$ such that*

$$\psi(\lambda) = V^*(1 + U\lambda)\,(1 - U\lambda)^{-1}V \tag{1.5}$$

for all $\lambda \subset \mathbb{D}$.

Theorem 1.4, while treating the somewhat spacious generality of an $\mathcal{L}(\mathcal{M})$-valued map, nevertheless contains whatever function theory is present in Theorem 1.3. Indeed, given the spectral theorem, Theorem 1.3 and Theorem 1.4 are equivalent.

We now describe our generalization of Theorem 1.4 to the polydisc $\mathbb{D}^d$. If $\mathcal{M}$ is Hilbert space let us agree to say that $D = (P_1, ..., P_d)$ is a *partition* of $\mathcal{M}$ if D is a d-tuple of orthogonal projections on $\mathcal{M}$ that are pairwise orthogonal and sum to one (equivalently, $\mathcal{M} = \oplus_{r=1}^d \text{ran } P_r$). If D is a partition of $\mathcal{M}$, then $D(\lambda)$, an $\mathcal{L}(\mathcal{M})$-valued holomorphic map on $\mathbb{D}^d$ can be defined by the formula

$$D(\lambda) = \sum_{r=1}^{d} \lambda_r P_r.$$

Observe that if $d = 1$, then (1.5) takes the form

$$\psi(\lambda) = V^*(1 + UD(\lambda)\,(1 - UD(\lambda))^{-1}V.$$

If ϕ is a holomorphic $\mathcal{L}(\mathcal{C})$-valued on $\mathbb{D}^d$ let us agree to say that ϕ is $\mathcal{F}_d$*-contractive* if

$$\|\phi(rT)\| \leq 1 \;\; \textit{whenever} \;\; T \in \mathcal{F}_d \;\; \textit{and} \;\; r < 1. \tag{1.6}$$

Evidently, if $dim\ \mathcal{C} = 1$, then a tautology is that ϕ is $\mathcal{F}_d$-contractive if and only if $\|\phi\|_{d,\infty} \leq 1$. Also, the work of Arveson ([**Ar1**] and [**Ar2**]) implies that if $d = 1$ or $d = 2$ then ϕ is $\mathcal{F}_d$-contractive if and only if

$$\|\phi(\lambda)\| \leq 1 \;\; \textit{for all} \;\; \lambda \in \mathbb{D}^d,$$

i.e. ϕ is contractive in the usual sense.

Following (1.6) let us agree to say that a holomorphic $\mathcal{L}(\mathcal{C})$-valued function ψ on $\mathbb{D}^d$ is $\mathcal{F}_d$-*Herglotz* if

$$(1.7) \qquad Re\ \psi(rT) \geq 0 \quad \textit{whenever} \quad T \in \mathcal{F}_d \quad \textit{and} \quad r < 1.$$

Reasoning similar to that in the previous paragraph guarantees that if $d = 1$ or $d = 2$ then ψ is $\mathcal{F}_d$-Herglotz if and only if $Re\ \psi(\lambda) \geq 0$ for all $\lambda \in \mathbb{D}^d$. Furthermore, it is easy check that in general ψ is $\mathcal{F}_d$-Herglotz if and only if $\frac{\psi-1}{\psi+1}$ is $\mathcal{F}_d$-contractive.

We now are able to state our generalization of the classical Herglotz representation theorem.

Theorem 1.8. *Let $\mathcal{C}$ be a Hilbert space. A holomorphic $\mathcal{L}(\mathcal{C})$-valued function ψ on $\mathbb{D}^d$ with $\psi(0) = 1$ is $\mathcal{F}_d$-Herglotz if and only if there exist a Hilbert space $\mathcal{M}$, a decomposition D of $\mathcal{M}$ (into d summands), an isometry $V : \mathcal{C} \to \mathcal{M}$ and a unitary operator $U \in \mathcal{L}(\mathcal{M})$ such that*

$$(1.9) \qquad \psi(\lambda) = V^*(1 + UD(\lambda))\,(1 - UD(\lambda))^{-1}V$$

for all $\lambda \in \mathbb{D}^d$.

We now turn to the problem of representing $\mathcal{F}_d$-contractive holomorphic functions on $\mathbb{D}^d$. Here the classical analog of Theorem 1.4 is the following result.

Theorem 1.10. *Let $\mathcal{C}$ be a Hilbert space. ϕ is a holomorphic $\mathcal{L}(\mathcal{C})$-valued function on $\mathbb{D}$ with $\|\phi(\lambda)\| \leq 1$ for all $\lambda \in \mathbb{D}$ if and only if there exist a Hilbert space $\mathcal{M}$ and a unitary operator*

$$U = \begin{bmatrix} U_{00} & U_{01} \\ U_{10} & U_{11} \end{bmatrix}$$

acting on $\mathcal{C} \oplus \mathcal{M}$ *such that*

$$\phi(\lambda) = U_{00} + U_{01}\lambda(1 - U_{11}\lambda)^{-1}U_{10} \tag{1.11}$$

for all $\lambda \in \mathbb{D}$.

Theorem 1.10 when applied to the characteristic function of a contraction represents an important component of the Sz-Nagy-Foias theory of contractions (see **[Sz.-N,F]** for the theorem's proof and an elaboration on this theme as well as for historical comments on the theorem). Also, formula (1.11), which is referred to as the realization of ϕ in engineering circles plays an important role in systems theory (see [**H**]).

Our generalization of Theorem 1.10 to the polydisc is the following theorem.

Theorem 1.12. *Let* $\mathcal{C}$ *be a Hilbert space.* ϕ *is a holomorphic* $\mathcal{L}(\mathcal{C})$*-valued function on* $\mathbb{C}^d$ *which is* $\mathcal{F}_d$*-contractive if and only if there exist a Hilbert space* $\mathcal{M}$*, a unitary operator*

$$U = \begin{bmatrix} U_{00} & U_{01} \\ U_{10} & U_{11} \end{bmatrix}$$

acting on $\mathcal{C} \oplus \mathcal{M}$ *and a decomposition* D *of* $\mathcal{M}$ *(with* d *summands) such that*

$$\phi(\lambda) = U_{00} + U_{01}D(\lambda)(1 - U_{11}D(\lambda))^{-1}U_{10} \tag{1.13}$$

for all $\lambda \in \mathbb{D}^d$.

2. THE HEREDITARY FUNCTIONAL CALCULUS ON $\mathcal{F}_d$.

In this section we shall derive a basic fact (Theorem 2.6 below) about the hereditary functional calculus ([**A1**]) for elements of $\mathcal{F}_d$. The proof of Theorem 2.6 is based on the first part of the proof of Theorem 3.16 in [**A2**].

Let $\mathcal{C}$ be a Hilbert space and let U be a domain $\mathbb{C}^d$. If $h(\lambda,\mu)$ a $\mathcal{L}(\mathcal{C})$-valued holomorphic function on $U\times\overline{U}$, we say that h is *positive semi-definite on U* if

$$\sum_{i=1}^{n} < h(\lambda_i,\overline{\lambda}_j)c_i,c_j > \geq 0$$

whenever n is a positive integer $\lambda_1,...,\lambda_n\in\mathbb{D}^d$, and $c_1,...,c_n\in\mathcal{C}$. The following theorem which gives a representation for $\mathcal{L}(\mathcal{C})$-valued positive definite holomorphic functions dates back to Aronsjan [**Aro**].

Theorem 2.1. *Let $\mathcal{C}$ be a Hilbert space, let U be a domain in $\mathbb{C}^d$, and let h be a holomorphic $\mathcal{L}(\mathcal{C})$-valued function on $U\times\overline{U}$. h is positive semidefinite if and only if there exist a Hilbert space $\mathcal{H}$ and a holomorphic $\mathcal{L}(\mathcal{C},\mathcal{H})$-valued function F on U such that*

$$h(\lambda,\overline{\mu}) = F(\mu)^*F(\lambda)$$

for all $\lambda,\mu\in U$.

Now, suppose $T=(T_1,...,T_d)$ is a pairwise commuting d-tuple of operators in $\mathcal{L}(H)$ with $\sigma(T)\subseteq\mathbb{D}^d$ (equivalently, $\sigma(T_r)\subseteq\mathbb{D}$ for each r). If $m=(m_1,...,m_d)$ is a multi-index define $T^m=T_1^{m_1}\cdot\ldots\cdot T_d^{m_d}$. If h is a holomorphic $\mathcal{L}(\mathcal{C},\mathcal{M})$-valued function on $\mathbb{D}^d$, then h possesses a power series representation

$$h(\lambda)=\sum_m \hat{h}(m)\lambda^m \tag{2.2}$$

which converges uniformly on compact subsets of $\mathbb{D}^d$. Since $\sigma(T) \subseteq \mathbb{D}^d$ it follows that

$$h(T) = \sum_m \hat{h}(m) \otimes T^m$$

defines an operator $h(T) \in \mathcal{L}(\mathcal{C}, \mathcal{M}) \otimes \mathcal{L}(\mathcal{H})$. More generally, if h is a holomorphic $\mathcal{L}(\mathcal{C}, \mathcal{M})$-valued function defined on $\mathbb{D}^d \times \mathbb{D}^d$, then $h(T) \in \mathcal{L}(\mathcal{C}, \mathcal{M}) \otimes \mathcal{L}(\mathcal{H})$ can be defined by

$$h(T) = \sum_{m,n} \hat{h}(m,n) \otimes T^{*n} T^m \tag{2.3}$$

where

$$h(\lambda, \overline{\mu}) = \sum_{m,n} \hat{h}(m,n) \overline{\mu}^n \lambda^m.$$

We note two algebraic identities of the hereditary functional calculus. Let $\mathcal{H}, \mathcal{H}_1, \mathcal{H}_2$, $\mathcal{H}_3$ and $\mathcal{H}_4$, be Hilbert spaces, assume that f, g are holomorphic functions on $\mathbb{D}^d$ taking values in $\mathcal{L}(\mathcal{H}_1, \mathcal{H}_2)$ and $\mathcal{L}(\mathcal{H}_3, \mathcal{H}_4)$ respectively, and assume that h is a holomorphic function on $\mathbb{D}^d \times \mathbb{D}^d$ taking values in $\mathcal{L}(\mathcal{H}_2, \mathcal{H}_3)$. Then naturally, $g(\overline{\mu}) h(\lambda, \overline{\mu}) f(\lambda)$ is a holomorphic function on $\mathbb{D}^d \times \mathbb{D}^d$ taking values in $\mathcal{L}(\mathcal{H}_1, \mathcal{H}_4)$. Furthermore, if $T \in \mathcal{L}(\mathcal{H})$ then

$$(g(\overline{\mu}) h(\lambda, \overline{\mu}) f(\lambda))(T) = g(T^*) h(T) f(T). \tag{2.4}$$

If h is as in the previous paragraph, we define a function $\check{h}$ by the formula

$$\check{h}(\lambda, \overline{\mu}) = h(\mu, \overline{\lambda})^* \quad , \quad \lambda, \mu \in \mathbb{D}^d.$$

With this definition $\check{h}$ is a holomorphic $\mathcal{L}(\mathcal{H}_3, \mathcal{H}_2)$ valued function and

$$h(T)^* = \check{h}(T). \tag{2.5}$$

Theorem 2.6. *Let $\mathcal{C}$ be a Hilbert space and let h be a holomorphic $\mathcal{L}(\mathcal{C})$-valued function defined on $\mathbb{D}^d \times \mathbb{D}^d$. $h(\rho T) \geq 0$ for all $T \in \mathcal{F}_d$ and all $\rho < 1$ if and only if there exist d Hilbert spaces $\mathcal{M}_1, ..., \mathcal{M}_d$ and d holomorphic maps $f_1, ..., f_d$ defined on $\mathbb{D}^d$ with f_r $L(\mathcal{C}, \mathcal{M}_r)$-valued and such that*

$$h(\lambda, \overline{\mu}) = \sum_{r=1}^{d} (1 - \overline{\mu}_r \lambda_r) f_r(\mu)^* f_r(\lambda) \tag{2.7}$$

for all $\lambda, \mu \in \mathbb{D}^d$.

PROOF. First assume that h has the form in (2.7). Fix $T \in \mathcal{F}_d$ and let $\rho < 1$. Using (2.4) and (2.5) we find that

$$h(\rho T) = \sum_{r=1}^{d} f_r(\rho T)^* (1 - \rho^2 T_r^* T_r) f_r(\rho T)$$

Since $T \in \mathcal{F}_d$, $1 - \rho^2 T_r^* T_r \geq 0$ and we see that $h(\rho T) \geq 0$ as was to be proved.

We now establish the converse direction of Theorem 2.6. We shall assume for convenience that $\mathcal{C}$ is separable. Fix a basis $\{e_i\}$ of $\mathcal{C}$. Let H denote the topological vector space of holomorphic $\mathcal{L}(\mathcal{C})$-valued functions defined on $\mathbb{D}^d \times \mathbb{D}^d$ with the topology induced by the countable sequence of seminorms

$$\|h\|_n = \max_{\lambda, \mu \in \left(\frac{n}{n+1} \mathbb{D}^-\right)^d} \|P_n h(\lambda, \overline{\mu}) P_n\| .$$

Here, P_n denotes the orthogonal projection of $\mathcal{C}$ onto $[e_i : 1 \leq i \leq n]$. H carries a locally convex Hausdorf topology. It is metrizable but is not complete. Define a convex cone C in H as follows.

$$C = \left\{ h \in H : \ h(\lambda, \overline{\mu}) = \sum_{r=1}^{d} (1 - \overline{\mu}_r \lambda_r) f_r(\lambda, \overline{\mu}), \ \ where \right.$$

$$\left. f_r \ \ is \ positive \ semidefinite \ holomorphic \ on \ \mathbb{C}^d \times \mathbb{C}^d \ \ for \ each \ \ r \right\}$$

We claim that C is closed in H. To see this assume that

$$h^j(\lambda, \overline{\mu}) = \sum (1 - \overline{\mu}_r \lambda_r) f_r^j(\lambda, \overline{\mu}) \in C$$

and that $h^j(\lambda, \overline{\mu}) \to h(\lambda, \overline{\mu}) \in H$. For each $n \geq 1$ inductively construct a sequence j_1^n, $j_2^n, \ldots$ as follows. For $n = 1$ note that since $h^j \to h$ in particular we have that

$$\lim_{j \to \infty} \sum (1 - |\lambda_r|^2) \ P_1 f_1^j(\lambda, \overline{\lambda}) P_1 = P_1 h(\lambda, \overline{\lambda}) P_1.$$

for $\lambda \in \left(\frac{1}{2} \mathbb{D}\right)^d$. Since f_r^j is positive semidefinite, $f_r^j(\lambda, \overline{\lambda}) \geq 0$, and consequently there exists a positive bounded function $g : \left(\frac{1}{2}\mathbb{D}\right)^d \to \mathcal{L}(P_1 \mathcal{C})$ such that

$$P_1 f_r^j(\lambda, \overline{\lambda}) P_1 \leq g(\lambda)$$

for all j, each r, and all $\lambda \in \left(\frac{1}{2}\mathbb{D}\right)^d$. Finally, since f_r^j is holomorphic and positive semidefinite we conclude there exists a subsequence $\{j_l^1\}$ and d $\mathcal{L}(P_1 \mathcal{C})$-valued positive semidefinite holomorphic functions $g_r^1(\lambda, \overline{\mu})$ such that

$$P_1 f_r^{j_l^1}(\lambda, \overline{\mu}) P_1 \to g_r^1(\lambda, \overline{\mu})$$

uniformly on compact subsets of $\left(\frac{1}{2}\mathbb{D}\right)^d \times \left(\frac{1}{2}\mathbb{D}\right)^d$.

Now suppose that $\{j_l^1\}, \ldots, \{j_l^{n-1}\}$ have been defined. The argument in the preceding paragraph guarantees that there exists a subsequence $\{j_l^n\}$ of $\{j_l^{n-1}\}$ and that there exist d $\mathcal{L}(P_n \mathcal{C})$-valued positive semidefinite holomorphic functions $g_r^n(\lambda, \overline{\mu})$ such that

$$P_n f_r^{j_l^n}(\lambda, \overline{\mu}) P_n \to g_r^n(\lambda, \overline{\mu})$$

uniformly on compact subsets of $\left(\frac{n}{n+1}\mathbb{D}\right)^d \times \left(\frac{n}{n+1}\mathbb{D}\right)^d$.

Now define d holomorphic l^2 matrix-valued functions $G_r(\lambda, \overline{\mu})$ on $\mathbb{D}^d \times \mathbb{D}^d$ by the formula

$$G_r(\lambda, \overline{\mu})_{ij} = \lim_{h \to \infty} < g_r^n(\lambda, \overline{\mu})e_j, e_i > .$$

G_r is well defined since by construction if $m \leq n$, $g_r^m(\lambda, \overline{\mu}) = P_m g_r^n(\lambda, \overline{\mu}) P_m$. Furthermore since by construction,

$$(2.9) \qquad < h(\lambda, \overline{\mu})e_j, e_i > = \sum (1 - \overline{\mu}_r \lambda_r) G_r(\lambda, \overline{\mu})_{ij} \quad ,$$

$h \in C$ provided that there exist $\mathcal{L}(\mathcal{C})$-valued holomorphic maps g_r with

$$(2.10) \qquad < g_r(\lambda, \overline{\mu})e_j, e_i > = G_r(\lambda, \overline{\mu})_{ij}$$

(i.e. $G_r(\lambda, \overline{\mu})$ is a bounded operator on l^2). Since $G_r(\lambda, \overline{\lambda})$ is positive semidefinite, (2.9) implies that $G_r(\lambda, \overline{\lambda})$ is bounded. Since $G_r(\lambda, \overline{\mu})$ is positive semidefinite $G_r(\lambda, \overline{\mu})$ is bounded. Hence (2.10) defines $L(\mathcal{C})$-valued maps g_r and we obtain from (2.9) that

$$h(\lambda, \overline{\mu}) = \sum (1 - \overline{\mu}_r \lambda_r) g_r(\lambda, \overline{\mu})$$

which establishes that C is closed.

Now assume that h_0 is a holomorphic $\mathcal{L}(\mathcal{C})$-valued function on $\mathbb{D}^d \times \mathbb{D}^d$ with the property that $h_0(\rho T) \geq 0$ whenever $\rho < 1$. By Theorem 2.1, Theorem 2.6 will follow if we can show that $h_o \in C$. Since C is closed, the Hahn-Banach separation principle implies that $h_o \in C$ if and only if

$$(2.11) \qquad Re\ L(h_o) \geq 0$$

whenever $L \in H^*$ has the property that

$$\text{(2.12)} \qquad Re\ L(h) \geq 0 \ \ for\ all \ \ h \in C\ .$$

Accordingly assume that $L \in H^*$ and that (2.12) holds. The proof of Theorem 2.6 will be complete if we can show that (2.11) holds. Define $L_1 \in H^*$ by the formula

$$L_1(h) = \frac{1}{2}\left(L(h) + \overline{L(\check{h})}\right).$$

Let H_o denote the vector space of holomorphic $\mathcal{L}(\mathcal{C},\mathbb{C})$-valued maps defined on $\mathbb{D}^d$. Define a sesquilinear form $[\cdot,\cdot]$ on H_o by the formula

$$[f,g] = L_1(g(\mu)^* f(\lambda))\ .$$

Observe that if $h \in H$ and $h = \check{h}$ then

$$L_1(h) = Re\ L(h)\ .$$

Hence since $(g(\mu)^* f(\lambda))\check{} = g(\mu)^* f(\lambda)$, we deduce from (2.12) and the fact that $g(\mu)^* f(\lambda) \in C$ that [,] is positive semidefinite on H_0. Letting $N = \{f \in H_0 : [f,f] = 0\}$ we deduce via Cauchy's inequality that N is a subspace of H_o and that [,] induces an inner product on H_o/N. Let $H^2(L_1)$ denote the Hilbert space obtained by completing H_o/N with respect to this inner product. Attempt to densely define a d-tuple of operators $M = (M_1, ..., M_d)$ on $H^2(L_1)$ by the formulas

$$(M_r f)(\lambda) = \lambda_r f(\lambda)\ .$$

Now, let $f \in H_o$ and fix r. We have

$$\begin{aligned}\|f\|^2 - \|M_r f\|^2 &= [f,f] - [M_r f, M_r f] \\ &= L_1(f(\mu)^* f(\lambda)) - L_1(\overline{\mu}_r \lambda_r f(\mu)^* f(\lambda)) \\ &= L_1((1 - \overline{\mu}_r \lambda_r) f(\mu)^* f(\lambda)) \geq 0\end{aligned}$$

since $(1-\overline{\mu}_r\lambda_r)\ f(\mu)^*f(\lambda) \in C$. Hence M_r is not only well defined on H_o/N but extends by continuity to a contraction defined on $H^2(L_1)$. Thus, $M \in F_d$.

Fix $h = \sum_{m,n} c_{mn}\overline{\mu}^n\lambda^m \in H$, let $\rho < 1$, and let $f = \sum e_j \otimes f_j \in C \otimes H^2(L_1)$. We wish to derive a formula for $< h(\rho M)f, f >$.

$$\begin{aligned}
\langle h(\rho M)f, f\rangle &= \sum_{m,n} \langle c_{mn} \otimes (\rho M)^{*n}(\rho M)^m f, f\rangle \\
&= \sum_{m,n}\sum_{i,j} \langle c_{mn} \otimes (\rho M)^{*n}(\rho M)^m\ (e_j \otimes f_j),\ e_i \otimes f_i\rangle \\
&= \sum_{m,n}\sum_{i,j} \langle c_{mn}e_j, e_i\rangle\ [(\rho M)^m f_j, (\rho M)^n f_i] \\
&= \sum_{m,n}\sum_{i,j} \langle c_{mn}e_j, e_i\rangle\ L_1\left((\overline{\rho\mu})^n(\rho\lambda)^m f_i(\mu)^* f_j(\lambda)\right) \\
&= L_1\left(\sum_{i,j}\langle\sum_{m,n} c_{mn}(\overline{\rho\mu})^n(\rho\lambda)^m e_j, e_i\rangle\ f_i(\mu)^* f_j(\lambda)\right) \\
&= L_1\left(\sum_{i,j}\langle h(\rho\lambda, \overline{\rho\mu})e_j, e_i\rangle f_i(\mu)^* f_j(\lambda)\right).
\end{aligned}$$

Letting $f = \sum_{j=1}^n e_j \otimes (1 \otimes e_j)$ in the above calculation and noting in that event that $f_i(\mu)^* f_j(\lambda) = e_i \otimes e_j$. We thus obtain that

$$[h(\rho M)f, f] = L_1(P_n h(\rho\lambda, \overline{\rho\mu})P_n)\ .$$

Since $M \in \mathcal{F}_d$, $[h_o(\rho M)f, f] \geq 0$ *for all* $\rho < 1$. Hence

$$L_1(P_n h(\rho\lambda, \overline{\rho\mu})P_n) \geq 0$$

whenever $\rho < 1$ and $n \geq 1$. By the continuity of L_1 we deduce that

$$L_1(h) \geq 0. \tag{2.13}$$

Finally, since $h_o(\rho M) \geq 0$ whenever $M \in \mathcal{F}_d$ and $\rho < 1$ we claim that $h_o = \check{h}_o$. To see this simply take $M = (\lambda_1, ..., \lambda_d)$ with $\lambda \in \mathbb{D}^d$. In particular, we have that

$L_1(h_o) = Re\ L(h_o)$ and consequently (2.13) implies that (2.11) holds. This establishes Theorem 2.6.

3. THE REPRESENTATION

In this section we shall derive the representation theorems stated in the introduction. We shall require the following elementary lemma.

Lemma 3.1. *Assume that $f : \mathbb{D}^d \to \mathcal{L}(\mathcal{C}, \mathcal{H})$ and $g : \mathbb{D}^d \to \mathcal{L}(\mathcal{C}, \mathcal{K})$ are holomorphic operator-valued functions and define $F \subseteq \mathcal{H}$ and $G \subseteq \mathcal{K}$ by*

$$F = [f(\lambda)c : \ \lambda \in \mathbb{D}^d, \ c \in \mathcal{C}]$$

and

$$G = [g(\lambda)c : \ \lambda \in \mathbb{D}^d, \ c \in \mathcal{C}].$$

If the $\mathcal{L}(\mathcal{C})$-valued kernel over $\mathbb{D}^d$,

$$g(\mu)^* g(\lambda) - f(\mu)^* f(\lambda) \tag{3.2}$$

is positive semidefinite, then there exists a unique contractive linear mapping $L : \ G \to F$ with

$$f(\lambda) = Lg(\lambda) \ \ \text{for all} \ \ \lambda \in \mathbb{D}^d. \tag{3.3}$$

Proof Define L on vectors in G of the form $g(\lambda)c$ by the formula

$$L(g(\lambda)c) = f(\lambda)c.$$

Using positive semidefiniteness of (3.2) we observe that

$$\begin{aligned}\|L(\sum g(\lambda_i)e_i)\|^2 &= \|\sum f(\lambda_i)c_i\|^2 \\ &= \sum_{i,j} < f(\lambda_j)^* f(\lambda_i)c_i, c_j > \\ &\leq \sum_{i,j} < g(\lambda_j)^* g(\lambda_i)c_i, c_j > \\ &= \|\sum g(\lambda_i)c_i\|^2 ,\end{aligned}$$

an inequality that both shows L is well defined as a linear operator and extends by continuity to a contraction defined on all of G. L obviously satisfies (3.3) and (3.3) obviously determines L uniquely. This establishes Lemma 3.1.

As a corollary of Lemma 3.1 we obtain the following result.

Lemma 3.4. *In the setup of Lemma 3.1 if we assume in place of (3.2) that*

$$g(\mu)^* g(\lambda) = f(\mu)^* f(\lambda) \text{ for all } \lambda, \mu \in \mathbb{D}^d, \tag{3.5}$$

then there exists a unique Hilbert space isomorphism $L: G \to F$ *with*

$$f(\lambda) = Lg(\lambda) \text{ for all } \lambda \in \mathbb{D}^d.$$

Proof Using (3.5) and applying Lemma 3.1 twice we deduce the existence of contractive linear maps $L_1 : G \to F$ and $L_2 : F \to G$ such that $f(\lambda) = L_1 g(\lambda)$ and $g(\lambda) = L_2 f(\lambda)$ for all $\lambda \in \mathbb{D}^d$. In particular $L_1 L_2 = id$ on F and $L_2 L_1 = id$ on G. Hence $L_2 = L_1^{-1}$. Since $\|L_1\| \leq 1$ and $\|L_1^{-1}\| \leq 1$ we deduce that L_1 is a Hilbert space isomorphism. Lemma 3.4 now follows by setting $L = L_1$.

We now turn to the proof of Theorem 1.12 from the introduction. Accordingly, assume that

$$\phi \in \mathcal{L}(\mathcal{C}) \otimes H_d^\infty \text{ with } \|\phi\|_{d,\infty,\mathcal{C}} \leq 1.$$

By Theorem 2.1 and Theorem 2.6 we deduce there exists for each r in the range $1 \le r \le d$ a holomorphic mapping $f_r : \mathbb{D}^d \to \mathcal{L}(\mathcal{C}, \mathcal{H}_r)$ with

$$1 - \phi(\mu)^* \phi(\lambda) = \sum_{r=1}^{d} (1 - \overline{\mu}_r \lambda_r) f_r(\mu)^* f_r(\lambda) \tag{3.6}$$

for all $\lambda, \mu \in \mathbb{D}^d$. Rewrite (3.6) in the form

$$1 + \sum_r (\mu_r f_r(\mu))^* \lambda_r f_r(\lambda) = \phi(\mu)^* \phi(\lambda) + \sum_{r=1}^{d} f_r(\mu)^* f_r(\lambda), \tag{3.7}$$

set $\mathcal{H} = \mathcal{K} = \mathcal{C} \oplus \oplus_{r=1}^{d} \mathcal{H}_r$, and define holomorphic maps $f : \mathbb{D}^d \to \mathcal{L}(\mathcal{C}, \mathcal{H})$ and $g : \mathbb{D}^d \to \mathcal{L}(\mathcal{C}, \mathcal{K})$ by

$$f(\lambda) = \begin{bmatrix} \phi(\lambda) \\ f_1(\lambda) \\ \vdots \\ f_d(\lambda) \end{bmatrix}$$

and

$$g(\lambda) = \begin{bmatrix} 1 \\ \lambda_1 f_1(\lambda) \\ \vdots \\ \lambda_d f_d(\lambda) \end{bmatrix}$$

It should be clear from (3.7) that $(f(\mu)^* f(\lambda) = g(\mu)^* g(\lambda)$ for all $\lambda, \mu \in \mathbb{D}^d$ and hence by Lemma 3.4 there exists a Hilbert space isomorphism $L : G \to F$ such that

$$f(\lambda) = Lg(\lambda) \tag{3.8}$$

for all $\lambda \in \mathbb{D}^d$. Since $\mathcal{H} = \mathcal{K}$ and since dim $F =$ dim G there exists a unitary $U \in \mathcal{L}(\mathcal{H})$ with the property that $U \,|\, G = L$; if necessary the spaces $\mathcal{H}_r$ can be replaced by Hilbert spaces containing $\mathcal{H}_r$ to guarantee that $\dim(\mathcal{H} \ominus F) = \dim(\mathcal{H} \ominus G)$). In particular (3.8) implies that

$$f(\lambda) = Ug(\lambda) \tag{3.9}$$

for all $\lambda \in \mathbb{D}^d$. Set $\mathcal{M} = \oplus_r \mathcal{H}_r$ and decompose U with respect to the decomposition $\mathcal{H} = \mathcal{C} \oplus \mathcal{M}$. Thus

$$U = \begin{bmatrix} U_{00} & U_{01} \\ U_{10} & U_{11} \end{bmatrix}$$

with $U_{00} \in \mathcal{L}(\mathcal{C})$, $U_{01} \in \mathcal{L}(\mathcal{M}, \mathcal{C})$, $U_{10} \in \mathcal{L}(\mathcal{C}, \mathcal{M})$, and $U_{11} \in \mathcal{L}(\mathcal{M}, \mathcal{M})$. Furthermore if we decompose

$$f(\lambda) = \begin{bmatrix} \phi(\lambda) \\ F(\lambda) \end{bmatrix}$$

and

$$g(\lambda) = \begin{bmatrix} 1 \\ D(\lambda)F(\lambda) \end{bmatrix}$$

where $F(\lambda) \in \mathcal{L}(\mathcal{C}, \mathcal{M})$ and $D(\lambda) \in \mathcal{L}(\mathcal{M})$ is defined by

$$D(\lambda)(\oplus h_r) = \lambda_r h_r \ ,$$

then (3.9) becomes the pair of equations

$$\phi(\lambda) = U_{00} + U_{01} D(\lambda) F(\lambda)$$

and

$$F(\lambda) = U_{10} + U_{11} D(\lambda) F(\lambda) \ ,$$

valid for all $\lambda \in \mathbb{D}^d$. Finally, the above equations are solved to obtain

$$\phi(\lambda) = U_{00} + U_{01} D(\lambda)(1 - U_{11} D(\lambda))^{-1} U_{10} \ ,$$

which establishes (1.13). To establish the converse direction of Theorem 1.12 assume that (1.13) holds. Directly calculating with the relations

$$U_{00}^* U_{00} + U_{10}^* U_{10} = 1 \qquad U_{00}^* U_{01} + U_{10}^* U_{11} = 0$$

$$U_{01}^*U_{00} + U_{11}^*U_{10} = 0 \qquad U_{01}^*U_{01} + U_{11}^*U_{11} = 1$$

yields the identity

$$(3.10)\quad 1 - \phi(\mu)^*\phi(\lambda) = ((1 - U_{11}D(\mu))^{-1}U_{10})^* \ (1 - D(\mu)^*D(\lambda))(1 - U_{11}D(\lambda))^{-1}U_{10} \ ,$$

which can be rewritten in the form

$$(3.11)\qquad 1 - \phi(\mu)^*\phi(\lambda) = \sum_{r=1}^{d}(1 - \overline{\mu}_r\lambda_r)f_r(\mu)^*f_r(\lambda)$$

where $f_r(\lambda) : \mathcal{C} \to \mathcal{M}$ is defined by

$$f_r(\lambda) = P_r(1 - U_{11}D(\lambda))^{-1}U_{10}.$$

From (3.11) it now follows immediately via Theorem 2.6 that ϕ is F_d-contractive. This concludes our proof of Theorem 1.12.

We now turn to the proof of Theorem 1.8 in the introduction. Accordingly assume that ψ is a $\mathcal{F}_d$-Herglotz $\mathcal{L}(\mathcal{C})$-valued holomorphic function on $\mathbb{D}^d$ which is normalized so that $\psi(0) = 1$. By Theorem 2.6 we obtain that

$$(3.12)\qquad 2(\psi(\mu)^* + \psi(\lambda)) = \sum_{r=1}^{d}(1 - \overline{\mu}_r\lambda_r)f_r(\mu)^*f_r(\lambda).$$

Here, for each r, $f_r : \mathbb{D}^d \to \mathcal{L}(\mathcal{C}, \mathcal{H}_r)$ is holomorphic. Rewriting (3.10) in the form

$$(1 + \psi(\mu))^*(1 + \psi(\lambda)) + \sum_r(\mu_r f_r(\mu))^*\lambda_r f_r(\lambda) = (1 - \psi(\mu))^*(1 - \psi(\lambda)) + \sum_r f_r(\mu)^*f_r(\lambda)$$

and continuing as in the proof of Theorem 1.12 we obtain a model space $\mathcal{M}$ and a unitary operator

$$W = \begin{bmatrix} W_{00} & W_{01} \\ W_{10} & W_{11} \end{bmatrix}$$

defined on $\mathcal{C} \oplus \mathcal{M}$ with properties that

$$(3.13) \qquad \begin{cases} 1 - \psi(\lambda) = W_{00}(1 + \psi(\lambda)) + W_{01}D(\lambda)F(\lambda) & \& \\ \\ F(\lambda) = W_{10}(1 + \psi(\lambda)) + W_{11}D(\lambda))F(\lambda) \end{cases}$$

for all $\lambda \in \mathbb{D}^d$. Here,

$$F(\lambda) = \begin{bmatrix} f_1(\lambda) \\ \vdots \\ f_d(\lambda) \end{bmatrix}.$$

We shall require the following lemma whose proof is a straight-forward calculation.

Lemma 3.14. *Let $\mathcal{H}_0$ and $\mathcal{H}_1$ be Hilbert spaces, let*

$$W = \begin{bmatrix} W_{00} & W_{01} \\ W_{10} & W_{11} \end{bmatrix}$$

be a bounded operator on $\mathcal{H}_0 \oplus \mathcal{H}_1$, assume that $-1 \notin \sigma(W_{00})$ and set $U = W_{11} - W_{10}(1 + W_{00})^{-1}W_{01} \in \mathcal{L}(\mathcal{H}_1)$. If W is an isometry, then U is an isometry and if W is unitary, then U is unitary.

Theorem 1.8 can now be proven by solving equations (3.13) for ψ and then using Lemma 3.14 to express ψ in the necessary form. As the algebra of this computation is somewhat tedious we shall adopt a lazier approach. Differentiate the equations (3.13) with respect to λ_r and eliminate ψ_r to obtain

$$(3.15) \qquad \frac{\partial}{\partial \lambda_r} F(\lambda) = U \frac{\partial}{\partial \lambda_r}(D(\lambda)F(\lambda))$$

where $U = W_{11} - W_{10}(1 + W_{00})^{-1}W_{01}$ is a unitary operator on $\mathcal{M}$ by Lemma 3.14 (here, since $\psi(0) = 1$, $W_{00} = 0$, so there is no difficulty in meeting the condition $-1 \notin \sigma(W_{00})$). Now, solving the system of equations in (3.15) gives that there exists $A \in \mathcal{L}(\mathcal{C}, \mathcal{M})$ such that

$$(3.16) \qquad F(\lambda) = UD(\lambda)F(\lambda) + A$$

for all $\lambda \in \mathbb{D}^d$. Hence

$$F(\lambda) = (1 - UD(\lambda))^{-1}A. \tag{3.17}$$

Now, (3.12) may be rewritten in the form

$$2(\psi(\mu)^* + \psi(\lambda)) = F(\mu)^*F(\lambda) - (D(\mu)F(\mu))^*D(\lambda)F(\lambda).$$

Thus, we obtain from (3.16), that

$$2(\psi(\mu)^* + \psi(\lambda)) = (UD(\mu)F(\mu))^*A + A^*UD(\lambda)F(\lambda) + A^*A,$$

and since $\psi(0) = 1$, we have that

$$\psi(\lambda) = \frac{1}{2}A^*UD(\lambda)F(\lambda) + 1$$

where $A^*A = 4$. Finally, getting $V = \frac{1}{2}A$ and using (3.17) we see that

$$\begin{aligned}\psi(\lambda) &= 2V^*UD(\lambda)(1 - UD(\lambda))^{-1}V + 1\\ &= V^*(2UD(\lambda)(1 - UD(\lambda))^{-1} + 1)V\\ &= V^*(1 + UD(\lambda))(1 - UD(\lambda))^{-1}V\end{aligned}$$

which establishes (1.9).

To establish the converse direction of Theorem 1.8 assume that ψ has the form in (1.9) and compute directly that

$$\frac{1}{2}(\psi(\mu)^* + \psi(\lambda)) = V^*(1 - UD(\mu))^{-1*}(1 - D(\mu)^*D(\lambda))(1 - UD(\lambda))^{-1}V. \tag{3.19}$$

Setting $f_r(\lambda) = P_r(1 - UD(\lambda))^{-1}V$ and rewriting (3.18) in the form

$$\frac{1}{2}(\psi(\mu)^* + \psi(\lambda)) = \sum_{r=1}^{d}(1 - \overline{\mu}_r\lambda_r)f_r(\mu)^*f_r(\lambda)$$

we deduce via Theorem 2.6 that ψ is $\mathcal{F}_d$-Herglotz. This concludes the proof of Theorem 1.8.

References

[A1] Agler, J., "An abstract approach to model theory," *Surveys of some recent results in operator theory,* **Vol. 2** J. B. Conway and B. B. Morrel (editors), Longman, Essex, 1988.

[A2] Agler, J., "Some interpolation theorems of Nevanlinna-Pick type," preprint.

[An] Ando, T., "On a pair of commutative contractions," *Acta Sci. Math.* **121** (1963), 88-90.

[Aro] Aronszajn, N., "Theory of reproducing kernels," *Trans. Am. Math. Soc.* **68** (1950), 337-404.

[Ar1] Arveson, W. B., "Subalgebras of C^*-algebras," *Acta Math.* **123** (1969), 141-224.

[Ar2] Arveson, W. B., "Subalgebras of C^*-algebras II," *Acta Math.* **128** (1972), 271-308.

[F] Fillmore, P. A., *Notes on Operator Theory*, Van Nostrand Reinhold, New York, 1970.

[H] Helton, J. W., *Operator theory, analytic functions, matrices, and electrical engineering*, CBMS Regional Conference Series in Mathematics No. 68, AMS, 1987.

[N] Nevanlinna, R., "Über Beschränkte Funktionen die in gegebene Punkten vorgeschriebene Werte annehmen," *Ann. Acad. Sci. Fenn. Ser.* **A13** (1919), No. 1.

[P] Parrott, S., "Unitary dilations for commuting contractions," *Pac. Jour. Math.* **34** (1970), 481-490.

[Pi] Pick, G., "Über die Beschränkungen analytischer Funktionen, welche durch vorgegebene Funktionswerte bewirkt werden," *Math. Ann.* **77** (1916), 7-23.

[Sz.-N,F] Sz.-Nagy, B. and Foias, C., *Harmonic Analysis of Operators on Hilbert Space*, North Holland, Amsterdam-London, 1970.

[V] Varopoulos, N., "On an inequality of von Neumann and an application of the metric theory of tensor products to operator theory," *Jour. Func. Anal.* **16** (1974), 83-100.

[vN] von Neumann, J., "Eine Spectraltheorie für allgemeine Operatoren eines unitären Ranmes," *Math. Nach.* **4** (1951), 258-281.

Department of Mathematics, C-012
University of California, San Diego
La Jolla, CA 92093-0112

Operator Theory:
Advances and Applications, Vol. 48

INVARIANT HILBERT SPACES OF ANALYTIC FUNCTIONS ON BOUNDED SYMMETRIC DOMAINS

* Jonathan Arazy and Stephen D. Fisher

We determine those Hilbert spaces of analytic functions on a bounded symmetric domain in $\mathbf{C}^N$ which are mapped unitarily onto themselves by composition with each automorphism of the domain. We give several extensions of this result to cases when the action of the automorphism group is weighted or when the action is only isomorphic (and not isometric) or when the space is only a semi-Hilbert space; that is, the "norm" has a non-trivial kernel.

INTRODUCTION. Let Ω be a bounded symmetric domain in $\mathbf{C}^N$ in its Harish-Chandra realization. That is, the origin lies in Ω, Ω is convex and circular, and $Aut(\Omega)$, the group of biholomorphic automorphisms of Ω, acts transitively on Ω. We shall also assume that Ω is irreducible; that is, Ω is not biholomorphically equivalent to a Cartesian product of two or more bounded symmetric domains. We shall generally not have need in this paper for the whole group $Aut(\Omega)$, but only for the connected component of the identity, which we denote by G. Let dV be Lebesgue measure on Ω normalized so that Ω has measure one and let $L_a^2 = L_a^2(\Omega, dV)$ be the space of complex-valued analytic functions on Ω which lie in $L^2(\Omega, dV)$; L_a^2 is the *Bergman space* of Ω. For a point $w \in \Omega$, the linear functional $f \to f(w)$ is continuous on L_a^2; the function in L_a^2 which represents this functional is the *Bergman kernel for* w:

$$f(w) = \int_\Omega f(z)\overline{K(z;w)}dV(z) \quad ; \quad f \in L_a^2 \tag{0.1}$$

*Research supported in part by a grant from the U.S.-Israel Bi-national Science Foundation.

It is elementary that

$$K(z;w) \quad \text{is analytic in } z \text{ and conjugate analytic in } w \tag{0.2(a)}$$

$$K(z;w) = \overline{K(w;z)} \quad ; \quad z,w \in \Omega \tag{0.2(b)}$$

$$K(z;w) \neq 0 \quad ; \quad z,w \in \Omega \tag{0.2(c)}$$

As is usual, we write $K_w(z)$ for $K(z;w)$.

Let p be the genus of Ω (see [L] or [U1]) and let $\Lambda = \Lambda(\Omega)$ be those non-negative real numbers λ for which

$$[K(z;w)]^{\lambda/p} \quad \text{is non-negative definite.}$$

Λ is called the Wallach set of Ω; see [V-R]. For each irreducible bounded symmetric domain Ω, the Wallach set $\Lambda(\Omega)$ consists of a finite set and a ray, (λ_0, ∞); see [B], [FK], and the summary discussion in Section 1.

For $\lambda \in \Lambda(\Omega)$, we define an inner product on the linear span of $\mathcal{K} = \{K_w^{\lambda/p} : w \in \Omega\}$ by the rule

$$(K_w^{\lambda/p}, K_\zeta^{\lambda/p}) = [K(\zeta;w)]^{\lambda/p} \tag{0.3}$$

and then extend this by linearity to the linear span of $\mathcal{K}$. Define $\mathcal{H}_\lambda$ to be closure of the linear span of $\mathcal{K}$ with respect to this inner product. Thus, $\mathcal{H}_\lambda$ is a Hilbert space of analytic functions on Ω and $K(z;w)^{\lambda/p}$ is its reproducing kernel.

When $\lambda > p-1$, the integral

$$\int_\Omega [K(z;z)]^{1-\lambda/p} dV(z) \tag{0.4}$$

converges. For such λ, let c_λ be the value of the integral in (0.4) and define

$$d\mu_\lambda(z) = \frac{1}{c_\lambda}[K(z;z)]^{1-\lambda/p} dV(z).$$

Then, for these λ, $\mathcal{H}_\lambda$ is nothing but those analytic functions on Ω which lie in $L^2(\Omega;\mu_\lambda)$. (This is a consequence of our Theorem 3, for example, but may also be derived directly.)

Let $\varphi \in Aut(\Omega)$. A change of variables argument shows that the Bergman kernel satisfies

$$K(z;a) = (J\varphi)(z)K(\varphi(z);\varphi(a))\overline{(J\varphi)(a)}$$

where $J\varphi$ is the complex Jacobian of φ. Hence, both of the Hilbert spaces $\mathcal{H}_\lambda$ (for all $\lambda \in \Lambda$) and $L^2(\Omega;\mu_\lambda)$ (for $\lambda > p-1$) are mapped unitarily onto themselves by each of the operators $U_\varphi^{(\lambda)}$ given by the formula

$$(U_\varphi^{(\lambda)} f)(z) = f(\varphi(z))[(J\varphi)(z)]^{\lambda/p}; z \in \Omega \tag{0.5}$$

for $\varphi \in Aut(\Omega)$. Note that $U_{\psi\varphi}^{(\lambda)} = U_\varphi^{(\lambda)} U_\psi^{(\lambda)}$ for $\varphi, \psi \in Aut(\Omega)$. A canonical set of semi-Hilbert spaces on which the operators $U_\varphi^{(\lambda)}$ are unitary is described by J. Faraut and A. Koranyi in [FK] and is summarized here in Section 1. B. Orsted [O] had earlier described these spaces in an important special case. It is the purpose of this paper to show that for $\lambda \in \Lambda(\Omega)$, there are in essence no other semi-Hilbert spaces $\mathbf{H}$ on which the operators $U_\varphi^{(\lambda)}$ are unitary or, in some cases, even uniformly bounded.

<u>Assumptions on H</u>

Let $\lambda \in \Lambda(\Omega)$ be fixed. We shall make the following assumptions on $\mathbf{H}$:

$\mathbf{H}$ is a linear space of analytic functions on Ω which has a hermitian sesqui-linear form $(f,g)_\mathbf{H}$ satisfying

$$(f,f)_\mathbf{H} \geqslant 0 \tag{0.6}$$

The semi-norm $\|f\|_\mathbf{H} = (f,f)_\mathbf{H}^{\frac{1}{2}}$ may be zero without f being identically zero. We assume that the quotient space

$$\mathbf{H}/\{f \in \mathbf{H} : \|f\|_\mathbf{H} = 0\} \tag{0.7}$$

is complete.

Let K consist of those $\varphi \in G$ which send 0 to 0. It is well-known that each $\varphi \in K$ extends to a linear mapping of $\mathbf{C}^N$ onto itself and that K is a maximal compact subgroup of G. We shall assume that for each finite Borel measure μ on K, the operator

$$f \mapsto \int_K f(kz)\,d\mu(k) \tag{0.8}$$

maps $\mathbf{H}$ continuously into itself and that the integral in (0.8) converges in the semi-norm of $\mathbf{H}$.

Finally, we shall suppose that the operators $U_\varphi^{(\lambda)}$ map $\mathbf{H}$ onto itself with uniformly bounded norms:

$$\sup_{\varphi\in G}\sup_{\|f\|_{\mathbf{H}}\leq 1}\|U_\varphi^{(\lambda)}f\|_{\mathbf{H}} = C < \infty \tag{0.9}$$

In [AF] the present authors showed by ad hoc methods that when Ω is the open unit disc in the complex plane and $\lambda = 0$, there is just one semi-Hilbert space satisfying (0.6)–(0.9), namely the Dirichlet space consisting of all analytic functions $f(z) = \sum_0^\infty a_n z^n$ on $|z| < 1$ satisfying

$$\|f\|^2 = \sum_1^\infty n|a_n|^2 \quad < \infty.$$

In [P1], J. Peetre extended this result when $\lambda = 0$ and when $C = 1$ in (0.9) to the unit ball of complex N-space. He showed that the only space satisfying (0.6)–(0.9) consists of those analytic functions $f(z) = \sum a_\alpha z^\alpha$ for which

$$\|f\|^2 = \sum_\alpha |\alpha|\frac{\alpha!}{|\alpha|!}|a_\alpha|^2 \quad < \infty; \tag{0.10}$$

K. Zhu [Z] rediscovered this result in 1988. In [P2] Peetre extended the result once again, this time to domains of type I (c.f. Section 1) for most λ, again under the assumption that $C = 1$ in (0.9). The authors wish to thank Jaak Peetre for several helpful conversations. In particular, he suggested averaging over a minimal parabolic subgroup, a technique we employ heavily in Section 3. This extends the authors' use, in [AF1], of averaging over an amenable subgroup. Finally, the present authors developed, but never published, much of the contents of the present paper for the special case when Ω is the open unit disc in the complex plane; this is in [AF2].

We have tried to make the paper accessible to readers with no knowledge of bounded symmetric domains. Therefore, in our presentation of the background material, and in particular of [FK], we have tried to simplify matters and to avoid as much as we can the use of techniques from Lie algebras and Jordan triple systems. See [Helgason] for the connection between bounded symmetric domains and Lie algebras and [L], [UI] and [U2] for the connections between bounded symmetric domains and Jordan triple systems. We will make use of, but not explain, parameters p, a, r; we also avoid the use of Gindikin's gamma function.

SECTION 1. BACKGROUND.

PROPOSITION 1. *Suppose that* **H** *satisfies (0.6)–(0.9). Then* **H** *contains polynomials and is the closure of the polynomials which lie in it.*

PROOF. Let $f \in \mathbf{H}$; the function

$$f_r(z) = \frac{1}{2\pi}\int_0^{2\pi} f(e^{it}z)\frac{1-r^2}{1-2r\cos t+r^2}\,dt; \quad 0<r<1 \quad , \quad z \in \Omega$$

lies in **H** by (0.8). This function is analytic on a neighborhood of the closure of Ω and $f_r \to f$ in **H** as $r \to 1$ by Poisson summability arguments. Further,

$$f_N(z) = \frac{1}{2\pi}\int_0^{2\pi} f(e^{it}z)\Big[\sum_{k=0}^{N}(1-\frac{k}{N+1})e^{-ikt}\Big]\,dt$$

also lies in **H**, is a polynomial, and $f_N \to f$ in **H** as $N \to \infty$, by arguments with the Fejer kernel and Cesaro summability.

The irreducible bounded symmetric domains were completely classified up to biholomorphic automorphisms by E. Cartan; see [C]. For the reader's convenience and our future reference, we list below the Cartan classification of the various irreducible bounded symmetric domains Ω, the values of λ which comprise $\Lambda(\Omega)$, the genus p and rank r of Ω, and the value of a parameter a which will be needed later.

$\mathrm{I}_{n,m}$: $1 \leq n \leq m$: $n \times m$ matrices with complex entries and matrix (=operator) norm less than one.

$$\Lambda = \{0,1,2,\ldots,n-1\} \cup (n-1,\infty)$$

$$p = n+m \quad ; \quad r = n; \quad a = 2.$$

II_m: $m \geqslant 5$, $m \times m$ anti-symmetric matrices with complex entries and matrix norm less than one.

$$\Lambda = \{0,2,\ldots,m-2\} \cup (m-2,\infty) \qquad \text{if} \quad m \quad \text{is even}$$

$$\Lambda = \{0,2,\ldots,m-3\} \cup (m-3,\infty) \qquad \text{if} \quad m \quad \text{is odd}$$

$$p = 2m-2 \quad ; \quad r = [m/2]; \quad a = 4.$$

III_m: $m \geqslant 2$, $m \times m$ symmetric matrices with complex entries and matrix norm less than one.

$$\Lambda = \{0, \frac{1}{2}, 1, \frac{3}{2}, \ldots, \frac{1}{2}(m-1)\} \cup (\frac{1}{2}(m-1), \infty)$$
$$p = m+1 \quad ; \quad r = m; \quad a = 1.$$

IV_m : $m \geqslant 3$, $1 \times m$ matrices $z = (z_1, \ldots, z_m)$ satisfying

$$\Big[(\sum_1^m |z_j|^2)^2 - |\sum_1^m z_j^2|^2\Big]^{\frac{1}{2}} < 1 - \sum_1^m |z_j|^2;$$
$$\Lambda = \{0, \frac{1}{2}(m-2)\} \cup (\frac{1}{2}(m-2), \infty)$$
$$p = m \quad ; \quad r = 2; \quad a = m-2.$$

V: 1×2 matrices with entries from the complex 8-dimensional Cayley algebra with spectral norm less than one

$$\Lambda = \{0, 3\} \cup (3, \infty)$$
$$p = 12 \quad ; \quad r = 2; \quad a = 6.$$

VI: 3×3 matrices with entries from the complex 8- dimensional Cayley algebra which are hermitian with respect to the Cayley conjugation, [L;4.11], with spectral norm less than one

$$\Lambda = \{0, 4, 8\} \cup (8, \infty)$$
$$p = 18 \quad ; \quad r = 3; \quad a = 8.$$

Spaces of type I–IV are called the classical domains. Note that the open unit ball B of $\mathbf{C}^N$ is a type $\mathrm{I}_{1,N}$ domain; this is the only Cartan domain of rank 1.

Each irreducible bounded symmetric domain Ω is the open ball of $\mathbf{C}^N$ (for an appropriate N) with some norm; we shall denote this space by Z. The (complex) dimension N for the respective spaces are as follows: $\mathrm{I}_{n,m}$ has dimension nm; II_m has dimension $m(m-1)/2$; III_m has dimension $m(m+1)/2$; and VI_m has dimension m; V has dimension 16; and VI has dimension 27.

We now give a summary of some of the results in the recent paper of J. Faraut and A. Koranyi [FK]; we shall use their notation and numbering of theorems.

The space of all analytic polynomials on Ω is P. As above, r is the rank of Ω. A signature $\mathbf{m}$ is an r-tuple of integers $\mathbf{m} = (m_1, \ldots, m_r)$ with $m_1 \geqslant \cdots \geqslant m_r \geqslant 0$. $\Delta_1, \ldots, \Delta_r$ are certain cannonical polynomials on Ω. (Δ_q is the Koecher "norm" polynomial corresponding to $G^{(q)}$; [U1; p. 47].) When Ω is of type I, or III, Δ_q is nothing but the determinant of the $q \times q$ submatrix in the upper left corner of z and hence is a polynomial of degree q in the coordinate functions $\{z_{jk} : 1 \leq j, k \leq q\}$; the case of type II_m domains, Δ_r is the square root of $\det z$ and $\Delta_j, 1 \leq j < r$, is defined analogously; see [U1]. For a signature $\mathbf{m}$, we set

$$\Delta_{\mathbf{m}} = \Delta_1^{m_1 - m_2} \ldots \Delta_{r-1}^{m_{r-1} - m_r} \Delta_r^{m_r} \tag{1.1}$$

and

$$P_{\mathbf{m}} = \text{linear span } \{\Delta_{\mathbf{m}}(kz) : \quad k \in K\}. \tag{1.2}$$

If q is a polynomial in the coordinate functions, we set

$$q^*(x) = \overline{q(\overline{x})}.$$

The Fischer inner product on P is defined by

$$(p \mid q)_F = p(\frac{\partial}{\partial x})q^* \mid_{x=0} \tag{1.3}$$

Equivalently,

$$(p \mid q)_F = \frac{1}{\pi^N} \int_\Omega p(x)\overline{q(x)} e^{-\|x\|^2} dV(x) \tag{1.4}$$

where $\|x\|$ is the unique normalized K-invariant inner product norm on Z. For instance, for type I domains, the normalization is $\|z_{ij}\| = 1$.

THEOREM 2.1 [FK]. *$P = \sum_{\mathbf{m}} \oplus P_{\mathbf{m}}$ is the orthogonal decomposition of P into K-irreducible, F-orthogonal subspaces.*

This is the Peter-Weyl decomposition of P; see [FK], [S], and [U3].

$P_{\mathbf{m}}$ is a finite dimensional, reproducing kernel Hilbert space with the Fischer inner product. Let $K^{\mathbf{m}}(z; w)$ be the reproducing kernel for $w \in \Omega$ on $P_{\mathbf{m}}$.

Let $\mathbf{m}$ be a signature and $w \in \mathbf{C}$; we write

$$(w)_{\mathbf{m}} = \prod_{j=1}^{r} \prod_{l=0}^{m_j-1} (w + l - (j-1)\frac{a}{2}) \tag{1.5}$$

where r is the rank of Ω, and a is a certain number associated with Ω (see the list of Cartan domains).

THEOREM 3.8 [FK]. *For all $\lambda \in \mathbf{C}$ and all $z, w \in \Omega$, we have*

$$[K(z;w)]^{\lambda/p} = \sum_{\mathbf{m}} (\lambda)_{\mathbf{m}} K^{\mathbf{m}}(z;w) \tag{1.6}$$

Since $(w)_{\mathbf{m}}$ is a polynomial in w of degree $m_1 + \cdots + m_r$, Theorem 3.8 provides those values of λ for which $(\lambda)_{\mathbf{m}}$ is non-negative for all signatures $\mathbf{m}$ and hence gives the Wallach set $\Lambda(\Omega)$.

COROLLARY 5.2 [FK]. *$K^{\lambda/p}(z;w)$ is positive definite if and only if either $\lambda > \frac{a}{2}(r-1)$ or $\lambda = \frac{a}{2}j$ for some integer j, $0 \leq j \leq r-1$. Consequently, $\Lambda(\Omega) = \{j\frac{a}{2} : 0 \leq j \leq r-1\} \cup (\frac{a}{2}(r-1), \infty)$.*

$\Lambda(\Omega)$ is the Wallach set of Ω; $\Lambda(\Omega)$ has a discrete portion, $\{j\frac{a}{2} : 0 \leq j \leq r-1\}$, and a continuous portion, $(\frac{a}{2}(r-1), \infty)$.

Let w be a complex number; a sesqui-linear form is defined on P by

$$(f|g)_w = \sum{}' \frac{1}{(w)_{\mathbf{m}}} (f_{\mathbf{m}}|g_{\mathbf{m}})_F \tag{1.7}$$

where $f = \sum_{\mathbf{m}} f_{\mathbf{m}}$ and $g = \sum_{\mathbf{m}} g_{\mathbf{m}}$ are the orthogonal decompositions of f and g (c.f. Theorem 2.1) and the prime indicates that the sum is taken over all signatures $\mathbf{m}$ for which $(w)_{\mathbf{m}} \neq 0$. This form is hermitian if and only if w is real; it need not be definite (either positive or negative) even if $w \in \mathbf{R}$.

It is easy to see that for $\lambda \in \Lambda(\Omega)$ the positive semidefinite form $(\cdot,\cdot)_\lambda$ defined by (1.7) coincides with the one defined via (0.3). Therefore, the action $U^{(\lambda)}$ defined by (0.5) is a unitary action of G (and can be extended to a unitary representation of the covering group $\widetilde{G}$ of G) on $\mathcal{H}_\lambda$. More generally, the map $\lambda \mapsto U^{(\lambda)}$ can be extended homorphically to all $\lambda \in \mathbf{C}$, so that $(\cdot,\cdot)_\lambda$ becomes invariant under each of the operators $U_\varphi^{(\lambda)}$, $\varphi \in G$; see [FK, Section 5].

For $\lambda \in \mathbf{C}$ and a signature $\mathbf{m}$, let

$$q(\lambda, \mathbf{m}) = \text{multiplicity of } \lambda \text{ as a zero of } (w)_{\mathbf{m}}$$

and

$$q(\lambda) = \max\{q(\lambda, \mathbf{m}) : \mathbf{m} \text{ a signature}\}.$$

It is easy to see from the form of $(w)_{\mathbf{m}}$ that $q(\lambda, \mathbf{m}) \leq r$ for every $\lambda, \mathbf{m}$. Further, if $0 < q(\lambda, \mathbf{m})$, then λ lies in the discrete portion of $\Lambda(\Omega)$. For $0 \leq j \leq q(\lambda)$, let

$$S_j(\lambda) = \{\mathbf{m} : q(\lambda, \mathbf{m}) \leq j\}$$

and

$$M_j(\lambda) = \sum_{\mathbf{m} \in S_j(\lambda)} P_{\mathbf{m}}.$$

$M_j(\lambda)$ is given a Hermitian form by

$$(f, g)_{M_j(\lambda)} = \lim_{w \to \lambda} (w - \lambda)^j (f \mid g)_w. \tag{1.8}$$

Let $P^{(\lambda)}$ be the linear span of $\{U_\varphi^{(\lambda)} p : p \in P, \varphi \in G\}$.

THEOREM 5.3 [FK]. *For any fixed $\lambda \in \mathbf{C}$, $q = q(\lambda)$ is equal to the number of non-positive integers among $\lambda, \lambda - a/2, \ldots, \lambda - (r-1)a/2$. In particular, $q > 0$ if and only if $\lambda - (r-1)a/2$ or $\lambda - (r-2)a/2$ is a non-positive integer. In any case,*

$$M_0(\lambda) \subset M_1(\lambda) \subset \cdots \subset M_q(\lambda) = P^{(\lambda)}$$

is a composition series of $P^{(\lambda)}$; that is, for each j, $1 \leq j \leq q$, $M_j(\lambda)$ is invariant under the operators $U_\varphi^{(\lambda)}$ defined by (0.5) and

$$(U_\varphi^{(\lambda)} f, U_\varphi^{(\lambda)} g)_{M_j(\lambda)} = (f, g)_{M_j(\lambda)}$$

for all $f, g \in P, \varphi \in G$. Moreover, the quotient $M_j(\lambda)/M_{j-1}(\lambda)$ is irreducible under the action (0.5) of G and the form (1.8) is G-invariant. Finally, every G-invariant subspace of $P^{(\lambda)}$ is one of the $M_j(\lambda)$.

THEOREM 5.4 [FK]. *The Hermitian form (1.8) is definite (positive or negative) on $M_0(\lambda)$ if and only if $\lambda > (r-1)a/2$ or $\lambda = ja/2$ for some integer $j = 0, \ldots, r-1$.*

For $1 \leq j \leq q$*, the form (1.8) is definite on* $M_j(\lambda)/M_{j-1}(\lambda)$ *if and only if* $j = q$ *and* $\lambda - (r-1)a/2$ *is a non-positive integer.*

SECTION 2. ISOMETRIC INVARIANCE. We fix $\lambda \in \Lambda(\Omega)$ and assume that $\mathbf{H}$ satisfies (0.6)–(0.9) with the constant C appearing in (0.9) equal to 1. This is equivalent to the assumption that $U_\varphi^{(\lambda)}$ is a unitary mapping of $\mathbf{H}$ onto itself for every $\varphi \in G$. With this assumption, we shall show that $\mathbf{H}$ is one of the cannonical spaces identified in [FK]. We always assume that the semi-norm in (0.6) is not identically zero on $\mathbf{H}$.

In Section 3, we relax condition (0.9) to allow C to be larger than 1. This, then, gives the case of isomorphic, in contrast to isometric, action. With this weaker assumption, we still are able to draw the same conclusions in two cases: when $\|1\|_{\mathbf{H}} > 0$ and when rank $(\Omega) = 1$.

We begin with several definitions.

DEFINITION

(i) Let S be the set of those signatures $\mathbf{m}$ for which there is some $f \in \mathbf{H}$ with $f_{\mathbf{m}} \neq 0$.

(ii) Let M be the linear span of $\{U_\varphi^{(\lambda)} p : p \in P_{\mathbf{m}}, \mathbf{m} \in S$, and $\varphi \in G\}$.

(iii) Let s be the largest integer ℓ such that $S \cap (S_\ell(\lambda) - S_{\ell-1}(\lambda)) \neq 0$.

These definitions prepare for the following results.

PROPOSITION 1. *(a) If* $\mathbf{m} \in S$*, then* $P_{\mathbf{m}} \subset \mathbf{H}$ *(b)* $S = \{\mathbf{m} : q(\lambda; \mathbf{m}) \leq s\}$ *(c)* $M = M_s(\lambda)$ *(d) for any choice of* $\mathbf{m} \in S_s(\lambda) - S_{s-1}(\lambda)$*, the linear span of the G-orbit of* $\Delta_{\mathbf{m}}$ *is equal to* M *and* $\|\Delta_{\mathbf{m}}\|_{\mathbf{H}} > 0$.

PROOF. (a) Let $\mathbf{m} \in S$; there is a character $\chi_{\mathbf{m}}$ on K such that

$$g_{\mathbf{m}}(z) = \int_K g(kz)\overline{\chi_{\mathbf{m}}(k)}\,dk$$

for all $g \in P$ where dk is the Haar measure on K. Since $f \in \mathbf{H}$, the function $U_k^{(\lambda)} f$ also lies in $\mathbf{H}$. However, $(U_k^{(\lambda)} f)(z) = f(kz)c(k)$, where $c(k)$ is unimodular complex number.

By (0.8), $f_{\mathbf{m}}$ lies in $\mathbf{H}$. Since $f_{\mathbf{m}}$ also lies in $P_{\mathbf{m}}$ and since the action of K is irreducible on $P_{\mathbf{m}}$, it follows that all of $P_{\mathbf{m}}$ lies in $\mathbf{H}$ whenever $\mathbf{m} \in S$. This proves (a).

To see (b), (c) and (d), note that by the definition of s, we must have $S \subseteq \{\mathbf{m} : q(\lambda; \mathbf{m}) \leq s\}$. On the other hand, again by the definition of s, there must be some $F \in \mathbf{H}$ with $F_{\mathbf{m}} \neq 0$ and $q(\lambda; \mathbf{m}) = s$. By (a), we may take $F = F_{\mathbf{m}}$. The G-orbit of F lies in $\mathbf{H}$ and by Theorem 5.3 of Faraut and Koranyi, this orbit must be $M_s(\lambda)$. The only point remaining is to see that $\|\Delta_{\mathbf{m}}\| > 0$. If this were not the case, then $\|f\|_{\mathbf{H}} = 0$ for all f in the G-orbit of $\Delta_{\mathbf{m}}$. But this orbit is dense in $\mathbf{H}$. This would contradict our assumption that the norm is not identically zero on $\mathbf{H}$.

THEOREM 2. *Suppose that Ω is an irreducible bounded symmetric domain and $\lambda \in \Lambda(\Omega)$. Let $\mathbf{H}$ satisfy (0.6)–(0.9) with $C = 1$ and let s be as defined above. Then the norms in $\mathbf{H}$ and $M_s(\lambda)$ are proportional and $M_s(\lambda)$ is a dense subset of $\mathbf{H}$. In particular, $s = 0$ or $s = q$.*

PROOF. The K-invariance of $(\cdot,\cdot)_{\mathbf{H}}$ implies that the spaces $\{P_{\mathbf{m}} : \mathbf{m} \in S\}$ are mutually orthogonal in $\mathbf{H}$. Fix any signature $\mathbf{m} \in S_s(\lambda) - S_{s-1}(\lambda)$. If $g \in P_{\mathbf{m}}$ and $f \in M_s(\lambda)$, then

$$(f, g)_{\mathbf{H}} = (f_{\mathbf{m}}, g)_{\mathbf{H}}.$$

However, $(\cdot,\cdot)_{\mathbf{H}}$ is a K-invariant inner product on $P_{\mathbf{m}}$ and so the K-irreducibility of $P_{\mathbf{m}}$ implies that this inner product must be a constant multiple of the $M_s(\lambda)$-inner product. If we denote this constant by β, then

$$(f, g)_{\mathbf{H}} = \beta (f, g)_{M_s(\lambda)} \quad , g \in P_{\mathbf{m}}, f \in M_s(\lambda) \tag{2.2}$$

Now let $\varphi, \psi \in G$ and denote ψ^{-1} by γ. Using (0.9) with $C = 1$ and the (anti-)homomorphism property of $U^{(\lambda)}$, we obtain

$$\begin{aligned}(U_{\varphi}^{(\lambda)} \Delta_{\mathbf{m}}, U_{\psi}^{(\lambda)} \Delta_{\mathbf{m}})_{\mathbf{H}} &= (U_{\varphi\gamma}^{(\lambda)} \Delta_{\mathbf{m}}, \Delta_{\mathbf{m}})_{\mathbf{H}} \\ &= \beta (U_{\varphi\gamma}^{(\lambda)} \Delta_{\mathbf{m}}, \Delta_{\mathbf{m}})_{M_s(\lambda)} \\ &= \beta (U_{\varphi}^{(\lambda)} \Delta_{\mathbf{m}}, U_{\psi}^{(\lambda)} \Delta_{\mathbf{m}})_{M_s(\lambda)}.\end{aligned}$$

Since the G-orbit of $\Delta_{\mathbf{m}}$ is dense in $\mathbf{H}$ and in $M_s(\lambda)$, we obtain

$$(f, g)_H = \beta (f, g)_{M_s(\lambda)}. \tag{2.3}$$

The final conclusion, that $s = 0$ or $s = q$, follows from Theorem 5.4 of Faraut and Koranyi and our assumption (0.6).

SECTION 3. ISOMORPHIC RESULTS.

In this section we assume that $\mathbf{H}$ satisfies (0.6)–(0.9) and we allow C to be any constant (larger than 1, of course) in (0.9). In two cases—when $\|1\|_{\mathbf{H}} > 0$ and when Ω is the unit ball complex N-space—we can show that $\mathbf{H}$ is one of the cannonical spaces described in [FK] and outlined in Section 1.

We begin with a general construction valid for any bounded symmetric domain. Let P be an amenable subgroup of G; see [G] and [Pi] for an extensive treatment of amenable groups. For instance, P could be the minimal parabolic subgroup of G. We define another inner product on $\mathbf{H}$ by

$$[f, g] = m_{\varphi \in P}(U_{\varphi}^{(\lambda)} f, U_{\varphi}^{(\lambda)} g)_{\mathbf{H}}$$

where m is an invariant mean on P. Then $[f, f]$ is equivalent to (f, f):

$$(1/C^2)(f, f)_{\mathbf{H}} \leq [f, f] \leq C^2 (f, f)_{\mathbf{H}}$$

by (0.9). Since m is invariant on the group P it also follows that

$$[U_{\psi}^{(\lambda)} f, U_{\psi}^{(\lambda)} g] = [f, g]; \psi \in P, f, g \in \mathbf{H}.$$

We may therefore assume without further comment that operators $U_{\psi}^{(\lambda)}$, $\psi \in P$, are unitary on $\mathbf{H}$. We shall also continue to write $(f, g)_{\mathbf{H}}$ for the semi-inner product in $\mathbf{H}$. The reader should note that the group G is not amenable and the construction above cannot be applied to G.

We will generally assume that the amenable subgroup P is also transitive on Ω. For instance, if $G = KAN$ is the Iwasawa decomposition of G, then AN is (maximal) solvable and so amenable. Further, it is transitive since G is and since each element of K fixes the origin. Moreover, if M is the normalizer of AN in K, then MAN is another amenable, transitive subgroup of G.

With this background we are ready for the first of our results, the case when $\|1\|_{\mathbf{H}} > 0$.

THEOREM 3. *Let Ω be an irreducible bounded symmetric domain, let $\lambda \in \Lambda(\Omega)$, and let* $\mathbf{H}$ *satisfy (0.6)–(0.9). If* $\|1\|_{\mathbf{H}} > 0$, *then* $\lambda > 0$, $\mathbf{H}$ *is a reproducing kernel Hilbert space, and* $\mathbf{H} = \mathcal{H}_\lambda$ *with equivalent norms where* $\mathcal{H}_\lambda$ *is given by (0.3).*

PROOF. Suppose that $f \in \mathbf{H}$; the Bochner integral

$$\frac{1}{2\pi}\int_0^{2\pi} f(e^{it}z)dt$$

converges by (0.8) and defines an element of $\mathbf{H}$. But the value of this integral is $f(0)\cdot 1$. Hence,

$$\begin{aligned}|f(0)|\|1\|_{\mathbf{H}} &= \|\frac{1}{2\pi}\int_0^{2\pi} f(e^{it}z)dt\|_{\mathbf{H}} \\ &\leq C\|f\|_{\mathbf{H}}.\end{aligned}$$

Now replace f by $U_\varphi^{(\lambda)}f$, $\varphi \in P$. This gives

$$|f(\varphi(0))||J\varphi(0)|^{\lambda/p}\|1\|_{\mathbf{H}} \leq C\|f\|_{\mathbf{H}}.$$

That is, for each $a \in \Omega$ there is a constant $c(a)$ depending on a but not f with

$$|f(a)| \leq C(a)\|f\|_{\mathbf{H}}.$$

This shows that point evaluation at any point of Ω is a bounded linear functional on $\mathbf{H}$; that is, $\mathbf{H}$ is a reproducing kernel Hilbert space.

Let $L(z;a)$ be the reproducing kernel for a on $\mathbf{H}$:

$$f(a) = (f, L(\quad;a))_{\mathbf{H}}.$$

Let $\psi \in P$ satisfy $\psi(a) = 0$ and set $\varphi = \psi^{-1}$. Then

$$(U_\varphi^{(\lambda)}g)(0) = (U_\varphi^{(\lambda)}g, L(\quad;0))_{\mathbf{H}}$$

and so

$$\begin{aligned}g(a)(J\varphi)^{\lambda/p}(0) &= (U_\varphi^{(\lambda)}g, L(\quad;0))_{\mathbf{H}} \\ &= (g, U_\psi^{(\lambda)}L(\quad;0))_{\mathbf{H}} \\ &= (g(z), L(\psi(z);0)(J\psi)^{\lambda/p}(z))_{\mathbf{H}}.\end{aligned}$$

However, $(J\varphi)(0) = (J\psi)^{-1}(a)$ since $\psi(a) = 0$ and $\varphi = \psi^{-1}$. This then gives

$$\overline{(J\psi)^{\lambda/p}(a)}L(\psi(z);\psi(a))(J\psi)^{\lambda/p}(z) = L(z;a) \tag{3.1}$$

Because P is transitive (3.1) holds for every $a, z \in \Omega$.

We now show that $\lambda > 0$. Suppose, to the contrary, that $\lambda = 0$. The transformation rule of the reproducing kernel (3.1) then becomes

$$L(\psi(z);\psi(a)) = L(z;a) \qquad z;a \in \Omega; \psi \in P.$$

Since P is transitive, $L(z;z) = L(0;0)$ for all $z \in \Omega$. Hence, $L(z;w)$ is analytic in z, conjugate analytic in W and constant on the diagonal, $z = w$. Therefore, by an elementary power series argument

$$L(z;w) = L(0;0) \qquad ; z, w \in \Omega.$$

Since the linear span of $\{L(\cdot;a) : a \in \Omega\}$ is dense in $\mathbf{H}$, a simple argument now shows that every function in $\mathbf{H}$ is constant, contrary to our assumption.

Since $\lambda > 0$, we may consider the space $\mathcal{H}_\lambda$ whose inner product is defined in (0.3). That is $\mathcal{H}_\lambda$ is the closure of the linear span of $\{K_z^{\lambda/p} : z \in \Omega\}$ with norm determined by the inner product of the kernel functions

$$(K_z^{\lambda/p}, K_w^{\lambda/p}) = [K(w;z)]^{\lambda/p}.$$

The space $\mathcal{H}_\lambda$ is a reproducing kernel Hilbert space, whose reproducing kernel for $w \in \Omega$ is $K_w^{\lambda/p}$. This kernel then satisfies

$$\overline{(J\psi)^{\lambda/p}(a)}K^{\lambda/p}(\psi(z);\psi(a))(J\psi)^{\lambda/p}(z) = K^{\lambda/p}(z;a)$$

for all $z, a \in \Omega$ and all $\psi \in P$; indeed, for all $\psi \in G$.

The rest of the argument is now like that of R. Kunze [K] and [Ko1], [Ko2]. Let $\beta = L(0;0)$; then

$$\begin{aligned} L(z;z) &= |(J\psi)(z)|^{2\lambda/p} L(0;0) \quad , \quad \psi(z) = 0 \\ &= \beta K^{\lambda/p}(z;z). \end{aligned}$$

The function

$$F(z;w) = \beta K^{\lambda/p}(z;w) - L(z;w)$$

is then analytic in z, conjugate analytic in w, and $F(z;z) = 0$. Hence, F is identically zero. Therefore,

$$L(z;w) = \beta K^{\lambda/p}(z;w). \tag{3.2}$$

Let $f(z) = \sum_1^m c_j L(z;a_j)$. Then

$$\begin{aligned}\|f\|_{\mathbf{H}}^2 = (f,f)_{\mathbf{H}} &= \sum_{j,l} c_j \bar{c}_l L(a_\ell;a_j) \\ &= \beta \sum_{j,l} c_j \bar{c}_l K^{(\lambda)}(a_\ell;a_j) \\ &= \beta (f,f)_{\mathcal{H}_\lambda} = \beta \|f\|_{\mathcal{H}_\lambda}^2.\end{aligned}$$

Functions of the form of f are dense in both $\mathbf{H}$ and $\mathcal{H}_\lambda$ and so $\mathbf{H} = \mathcal{H}_\lambda$ and the $\mathbf{H}$ norm is a multiple of that in $\mathcal{H}_\lambda$. We are done.

COROLLARY. *If* $\|1\|_{\mathbf{H}} > 0$, *then* $\mathbf{H}$ *contains* $M_0(\lambda)$ *as a dense subspace.*

PROOF. Since $\lambda > 0$, we know that $M_0(\lambda)$ is non-trivial since if $\lambda = (\nu-1)\frac{a}{2}$, $2 \le \nu \le r$, then

$$M_0(\lambda) = \sum_{m_\nu=0} \oplus P_{\mathbf{m}} = \sum_{m_1 \geqslant \cdots \geqslant m_{\nu-1} \geqslant 0} \oplus P_{(m_1,\ldots,m_{\nu-1},0,\ldots,0)}.$$

Hence, $1 \in M_0(\lambda)$ and $\|1\|_{M_0(\lambda)} > 0$. Thus, exactly as above,the closure of $M_0(\lambda)$ is reproducing kernel Hilbert space whose reproducing kernel $K_z^{(\lambda)}$ satisfies a transformation rule just like that of $\mathbf{H}$ or $\mathcal{H}_\lambda$; namely,

$$\overline{(J\psi)^{\lambda/p}(a)} K^{(\lambda)}(\psi(z);\psi(a))(J\psi)^{\lambda/p}(z) = K^{(\lambda)}(z;a); z,a \in \Omega$$

and $\psi \in P$. We may now repeat the argument given above which culminated in (3.2) to conclude that the closure of $M_0(\lambda)$ is $\mathbf{H} = \mathcal{H}_\lambda$.

PROPOSITION 4. *Let* B *be the open unit ball of* $\mathbf{C}^N$ *with the usual norm and let* $e_1 = (1,0,\ldots,0)$. *The subgroup*

$$P = \{\varphi \in G : \varphi(e_1) = e_1\}$$

of $Aut(B)$ is both amenable and transitive on B.

PROOF. It is easier to work in the upper half-plane

$$U = \{(w_1, w') : Im\, w_1 > |w'|^2\}$$

where we have adopted the conventional notation $w_1 \in \mathbf{C}$ and $w' \in \mathbf{C}^{N-1}$; see [R]. Let

$$\Phi(z_1, z') = i(\frac{1+z_1}{1-z_1}, \frac{z'}{1-z_1})$$

be the Cayley transform of B onto U. The subgroup P is transformed to

$$Q = \{\psi = \Phi \circ \varphi \circ \Phi^{-1} : \varphi \in P\},$$

the subgroup of $Aut(U)$ which fixes "∞". We now define three subgroups of Q.

(i) $\quad \mathrm{J} = \{\psi \in Q : \varphi \in P$ and φ is linear $\}$

That is $\psi \in Q$ if and only if φ is a member of the compact subgroup of K consisting of those linear automorphisms of B which fix e_1.

(ii) $\quad$ for $0 < t < \infty$, let

$$\delta_t(w_1, w') = (t^2 w_1, tw')$$

and

$$D = \{\delta_t : 0 < t < \infty\}$$

(iii) $\quad$ for $a \in \partial U$, let

$$h_a(w_1, w') = (w_1 + a_1 + 2i < w', a' >, w' + a')$$

and

$$H = \{h_a : a \in \partial U\}.$$

H is known as the Heisenberg group; see [R] and [Ta]. Clearly, D and H are subgroups of Q.

To show that Q is transitive on U, let $b = (b_1, b')$ be any point of U and set

$$t = (Im\, b_1 - |b'|^2)^{\frac{1}{2}}, \quad a_1 = -i + t^{-2} b_1, \quad a' = t^{-1} b'.$$

Then $a = (a_1, a') \in \partial U$ and a simple computation establishes that

$$\delta_t(h_a(i, 0')) = b.$$

We shall now establish that each $\psi \in Q$ has a unique factorization

$$\psi = h_a \delta_t l \quad ; \quad h_a \in H, \delta_t \in D, l \in J. \tag{3.3}$$

To see this, let $\psi \in Q$ and write $\psi(i, 0') = (b_1, b') \in U$. Put

$$c = (i|b'|^2 - Re b_1, -b').$$

Then $c \in \partial U$ and

$$\begin{aligned} h_c(\psi(i, 0')) &= h_c(b_1, b') \\ &= (b_1 + i|b'|^2 - Re b_1 + 2i < b', -b' >, b' - b') \\ &= (i Im\, b_1 - i|b'|^2, 0'). \end{aligned}$$

Let $t = (Im\, b_1 - |b'|^2)^{\frac{1}{2}}$ and put

$$\zeta = \delta_t h_c \psi.$$

Then $\zeta \in Q$ since each of ψ, h_c, δ_t lie in Q and further

$$\zeta(i, 0') = (i, 0'). \tag{3.4}$$

(3.4) implies that

$$\varphi = \Phi^{-1} \circ \zeta \circ \Phi$$

is an automorphism of B which fixes the origin. Hence, φ is linear (by Cartan's linearity theorem) and so by definition, $\varphi \in J$. This establishes the factorization in (3.3).

To see that the factorization is unique, we note first that each $l \in J$ maps 0 to 0. This is because $l = \Phi \circ \varphi \circ \Phi^{-1}$ where φ is both linear and in P, and so

$$\begin{aligned} l(0) &= \Phi(\varphi(\Phi^{-1}(0))) \\ &= \Phi(\varphi(-e_1)) \\ &= \Phi(-e_1) \\ &= 0. \end{aligned}$$

Suppose, then, that

$$h_a \delta_t l_1 = h_b \delta_s l_2 \qquad , \quad l_1, l_2 \in J.$$

Then

$$a = h_a(\delta_t(l_1(0))) = h_b(\delta_s(l_2(0))) = b.$$

Further, at the point $(is^{-2}, 0') \in U$, we have

$$\begin{aligned} (it^2 s^{-2}, 0') &= \delta_t(is^{-2}, 0') \\ &= \delta_s(l_2(l_1^{-1}(is^{-2}, 0'))) \\ &= \delta_s(is^{-2}, 0') \\ &= (i, 0') \end{aligned}$$

and so $t = s$. This further implies that $l_1 = l_2$. Therefore, (3.3) is proved.

To prove that Q is amenable, we note the following easily established relations:

$$l h_a l^{-1} = h_{l(a)} \qquad , l \in J, a \in \partial U \tag{3.5}$$

$$\delta_t h_a \delta_t^{-1} = h_{\delta_t(a)} \quad , t \in (0, \infty), a \in \partial U \tag{3.6}$$

$$\delta_t l = l \delta_t \qquad , t \in (0, \infty), l \in J. \tag{3.7}$$

(We have used (3.7) already in proving the uniqueness of the factorization in (3.3).) The group H is nilpotent [Ta] and so solvable, D is abelian, and J is compact. Hence, each is amenable. By (3.7), $DJ = JD$ so DJ is amenable. By (3.5) and (3.6) JD normalizes

H and so $Q = HJD$ is amenable. The reader is referred to [G] for these results about amenable groups.

REMARK. A shorter, but indirect, proof of Proposition 4 depends on the fact that any minimal parabolic subgroup of G is amenable (see [M]) and the fact that in the case of the unit ball of C^N (i.e. $r = 1$) the stabilizer subgroup of any boundary point is minimal parbolic. This follows for instance, by a general description of the parabolic subgroups of G in terms of its action on the boundary components of $\overline{\Omega}$, see [Loos, 9.17–9.18].

THEOREM 5. *Let* $\mathbf{H}$ *be a semi-inner product space of analytic functions on* B*, the unit ball of complex* N *space. Suppose* H *satisfies (0.6)–(0.9) for* $\lambda = 0$*. Then* $\mathbf{H}$ *coincides with the space* $\mathcal{H}$ *of those analytic functions*

$$f(z) = \sum_{\alpha} a_\alpha z^\alpha$$

which satisfy

$$\|f\|_{\mathcal{H}}^2 = \sum_{\alpha} |a_\alpha|^2 \frac{\alpha!}{|\alpha|!} |\alpha| \quad < \infty \tag{3.8}$$

and the semi-norm of $\mathbf{H}$ *is equivalent to that of* $\mathcal{H}$*.*

PROOF. We take the subgroup P to be the stabilizer group of $e_1 = (1, 0, \ldots, 0)$. P is, indeed, amenable and transitive by Proposition 4.

We note at the outset of the proof that since B has rank 1, the composition series from Theorem 5.3 of Faraut and Koranyi has length 1 and so the Hilbert space $\mathbf{H}$ must contain all analytic polynomials; see the proof of (a) of Proposition 1.

Let $aut(B)$ be the set of generators of one parameter subgroups of $Aut(B)$; the elements of $aut(B)$ are complete holomorphic vector fields of a special type; see [L]. Let $h \in aut(B)$ generate a 1-parameter subgroup $\{\varphi_t\}_{t \in \mathbf{R}}$ of elements of P. Then for $f, g \in \mathbf{H}$

$$(f, g)_{\mathbf{H}} = (f \circ \varphi_t, g \circ \varphi_t)_{\mathbf{H}}, \quad -\infty < t < \infty. \tag{3.9}$$

If f, g are analytic on a neighborhood of the closure of Ω, then we may differentiate both sides of (3.9) with respect to t at $t = 0$ and obtain

$$0 = ([f', h], g)_{\mathbf{H}} + (f, [g', h])_{\mathbf{H}} \tag{3.10}$$

where

$$f' = gradient f = (\frac{\partial f}{\partial z_1}, \dots, \frac{\partial f}{\partial z_N})$$

$[\ ,\ ]$ is the **real** inner product on $\mathbf{C}^N$:

$$[v, w] = \sum_1^N v_j w_j.$$

The choice of P as the stabilizer group of e_1 forces the elements $h \in aut(\Omega)$ which appear in (3.10) to satisfy $h(e_1) = 0$. Conversely, if $h \in aut(\Omega)$ and $h(e_1) = 0$, then the one parameter subgroup G generated by h belongs to P. Each such h has the form

$$h(z) = Az + a - < z, a > z \tag{3.11}$$

for a skew hermitian matrix A and $a \in B$; see [L; Chapter 2]. Since $h(e_1) = 0$ we find $A = A_1 + A_2$ where

$$A_1 = \begin{bmatrix} \overline{a}_1 - a_1 & \overline{a}_2 & \dots & \overline{a}_N \\ -a_2 & & & \\ \vdots & & 0 & \\ -a_N & & & \end{bmatrix}$$

and

$$A_2 = \begin{bmatrix} 0 & 0 \\ 0 & W \end{bmatrix}$$

where W is any $(N-1) \times (N-1)$ skew-hermitian matrix. Since W and a are independent of one another, (3.10) holds with $A = A_2, a = 0$, and also for $A = A_1, a \in B$. In the former case we obtain

$$h(z) = A_2 z = (0, \sum_{k=2}^N w_{2k} z_k, \dots, \sum_{k=2}^N w_{Nk} z_k)$$

and, from (3.10)

$$0 = \sum_{j=2}^N (\frac{\partial f}{\partial z_j} \sum_{k=2}^N w_{jk} z_k, g)_{\mathbf{H}} + \sum_{k=2}^N (f, \frac{\partial g}{\partial z_k} \sum_{j=2}^N w_{kj}, z_j)_{\mathbf{H}}$$

Since $-\overline{w}_{kj} = w_{jk}$, this yields

$$(\frac{\partial f}{\partial z_j} z_k, g)_{\mathbf{H}} = (f, \frac{\partial g}{\partial z_k} z_j)_{\mathbf{H}}, \quad 2 \le j, k \le N \tag{3.12}$$

Suppose first that

$$f(z) = z_1^{\alpha_1}(z')^{\alpha'}, g(z) = z_1^{\beta_1}(z')^{\beta'} \quad \text{and } \alpha' \neq \beta'.$$

Choose $j \geqslant 2$ so that $\alpha'_j \neq \beta'_j$ and in (3.12) take $k = j$. Then

$$\alpha'_j(f,g)_{\mathbf{H}} = \beta'_j(f,g)_{\mathbf{H}}.$$

Hence,

$$(z^\alpha, z^\beta)_{\mathbf{H}} = 0 \quad \text{if } \alpha' \neq \beta' \tag{3.13}$$

Suppose next that $\alpha' = \beta'$; in (3.12) take

$$k = N, \quad j = 2, \qquad f = z_1^{\alpha_1} z_2^{\alpha_2+1} \dots z_{N-1}^{\alpha_{N-1}} z_N^{\alpha_N - 1}, g = z_1^{\beta_1} z_2^{\alpha_2} \dots z_{N-1}^{\alpha_{N-1}} z_N^{\alpha_N}.$$

This will give

$$(\alpha_2+1)(z_1^{\alpha_1} z_2^{\alpha_2} \dots z_N^{\alpha_N}, z_1^{\beta_1} z_2^{\alpha_2} \dots z_N^{\alpha_N})_{\mathbf{H}} = \alpha_N(z_1^{\alpha_1} z_2^{\alpha_2+1} \dots z_N^{\alpha_N-1}, z_1^{\beta_1} z_2^{\alpha_2+1} \dots z_N^{\alpha_N-1})_{\mathbf{H}}.$$

Repeat this step until the exponent of z_N has been reduced to zero. Thus,

$$(z_1^{\alpha_1} z_2^{\alpha_2} \dots z_N^{\alpha_N}, z_1^{\beta_1} z_2^{\alpha_2} \dots z_N^{\alpha_N})_{\mathbf{H}} = \frac{\alpha_N!}{(\alpha_2+1)(\alpha_2+2)\cdots(\alpha_2+\alpha_N)}(z^{\alpha''}, z^{\beta''})_{\mathbf{H}}$$

where $\alpha'' = (\alpha_1, \alpha_2 + \alpha_N, \alpha_3, \dots, \alpha_{N-1}, 0)$, $\beta'' = (\beta_1, \alpha_2 + \alpha_N, \alpha_3, \dots, \alpha_{N-1}, 0)$. Now repeat again, this time with $k = N-1$, $j = 2$. Continue this process until we obtain

$$\begin{aligned}(z^\alpha, z^\beta)_{\mathbf{H}} &= \frac{\alpha_N! \cdots \alpha_3!}{(\alpha_2+1)\cdots(\alpha_2+\alpha_3+\cdots+\alpha_N)}(z_1^{\alpha_1} z_2^{\alpha_2+\alpha_3+\cdots+\alpha_N}, z_1^{\beta_1} z_2^{\alpha_2+\cdots+\alpha_N})_{\mathbf{H}} \\ &= \frac{(\alpha')!}{|\alpha'|!}(z_1^{\alpha_1} z_2^{|\alpha'|}, z_1^{\beta_1} z_2^{|\alpha'|})_{\mathbf{H}} \end{aligned} \tag{3.14}$$

This is almost what is desired; we need only work with the z_1-terms now.

We return to (3.11) and take $a \in B$ and $A = A_1$. This yields

$$h(z) = (C_a - C_a^*)z + a - \langle z, a \rangle z$$

where

$$C_a = \begin{pmatrix} -a_1 & 0 & \dots & 0 \\ \vdots & \vdots & & \\ -a_N & 0 & \dots & 0 \end{pmatrix}.$$

Substitute this into (3.10) and separate those terms which are linear in a and those which are conjugate linear in a. Since a is any point of B, this yields

$$((-z_1+1)\frac{\partial f}{\partial z_j}, g)_{\mathbf{H}} = (f, z_j(Rg - \frac{\partial g}{\partial z_1}))_{\mathbf{H}} \quad ,j=1,\ldots,N \tag{3.15}$$

where $Rg = \sum_1^N z_k \partial g/\partial z_k$ is the radial derivative of g.

In (3.15) take $j=1$, $f = z_1^\alpha z_2^m$, $g = z_1^\beta z_2^m$, α, β, m all non-negative integers. This gives

$$(\alpha-\beta)(z_1^\alpha z_2^m, z_1^\beta z_2^m)_{\mathbf{H}} = \alpha(z_1^{\alpha-1} z_2^m, z_1^\beta z_2^m)_{\mathbf{H}} - (\beta+m)(z_1^\alpha z_2^m, z_1^{\beta+1} z_2^m)_{\mathbf{H}} \tag{3.16}$$

We proceed by induction. If $\alpha=\beta=0$, then

$$(z_2^m, z_1 z_2^m)_{\mathbf{H}} = 0.$$

Assume $(z_2^m, z_1^\beta z_2^m)_{\mathbf{H}} = 0$, then (3.16) gives $(z_2^m, z_1^{\beta+1} z_2^m)_{\mathbf{H}} = 0$. Hence,

$$(z_2^m, z_1^\beta z_2^m)_{\mathbf{H}} = 0 \quad \text{for all } \beta \geqslant 1.$$

Now take $\alpha=\beta=1$ in (3.16); this gives

$$(z_1 z_2^m, z_1^2 z_2^m)_{\mathbf{H}} = 0.$$

Once again proceed by induction on α, and for each α by induction on β, using (3.16). This will give

$$(z_1^\alpha z_2^m, z_1^\beta z_2^m)_{\mathbf{H}} = 0 \quad \text{if } \beta \geqslant \alpha+1. \tag{3.17}$$

In particular, $(z_1^\alpha z_2^m, z_1^\beta z_2^m) = 0$ if $\alpha \neq \beta$.

We next use (3.15) with $j=2$, $f = z_2^m$, $g = z_2^{m-1}$. We obtain

$$((z_1+1)m z_2^{m-1}, z_2^{m-1})_{\mathbf{H}} = (z_2^m, z_2(m-1)z_2^{m-1})_{\mathbf{H}}$$

or

$$(z_2^m, z_2^m)_{\mathbf{H}} = \frac{m}{m-1}(z_2^{m-1}, z_2^{m-1})_{\mathbf{H}}.$$

This yields (after using (3.17))

$$(z_2^m, z_2^m)_{\mathbf{H}} = m(z_2, z_2)_{\mathbf{H}}. \tag{3.18}$$

Finally, we again use (3.15) with $j = 1$, $f = z_1^\alpha z_2^m$ and $g = z_1^{\alpha-1} z_2^m$. This gives (using (3.17))

$$\alpha(z_1^{\alpha-1}z_2^m, z_1^{\alpha-1}z_2^m)_{\mathbf{H}} = (\alpha - 1 + m)(z_1^\alpha z_2^m, z_1^\alpha z_2^m)_{\mathbf{H}}$$

or

$$(z_1^\alpha z_2^m, z_1^\alpha z_2^m)_{\mathbf{H}} = \frac{\alpha}{\alpha - 1 + m}(z_1^{\alpha-1}z_2^m, z_1^{\alpha-1}z_2^m)_{\mathbf{H}}.$$

Continuing

$$\begin{aligned}(z_1^\alpha z_2^m, z_1^\alpha z_2^m)_{\mathbf{H}} &= \frac{\alpha!}{(\alpha-1+m)\cdots m}(z_2^m, z_2^m)_{\mathbf{H}} \\ &= \frac{\alpha!(m-1)!}{(\alpha+m-1)!}m(z_2, z_2)_{\mathbf{H}} \\ &= \frac{\alpha! m!}{(\alpha+m-1)!}(z_2, z_2)_{\mathbf{H}}.\end{aligned}$$

When this is combined with (3.12) we obtain

$$(z^\alpha, z^\beta)_{\mathbf{H}} = \delta_{\alpha,\beta} \quad \frac{\alpha!}{|\alpha|!}|\alpha|(z_2, z_2)_{\mathbf{H}}. \tag{3.19}$$

Set $\lambda = (z_2, z_2)_{\mathbf{H}}$. Then (3.19) gives the desired conclusion: $\mathbf{H} = \mathcal{H}$ and

$$\|f\|_{\mathbf{H}}^2 = \lambda \|f\|_{\mathcal{H}}^2.$$

FINAL REMARKS. The result in [AF1] for unweighted action, $\lambda = 0$, which now becomes a very special case of the results here, was derived under the additional assumption that the functions in the semi-Hilbert space were also Bloch functions. As it has turned out, that particular assumption can now be seen to be redundant. However, if we assume that the semi-Hilbert space $\mathbf{H}$ of analytic functions on the bounded symmetric domain Ω contains only Bloch functions (see [T2], [T3]), then the results of the present paper show that the number of such spaces is quite small. Indeed, when the rank of Ω is one, that is, Ω is the unit ball of $\mathbf{C}^N$ with its usual norm, then the space $\mathbf{H}$ must be that of Theorem 5. However, when the rank of Ω is 2 or more and the group G acts isometrically (the assumption in Theorem 2) with $\lambda = 0$, then there are no (non-trivial) semi-Hilbert spaces of analytic functions which lie within the Bloch functions. A different way of stating this is that if Ω has rank 2 or more and if $Aut(\Omega)$ acts isometrically by composition on **H**, then either **H** contains only constants or **H** has no "decent" linear functional; see [T3].

Theorems 3 and 5 say that certain uniformly bounded actions of G are similar to unitary actions. One interpretation of this is a a solution in special cases of the celebrated "similarity problem" of Dixmier.

REFERENCES

[AF1] J. Arazy and S. D. Fisher, *The uniqueness of the Dirichlet space among Möbius-invariant Hilbert spaces*, Ill. J. Math vol. 29 (1985), 449–462.

[AF2] __________, *Weighted actions of the Möbius group and their invariant Hilbert spaces*, Univ. of Haifa preprint (1987).

[B] F. A. Berezin, *Quantization in complex symmetric spaces*, Izv. Akad. Nauk SSSR vol. 9 (1975), 341–379.

[C] E. Cartan, *Sur les domaines bornés homogènes de l'espace de n variables complexes*, Abh. Math. Sem. Univ. Hamburg vol. 11 (1935), 116–162.

[FK] J. Faraut and A. Koranyi, *Function spaces and reproducing kernels on bounded symmetric domains*, to appear in J. Functional Analysis.

[G] F P. Greenleaf, "Invariant Means and Topological Groups," Van Nostrand, New York, 1969.

[H] L. K. Hua, "Harmonic Analysis of Functions of Several Complex Variables in the Classical Domains," Amer. Math. Soc. Translations, vol. 6, Amer. Math. Soc., Providence, 1963.

[Ko1] S. Kobayashi, *On automorphism groups of homogeneous complex manifolds*, Proc. Amer. Math. Soc. vol. 12 (1961), 359–361.

[Ko2] __________, *Irreducibility of certain unitary representations*, J. Math. Soc. Japan vol. 20 (1968), 638–642.

[Ku] R. A. Kunze, *On the irreducibility of certain multiplier representations*, Bull. Amer. Math. Soc. vol. 68 (1962), 93–94.

[L] O. Loos, *Bounded symmetric domains and Jordan pairs*, Dep't. of Mathematics, Univ. of California, Irvine, CA (1977).

[M] C. C. Moore, *Amenable subgroups of semi-simple groups and proximal flows*, Israel J. Math vol. 34 (1979), 121–138.

[O] B. Orsted, *Composition series for analytic continuations of holomorphic discrete series representations of $SU(n,n)$*, Trans. Am. Math. Soc. vol. 260 (1980), 563–573.

[P1] J. Peetre, *Möbius invariant function spaces in several variables*, preprint (1982).

[P2] __________, *Type I*, manuscript, 1988.

[Pi] J.-P. Pier, "Amenable Locally Compact Groups," Wiley, New York, 1984.

[P-S] I. I. Pyatetskii-Shapiro, "Auatomorphic Functions and the Geometry of Classical Domains," Gordon Breach, New York, 1969.

[R] W. Rudin, "Function Theory on the Unit Ball of $\mathbf{C}^n$," Grundlehren der natuematischen Wissenschaften 241, Springer-Verlag, New York, 1980.

[S] W. Schmid, *Die Randwerte holomorpher Funktionen auf hermitesch symmetrischen Räumen*, Inventiones Math vol. 9 (1969), 61–80.

[Ta] M. Taylor, "Non-commutative Harmonic Analysis," American Math. Soc., Providence, RI, 1986.

[T1] R. Timoney, *Bloch functions in several complex variables, I*, Bull. London Math. Soc. vol. 12 (1980), 241–267.

[T2] __________, *Bloch functions in several complex variables, II*, J. Reine Angew. Math vol. 319 (1980), 1–22.

[T3] __________, *Maximal invariant spaces of analytic functions*, Indiana Univ. Mathematics Journal vol. 31 (1982), 651–663.

[U1] H. Upmeier, *Jordan algebras in analysis, operator theory and quantum mechanics*, CBMS Regional Conference Series in Math no. 67 (1987), Amer. Math. Soc., Providence, RI.

[U2] __________, "Symmetric Banach Manifolds and Jordan $C*$-alagebras," North Holland Math Studies no. 104 (1985), Elsevier, New York, N.Y.

[U3] __________, *Jordan algebras and harmonic analysis on symmetric spaces*, Amer. J. Math vol. 108 (1986), 1–25.

[V-R] M. Vergne and H. Rossi, *Analytic continuation of the holomorphic discrete series of a semi-simple Lie group*, Acta Math vol. 136 (1976), 1–59.

[W] N. Wallach, *The analytic continuation of the discrete series, I, II*, Trans. Am. Math. Soc., vol. 251 (1979), 1–17, 19–37.

Jonathan Arazy
Department of Mathematics
University of Haifa
Haifa, 31999 Israel
and
Department of Mathematics
University of Kansas
Lawrence, Kansas 66045

Stephen D. Fisher
Department of Mathematics
Northwestern University
Evanston, Illinois 60208

Operator Theory:
Advances and Applications, Vol. 48

CLOSABILITY OF DIFFERENTIAL OPERATORS AND REAL SUBJORDAN OPERATORS

JOSEPH A. BALL[1] AND THOMAS R. FANNEY

Dedicated to the memory of E. Hellinger, in admiration of his fundamental contributions to operator theory

A (bounded linear) operator J on a Hilbert space is said to be jordan if $J = S + N$ where $S = S^*$ and $N^2 = 0$. The operator T is subjordan if T is the restriction of a jordan operator to an invariant subspace, and pure subjordan if no nonzero restriction of T to an invariant subspace is jordan. The main operator theoretic result of the paper is that a compact subset of the real line is the spectrum of some pure subjordan operator if and only if it is the closure of its interior. The result depends on understanding when the operator $D = \theta + \frac{d}{dx} : L^2(\mu) \to L^2(\nu)$ is closable. Here θ is an $L^2(\nu)$ function, μ and ν are two finite regular Borel measures with compact support on the real line, and the domain of D is taken to be the polynomials. More generally an explicit characterization of the closure of the graph of D for a large class of (θ, μ, ν) is obtained.

1. Introduction.

A (bounded, linear) operator J on a Hilbert space $\mathcal{K}$ is said to be *jordan* (of order two) if $J = S + N$ where $S = S^*$, $SN = NS$ and $N^2 = 0$. We say that the operator T on $\mathcal{H}$ is *subjordan* if there is a Hilbert space $\mathcal{K} \supset \mathcal{H}$ and a jordan operator J on $\mathcal{K}$ such that $\mathcal{H}$ is invariant for J and $T = J|\mathcal{H}$. The subjordan operator T is said to be *pure* if there is no nonzero invariant subspace $\mathcal{H}_0$ for T for which the restriction $T_0 = T|\mathcal{H}_0$ is jordan. We say that a compact subset K of the real line $\mathbb{R}$ is *regularly closed* if and only if K is the closure of its interior. We can now state one of the main results of this paper.

THEOREM A. *(See Theorem 3.17) A compact subset K of the real line arises as the spectrum $\sigma(T)$ of a pure subjordan operator T if and only if K is regularly closed.*

For the proof of Theorem A we use the model for subjordan operators given by Ball and Helton [BH] to deduce a model for pure subjordan operators. Characterizing these models requires determining when a differential operator of a certain class is closable. We actually obtain results

[1]Research of the first author was partially supported by the National Science Foundation.

on the closure of the graph of such a differential operator more general than what are required for the proof of Theorem A which should be of interest in their own right.

We now discuss the approximation question which comprises a large portion of the paper. We consider two finite compactly supported measures μ and ν on the real line and a complex valued function θ in $L^2(\nu)$. Define an operator D from $L^2(\mu)$ into $L^2(\nu)$ with domain equal to the polynomials in $L^2(\mu)$ by

$$Dp = \theta p + \frac{d}{dx}p.$$

A main interest of this paper is the study of the closure of the graph of D

$$\mathcal{G} = \{p \oplus Dp : \; p \text{ a polynomial}\}$$

in $L^2(\mu) \oplus L^2(\nu)$. For a large class of (μ, ν, θ) we obtain a complete explicit description of the closure $\mathcal{G}^-$ of $\mathcal{G}$. As corollaries we are able to give complete characterizations of when D is closable (i.e. $\mathcal{G}^-$ contains no nonzero elements of the form $0 \oplus g$) as well as of when $\mathcal{G}$ is dense in $L^2(\mu) \oplus L^2(\nu)$.

To describe our results in detail, let $d\nu = wdx + \nu_{sing}$ be the Lebesgue decomposition of ν with respect to Lebesgue measure dx on $\mathbb{R}$. When $w(x) = 0$ define $1/w(x) = \infty$. An interval I is said to be an interval of local integrability for $1/w$ if $\int_a^b \frac{1}{w(x)}dx < \infty$ for every compact subinterval $[a, b] \subset I$; here none, either one or both endpoints of I may belong to I. An interval I (possibly open, closed or half open - half closed) is said to be a *maximal interval of local integrability* (MILI) for $1/w$ if $1/w$ is locally integrable over I but no interval strictly containing I has this property. The collection of all MILI's for $1/w$ forms a partitioning of a subset of the support *spptν* of ν. For I an MILI, denote by $(\mathcal{G}|I)^-$ the closure of $\mathcal{G}$ in $L^2(\mu|I) \oplus L^2(\nu|I)$. Our first result shows how the characterization of $\mathcal{G}^-$ can be localized to MILI's.

THEOREM B. *(see Lemma 2.16 and Corollary 2.8) A given $h \oplus k \in L^2(\mu) \oplus L^2(\nu)$ is in $\mathcal{G}^-$ if and only if $(h \oplus k)|I$ is in $(\mathcal{G}|I)^-$ for each MILI I for $1/w$. Moreover, $h \oplus k$ is in $\mathcal{G}^-$ if and only if $h \oplus k_a$ is in the closure of $\mathcal{G}$ in $L^2(\mu) \oplus L^2(wdx)$, where $k = k_a + k_s$ is the decomposition of $k \in L^2(\nu)$ with $k_a \in L^2(wdx)$ and $k_s \in L^2(\nu_{sing})$.*

As a consequence of this localization result, characterizing $\mathcal{G}^-$ in $L^2(\mu) \oplus L^2(\nu)$ reduces to the case where $\nu = wdx$ and $sppt\nu$ is a MILI for $1/w$. The analysis for this reduced case splits into eight cases summarized in Table 2.1. For five of the eight cases, we have a complete explicit characterization of $\mathcal{G}^-$; the remaining three remain open, although we have a natural conjecture for the form of $\mathcal{G}^-$ for these cases as well which we can verify for some simple examples. We also give examples to verify that all cases are not vacuous. We present here the simplest sample result.

THEOREM C. *(Type 1 in Table 2.1) Suppose* $\nu = wdx$, $sppt\nu$ *is the closed interval* $I = [a,b]$ *and* $\int_a^b \frac{1}{w(x)}dx < \infty$. *Then a function* $h \oplus k \in L^2(\mu) \oplus L^2(wdx)$ *is in* $\mathcal{G}^-$ *if and only if there exists a function* h_1 *such that*

(i) h_1 *is absolutely continuous on* I

(ii) $h = h_1 \;\mu - a.e.$

(iii) $e^{\Theta(x)}h_1(x) = h_1(a) + \int_a^x (e^{\Theta}k)(t)dt$

where $\Theta(x) = \int_a^x \theta(t)dt$.

For the general case the following partial result is sufficient for many purposes.

THEOREM D. *(see Corollary 2.18) Suppose* $\nu = wdx$ *and* $sppt\nu$ *is a MILI* I *for* $1/w$. *Suppose that a function* $h \oplus k \in L^2(\mu) \oplus L^2(wdx)$ *satisfies (i), (ii) and (iii) in Theorem B, where* a *is some point in* I, *as well as*

(iv) $h \oplus k$ *has compact support in* I.

Then $h \oplus k \in \mathcal{G}^-$.

Using these general results concerning $\mathcal{G}^-$ we are able to obtain complete answers to when D is closable and (the opposite extreme) when $\mathcal{G}$ is dense.

THEOREM E. *(see Theorem 2.24) The operator* $D = \theta + \frac{d}{dx} : L^2(\mu) \to L^2(\nu)$ *is closable if and only if* ν *is absolutely continuous (so* $d\nu = wdx$*), the union of the MILI's for* $1/w$ *is a carrier for* wdx, *and the complement of* $sppt\mu$ *contains no intervals in a MILI for* $1/w$.

To give the characterization of when $\mathcal{G}$ is dense in $L^2(\mu) \oplus L^2(\nu)$ we need a definition. If I is a MILI for $1/w$ with endpoints a and b (which may or may not be in I), we define an auxiliary interval $J(= J(I))$ to be an interval with the same endpoints as I such that

(i) $a \in J$ if and only if $\int_a^{a+\delta} e^{2Re\Theta(t)} w(t)^{-1} dt < \infty$ for some $\delta > 0$

and

(ii) $b \in J$ if and only if $\int_{b-\delta}^{b} e^{2Re\Theta(t)} w(t)^{-1} dt < \infty$ for some $\delta > 0$.

Here $\Theta(t) = \int_{x_0}^{t} \theta(s) ds$ where x_0 is an arbitrary point of I. It turns out that always $J \supset I$. We can now state the result concerning the density of $\mathcal{G}$.

THEOREM F. *(see Theorem 2.31)* $\mathcal{G}^- = L^2(\mu) \oplus L^2(\nu)$ *(where* $d\nu = w dx + \nu_{sing}$*) if and only if, for each MILI* I *for* $1/w$*, either*

(i) $\mu | I$ *is a single point mass and* $J = I$

or

(ii) $\mu | I = 0$ *and* $J \setminus I$ *has at most one point.*

Special cases of Theorem F were suggested by J. Agler (private communication).

Similar results hold for the closure of $\mathcal{G}$ in $L^p(\mu) \oplus L^p(\nu)$ where $1 \le p < \infty$. In the case $1 < p < \infty$ one works with MILI's for w^{1-q} (where $\frac{1}{p} + \frac{1}{q} = 1$) rather than for w^{-1}; if $p = 1$ one works with maximal intervals on which $1/w$ is locally essentially bounded. We do not give the details of this extension here but rather refer to [F]. One may also consider the closure of $\mathcal{G}$ (for the special case $\theta = 0$) in the uniform norm of $C(E) \oplus C(E)$, where E is a compact subset of $\mathbb{R}$ and $C(E)$ is the space of continuous functions on E; this is also discussed in [F].

The study of subjordan operators was initiated by Helton in [H1], [H2], [H3] and has continued in [BH], [Ag1], [Ag2]. Included in some of this work are some results concerning such generalizations as complex subjordan operators of higher order (T is the restriction of $M + N$ to an invariant subspace where M is normal, M commutes with N and $N^k = 0$ for some positive integer k) and commuting n-tuples of complex subjordan operators. Our result Theorem A is an analogue for pure subjordan operators of the result of Clancey and Putnam [CP] which characterizes the spectrum of a pure subnormal operator. We remark that much earlier Dixmier and Foias [DF] used multiplication operators on Sobolev spaces to construct an operator having a preassigned F_σ–set as its point spectrum and a preassigned compact superset as its spectrum.

The closability of a differential operator as discussed above is closely connected with the closability of a Dirichlet form as discussed by Fukushima in [Fu]; the motivation there is the Beurling-Deny theory of Markovian semigroups. In particular our Theorem E has a large intersection with Theorem 2.1.4 in [Fu] (see also [Ham]).

Our technique for all the different Banach space norms considered is a straightforward duality analysis. We first obtain an explicit characterization of the annihilator $\mathcal{G}^{\perp}$ of $\mathcal{G}$ in the dual space. We then compute $\mathcal{G}^{-}$ as the preannihilator ${}^{\perp}\mathcal{G}^{\perp}$ of the annihilator. By restricting to the real line setting, all the calculations can be carried out quite far in an explicit form. This enables us to obtain an explicit characterization of the closure $\mathcal{G}^{-}$ for most cases.

The organization of the paper is as follows. Section 2 developes the needed duality machinery and proves the general results on the closure of $\mathcal{G}$ in $L^2(\mu) \oplus L^2(\nu)$ as listed in Theorems B-D and in Table 2.2. Section 2.a uses these results to characterize when D is closable (Theorem E) while Section 2.b characterizes when $\mathcal{G}$ is dense. Section 3 sets down the model for pure subjordan operators and uses the results of Section 2 to prove Theorem A. The results of this paper form a partion of the second authors Ph.D. dissertation [F] written under the direction of the first author.

2. The Closure of the Graph of a Differential Operator.

We begin by posing the problem and introducing the notation and terminology that will be used. Let μ and ν be two finite, positive measures both with support contained in $[0,1]$. Let m (or equivalently dx) represent Lebesgue measure on $[0,1]$. Denote by $\nu = \nu_a + \nu_s$ the Lebesgue decomposition of ν with respect to m, where $\nu_a << m$ and $\nu_s \perp m$. (Note: throughout this paper the symbol "$<<$" reads "is absolutely continuous with respect to", while "$\perp$" reads "is singular to" when referring to measures.) If we let $d\nu/dm = w \in L^1(dm)$, then $L^2(\nu)$ splits as $L^2(wdm) \oplus L^2(\nu_s)$, and for every $k \in L^2(\nu), k = k_a \oplus k_s$ where $k_a \in L^2(wdm)$ and $k_s \in L^2(\nu_s)$.

For $\theta \in L^2(\nu)$ define $D : L^2(\mu) \to L^2(\nu)$ via

$$Dp = \theta p + p'$$

for all polynomials p, and let

$$\mathcal{G} = \{p \oplus Dp : \ p \text{ a polynomial}\}$$

denote the graph of D in $L^2(\mu) \oplus L^2(\nu)$.

In the proofs of the following results a key observation is that

$$\int_a^b dm = \int_a^b \frac{1}{w} d\nu_a$$

since $wdm = d\nu_a$. Thus $1/w \in L^1(m|[a,b])$ if and only if $1/w \in L^2(\nu_a|[a,b])$. Now let $F \in L^2(\nu)$. Then, as mentioned above, F can be written as $F_a \oplus F_s$ with $F_a \in L^2(wdm)$ and $F_s \in L^2(\nu_s)$. Since $\nu_s \perp m, |F|dm = |F_a|dm$ and it follows that

$$\int_a^b |F|dm = \int_a^b |F|\frac{1}{w} d\nu_a \le \|F\|_{L^2(\nu_a)} \|\frac{1}{w}\|_{L^2(\nu_a|[a,b])}. \tag{2.1}$$

Thus $L^2(\nu_a|[a,b])$ can be continuously embedded into $L^1(m|[a,b])$ when $\int_a^b \frac{1}{w} dm < \infty$.

We will see that for $h \oplus k \in \mathcal{G}^-$, if we let $k = k_a \oplus k_s$ be the Lebesgue decomposition of k, then k_s can be chosen independently of both h and k_a. This fact and (2.1) will enable us to restrict our attention to intervals over which $1/w$ is integrable in a local sense. We therefore define the following terms and notation.

Definition 2.2. A property P is said to be *local* for an interval I, written $P_{loc}(I)$, if it holds for every compact subinterval $[c,d] \subseteq I$. This will be applied specifically to $P : f$ is Lebesgue integrable over I, and $P : f$ is absolutely continuous over I. We write these as $f \in L^1_{loc}(I)$ and $f \in AC_{loc}(I)$, respectively.

Thus $f \in AC_{loc}(I)$ implies that

$$f(y) - f(x) = \int_x^y f'(t)dt$$

for all $x, y \in I$.

Definition 2.3. An interval I will be termed a *maximal interval of local integrability*, (*MILI*) for a function f, if f is $L^1_{loc}(I)$, and if J is any other interval such that $J \supset I$ with $f \in L^1_{loc}(J)$, then $J = I$.

The thrust of our analysis of the norm closure of $\mathcal{G}$ relies on partitioning the support of ν_a into MILI's for $1/w$, and directing our attention to one such interval. Towards this, let I be a MILI for

$1/w$ and for some point $t_0 \in I$, let

$$\Theta(x) = \int_{t_0}^{x} \theta(t)dt. \tag{2.4}$$

Then, by observation (2.1), $\Theta \in AC_{loc}(I)$. Later we shall see that Θ plays a vital role in the determination of $\mathcal{G}^-$ in certain circumstances.

Definition 2.4. For a point $p \in \mathbb{R}$ and for a function $f \geq 0$, we say p is *right integrable for* f, written p is $RI(f)$, if for sufficiently small $\delta > 0$,

$$\int_{p}^{p+\delta} f dm < \infty.$$

Similarly, we say p is *left integrable for* f, written p is $LI(f)$, if for sufficiently small $\delta > 0$,

$$\int_{p-\delta}^{p} f dm < \infty.$$

If p is not right integrable for f, we write p is $NRI(f)$. Similarly we define the notation $NLI(f)$.

Now suppose I, with endpoints a and b, is a MILI for $1/w$. We shall see that the characterization of $\mathcal{G}^-$ depends on whether or not the functions $1/w$ and $e^{2Re\Theta}(1/w)$ are right integrable at a, and similarly left integrable at b. Thus for each MILI I for $1/w$, we define a corresponding interval J, with the same endpoints a and b as I, but where $a \in J$ if a is $RI(e^{2Re\Theta}1/w)$, and $b \in J$ if b is $LI(e^{2Re\Theta}1/w)$.

As noted where $1/w$ is integrable, Θ is absolutely continuous by (2.1), and hence $e^{2Re\Theta}$ is bounded. That is, integrability of $1/w$ implies integrability of $e^{2Re\Theta}1/w$. Thus I is always a subset of J. From this and the symmetry of some cases, we have six qualitatively different types of intervals I which are MILI's for $1/w$:

(T.1): $I = [a,b] = J$

(T.2): $I = (a,b) = J$

(T.3): $I = [a,b) = J$

(T.4): $I = (a,b), J = [a,b]$

(T.5): $I = [a,b), J = [a,b]$

(T.6): $I = (a,b), J = [a,b)$

When I is of the first three types, the problem is totally solved. The next two types are handled under an additional assumption, and the last type remains an open question. A table of examples of all these cases, with or without the additional assumption, will be provided after the presentation and proofs of the results.

We first need the following easy lemma:

LEMMA 2.5. *Suppose I is a MILI for $1/w$ and J is the corresponding interval of integrability for $e^{2Re\Theta}1/w$. Then $e^{\Theta}k \in L^1_{loc}(J)$ whenever $k \in L^2(wdm|J)$.*

PROOF: Let C be a compact subinterval of J. Then since $e^{\Theta}/w \in L^2(wdm|C)$ and $k \in L^2(wdm|J)$ we have

$$\int_C |e^{\Theta}k|dx = \int_C |(e^{\Theta}/w)k|wdx$$
$$=< |e^{\Theta}/w|,\ |k| >_{L^2(wdm|C)} < \infty. \square$$

Definition 2.6. Let $\hat{\mathcal{G}}$ denote the submanifold of $L^2(\mu) \oplus L^2(\nu)$ consisting of all $h \oplus k$ such that h and k satisfy the following: if I is an interval with endpoints a and b which is a MILI for $1/w$, and J is the corresponding interval determined by the integrability of $e^{2Re\Theta}1/w$, then

there exists a function $h_1 \in AC_{loc}(I)$ so that $h_1 = h$ $\mu-$ a.e., and

$$e^{\Theta(x)}h_1(x) - e^{\Theta(y)}h_1(y) = \int_y^x (e^{\Theta}k)(t)dt; \tag{2.6.1}$$

if $a \in J \setminus I$, then for $m-$ a.e. $x \in J$,

$$(e^{\Theta}h_1)(x) = \int_a^x (e^{\Theta}k)(t)dt, \tag{2.6.2}$$

and similarly, if $b \in J \setminus I$, then for $m-$ a.e. $x \in J$,

$$-(e^{\Theta}h_1)(x) = \int_x^b (e^{\Theta}k)(t)dt. \tag{2.6.3}$$

(Note that 2.6.1 makes sense by Lemma 2.5).

The conjecture is that $\hat{\mathcal{G}} = \mathcal{G}^-$, and our first theorem gives the inclusion in one direction: $\mathcal{G}^- \subset \hat{\mathcal{G}}$. A few points should be made about the definition of $\hat{\mathcal{G}}$. First of all, if $k = k_a + k_s$, notice that no restriction has been placed on k_s. Secondly, for the absolutely continuous extension h_1 of h chosen, $k_a = \theta h_1 + h_1'$ $m-$ a.e. on I. Also given h there may be more than one absolutely continuous extension of h satisfying (2.6.2) of Definition 2.6 for some k. Thus for a given h there may not be a unique k_a so that $h \oplus k$ is in $\hat{\mathcal{G}}$.

For instance, if μ consists of exactly one point mass in J, (2.6.2) places no restriction on the pair $h \oplus k$. Similarly, if μ consists of two point masses in J, say x_0 and y_0, then the only restriction is that

$$e^{\Theta(x_0)}h_1(x_0) - e^{\Theta(y_0)}h_1(y_0) = \int_{y_0}^{x_0} (e^{\Theta}k)(t)dt$$

for an absolutely continuous extension h_1 of h. On the other hand, if $\mu = m$, (2.6.2) is a very stiff requirement. We also note that since $L^2(\mu)$ consists of equivalence classes, to show a pair $h \oplus k \in \hat{\mathcal{G}}$ is in $\mathcal{G}^-$, we may use the absolutely continuous extension h_1 of h in place of h, since $h_1 = h$ μ a.e.

Much of our analysis is based on considering $\mathcal{G}^-$ as a double orthogonal complement $(\mathcal{G}^{\perp})^{\perp}$ of $\mathcal{G}$. Towards this we characterize $\mathcal{G}^{\perp}$ in the following lemma. For convenience, we assume that μ and ν have support in the unit interval $[0,1]$.

LEMMA 2.7. *A pair $f \oplus g$ in $L^2(\mu) \oplus L^2(\nu)$ belongs to $\mathcal{G}^{\perp}$ if and only if the following conditions are satisfied:*

$$gd\nu = gd\nu_a = gwdm \tag{2.7.1}$$

$$\int_{[0,1]} f(x)d\mu(x) + \int_{[0,1]} \overline{\theta(x)}g(x)w(x)dx = 0 \tag{2.7.2}$$

$$g(x)w(x) = -(\int_{(x,1]} f(t)d\mu(t) + \int_{(x,1]} \overline{\theta(t)}g(t)w(t)dt). \tag{2.7.3}$$

PROOF: Let $f \oplus g = \mathcal{G}^{\perp}$. Then for all polynomials p,

$$\begin{aligned}
0 &= \int_{[0,1]} p(x)\overline{f(x)}d\mu(x) + \int_{[0,1]} (\theta(x)p(x) + p'(x))\overline{g(x)}d\nu(x) \\
&= p(0)[\int_{[0,1]} \overline{f(x)}d\mu(x) + \int_{[0,1]} \theta(x)\overline{g(x)}d\nu(x)] + \int_{[0,1]} (\int_{[0,x]} p'(t)dt)\overline{f(x)}d\mu(x) \\
&+ \int_{[0,1]} \theta(x)(\int_{[0,x]} p'(t)dt)\overline{g(x)}d\nu(x) + \int_{[0,1]} p'(x)\overline{g(x)}d\nu(x) \\
&= p(0)[\int_{[0,1]} \overline{f(x)}d\mu(x) + \int_{[0,1]} \theta(x)\overline{g(x)}d\nu(x)]
\end{aligned}$$

$$+\int_{[0,1]} p'(x)[(\int_{(x,1]} \overline{f(t)}d\mu(t) + \int_{(x,1]} \theta(t)\overline{g(t)}d\nu(t))dx + \overline{g(x)}d\nu(x)].$$

We justify the change in order of integration by applying Fubini's theorem to the measures $d\mu(t) \times dm(x)$ and $d\nu(t) \times dm(x)$; with integrands $\chi_{\{x\leq t\}}(x,t)p'(x)\overline{f(t)}$ and $\chi_{\{x\leq t\}}(x,t)p'(x)\theta(t)\overline{g(t)}$, respectively; here $\chi_S(x,t)$ denotes the characteristic function of the set S: $\chi_S(s,t) = 1$ if $(s,t) \in S$ and 0 otherwise. Now using $p = 1$, we get that

$$0 = \int_{(0,1]} f(x)d\mu(x) + \int_{[0,1]} \overline{\theta(x)}g(x)d\nu(x). \tag{2.7.4}$$

Next, considering all polynomials p' with $p(0) = 0$, a dense set, we get for $m-$ a.e. $x \in [0,1]$

$$g(x)d\nu(x) = -(\int_{(x,1]} f(t)d\mu(t) + \int_{(x,1]} \overline{\theta(t)}g(t)d\nu(t))dx. \tag{2.7.5}$$

Since the measure on the right is absolutely continuous with respect to m, we must have $gd\nu$ is also. Thus, if $d\nu(x) = w(x)dx + d\nu_s(x)$, then $g(x)d\nu_s(x) = 0$. This proves (2.7.1) of the lemma. Applying the result to equations (2.7.4) and (2.7.5) yields that $f\oplus g$ in $\mathcal{G}^\perp$ must satisfy parts (2.7.2) and (2.7.3).

To complete the proof, simply reverse the argument. □

Before we proceed to prove $\mathcal{G}^- \subset \hat{\mathcal{G}}$, we give two corollaries which exhibit the ramifications of (2.7.1).

Corollary 2.8. *For $h \oplus k \in L^2(\mu) \oplus L^2(\nu)$, let $k = k_a \oplus k_s$ be the Lebesgue decomposition of k with respect to $\nu_a \oplus \nu_s$. Then $h \oplus k \in \mathcal{G}^-$ if and only if, for all $f \oplus g \in \mathcal{G}^\perp$,*

$$0 = \int_{[0,1]} f(x)\overline{h(x)}d\mu(x) + \int_{[0,1]} g(x)\overline{k_a}(x)w(x)dx.$$

Proof: Since $(\mathcal{G}^\perp)^\perp = \mathcal{G}^-, h \oplus k \in \mathcal{G}^-$ if and only if for each $f \oplus g \in \mathcal{G}^\perp$

$$0 = \int_{[0,1]} f\bar{h}d\mu + \int_{[0,1]} g\bar{k}d\nu.$$

But, from (2.7.1), the second integral is equal to

$$\int_{[0,1]} g\bar{k}_a w dm.$$

The results follow.

COROLLARY 2.9. *If $\nu_s \neq 0$, then D is not closable.*

PROOF: If $\nu_s \neq 0$, then there is a set A so that $\nu_s([0,1] \setminus A) = \nu_a(A) = m(A) = 0$. It follows that $\chi_A(x)$ is a non-zero element of $L^2(\nu)$ and from Corollary 2.8, $0 \oplus \chi_A \in \mathcal{G}^\perp$. Thus D is not closable.

□

Notice we have shown that, as was the case for $\hat{\mathcal{G}}$, belonging to $\mathcal{G}^-$ places no restriction on k_s. For this reason we see the closure depends only on the properties of ν_a (that is, the integrability of $1/w$), and is independent of ν_s. Thus, without loss of generality, we may assume $d\nu = wdm$ at times in the following proofs.

THEOREM 2.10. $\mathcal{G}^- \subset \hat{\mathcal{G}}$.

PROOF: We need to show $h \oplus k \in \mathcal{G}^-$ satisfies (2.6.1)-(2.6.3) of Definition 2.6. Let I, with endpoints a and b, be a MILI for $1/w$. If Θ is defined by (2.4), then Θ is $AC_{loc}(I)$. Let $\{p_n\}$ be a sequence of polynomials such that $\{p_n \oplus Dp_n\}$ converges to $h \oplus k$ in $L^2(\mu) \oplus L^2(\nu)$. By choosing a subsequence if necessary we may assume that for $\mu-$ a.e. x in I

$$\lim_n p_n(x) = h(x)$$

Since for ν-a.e. x in I,

$$\{e^\Theta p_n\}' = e^\Theta (Dp_n)$$

we have for all $x, y \in I$,

$$e^{\Theta(y)} p_n(y) - e^{\Theta(x)} p_n(x) = \int_{[x,y]} e^{\Theta(t)} (Dp_n)(t) dt. \tag{2.10.1}$$

Now, $\lim_{n \to \infty} Dp_n = k$ in $L^2(\nu)$, so by (2.1), Dp_n approaches k in $L^1_{loc}(I)$. Thus, if $[c,d] \subset I$, we have for all $x, y \in [c,d]$

$$\lim_{n \to \infty} \int_{[x,y]} e^{\Theta(t)} (Dp_n)(t) dt = \int_{[x,y]} e^{\Theta(t)} k(t) dt,$$

since e^Θ is locally bounded in I. But the left hand side of (2.10.1) converges $\mu-$ a.e. to $e^{\Theta(y)} h(y) - e^{\Theta(x)} h(x)$. Thus, for $\mu-$ a.e. $x, y \in [c,d]$,

$$e^{\Theta(t)} h(y) - e^{\Theta(x)} h(x) = \int_{[x,y]} e^{\Theta(x)} k(t) dt.$$

But, since e^{Θ} has bounded inverse and is $AC_{loc}(I)$, we have there is an $AC_{loc}(I)$ extension h_1 of h, so that $h = h_1$ $\mu-$ a.e. It follows that if $h \oplus k \in \mathcal{G}^-$, then $h \oplus k$ satisfies (2.6.2) of Definition 2.6. Moreover, since

$$\int_{[x,y]} |e^{\Theta(t)}k(t)|dt = \int_{[x,y]} |e^{\Theta(t)}k(t)|1/w(t)d\nu(t) \leq \|k\|_{L^2(\nu)}\|e^{2Re\Theta}1/w\|^{1/2}_{L^1(m)}, \tag{2.10.2}$$

we see that $e^{\Theta}k$ is actually in $L^1_{loc}(J)$. Thus h and k satisfy property (2.6.1) of $\hat{\mathcal{G}}$.

For property (2.6.3), notice that if a is $NRI(1/w)$ but

$$\int_{[a,a+\delta]} e^{2Re\Theta}\frac{1}{w}dm < \infty,$$

it must be the case that $\lim_{\delta\to 0} inf\ e^{2Re\Theta(a+\delta)} = 0$. Thus, for every n, $\lim_{x\to a} inf\ e^{\Theta(x)}p_n(x) = 0$. But then from the general estimate (2.10.2) we may take limits to get

$$e^{\Theta(x)}p_n(x) = \int_{[a,x]}(e^{\Theta}p_n)'(t)dt = \int_{[a,x]}[e^{\Theta}Dp_n](t)dt.$$

Now again by (2.10.2) the right hand side converges to

$$\int_{[a,x]}(e^{\Theta}k)(t)dt,$$

while the left hand side converges for $\mu-$ a.e. x in I to $(e^{\Theta}h)(x)$.

Similarly, if $b \in J \setminus I$, we have

$$(e^{\Theta}h)(x) = -\int_{[x,b]}(e^{\Theta}k)(t)dt. \ \square$$

We next address the question of when $\hat{\mathcal{G}} \subset \mathcal{G}^-$. The first step is to prove five lemmas based on the characterization of $\mathcal{G}^{\perp}$ reached in Lemma 2.7. These results will enable us to localize our concentration to one MILI for $1/w$. We begin with two lemmas which determine the set where all elements of $\mathcal{G}^{\perp}$ vanish identically. If we consider the manifold consisting of all $h \oplus k$ that are zero off this set, we see that this manifold must be contained in $\mathcal{G}^-$ since it is perpendicular to $\mathcal{G}^{\perp}$.

LEMMA 2.11. *Let $f \oplus g \in \mathcal{G}^{\perp}$. If x is $NRI(1/w)$, then*

$$\lim_{\delta\to 0}\left(\int_{(x+\delta,1]} f(t)d\mu(t) + \int_{(x+\delta,1]}\overline{\theta(t)}g(t)w(t)dt\right) = \int_{(x,1]} f(t)d\mu(t) + \int_{(x,1]}\overline{\theta(t)}g(t)w(t)dt = 0.$$

Similarly, if x is $NLI(1/w)$, then

$$\lim_{\delta\to 0}\left(\int_{(x-\delta,1]} f(t)d\mu(t) + \int_{(x-\delta,1]} \overline{\theta(t)}g(t)w(t)dt\right) = \int_{[x,1]} f(t)d\mu(t) + \int_{[x,1]} \overline{\theta(t)}g(t)w(t)dt = 0.$$

PROOF: Since the measure

$$\lambda(A) = \int_A fd\mu + \int_A \bar{\theta}gwdm$$

is countably additive,

$$\lim_{\delta\to 0}(\int_{(x+\delta,1]} fd\mu + \int_{(x+\delta,1]} \bar{\theta}gwdm) = \int_{(x,1]} fd\mu + \int_{(x,1]} \bar{\theta}gwdm,$$

for all x. Now assume x is $NRI(1/w)$ and suppose the limit above is nonzero. Then there is an $\epsilon > 0$ and an interval I with left end point x, so that for all y in I,

$$\left|\int_{(y,1]} fd\mu + \int_{(y,1]} \bar{\theta}gwdm\right| > \epsilon.$$

But then from (2.7.3)

$$\begin{aligned}\|g\|^2_{L^2(\nu)} &= \int_{[0,1]} |g(y)w(y)|^2\frac{1}{w(y)}dy = \int_{[0,1]} |\int_{(y,1]} f(t)d\mu(t) + \int_{(y,1]} \overline{\theta(t)}g(t)w(t)dt|^2\frac{1}{w(y)}dy \\ &\geq \int_I |\int_{(y,1]} fd\mu + \int_{(y,1]} \bar{\theta}gw\ dm|^2\frac{1}{w(y)}dy > \epsilon^2\int_I \frac{1}{w(y)}dy = \infty,\end{aligned}$$

a contradiction. A similar argument applies if x is $NLI(1/w)$. □

We next define a set K by

$$K = \{x \in [0,1]\ : x \text{ is both } NRI(1/w) \text{ and } NLI(1/w)\}. \tag{2.12.1}$$

As the next lemma will point out, K is precisely the set where elements of $\mathcal{G}^\perp$ vanish identically. We note two other properties. First, K is the complement of the union of all MILI's for $1/w$. Secondly, for each $x \in K$

$$\int_{[x,1]} fd\mu + \int_{[x,1]} \bar{\theta}gwdm = \int_{(x,1]} fd\mu + \int_{(x,1]} \bar{\theta}gwdm = 0, \tag{2.12.2}$$

by Lemma 2.11.

Lemma 2.13. *If $f \oplus g \in \mathcal{G}^{\perp}$, then $f = 0$ $\mu-$ a.e. and $g = 0$ $\nu-$ a.e. on K.*

Proof: Let λ_1 and λ_2 denote the measures defined by

$$\lambda_1(A) = \int_A f d\mu, \qquad \lambda_2(A) = \int_A \bar{\theta} g w dm$$

for all subsets A of $[0,1]$. Then $\lambda = \lambda_1 + \lambda_2$ where λ was defined in the proof of Lemma 2.11. By (2.7.3) of Lemma 2.7 and Lemma 2.11, for $m-$ a.e. $x \in K$,

$$g(x)w(x) = -\lambda((x,1]) = 0.$$

That is, $gw = 0$, $m-$ a.e. on K, and hence $g = 0$, $\nu-$ a.e. on K. Moreover, this implies $\lambda_2(A) = 0$ for all subsets A of K.

To finish the proof we need only show $\lambda_1(A) = 0$ for all $A \subset K$ as this implies $f = 0$, $\mu-$ a.e. on K. But λ_2 is zero on K and $\lambda_1 = \lambda - \lambda_2$ so it suffices to show $\lambda(A) = 0$ for every subset A of K.

Since $\{x\} = [x,1] \setminus (x,1]$, by additivity of λ and (2.12.2), we have $\lambda(\{x\}) = 0$ for all $x \in K$. That is, λ can have no point mass in K. Now let $A \subset K$. By regularity of the total variation measure $|\lambda|$, we may assume that A is closed. Let $\epsilon > 0$ be given and choose an open set U containing A so that $|\lambda|(U \setminus A) < \epsilon$. Write

$$U = \bigcup_n (a_n, b_n),$$

as a disjoint union of open intervals. We construct a new open set U' as follows: Let

$$\mathbb{Z}' = \{n \in \mathbb{Z} : (a_n, b_n) \cap A \neq \phi\},$$

$$a'_n = inf\{x : x \in (a_n, b_n) \cap A, n \in \mathbb{Z}'\},$$

$$b'_n = sup\{x : x \in (a_n, b_n) \cap A, n \in \mathbb{Z}'\},$$

and let

$$U' = \bigcup_{n \in \mathbb{Z}} (a'_n, b'_n).$$

Then notice that since A is closed $a'_n, b'_n \in A$ for all n, and $A \subset U'$ modulo these endpoints a'_n and b'_n. But as these endpoints are a countable set and since λ has no point masses in K, this set has

λ-measure zero. Moreover, since each a'_n and $b'_n \in K$, by additivity and (2.12.1) we have

$$\lambda((a'_n, b'_n)) = \lambda((a'_n, 1]) - \lambda([b'_n, 1]) = 0.$$

Thus $\lambda(U') = 0$. Hence

$$\begin{aligned} |\lambda(A)| &= |\lambda(U') - \lambda(A)| \\ &= |\lambda(U' \setminus A)| \\ &\leq |\lambda|(U' \setminus A) \\ &\leq |\lambda|(U \setminus A) < \epsilon. \end{aligned}$$

So by the arbitrariness of ϵ, we conclude that $\lambda(A) = 0$. This completes the proof. □

The next result, a corollary to Lemma 2.13, does not in itself contribute to the analysis of when $\mathcal{G}^- = \hat{\mathcal{G}}$, but does introduce the question of the closability of D. The answer to this question will be given at the end of this section.

COROLLARY 2.14. *If $\nu << m$, and $\nu(K) > 0$, then $\mathcal{G}^-$ contains non-zero elements of the form $0 \oplus k$, and hence D is not closable.*

PROOF: Suppose there is a set $A \subset K$ so that $\nu(A) > 0$. Let $k = \chi_A$, a nonzero element of $L^2(\nu)$. By Lemma 2.13, if $f \oplus g \in \mathcal{G}^\perp$, then

$$\int_{[0,1]} g\bar{k}d\nu = \int_{[0,1]\setminus K} g\bar{k}d\nu = 0.$$

Thus $0 \oplus k \in (\mathcal{G}^\perp)^\perp = \mathcal{G}^-$. □

We next investigate the complement of the set K given above, namely the set of points where elements of $\mathcal{G}^\perp$ can be nonzero, and thus where the elements of $\mathcal{G}^-$ are constrained. We do this by localizing to intervals I which are MILI's for $1/w$.

LEMMA 2.15. *If I is a MILI for $1/w$, then for $f \oplus g \in \mathcal{G}^\perp$,*

$$\int_I f(t)d\mu(t) + \int_I \overline{\theta(t)}g(t)w(t)dt = 0.$$

Proof: Let I have endpoints a and b. Then I has four forms depending on whether or not a and b belong to I. Suppose $a \notin I$, that is, a is $NRI(1/w)$. Then for small $\delta > 0$,

$$\int_a^{a+\delta} 1/w dm = \infty.$$

If $a \in I$, then by the maximality of I, it must be the case that for small $\delta > 0$,

$$\int_{a-\delta}^{a} 1/w dm = \infty.$$

By Lemma 2.11, in the first case

$$\lim_{\delta \to 0} \left(\int_{(a+\delta,1]} f(t)d\mu(t) + \int_{(a+\delta,1]} \overline{\theta(t)}g(t)w(t)dt \right) = \int_{(a,1]} f(t)d\mu(t) + \int_{(a,1]} \overline{\theta(t)}g(t)w(t)dt = 0,$$

while in the second,

$$\lim_{\delta \to 0} \left(\int_{(a-\delta,1]} f(t)d\mu(t) + \int_{(a-\delta,1]} \overline{\theta(t)}g(t)w(t)dt \right) = \int_{[a,1]} f(t)d\mu(t) + \int_{[a,1]} \overline{\theta(t)}g(t)w(t)dt = 0.$$

Similar results hold for the endpoint b. Thus, if $b \in I$, then

$$\int_{(b,1]} f(t)d\mu(t) + \int_{(b,1]} \overline{\theta(t)}g(t)w(t)dt = 0,$$

and if $b \notin I$

$$\int_{[b,1]} f(t)d\mu(t) + \int_{[b,1]} \overline{\theta(t)}g(t)w(t)dt = 0.$$

In all cases, I can be obtained as the difference of the first interval and the second. Thus

$$\int_I f(t)d\mu(t) + \int_I \overline{\theta(t)}g(t)w(t)dt = 0$$

as claimed. □

Lemma 2.16. *Suppose that for each I which is a MILI for $1/w$ and for each $f \oplus g \in \mathcal{G}^\perp$, the element $h \oplus k$ of $L^2(\mu) \oplus L^2(\nu)$ satisfies*

$$\int_I f\bar{h}d\mu + \int_I g\bar{k}d\nu = 0.$$

Then $h \oplus k \in \mathcal{G}^-$.

Proof: The observation here is that if $\mathcal{M} = \cup\{I : I \text{ is a MILI for } 1/w\}$, then $[0,1] \setminus \mathcal{M} = K$. Since, by Lemma 2.13, $f \oplus g \in \mathcal{G}^\perp$ implies $f \oplus g = 0$ on K, we have

$$\int_{[0,1]} f\bar{h}d\mu + \int_{[0,1]} g\bar{k}d\nu = \int_{\mathcal{M}} f\bar{h}d\mu + \int_{\mathcal{M}} g\bar{k}d\nu.$$

Thus by countable additivity, our assumption implies that this term is zero as needed. □

LEMMA 2.17. *Let I be a MILI for $1/w$ and suppose that $h \oplus k \in L^2(\mu) \oplus L^2(\nu)$ satisfies*

$$(h \oplus k)(x) = 0 \text{ for } x \in [0,1] \setminus I, \tag{2.17.1}$$

$$\text{there exists a function } h_1 \in AC_{loc}(I) \text{ with } h_1 = h \ \mu \text{ a.e. for } h_1 \text{ as in (2.6.1)}, \tag{2.17.2}$$

$$k_a = \theta h_1 + h_1' \quad \nu_a - \text{ a.e. on } I, \tag{2.17.3}$$

and for some sequences $\{c_n\}$ and $\{d_n\}$ such that $\{c_n\}$ approaches a from the right and $\{d_n\}$ approaches b from the left and for h_1 as in (2.17.2)

$$\lim_{n\to\infty} [\overline{h_1(c_n)}(\int_{[c_n,1]} f d\mu + \int_{[c_n,1]} \bar{\theta} g d\nu) - \overline{h_1(d_n)}\left(\int_{(d_n,1]} f d\mu + \int_{(d_n,1]} \bar{\theta} g d\nu\right)] = 0 \tag{2.17.4}$$

for all $f \oplus g \in \mathcal{G}^{\perp}$.

Then $h \oplus k \in \mathcal{G}^{-}$.

PROOF: It suffices to show that $h \oplus k$ is orthogonal to $\mathcal{G}^{\perp}$. Since $(h \oplus k)(x) = 0$ for $x \notin I$, by Lemma 2.7 this amounts to showing

$$0 = \int_I f\bar{h} d\mu + \int_I g\bar{k} w dx \tag{2.17.5}$$

for all $f \oplus g$ in $\mathcal{G}^{\perp}$. But also by Lemma 2.7 and the assumptions on $h \oplus k$, the second integral in (2.17.5) becomes

$$\int_I g\bar{k} w dx = -\int_I \{\int_{(x,1]} f(t) d\mu(t) + \int_{(x,1]} \overline{\theta(t)} g(t) w(t) dt\} \overline{k(x)} dx. \tag{2.17.6}$$

Now suppose $[c,d]$ is a compact subinterval of I. Then $1/w \in L^1(dx|[c,d])$. Noting that I can be written as the increasing union of such intervals, we have

$$\begin{aligned}\int_{[c,d]} g(x)\overline{k(x)} w(x) dx &= -\int_{[c,d]} \{\int_{(x,1]} f(t) d\mu(t) + \int_{(x,1]} \overline{\theta(t)} g(t) w(t) dt\} \overline{k(x)} dx \\ &= -\int_{[c,d]} \int_{[0,1]} \chi_{\{x<t\}}(x,t) f(t) \overline{k(x)} d\mu(t) dx - \int_{[c,d]} \int_{[0,1]} \chi_{\{x<t\}}(x,t) \overline{\theta(t)} g(t) w(t) \overline{k(x)} dt dx.\end{aligned} \tag{2.17.7}$$

Now $f(t) \in L^2(\mu) \subset L^1(\mu)$, and $k \in L^1(m|[c,d])$, so the integrand in the first integral in (2.17.7) is in $L^1(\mu(t) \times m(x)|[0,1] \times [c,d])$. Thus we may apply Fubini's theorem to get this first integral equal to

$$-\int_{[0,1]} \int_{[c,d]} \chi_{\{x<t\}}(x,t) f(t) \overline{k(x)} dx d\mu(t) =$$

$$- \int_{[c,d]} f(t)(\int_{[c,t]} \overline{k(x)}dx)d\mu(t) - \int_{(d,1]} f(t)(\int_{[c,d]} \overline{k(x)}dx)d\mu(t). \qquad (2.17.8)$$

In the second integral in (2.17.7), $\overline{\theta(t)}g(t) \in L^2(wdt) \subset L^1(wdt)$. Thus $\overline{\theta(t)}g(t)w(t) \subset L^1(dt)$. Moreover, k is $L^1_{loc}(I)$, so k is $L^1(m|[c,d])$. Therefore the integrand is in $L^1(m \times m|[0,1] \times [c,d])$, so using Fubini's theorem again we get this second integral is equal to

$$- \int_{[0,1]} \int_{[c,d]} \chi_{\{x<t\}}(x,t)\overline{\theta(t)}g(t)w(t)\overline{k(x)}dxdt =$$

$$- \int_{[c,d]} \overline{\theta(t)}g(t)w(t) \left(\int_{[c,t]} \overline{k(x)}dx \right) dt - \int_{(d,1]} \overline{\theta(t)}g(t)w(t) \left(\int_{[c,d]} \bar{k}dx \right) dt. \qquad (2.17.9)$$

But by (2.17.2), for some absolutely continuous extension h_1 of h,

$$\int_{[c,t]} \overline{k(x)}dx = \overline{h_1(t)} - \overline{h_1(c)} + \int_{[c,t]} \overline{(\theta h_1)(x)}dx.$$

Using this, and $h_1 = h$ $\mu-$ a.e., we combine (2.17.7)-(2.17.9) to get

$$\int_{[c,d]} g(t)\overline{k(t)}w(t)dt = - \int_{[c,d]} f(t)\overline{h_1(t)}d\mu(t) - \int_{[c,d]} \overline{\theta(t)}g(t)\overline{h_1(t)}w(t)dt$$
$$+ \overline{h_1(c)} \left[\int_{[c,d]} f(t)d\mu(t) + \int_{[c,d]} \overline{\theta(t)}g(t)w(t)dt \right] - \int_{[c,d]} f(t) \left(\int_{[c,t]} \overline{\theta(x)h_1(x)}dx \right) d\mu(t)$$
$$- \int_{[c,d]} \overline{\theta(t)}g(t)w(t) \left(\int_{[c,t]} \overline{\theta(x)h_1(x)}dx \right) dt - \overline{h_1(d)} \left[\int_{(d,1]} f(t)d\mu(t) + \int_{(d,1]} \overline{\theta(t)}g(t)w(t)dt \right]$$
$$+ \overline{h_1(c)} \left[\int_{(d,1]} f(t)d\mu(t) + \int_{(d,1]} \overline{\theta(t)}g(t)w(t)dt \right] - \int_{(d,1]} f(t) \left(\int_{[c,d]} \overline{\theta(x)h_1(x)}dx \right) d\mu(t)$$
$$- \int_{(d,1]} \overline{\theta(t)}g(t)w(t) \left(\int_{[c,d]} \overline{\theta(x)h_1(x)}dx \right) dt.$$

Thus

$$\int_I f\bar{h}d\mu + \int_I g\bar{k}d\nu = \int_{I/[c,d]} f\overline{h_1}d\mu + \int_{[c,d]} f\overline{h_1}d\mu + \int_{I\setminus[c,d]} g\bar{k}wdx - \int_{[c,d]} f\overline{h_1}d\mu - \int_{[c,d]} \bar{\theta}g\bar{h}_1wdx$$
$$+ \overline{h_1(c)} \left[\int_{[c,d]} fd\mu + \int_{[c,d]} \bar{\theta}gwdx \right] - \int_{[c,d]} f(t) \left(\int_{[c,t]} \overline{\theta(x)h_1(x)}dx \right) d\mu(t)$$
$$- \int_{[c,d]} \overline{\theta(t)}g(t)w(t) \left(\int_{[c,t]} \overline{\theta(x)h_1(x)}dx \right) dt - \overline{h_1(d)} \left[\int_{(d,1]} fd\mu + \int_{(d,1]} \bar{\theta}gwdx \right]$$
$$+ h_1\overline{(c)} \left[\int_{(d,1]} fd\mu + \int_{(d,1]} \bar{\theta}gwdx \right] - \left(\int_{[c,d]} \overline{\theta h_1}dx \right) \left(\int_{(d,1]} fd\mu + \int_{(d,1]} \bar{\theta}gwdx \right)$$

$$= \int_{I\setminus[c,d]} f\bar{h}_1 d\mu + \int_{I\setminus[c,d]} g\bar{k} d\nu + \overline{h_1(c)} \left[\int_{[c,1]} f d\mu + \int_{[c,1]} \bar{\theta} g w dx \right]$$
$$- \overline{h_1(d)} \left[\int_{(d,1]} f d\mu + \int_{(d,1]} \bar{\theta} g w dx \right] - \int_{[c,d]} \bar{\theta} g \bar{h}_1 w dx - \int_{[c,d]} f(t) \left(\int_{[c,t]} \overline{\theta(x) h_1(x)} dx \right) d\mu(t)$$
$$- \int_{[c,d]} \overline{\theta(t)} g(t) w(t) \left(\int_{[c,t]} \overline{\theta(x) h_1(x)} dx \right) dt - \left(\int_{[c,d]} \overline{\theta h_1} dx \right) \left(\int_{(d,1]} f d\mu + \int_{(d,1]} \bar{\theta} g w dx \right). \tag{2.17.10}$$

Now, carrying out another interchange of the order of integration, we get

$$- \int_{[c,d]} \int_{[c,d]} \chi_{\{x<t\}}(x,t) f(t) \overline{(\theta h_1)(x)} dx d\mu(t) - \int_{[c,d]} \int_{[c,d]} \chi_{\{x<t\}}(x,t) \overline{\theta(t)} g(t) w(t) \overline{(\theta h_1)(x)} dx dt$$
$$= - \int_{[c,d]} \left(\int_{(x,d]} f(t) d\mu(t) \right) \overline{(\theta h_1)(x)} dx - \int_{[c,d]} \left(\int_{(x,d]} (\bar{\theta} g w)(t) dt \right) \overline{(\theta h_1)(x)} dx. \tag{2.17.11}$$

But by Lemma 2.7 (2.7.3),

$$\int_{(x,d]} f(t) d\mu(t) + \int_{(x,d]} (\bar{\theta} g w)(t) dt = \left[\int_{(x,1]} f(x) d\mu(t) + \int_{(x,1]} (\bar{\theta} g w)(t) dt \right]$$
$$- \left[\int_{(d,1]} f(t) d\mu(t) + \int_{(d,1]} (\bar{\theta} g w)(t) dt \right] = -(gw)(x) - \int_{(d,1]} f(t) d\mu(t) - \int_{(d,1]} (\bar{\theta} g w)(t) dt.$$

Thus (2.17.11) becomes

$$- \int_{[c,d]} f(t) \left(\int_{[c,t]} \overline{(\theta h_1)(x)} dx \right) d\mu(t)$$
$$- \int_{[c,d]} (\bar{\theta} g w)(t) \left(\int_{[c,t]} \overline{(\theta h_1)(x)} dx \right) dt$$
$$= \int_{[c,d]} (gw)(x) \overline{(\theta h_1)(x)} dx$$
$$+ \left[\int_{(d,1]} f(t) d\mu(t) + \int_{(d,1]} (\bar{\theta} g w)(x) dx \right] \left[\int_{[c,d]} \overline{(\theta h_1)(x)} dx \right].$$

Substituting this expression into (2.17.10) results in

$$\int_I f\bar{h} d\mu + \int_I g\bar{k} d\nu = \int_{I\setminus[c,d]} f\bar{h} d\mu + \int_{I\setminus[c,d]} g\bar{k} d\nu$$
$$+ \overline{h_1(c)} \left(\int_{[c,1]} f(t) d\mu(t) + \int_{[c,1]} (\bar{\theta} g w)(t) dt \right) - \overline{h_1(d)} \left(\int_{(d,1]} f(t) d\mu(t) + \int_{(d,1]} (\bar{\theta} g w)(t) dt \right).$$

Finally, notice that as c approaches a from the right, d approaches b from the left, and both $\int_{I\setminus[c,d]} f\bar{h}d\mu$ and $\int_{I\setminus[c,d]} g\bar{k}d\nu$ converge to zero by hypothesis (2.17.1) and the last two terms tend to zero by (2.17.4). Thus (2.17.5) is verified and the proof is complete. □

COROLLARY 2.18. *Suppose $h \oplus k \in \hat{\mathcal{G}}$ and for each MILI I for $1/w$, $(h\oplus k)|I$ has compact support in I. Then $h \oplus k$ is in $\mathcal{G}^-$.*

PROOF: The result follows immediately from Lemmas 2.16 and 2.17. □

We now turn to the consideration of cases (T.1) through (T.5). The first three types rely on groundwork laid in the preceding lemmas. The analysis of (T.4) and (T.5) rely on Lemmas 2.11 through 2.16, an additional assumption, and on a transformation, by the function e^{Θ}, to another direct sum space

$$L^2(e^{-2Re\Theta}d\mu|I) \oplus L^2(e^{-2Re\Theta}wdx|I).$$

Type (T.6) remains open.

THEOREM 2.19. *(T.1). Let I be the MILI for $1/w$ which is of type (T.1). If $h \oplus k \in \hat{\mathcal{G}}$ and is zero on $[0,1]\setminus I$, then $h \oplus k \in \mathcal{G}^-$.*

PROOF: By assumption $I = [a,b]$ is compact, and by maximality a is $NLI(1/w)$ and b is $NRI(1/w)$. Thus by Lemma 2.11 both

$$\left(\int_{[a,1]} fd\mu + \int_{[a,1]} \bar{\theta}gwdt\right) = \left(\int_{(b,1]} fd\mu + \int_{(b,1]} \bar{\theta}gwdt\right) = 0. \qquad (2.19.1)$$

Since, we may take $[c,d] = [a,b]$ in the proof of Lemma 2.17 and since by assumption there is an $AC(I)$ extension $\bar{h}_1$ of h, it suffices only to show

$$\overline{h_1(a)}\left(\int_{[a,1]} fd\mu + \int_{[a,1]} \bar{\theta}gwdt\right) - \overline{h_1(b)}\left(\int_{(b,1]} fd\mu + \int_{(b,1]} \bar{\theta}gwdt\right) = 0.$$

But this follows immediately from (2.19.1). □

THEOREM 2.20. *(T.2). Let I be a MILI for $1/w$ which is of type (T.2). If $h \oplus k \in \hat{\mathcal{G}}$ and is zero on $[0,I]\setminus I$, then $h \oplus k \in \mathcal{G}^-$.*

PROOF: Here we have $I = (a, b)$. By Lemma 2.17, we need only show there are sequences $\{c_n\}$ and $\{d_n\}$, so that c_n approaches a from the right, d_n approaches b from the left and

$$\lim_{n\to\infty}[\overline{h_1(c_n)}\left(\int_{[c_n,1]} fd\mu + \int_{[c_n,1]} \bar{\theta} gwdt\right) - \overline{h_1(d_n)}\left(\int_{(d_n,1]} fd\mu + \int_{(d_n,1]} \bar{\theta} gwdt\right)] = 0$$

for all $f \oplus g \in \mathcal{G}^{\perp}$. Suppose there is no sequence $\{d_n\}$ approaching b from the left, so that

$$\lim_{n\to\infty}\left[\overline{h_1(d_n)}\left(\int_{(d_n,1]} fd\mu + \int_{(d_n,1]} \bar{\theta} gwdt\right)\right] = 0.$$

Then

$$\lim_{d\to b} inf\left|\overline{h_1(d)}\left(\int_{(d,1]} fd\mu + \int_{(d,1]} \bar{\theta} gwdt\right)\right| \neq 0.$$

That is, there is an $\epsilon > 0$, and $\delta > 0$, so that

$$\left|h_1(x)\left(\int_{(x,1]} fd\mu + \int_{(x,1]} \bar{\theta} gwdt\right)\right| > \epsilon \tag{2.20.1}$$

for all x, with $b - \delta < x < b$. Let $t_0 \in I$, and define

$$\Theta(x) = \int_{t_0}^{x} \theta(t)dt.$$

Now, squaring both sides of the inequality (2.20.1) and multiplying by $e^{2Re\Theta(x)}$ results in

$$\left|e^{\Theta(x)}h_1(x)\left(\int_{(x,1]} fd\mu + \int_{(x,1]} \bar{\theta} gwdt\right)\right|^2 > \epsilon^2 e^{2Re\Theta(x)}.$$

But $(e^{\Theta}h_1)(x) = \int_{[t_0,x]} e^{\Theta(t)}k(t)dt + e^{\Theta(t_0)}h_1(t_0)$.

Using this we get,

$$\begin{aligned}|&\left(\int_{[t_0,x]} e^{\Theta(t)}k(t)dt\right)\left(\int_{[x,1]} fd\mu + \int_{(x,1]} \bar{\theta} gwdt\right)\\ &+ e^{\Theta(t_0)}h_1(t_0)\left(\int_{(x,1]} fd\mu + \int_{(x,1]} \bar{\theta} gwdt\right)|^2 > \epsilon^2 e^{2Re\Theta(x)}.\end{aligned}$$

However, b is NLI $(1/w)$, so by Lemma 2.11,

$$\lim_{x\to b}\left(\int_{(x,1]} fd\mu + \int_{(x,1]} \bar{\theta} gwdt\right) = \left(\int_{[b,1]} fd\mu + \int_{[b,1]} \bar{\theta} gwdt\right) = 0.$$

So it suffices to only consider the situation where

$$\left|\left(\int_{[t_0,x]} e^{\Theta(t)}k(t)dt\right)\left(\int_{(x,1]} fd\mu + \int_{(x,1]} \bar{\theta}gwdt\right)\right|^2 > \epsilon^2 e^{2Re\Theta(x)},$$

or

$$\left|\int_{(x,1]} fd\mu + \int_{(x,1]} \bar{\theta}gwdt\right|^2 > \epsilon^2 e^{2Re\Theta(x)} \left|\int_{[t_0,x]} e^{\Theta(t)}k(t)dt\right|^{-2}. \tag{2.20.2}$$

Now

$$\begin{aligned}\left|\int_{[t_0,x]} e^{\Theta(t)}k(t)dt\right|^2 &\le \left[\int_{[t_0,x]} e^{Re\Theta(t)}|k(t)|\frac{1}{w(t)}d\nu_a(t)\right]^2 \\ &\le \|k\|^2_{L^2(\nu)} \int_{[t_0,x]} e^{2Re\Theta(t)}\left(\frac{1}{w(t)}\right)^2 d\nu_a(t) \\ &\le \|k\|^2_{L^2(\nu)} \int_{[t_0,x]} e^{2Re\Theta(t)}1/w(t)dt.\end{aligned}$$

Thus, (2.20.2) may be replaced by the inequality

$$|(gw)(x)|^2 \ge \epsilon^2\|k\|^{-2}e^{2Re\Theta(x)}\left\{\int_{[t_0,x]} e^{2Re\Theta(t)}\frac{1}{w(t)}dt\right\}^{-1}. \tag{2.20.3}$$

Let $b-\delta < t_1 < T < b$, and $t_1 \ne t_0$. Then

$$\int_{[0,1]} |(gw)(x)|^2\frac{1}{w(x)}dx = \|g\|^2_{L^2(\nu)} \ge \int_{[t_1,T]} |g|^2 d\nu.$$

So, integrating both sides of (2.20.3) with respect to $1/w(x)dx$ from t_1 to y, where $b-\delta < y < b$, we get

$$\begin{aligned}\|g\|^2_{L^2(\nu)} &\ge \left(\frac{\epsilon}{\|k\|}\right)^2 \int_{[t_1,y]} e^{2Re\Theta(x)}\frac{1}{w(x)}\left(\int_{[t_0,x]} e^{2Re\Theta(t)}\frac{1}{w}(t)dt\right)^{-1} dx \\ &= \left(\frac{\epsilon}{\|k\|}\right)^2 \ell n\left(\int_{[t_0,x]} e^{2Re\Theta(t)}1/w(t)dt\right)\Big|_{t_1}^{y} \\ &= \left(\frac{\epsilon}{\|k\|}\right)^2\left[\ell n\int_{[t_0,y]} e^{2Re\Theta(t)}/w(t)dt - \ell n\left(\int_{[t_0,t_1]} e^{2Re\Theta(t)}/w(t)dt\right)\right].\end{aligned}$$

Finally, letting y tend to b, and using the assumption that b is NLI $(e^{2Re\Theta}1/w)$, we get the last expression goes to infinity, a contradiction. We conclude that there is a sequence $\{d_n\}$ tending to b from the left such that

$$\lim_{n\to\infty} \overline{h_1(d_n)}\left(\int_{(d_n,1]} fd\mu + \int_{(d_n,1]} \bar{\theta}gwdt\right) = 0.$$

Since a is $NRI(1/w)$ we have

$$\lim_{x\searrow a}\left(\int_{[x,1]} f d\mu + \int_{[c_n,1]} \bar{\theta} g w dt\right) = \int_{(a,1]} f d\mu + \int_{(a,1]} \bar{\theta} g w d = 0.$$

In addition we have that a is $NRI(e^{2Re\Theta}1/w)$, so an argument similar to the one above gives that there must be a sequence $\{c_n\}$ approaching a from the right, so that

$$\lim_{n\to\infty} h_1\overline{(c_n)}\left(\int_{[c_n,1]} f d\mu + \int_{[c_n,1]} \bar{\theta} g w dt\right) = 0. \square$$

THEOREM 2.21. *(T.3). Let I be a MILI for $1/w$ which is of type (T.3). If $h \oplus k \in \hat{\mathcal{G}}$ and is zero on $[0,1]\setminus I$, then $h \oplus k \in \mathcal{G}^-$.*

PROOF: In this case, $I = [a,b)$, so we may take $c_n = a$, for all n. By maximality of I, a is $NLI(1/w)$, so by Lemma 2.11, for all $f \oplus g \in \mathcal{G}^\perp$,

$$\int_{[a,1]} f d\mu + \int_{[a,1]} \bar{\theta} g w dt = 0.$$

Moreover, h_1 is $AC_{loc}[a,b)$, so $h_1(a)$ is defined and finite. By Lemma 2.17, it suffices to show

$$\lim_{n\to\infty} \overline{h_1(d_n)}\left(\int_{(d_n,1]} f d\mu + \int_{(d_n,1]} \bar{\theta} g w dt\right) = 0$$

for some sequence $\{d_n\}$ approaching b from the left. But since the endpoint b is as in Theorem 2.20, the same argument shows that such a sequence exists. $\square$

In the consideration of cases where I is of the type (T.4) and (T.5), notice that

$$\int_{[a,b]} e^{2Re\Theta} 1/w dt < \infty$$

in both cases. From this we see that for any polynomial $p, (e^\Theta p)'$ is $L^1[a,b]$, since

$$\int_{[a,b]} |(e^\Theta p)'|\, dx = \int_{[a,b]} e^{Re\Theta}\, |\Theta p + p'| \frac{1}{w} d\nu_a \leq \left(\int_{[a,b]} e^{2Re\Theta}\frac{1}{w} dt\right)^{1/2} \left(\int_{[a,b]} |\theta p + p'|^2 d\nu_a\right)^{1/2} < \infty.$$

Thus $e^{\Theta}p \in AC[a,b]$ and

$$(e^{\Theta}p)(x) - (e^{\Theta}p)(y) = \int_y^x (e^{\Theta}p)'(t)dt$$

for all x in $[a,b]$. To utilize this we consider the direct sum space

$$L^2(e^{-2Re\Theta}d\mu|I) \oplus L^2(e^{-2Re\Theta}wdx|I)$$

and the submanifold $\mathcal{G}_{\Theta} = \{(e^{\Theta}p) \oplus (e^{\Theta}p)' : p \text{ a polynomial }\}$ Now notice that $h \oplus k \longrightarrow e^{\Theta}h \oplus e^{\Theta}k$ is a unitary transformation of $L^2(\mu) \oplus L^2(\nu_a)$ onto $L^2(e^{-2Re\Theta}d\mu) \oplus L^2(e^{-2Re\Theta}wdx)$ when localized to I, and that $(e^{\Theta}p)' = e^{\Theta}(\theta p + p')$. Therefore $h \oplus k \in L^2(d\mu|I) \oplus L^2(wdx|I)$ belongs to $\mathcal{G}^-$ if and only if $(e^{\Theta}h) \oplus (e^{\Theta}k)$ belongs to $\mathcal{G}_{\Theta}^-$. We show that, under the assumption that $(e^{-2Re\Theta}d\mu)|I$ is a finite measure, $h \oplus k \in \hat{\mathcal{G}}$ implies $(e^{\Theta}h) \oplus (e^{\Theta}k) \in \mathcal{G}_{\Theta}^-$.

THEOREM 2.22. *(T.4). Let I be a MILI for $1/w$ which is of type (T.4). Suppose that the measure $e^{-2Re\Theta}d\mu|I$ is finite. If $h \oplus k \in \hat{\mathcal{G}}$ and is zero on $[0,1] \setminus I$, then $h \oplus k \in \mathcal{G}^-$.*

PROOF: Here we have $I = (a,b)$ but $J = [a,b]$. We begin by investigating the orthogonal complement of $\mathcal{G}_{\Theta}$ in $L^2(e^{-2Re\Theta}d\mu) \oplus L^2(e^{-2Re\Theta}wdx)$. So let $f \oplus g \in \mathcal{G}_{\Theta}^{\perp}$. Then for all polynomials p,

$$0 = \int_{[a,b]} (e^{\Theta}p)(x)\overline{f(x)}e^{-2Re\Theta(x)}d\mu(x) + \int_{[a,b]} (e^{\Theta}p)'(x)\overline{g(x)}e^{-2Re\Theta(x)}w(x)dx. \tag{2.22.1}$$

Now note that $(e^{\Theta}p)'(x) = \int_{(a,x)}(e^{\Theta}p)'(t)dt + (e^{\Theta}p)(a)$ and that by assumption $a \in J$ but $a \notin I$ so that $(e^{\Theta}p)(a) = 0$ (see the proof of Theorem 2.10). Thus (2.22.1) becomes

$$0 = \int_{[a,b]} (\int_{[a,x]} (e^{\Theta}p)'(t)dt)\overline{f(x)}e^{-2Re\Theta(x)}d\mu(x) + \int_{[a,b]} (e^{\Theta}p)'(x)\overline{g(x)}e^{-2Re\Theta(x)}w(x)dx. \tag{2.22.2}$$

The hypothesis that $(e^{-2Re\Theta}d\mu)|I$ is a finite measure implies that $L^2((e^{-2Re\Theta}d\mu)|I) \subset L^1((e^{-2Re\Theta}d\mu)|I)$. Thus we may apply Fubini's theorem to the measure $[(e^{-2Re\Theta(x)}d\mu(x))|I] \times [(dt)|I]$ and the integrand $\chi_{\{t<x\}}\{t,x\}(e^{\Theta}p)'(t)\overline{f(x)}$ in the first integral of (2.22.2), and get

$$0 = \int_{[a,b]} (e^{\Theta}p)'(x) \left[\left(\int_{[x,b]} \overline{f(t)}e^{-2Re\Theta(t)}d\mu(t) \right) e^{2Re\Theta(x)}1/w(x) + \overline{g(x)} \right] e^{-2Re\Theta(x)}w(x)dx. \tag{2.22.3}$$

If $H(x)$ is the expression in the brackets in (2.22.3), then $H(x)$ is in $L^2(e^{-2Re\Theta(x)}w(x)dx)$, and satisfies, for every polynomial p,

$$0 = \int_{[a,b]} (e^{\Theta}p)'(x)H(x)e^{-2Re\Theta(x)}w(x)dx.$$

Thus,

$$\begin{aligned}0 &= \int_{[a,b]} e^{\Theta(x)}(\theta(x)p(x) + p'(x))H(x)e^{-2Re\Theta(x)}w(x)dx \\ &= \int_{[a,b]} e^{-\overline{\Theta(x)}}\theta(x)\left(\int_{[a,x]} p'(t)dt\right)H(x)w(x)dx \\ &+ \int_{[a,b]} e^{-\overline{\Theta(x)}}p'(x)H(x)w(x)dx + p(a)\int_{[a,b]} e^{-\overline{\Theta(x)}}\theta(x)H(x)w(x)dx.\end{aligned} \tag{2.22.4}$$

Now, considering polynomials p' with $p(a) = 0$ and interchanging the order of integration in the first integral of (2.22.4), we get

$$0 = \int_{[a,b]} p'(x)[\int_{[x,b]} e^{-\overline{\Theta(t)}}\theta(t)H(t)w(t)dt + e^{-\overline{\Theta(x)}}H(x)w(x))]dx.$$

The interchange is justified since $H \in L^2(e^{-2Re\Theta}wdx)$ and $e^{\Theta}\theta \in L^2(e^{-2Re\Theta}wdx)$ imply that $e^{-\bar{\Theta}}\theta Hw$ is integrable by the Schwarz inequality in the $L^2(e^{-2Re\Theta}wdx)$ inner product. But the class of the polynomials p' with $p(a) = 0$ is dense in $L^2(dx|I)$, so we conclude

$$\int_{[x,b]} e^{-\overline{\Theta(t)}}\theta(t)H(t)w(t)dt = -e^{-\overline{\Theta(x)}}H(x)w(x) \tag{2.22.5}$$

for $m - a.e.\ x \in I$. Denoting the lefthand side of (2.22.5) by Y, we see from (2.22.5) that Y satisfies the differential equation

$$Y' - \theta Y = 0,$$

which has the general solution $Y = ce^{\Theta(x)}$ (where c is a constant dependent on H). But by definition of Y and from (2.22.5),

$$Y = -H(x)e^{-\overline{\Theta(x)}}w(x).$$

Thus $H(x) = ce^{2Re\Theta(x)}/w(x)$. In summary, for $f \oplus g \in \mathcal{G}_{\Theta}^{\perp}$, for m $a.e.\ x \in I$,

$$\int_{[x,b]} \overline{f(t)}e^{-2Re\Theta(t)}d\mu(t) + \overline{g(x)}e^{-2Re\Theta(x)}w(x)$$

$$= \left[c e^{2Re\Theta(x)} \frac{1}{w(x)} \right] \left[e^{-2Re\Theta(x)} w(x) \right] = c \tag{2.22.6}$$

where c is a constant dependent on the pair $f \oplus g$.

Now, let $h \oplus k \in \hat{\mathcal{G}}$. We want to show $e^{\Theta} h \oplus e^{\Theta} k$ is in $(\mathcal{G}_{\Theta}^{-})$. That is, for every $f \oplus g \in \mathcal{G}_{\Theta}^{\perp}$,

$$0 = \int_{[a,b]} f(x)(e^{\bar{\Theta}} \bar{h})(x) e^{-2Re\Theta(x)} d\mu(x) + \int_{[a,b]} g(x)(e^{\bar{\Theta}} \bar{k})(x) e^{-2Re\Theta(x)} w(x) dx.$$

By the definition of $\hat{\mathcal{G}}$ in this case, $e^{\Theta} k = (e^{\Theta} h_1)'$. Also $e^{\Theta} h_1$ is $AC(J)$ with $(e^{\Theta} h_1)(a) = 0, e^{\Theta} k \in L^1(m|[a,b])$

and

$$(e^{\Theta} h_1)(x) = \int_{[a,x]} (e^{\Theta} k)(t) dt.$$

Thus, for $f \oplus g \in (\mathcal{G}_{\Theta})^{\perp}$,

$$\int_{[a,b]} f(x)(e^{\bar{\Theta}} \bar{h})(x) e^{-2Re\Theta(x)} d\mu(x) + \int_{[a,b]} g(x)(e^{\bar{\Theta}} \bar{k})(x) e^{-2Re\Theta(x)} w(x) dx$$
$$= \int_{[a,b]} f(x) \left(\int_{[a,x]} (e^{\bar{\Theta}} \bar{k})(t) dt \right) e^{-2Re\Theta(x)} d\mu(x) + \int_{[a,b]} g(x)(e^{\bar{\Theta}} \bar{k})(x) e^{-2Re\Theta(x)} w(x) dx$$
$$= \int_{[a,b]} (e^{\bar{\Theta}} \bar{k})(x) [\int_{[x,b]} f(t) e^{-2Re\Theta(t)} d\mu(t) + g(x) e^{-2Re\Theta(x)} w(x)] dx.$$

Finally, applying (2.22.6), we see that the last expression becomes

$$\bar{c} \int_{[a,b]} (e^{\bar{\Theta}} \bar{k})(t) dt = \bar{c}[(e^{\bar{\Theta}} \bar{h}_1)(b) - (e^{\bar{\Theta}} \bar{h}_1)(a)] = 0. \square$$

THEOREM 2.23 (T.5). *Let I be a MILI for $\frac{1}{w}$ which is of type (T.5) and suppose that the measure $(e^{-2Re\Theta} d\mu)|I$ is finite. If $h \oplus k \in \hat{\mathcal{G}}$ and is zero on $[0,1] \setminus I$, then $h \oplus k \in \mathcal{G}^{-}$.*

PROOF: Here we have $I = [a,b)$ and $J = [a,b]$. This proof follows the proof of the previous theorem with some modifications. First of all, note that $(e^{\Theta} p)(a)$ is well defined but not necessarily zero. In this case, we can take the normalization

$$\Theta(x) = \int_{[a,x]} \theta(t) dt,$$

so we do have $(e^{\Theta} p)(a) = p(a)$. Thus for $f \oplus g \in (\mathcal{G}_{\Theta})^{\perp}$, (2.22.3) becomes, for all polynomials p,

$$0 = p(a) \int_{[a,b]} \overline{f(x)} e^{-2Re\Theta(x)} d\mu(x) \tag{2.23.1}$$

$$+\int_{[a,b]}(e^{\Theta}p)'(x)[\left(\int_{[x,b]}\overline{f(t)}e^{-2Re\Theta(t)}d\mu(t)\right)e^{2Re\Theta(x)}\frac{1}{w(x)}+\overline{g(x)}]e^{-2Re\Theta(x)}w(x)dx.$$

Now, considering polynomials p' with $p(a)=0$, again a dense set in $L^1(dx|I)$, we get as in the proof of Theorem 2.22

$$\int_{[x,b]}\overline{f(t)}e^{-2Re\Theta(t)}d\mu(t)+\overline{g(x)}e^{-2Re\Theta(x)}w(x)=c \tag{2.22.6}$$

for $m-a.e.\ x\in I$ and $f\oplus g\in(\mathcal{G}_{\Theta}^{\perp})$, where the constant c depends on $f\oplus g$.

But considering the polynomial $p\equiv 1$, (2.23.1) yields

$$\begin{aligned}0=&\int_{[a,b]}\overline{f(x)}e^{-2Re\Theta(x)}d\mu(x)\\&+\int_{[a,b]}\theta(x)e^{\Theta(x)}[\left(\int_{[x,b]}\overline{f(t)}e^{-2Re\Theta(t)}d\mu(t)\right)e^{2Re\Theta(x)}\frac{1}{w(x)}\\&+\overline{g(x)}]e^{-2Re\Theta(x)}w(x)dx.\end{aligned}$$

So, from (2.22.6)

$$\int_{[a,b]}\overline{f(x)}e^{-2Re\Theta(x)}d\mu(x)=-c\int_{[a,b]}\theta(x)e^{\Theta(x)}dx=-c(e^{\Theta(b)}-e^{\Theta(a)})=c \tag{2.23.2}$$

since $e^{\Theta(a)}=1$ by our normalization, and $e^{\Theta(b)}=0$, because $b\in J$ but $b\notin I$. Now let $h\oplus k\in\hat{\mathcal{G}}$. Then we must show for each $f\oplus g\in(\mathcal{G}_{\Theta})^{\perp}$,

$$0=\int_{[a,b]}f(x)(e^{\bar{\Theta}}\bar{h})(x)e^{-2Re\Theta(x)}d\mu(x)+\int_{[a,b]}g(x)(e^{\bar{\Theta}}\bar{k})(x)e^{-2Re\Theta(x)}w(x)dx. \tag{2.23.3}$$

As before, we use the identity

$$(e^{\Theta}h_1)(x)=\int_{[a,x]}(e^{\Theta}k)(t)dt+(e^{\Theta}h_1)(a)=\int_{[a,x]}(e^{\Theta}k)(t)+h_1(a),$$

interchange the order of integration, and use (2.23.2), (2.22.6), to see that (2.23.3) reduces to showing

$$0=\overline{h_1(a)}\int_{[a,b]}f(x)e^{-2Re\Theta(x)}d\mu(x)=\bar{c}\int_{[a,b]}(e^{\bar{\Theta}}\bar{k})(x)dx.$$

But the right hand side becomes

$$\overline{ch_1(a)}+\bar{c}((e^{\bar{\Theta}}\bar{h}_1)(b))-(e^{\bar{\Theta}}\bar{h}_1)(a))=0,$$

as needed. □

The following table summarizes our results. Most of the table is self-explanatory. The fifth, sixth and seventh column give specific choices of μ, w, and θ which satisfy all the conditions required by the first, second, and fourth columns of the same row. This shows that the special cases corresponding to each of the eight rows are all nonvacuous. Note that in all cases $\mathcal{G}^- \subset \hat{\mathcal{G}}$ by Theorem 2.10. In five of eight cases, as indicated in the last column we have characterized $\mathcal{G}^-$ as exactly equal to $\hat{\mathcal{G}}$. In the table the interval I with endpoints 0 and 1 is assumed to be an MILI for $\frac{1}{w}$, and J is the associated interval with the same endpoints as in Definition 2.5.

TABLE 2.1

Type	I	J	Added assumption	μ	w	θ	$e^{2Re\Theta}\frac{1}{w}$	$\hat{\mathcal{G}} = \mathcal{G}^-$?
1	[0,1]	[0,1]	None	arbitrary	1	1	e^{2x}	yes
2	(0,1)	(0,1)	None	arbitrary	$x(1-x)$	1	e^{2x}	yes
3	[0,1)	[0,1)	None	arbitrary	$(1-x)$	1	e^{2x}	yes
4.a	(0,1)	[0,1]	the measure $e^{-2Re\Theta}d\mu$ is finite	wdx	$x^2(1-x)^2$	$\frac{1-2x}{x(1-x)}$	1	yes
4.b	(0,1)	[0,1]	the measure $e^{-2Re\Theta}d\mu$ is infinite	dx	$x^2(1-x)^2$	$\frac{1-2x}{x(1-x)}$	1	open
5.a	[0,1)	[0,1]	the measure $e^{-2Re\Theta}d\mu$ is finite	wdx	$(1-x)^2$	$\frac{-1}{1-x}$	1	yes
5.b	[0,1)	[0,1]	the measure $e^{-2Re\Theta}d\mu$ is infinite	dx	$(1-x)^2$	$\frac{-1}{1-x}$	1	open
6	(0,1)	(0,1]	None	arbitrary	$x(1-x)^2$	$\frac{-1}{1-x}$	$\frac{1}{x}$	open

As an illustration of the general ideas developed so far, we settle two questions: 1) when is D a closable operator, and 2) when is D the "most unclosable", i.e. when is $\mathcal{G}$ dense in $L^2(\mu) \oplus L^2(\nu)$. The answers are more subtle than one might at first expect.

2a. Closability of D

The following theorem gives a complete characterization (in terms of the data θ, μ, ν) of when $D = \theta + \frac{d}{dx} : L^2(\mu) \to L^2(\nu)$ is closable.

THEOREM 2.24. *$D = \theta + \frac{d}{dx} : L^2(\mu) \to L^2(\nu)$ is closable if and only if $\nu << m$, $\nu(K) = 0$ (where K is given by (2.11.1)) and the complement of $sppt\mu$ contains no intervals in a MILI for $\frac{1}{w}$.*

PROOF: Suppose $\nu << m$, $\nu(K) = 0$ and the complement of $sppt\mu$ contains no intervals in a MILI for $\frac{1}{w}$. We want to show that if $0 \oplus k \in \mathcal{G}^-$, then $k = 0$ $\nu - a.e.$. Since $\nu(K) = 0$, it suffices to show $k|I = 0$ for each MILI for $\frac{1}{w}$. Since $\mathcal{G}^- \subset \hat{\mathcal{G}}$ (Theorem 2.10), we know $e^{\Theta}k \in L^1_{loc}(J)$ and there is a $h_1 \in AC_{loc}(I)$ with $h_1 = 0$ $\mu - a.e.$ so that

$$e^{\Theta(x)}h_1(x) - e^{\Theta(y)}h_1(y) = \int_y^x (e^{\Theta}k)(t)dt.$$

Now $sppt\mu \cap I$ is dense in I by assumption, so $h_1 \equiv 0$ on I (since h_1 is continuous). Thus

$$\int_y^x (e^{\Theta}k)(t)dt = 0$$

for all $x, y \in I$. It follows that $(e^{\Theta}k)(t) = 0$ $m - a.e.$ on I and hence so is $k(t)$. Finally since we assume $\nu << m$, we have $k = 0$ $\nu - a.e.$ on I. We conclude that D is closable.

Conversely, if ν is not absolutely continuous with respect to m, we have D is not closable by Corollary 2.9. If $\nu << m$ and $\nu(K) > 0$ consider the non-zero element of $L^2(\mu) \oplus L^2(\nu)$, $0 \oplus \chi_K$. This element is orthogonal to $\mathcal{G}^{\perp}$ by Lemma 2.13, and hence is in $\mathcal{G}^-$, so D is not closable in this case. Finally, suppose $\nu << m$, $\nu(K) = 0$ but there is an interval $A = (a', b')$ in $I \setminus sppt\mu$ where I with endpoints a, b is some MILI for $1/w$. Choose a'', b'' with $a' < a'' < b'' < b'$. Set

$$k(t) = e^{-\Theta(t)}[\chi_{[a'',\xi]}(t) - \chi_{[\xi,b'']}(t)]$$

where $\xi = a'' + \frac{1}{2}(b'' - a'')$. Then

$$\int_{a''}^{b''} e^{\Theta(t)} k(t)dt = 0$$

and *sppt* $k \subset [a'', b'']$. If we define h_1 on $[a, b]$ by

$$h_1(x) = e^{-\Theta(x)} \int_{a''}^{x} e^{\Theta(t)} k(t)dt$$

then h_1 is absolutely continuously on $[a, b]$ with support contained in $[a'', b'']$. Thus in particular, $h_1 = 0$ $\mu - a.e.$ and $0 \oplus k$ has compact support inside I. By Corollary 2.18 we see that $0 \oplus k \in \mathcal{G}^-$ and hence D is not closable in this case either. □

We will need the following Corollary of Theorem 2.24 for our operator theory application in Chapter 3.

COROLLARY 2.24.1. *Suppose $D = \theta + \frac{d}{dx} : L^2(\mu) \to L^2(\nu)$ is closable. Then the support of ν is a regularly closed subset of* $\mathbb{R}$ *(i.e. spptν is the closure of its interior).*

PROOF: By Theorem 2.24, if D is closable then $\nu << m$ and $\nu(K) = 0$, i.e. ν is carried by $U = \cup\{int\ I :\ I \text{ is a MILI for } \frac{1}{w}\}$. Thus *sppt*$\nu$ is the closure of the open set U. Hence *sppt*$\nu \supset (int\ sppt\nu)^- \supset U^- = sppt\nu$ and thus *sppt*ν is regularly closed. □

2b. Density of $\mathcal{G}$

We finally consider the most degenerate case, that is when $\mathcal{G}^- = L^2(\mu) \oplus L^2(\nu)$. Before we proceed, though, we give another localization theorem about elements in $\mathcal{G}^\perp$.

LEMMA 2.25. *Suppose I is an MILI for $1/w$ and $f \oplus g \in L^2(\mu) \oplus L^2(\nu)$ has support in I. Then $f \oplus g \in \mathcal{G}^\perp$ if and only if*

$$g d\nu = g d\nu_a = g w dm \text{ on } I \tag{2.25.1}$$

$$\int_I f d\mu + \int_I \bar{\theta} g w dm = 0 \tag{2.25.2}$$

For m -a.e. x in I, (2.25.3)

$$g(x)w(x) = -\int_{(x,1]\cap I} f(t)d\mu(t) - \int_{(x,1]\cap I} \overline{\theta(t)} g(t)w(t)dt.$$

PROOF: It is easy to check that if $f \oplus g$ has support in I and satisfies (2.25.1)-(2.25.3) above, then $f \oplus g$ satisfies (2.7.1)-(2.7.3) in Lemma 2.7 and hence is in $\mathcal{G}^{\perp}$. Conversely, suppose that $f \oplus g$ has support in I and is in $\mathcal{G}^{\perp}$. Then condition (2.25.1) is an immediate consequence of (2.7.1). To discuss condition (2.25.2), suppose $a < b$ are the endpoints of I. If $a \in I$, by maximality of I necessarily a is a $NLI(1/w)$ so by Lemma 2.11

$$\int_{[a,1]} f d\mu + \int_{[a,1]} \bar{\theta} g w dm = 0. \tag{2.25.4}$$

On the other hand, if $a \notin I$, then again by maximality of I a is $NRI(1/w)$, so again by Lemma 2.11

$$\int_{(a,1]} f d\mu + \int_{(a,1]} \bar{\theta} g w \, dm = 0. \tag{2.25.5}$$

Similarly, if $b \in I$, then b is $NRI(1/w)$ so

$$\int_{(b,1]} f d\mu + \int_{(b,1]} \bar{\theta} g w dm = 0 \tag{2.25.6}$$

while if $b \notin I$, then b is $NLI(1/w)$ so

$$\int_{[b,1]} f d\mu + \int_{[b,1]} \bar{\theta} g w dm = 0. \tag{2.25.7}$$

When we appropriately subtract (2.25.6) or (2.25.7) from (2.25.4) or (2.25.5), we get precisely (2.25.2). Finally, (2.25.3) follows upon combining (2.25.6) or (2.25.7) with (2.7.3). □

To characterize the conditions for which $\mathcal{G}^{-} = L^2(\mu) \oplus L^2(\nu)$, we proceed by a series of lemmas. We begin by showing $\mu | I$ must be zero or a single point mass for each MILI for $\frac{1}{w}$. Throughout the discussion we utilize the local formulation given by Lemma 2.25 and the equivalence that $\mathcal{G}^{-} = L^2(\mu) \oplus L^2(\nu)$ if and only if $\mathcal{G}^{\perp} = (0)$.

LEMMA 2.26. *If for some MILI I for $\frac{1}{w}$, $\mu | I$ is not zero or a single point mass, then $\mathcal{G}^{\perp} \neq (0)$ (and thus $\mathcal{G}^{-} \neq L^2(\mu) \oplus L^2(\nu)$).*

PROOF: Let I be a MILI for $\frac{1}{w}$ with endpoints a and b, $a < b$, and suppose $\mu | I$ is non-zero and not a single point mass. For simplicity we first illustrate the argument for the special case where

$\theta = 0\ \nu - a.e.$ on I. Since $\mu|I$ is more than a single point mass, there is a set $S \subset sppt(\mu|I)$, so that $0 < \mu(S) < \mu(sppt(\mu|I))$. Let

$$f(x) = \begin{cases} \frac{-\mu(I\setminus S)}{\mu(S)}\chi_S(x) + \chi_{I\setminus S}(x), & x \in I \\ 0, & x \notin I. \end{cases}$$

Then $f \in L^2(\mu)$ and is nonzero. Set

$$g(x) = \begin{cases} -\frac{1}{w}(x)\int_{[x,b]} f(t)d\mu(t), & w(x) \neq 0 \\ 0, & elsewhere. \end{cases}$$

Then g is in $L^2(\nu)$ and satisfies (2.25.1) and (2.25.3) of Lemma 2.25. Finally we see (2.25.2) is satisfied, since

$$\int_{[a,b]} f(t)d\mu(t) = \frac{-\mu(I|S)}{\mu(S)}\int_{[a,b]}\chi_S d\mu + \int_{[a,b]}\chi_{I\setminus S}d\mu = \frac{-\mu(I\setminus S)}{\mu(S)}\mu(S) + \mu(I\setminus S) = 0.$$

Now we consider the case $\theta(x) \neq 0$ on a set of positive ν-measure in I. Let $[\alpha,\beta]$ be a compact subinterval of I, so that $\mu|[\alpha,\beta]$ is more than a single point mass. Since $1/w \in L^1_{loc}(I)$, we have $1/w \in L^1[\alpha,\beta]$, and thus by observation (2.1), $\theta \in L^1[\alpha,\beta]$. Further if $\Theta(x) \equiv -\int_{[x,\beta]}\theta(t)dt$ for $x \in [\alpha,\beta]$, then $\Theta \in AC[\alpha,\beta]$ and $\Theta' = \theta\ m - a.e.$ on $[\alpha,\beta]$.

Consider the equation (θ times equation (2.25.3) of Lemma 2.25) for $m - a.e.\ x \in [\alpha,\beta]$,

$$\overline{\theta(x)}g(x)w(x) = -\overline{\theta(x)}\int_{(x,\beta]} f(t)d\mu(t) - \overline{\theta(x)}\int_{(x,\beta]}\overline{\theta(t)}g(t)w(t)dt. \tag{2.26.1}$$

If we set $y = \int_{(x,\beta]}\overline{\theta(t)}g(t)w(t)dt$ for $x \in [\alpha,\beta]$, then we notice that $y' = -\bar{\theta}(x)g(x)w(x)\ m - a.e.$ on $[\alpha,\beta]$. Thus (2.26.1) becomes the initial value problem on $[\alpha,\beta]$

$$\begin{cases} y' - \bar{\theta}y = \overline{\theta(x)}\int_{(x,\beta]} f(t)d\mu(t) \\ y(\beta) = 0. \end{cases} \tag{2.26.2}$$

Now (2.26.2) has absolutely continuous solution on $[\alpha,\beta]$

$$y = e^{\overline{\Theta(x)}}\int_{[x,\beta]}\overline{\theta(t)}e^{-\overline{\Theta(t)}}\left(\int_{(t,\beta]} f(s)d\mu(s)\right)dt. \tag{2.26.3}$$

Choosing an $f \in L^2(\mu)$ with $f = 0$ off $[\alpha,\beta]$, let y be defined by (2.26.3) and set for ν_a-a.e. x in I,

$$g(x)w(x) = \begin{cases} -\int_{(x,\beta]} f(t)d\mu(t) - y(x), & x \in [\alpha,\beta] \\ 0, & otherwise \end{cases}. \tag{2.26.4}$$

Then for such an f and g, (2.25.1) is satisfied. The verification of (2.25.3) requires

$$-\int_{(x,\beta]} f(t)d\mu(t) - y(x) = -\int_{(x,\beta]} f(t)d\mu(t) - \int_{(x,\beta]} \overline{\theta(t)}[-\int_{(t,\beta]} f(s)d\mu(s) - y(t)]dt$$

or more simply

$$y(x) = -\int_{(x,\beta]} \overline{\theta(t)}[\int_{(t,\beta]} f(s)d\mu(s) + y(t)]dt.$$

To verify this note that both sides are 0 at β so it suffices to check that the derivatives of both sides agree:

$$y'(x) = \overline{\theta(x)}\int_{(x,\beta]} f(s)d\mu(s) + \overline{\theta(x)}y(x).$$

This in turn follows from the defining differential equation (2.26.2) for y, and hence (2.25.3) holds for $x \in [\alpha, b]$. Moreover since $f \in L^1(\mu|[\alpha,\beta])$ and $y \in C[\alpha,\beta]$, it is clear that

$$m = max\{|g(x)w(x)| : x \in [\alpha,\beta]\} < \infty.$$

Thus

$$\int_{[\alpha,\beta]} |g|^2 w dx = \int_{[\alpha,\beta]} |gw|^2 \frac{1}{w} dx \leq m^2 \int_{[\alpha,\beta]} 1/w dx < \infty$$

and $g \in L^2(\nu)$.

Our proof is then complete if we can show there is a nonzero $f \in L^2(\mu)$, $f = 0$ off $[\alpha,\beta]$, so that with f, and g defined by (2.26.4), (2.25.2) is satisfied. That is,

$$\int_{[a,b]} \overline{\theta(t)}g(t)w(t)dt = -\int_{[a,b]} f(t)d\mu(t).$$

But since f is assumed to be zero off $[\alpha,\beta]$, and gw, as defined, is also, this reduces to

$$\int_{[\alpha,\beta]} \overline{\theta(t)}g(t)w(t)dt = -\int_{[\alpha,\beta]} f(t)d\mu(t).$$

This then will verify (2.25.2) for $x < \alpha$ as well. Moreover, with gw defined by (2.26.4), we have

$$y(x) = \int_{[x,\beta]} \overline{\theta(t)}g(t)w(t)dt.$$

Thus, showing (2.25.2) is reduced to finding a nonzero $f \in L^2(\mu)$, zero off $[\alpha,\beta]$ and for which

$$y(\alpha) = -\int_{[\alpha,\beta]} f(t)d\mu(t). \tag{2.26.5}$$

Or, from (2.26.3)

$$y(\alpha) = e^{\overline{\Theta(\alpha)}} \int_{[\alpha,\beta]} \overline{\theta(t)} e^{-\overline{\Theta(t)}} \left(\int_{(t,\beta]} f(s) d\mu(s) \right) dt = - \int_{[\alpha,\beta]} f(t) d\mu(t),$$

or

$$\int_{[\alpha,\beta]} \overline{\theta(t)} e^{-\overline{\Theta(t)}} \left(\int_{(t,\beta]} f(s) d\mu(s) \right) dt = -e^{-\overline{\Theta(\alpha)}} \int_{[\alpha,\beta]} f(t) d\mu(s).$$

Now, interchanging the order of integration on the left gives

$$\int_{[\alpha,\beta]} f(s) \left[\int_{[\alpha,s]} \overline{\theta(t)} e^{\overline{-\Theta(t)}} dt \right] d\mu(s) = -e^{\overline{\Theta(\alpha)}} \int_{[\alpha,\beta]} f(t) d\mu(s),$$

or

$$\int_{[\alpha,\beta]} f(s) \left[\left(\int_{[\alpha,s]} \overline{\theta(t)} e^{-\overline{\Theta(t)}} dt \right) + e^{-\overline{\Theta(\alpha)}} \right] d\mu(s) = 0.$$

But notice

$$\int_{[\alpha,s]} \overline{\theta(t)} e^{-\overline{\Theta(t)}} dt = - \int_{[\alpha,s]} (e^{-\overline{\Theta(t)}})' dt = e^{-\overline{\Theta(\alpha)}} - e^{-\overline{\Theta(s)}}.$$

So let $F(s) = 2e^{-\Theta(\alpha)} - e^{-\Theta(s)}$, noting that $F \in AC[\alpha,\beta] \subset L^2(\mu|[\alpha,\beta])$, and defining the continuous linear functional $L_F : L^2(\mu|[\alpha,\beta]) \to \mathbb{C}$, by

$$L_F(f) = \int_{[\alpha,\beta]} f \bar{F} d\mu, \quad f \in L^2(\mu|[\alpha,\beta]).$$

Then (2.26.5) is satisfied if there exists a nonzero f in $ker(L_F)$. But since $\mu|[\alpha,\beta]$ is more than a single point mass, $Dim(L^2|[\alpha,\beta]) \geq 2$, and therefore $ker(L_F)$ is nontrivial.

Choose a nonzero $\hat{f} \in ker(L_F)$, and let

$$f(x) = \begin{cases} \hat{f}(x), & x \in [\alpha,\beta] \\ 0, & elsewhere. \end{cases}$$

Define y, using this $f \in L^2(\mu)$, via (2.26.3), and g by (2.26.4), and we get $f \oplus g \in L^2(\mu) \oplus L^2(\nu)$, is nonzero, zero off I, and on I satisfies Lemma 2.25. Thus $f \oplus g \in \mathcal{G}^\perp$, so $\mathcal{G}^\perp$ contains a nontrivial element. □

LEMMA 2.27. *Suppose $\theta = 0$ $\nu - a.e.$ on a MILI I for $\frac{1}{w}$. Then, if $\mu|I$ is zero or a single point mass, $\mathcal{G}^-|I = (L^2(\mu) \oplus L^2(\nu))|I$.*

PROOF: First suppose $\mu|I = 0$. Then $L^2(\mu)|I = (0)$. Let $a < b$ be the endpoints of I. Suppose $g \in L^2(\nu)$ is such that $0 \oplus g \in \mathcal{G}^{\perp}$. Then for $m - a.e.\ x$ in I,

$$g(x)w(x) = -\int_{[x,b]} \overline{\theta(t)}g(t)w(t)dt = 0,$$

by (2.25.3). Thus $gw = 0\ \nu - a.e.$ But by (2.25.1), this implies $g = 0\ \nu - a.e.$ Thus $\mathcal{G}^{\perp}|I = 0$. Now suppose $\mu|I$ is a single point mass at $x_0 \in I$. Let $f \oplus g \in \mathcal{G}^{\perp}$. Then (2.25.2) gives

$$-f(x_0)\mu(\{x_0\}) = \int_{[a,b]} \bar{\theta}gwdt = 0.$$

Thus $f = 0\ \mu - a.e.$ on I. It therefore follows from (2.25.3) that for $m - a.e.\ x \in I$

$$g(x)w(x) = -\int_{[x,b]} \overline{\theta(t)}g(t)w(t)dt = 0.$$

Thus, with (2.25.1), this gives $g = 0\ \nu - a.e.$ on I. We conclude $\mathcal{G}^{\perp}|I = 0$. □

PROPOSITION 2.28. *Suppose $\theta = 0\ \nu - a.e.$ on $\cup\{I : I$ is a MILI for $\frac{1}{w}\}$. Then $\mathcal{G}^{-} = L^2(\mu) \oplus L^2(\nu)$ if and only if $\mu|I$ is no more than a single point mass for each MILI I for $\frac{1}{w}$.*

PROOF: Combine Lemmas 2.26 and 2.27. □

It is left to consider when μ is at most a point mass on a MILI I for $\frac{1}{w}$ for a general θ. We begin with a result for when $\mu|I = 0$.

PROPOSITION 2.29. *Let I be a MILI for $1/w$ such that $\mu|I = 0$. Then $\mathcal{G}^{-}|I = (L^2(\mu) \oplus L^2(\nu))|I$ if and only if $J \setminus I$ is at most one point, where J is the interval associated with I as in Definition 2.6.*

PROOF: If I has endpoints a and b then by Lemma 2.25 $0 \oplus g \in \mathcal{G}^{\perp}$ implies

$$\int_{[a,b]} \bar{\theta}gwdt = 0 \tag{2.29.1}$$

and for m-a.e. $x \in I$

$$g(x)w(x) = -\int_{(x,b]} \bar{\theta}gwdt. \tag{2.29.2}$$

Let $Y(x) = \int_{[x,b]} \bar{\theta}gwdm \in AC[a,b]$. Then $Y' = -\bar{\theta}gw\ m - a.e.$ on I. Define $\overline{\Theta(x)} = \int_{x_0}^{x} \overline{\theta(t)}dt$ for some $x_0 \in I$. Then $\bar{\Theta} \in AC_{loc}(I)$ since $\theta \in L^1_{loc}(I)$. Consider the equation, from (2.29.2),

$$\overline{\theta(x)}g(x)w(x) = -\overline{\theta(x)}\int_{[x,b]} \bar{\theta}gwdt,$$

or

$$Y' - \bar{\theta}Y = 0,$$

with $Y(a) = Y(b) = 0$ (from (2.29.1) and the definition of Y). If we solve this first order equation on compact intervals containing x_0, we get $Y(x) = d\overline{e^{\Theta(x)}}$ for some d. If I contains the endpoint a, $\bar{\theta}$ is integrable on $[a, x_0]$ and the boundary condition $Y(a) = 0$ together with the uniqueness theorem for differential equations (see e.g. Problem #1f, p. 97 of [CL]) forces $d = 0$. By a similar argument, $d = 0$ if the endpoint b is in I. Thus if I contains either endpoint a or b, then $g = 0$ $\nu_a - a.e.$ on I.

If $I = (a, b)$ and $J = [a, b)$, then $\int_{b-\delta}^{b} e^{2Re\Theta}\frac{1}{w}dm = \infty$. But since

$$\int |g|^2 w dm = \int |gw|^2 \frac{1}{w} dm = \int |Y|^2 \frac{1}{w} dm = |d|^2 \int e^{2Re\Theta} \frac{1}{w} dm$$

and g by assumption is in $L^2(\nu)$, we must have $d = 0$. A similar argument shows d must be zero if $I = (a, b)$ and $J = (a, b]$.

Conversely if $I = (a, b)$ and $J = [a, b]$, then Θ is such that while $\int_a^{a+\delta} \frac{1}{w}dx = \infty$, we have $\int_a^{a+\delta} e^{2Re\Theta}\frac{1}{w}dx < \infty$. Thus $\lim_{x\to a+} \Theta(x) = -\infty$. Similarly, $\lim_{x\to b-} \Theta(x) = -\infty$. Therefore $Y = de^{\Theta(x)}$ satisfies $Y' - \bar{\theta}Y = 0$ and $Y(a) = Y(b) = 0$. Hence we can define g, nonzero in $L^2(\nu)$, via

$$g(x) = \begin{cases} -\frac{1}{w(x)}Y(x) \text{ if } x \in I, & w(x) \neq 0 \\ 0, & elsewhere \end{cases}$$

and get $0 \oplus g \in \mathcal{G}^{\perp}$. Thus $\mathcal{G}^{\perp} \neq (0)$. □

The last case to consider is that where μ consists of a point mass on I.

PROPOSITION 2.30. *Let I be a MILI for $\frac{1}{w}$ such that $\mu|I$ is a single point mass. Then $\mathcal{G}^{-}|I = (L^2(\mu) \oplus L^2(\nu))|I$ if and only if $J = I$.*

PROOF: Let I be a MILI for $\frac{1}{w}$ with endpoints a and b, such that $\mu|I$ is a single point mass at $x = x_0$. Define $\Theta(x) = \int_{x_0}^{x} \theta(t)dt$. Then $\Theta(x) \in AC_{loc}(I)$ and $\Theta(x_0) = 0$. Consider (2.25.3)

$$g(x)w(x) = -\int_{(x,b]} \overline{\theta(t)}g(t)w(t)dt - \int_{(x,b]} f(t)d\mu(t) = -\int_{(x,b]} \bar{\theta}gwdm - f(x_0)\mu(\{x_0\})\chi_{[a,x_0)}(x).$$

Multiplying this equation by $\theta(x)$ and considering $Y = \int_{(x,b]} \bar{\theta}gwdm \in AC[a, b]$, we get

$$Y' - \bar{\theta}Y = f(x_0)\mu(\{x_0\})\chi_{[a,x_0)}(x) \tag{2.30.1}$$

with boundary conditions $Y(a) = f(x_0)\mu(\{x_0\})$ from (2.25.2), and $Y(b) = 0$ by definition of Y. The idea is that $(f \oplus g)|I \in \mathcal{G}^{\perp}|I$ if and only if there is a $Y \in AC[a,b]$ satisfying (2.30.1) with the boundary conditions, and with

$$g(x)w(x) = -Y(x) - f(x_0)\mu\{x_0\}\chi_{[a,x_0)}(x)$$

for $m - a.e.\ x \in I$.

Now all solutions to the differential equation (2.30.1) (with boundary conditions ignored for the moment) have the form

$$Y(x) = -ce^{\overline{\Theta(x)}} \int_{x_0}^{x} e^{-\overline{\Theta(t)}}\overline{\theta(t)}\chi_{[x,x_0)}(t)dt + de^{\overline{\Theta(x)}}$$

where $c = -f(x_0)\mu(\{x_0\})$. To fit the boundary conditions we must choose d so that $Y(a) = c$ and $Y(b) = 0$. Using that $(e^{-\overline{\Theta(t)}})' = -\overline{\theta(t)}e^{-\overline{\Theta(t)}}$ and $e^{\overline{\Theta(x_0)}} = 1$, we then have

$$Y(x) = ce^{\overline{\Theta(x)}} \int_{x_0}^{x} (e^{-\overline{\Theta(t)}})'dt + de^{\overline{\Theta(x)}} = ce^{\overline{\Theta(x)}}(e^{-\overline{\Theta(x)}} - e^{-\overline{\Theta(x_0)}}) + de^{\overline{\Theta(x)}} = (d-c)e^{\overline{\Theta(x)}} + c,$$

for $x < x_0$. Further for $x > x_0$, $Y(x) = de^{\overline{\Theta(x)}}$, and $Y(x_0) = d$ by either formula. Thus

$$Y(x) = \begin{cases} (d-c)e^{\overline{\Theta(x)}} + c & x \leq x_0 \\ de^{\overline{\Theta(x)}} & x > x_0. \end{cases} \tag{2.30.2}$$

Notice that from (2.30.2)

i) $Y(a) = c$ implies either $d = c$ or $e^{\overline{\Theta(a)}} = 0$, that is, $d = c$ or

$$\lim_{t \to a+} \overline{\Theta(t)} = -\infty \tag{2.30.3}$$

ii) $Y(b) = 0$ implies $d = 0$ or $\lim_{t \to b-} \Theta(t) = -\infty$.

Now by (2.25.3) of Lemma 2.25, $g(x)w(x) = c\chi_{[a,x_0)}(x) - Y(x)$, or by (2.30.2)

$$g(x)w(x) = \begin{cases} (c-d)e^{\overline{\Theta(x)}} & x \leq x_0 \\ -de^{\overline{\Theta(x)}} & x > x_0. \end{cases} \tag{2.30.4}$$

Considering that $g \in L^2(\nu)$, we have

$$\int_{[a,b]} |g|^2 w dx = \int |gw|^2 \frac{1}{w} dx = \int_{[a,x_0]} |d-c|^2 e^{2Re\Theta} \frac{1}{w} dx + \int_{[x_0,b]} |d|^2 e^{2Re\Theta} \frac{1}{w} dx.$$

Since, $g \in L^2(\nu)$, we must have

i) $d = c$ or $\int_a^{a+\delta} e^{2Re\Theta}\frac{1}{w}dm < \infty (a \in J)$,

and (2.30.5)

ii) $d = 0$, or $\int e^{2Re\Theta}\frac{1}{w}dm < \infty (b \in J)$.

We finish by considering all cases for I and J:

1) $I = [a,b] = J$: we have $\Theta \in AC[a,b]$, thus both $e^{\overline{\Theta(a)}}$ and $e^{\overline{\Theta(b)}}$ are not zero. So by (2.30.3) $d = c$ and $d = 0$. Now $0 = c = -f(x_0)\mu(\{x_0\})$ implies $f = 0$ $\mu - a.e.$ on I, and $c = d = 0$ implies $g = 0$ $\nu_a - a.e.$ by (2.30.4). Thus $\mathcal{G}^{\perp}|I = (0)$.

2) $I = (a,b) = J$: Here $a, b \notin J$, so $\int_a^{a+\delta} e^{2Re\Theta}\frac{1}{w}\,dm = \infty$ so by (2.30.5) we have $c = d = 0$. Thus, as above, $\mathcal{G}^{\perp}|I = (0)$.

3) $I = (a,b] = J$: Since $b \in I$, θ is left integrable at b, so $e^{\overline{\Theta(b)}} \neq 0$. So $d = 0$ by (2.30.3). But a not in J implies $\int_a^{a+\delta} e^{2Re\Theta}\frac{1}{w}dm = \infty$, so by (2.30.5) $c = d = 0$. Thus $\mathcal{G}^{\perp}|I = (0)$.

4) $I = [a,b] = J$: Here $e^{\overline{\Theta(a)}} \neq 0$, so $d = c$ by (2.30.3). But b is not in J, so $d = 0$ by (2.30.5). Thus $\mathcal{G}^{\perp}|I = (0)$.

5) $I = (a,b), J = [a,b)$: Since b is not in J, $d = 0$ by (2.30.5), but there is no restriction on c. So we can let $f(x) \equiv -c/\mu(\{x_0\}) \neq 0$ on I, and define g $\nu - a.e.$ on I by (2.30.4). Since $\int_a^{a+\delta} e^{2Re\Theta}\frac{1}{w}dm < \infty$ but $\int_a^{a+\delta}\frac{1}{w}dm = \infty$, we must have $\lim_{x\to a+} e^{\Theta(x)} = 0$, that is $Y(a) = c$. Of course $Y(b) = 0$ since $d = 0$. We conclude that with f and g so defined $f \oplus g \in \mathcal{G}^{\perp}|I$ by Lemma 2.25. Hence $\mathcal{G}^{\perp} \neq (0)$.

6) $I = (a,b),\ J = (a,b]$: Here a not in J implies $d = c$, but d can be chosen freely. So by the same argument as above, a nonzero $f \oplus g \in \mathcal{G}^{\perp}|I$ can be constructed.

7) $I = (a,b),\ J = [a,b]$: Here both c and d are unrestricted, so $\mathcal{G}^{\perp}|I \neq (0)$.

8) $I = [a,b),\ J = [a,b]$: Since $a \in I$, $e^{\Theta(a)} \neq 0$, thus $d = c$. But d can be anything. So $\mathcal{G}^{\perp}|I \neq (0)$.

9) $I = (a,b],\ J = [a,b]$: Here $e^{\Theta(b)} \neq 0$, thus $d = 0$, but c is not restricted. Thus $\mathcal{G}^{\perp}|I \neq (0)$. This completes the proof of Proposition 2.30. □

We finally summarize results 2.26 through 2.30 in the following:

THEOREM 2.31. *$\mathcal{G}^- = L^2(\mu) \oplus L^2(\nu)$ if and only if for each MILI I for $\frac{1}{w}$ either*

$$\mu|I \text{ is a point mass and } J = I, \tag{2.31.1}$$

or

$$\mu|I = 0 \text{ and } J \setminus I \text{ is at most one point.} \tag{2.31.2}$$

We finish this discussion by showing that $\mathcal{G}^- = \hat{\mathcal{G}}$ in the case where $\mu|I = \delta_{x_0}$, a point mass at $x = x_0$ in I, and thus substantiating our conjecture that $\mathcal{G}^- = \hat{\mathcal{G}}$ for this case.

PROPOSITION 2.32. *If $\mu|I$ consists of a single point mass at $x = x_0$ in I for a MILI I for $1/w$, then $\hat{\mathcal{G}}|I = \mathcal{G}^-|I$.*

PROOF: Without loss of generality we assume that $\mu|I$ equals the unit point mass measure δ_{x_0} supported on $\{x_0\} \subset I$. For this simple μ the measure $e^{-2Re\Theta}d\mu$ is certainly finite. Thus, by the results summarized in Table 2.1, we already know $\hat{\mathcal{G}}|I = \mathcal{G}^-|I$ when (I, J) is of types 1-5. There remains only Type 6 where I is open and the set $J \setminus I$ consists of one point. We proceed by showing $\hat{\mathcal{G}}^\perp|I = \mathcal{G}^\perp|I$ in this case as well.

First of all, from Definition 2.6 for $\hat{\mathcal{G}}$, $h \oplus k \in \hat{\mathcal{G}}|I$ where I is a MILI for $\frac{1}{w}$ with endpoints a and b, means $h = h(x_0)$ $\mu - a.e.$ on I and for some $h_1 \in AC_{loc}(I)$, $h_1(x_0) = h(x_0)$,

$$e^{\Theta(x)}h_1(x) - e^{\Theta(y)}h_1(y) = \int_y^x (e^\Theta k)(t)dt$$

for m-a.e. $x \in I$, where we let $\Theta(x) = \int_{x_0}^x \theta(t)dt$. Letting $y = x_0$, and noting $e^{\Theta(x_0)} = 1$, we get

$$e^{\Theta(x)}h_1(x) = \int_{x_0}^x (e^\Theta k)(t)dt + h(x_0) \tag{2.32.1}$$

for $m - a.e. \in I$. Now if $a \in J \setminus I$, by (2.32.1) and (2.6.2), we get

$$\int_{x_0}^x (e^\Theta k)(t)dt + h(x_0) = (e^\Theta h_1)(x) = \int_a^x (e^\Theta k)(t)dt = \int_a^{x_0} (e^\Theta k)(t)dt + \int_{x_0}^x (e^\Theta k)(t)dt.$$

Thus, if $a \in J \setminus I$,

$$\int_a^{x_0} (e^\Theta k)(t) = h(x_0). \tag{2.32.2}$$

Similarly, if $b \in J \setminus I$, by (2.32.1) and (2.6.2) of Def. 2.6,

$$\int_{x_0}^{x} (e^{\Theta}k)(t)dt + h(x_0) = (e^{\Theta}h_1)(x) = -\int_{x}^{b} (e^{\Theta}k)(t)dt$$
$$= \int_{x_0}^{x} (e^{\Theta}k)(t)dt - \int_{x_0}^{b} (e^{\Theta}k)(t)dt.$$

That is, if $b \in J \setminus I$,

$$\int_{x_0}^{b} (e^{\Theta}k)(t)dt = -h(x_0). \tag{2.32.3}$$

We now consider explicitly the two cases where $(J \setminus I)$ consists of one point.

1) $I = (a,b),\ J = [a,b)$: By (2.32.2) we have

$$\hat{\mathcal{G}}^{\perp}|I = \{c(-1 \oplus -e^{\bar{\Theta}}\chi_{[a,x_0)}1/w) : c \in \mathbb{C}\}$$

which agrees with $\mathcal{G}^{\perp}$ by case (5) of the proof of Theorem 2.30, via (2.30.4).

2) $I = (a,b),\ J = (a,b]$: By (2.31.3),

$$\hat{\mathcal{G}}^{\perp}|I = \{c[(-1 \oplus -e^{\Theta}\chi_{[x_0,b]}1/w) : c \in \mathbb{C}\}$$

which, from (2.30.4), agrees with $\mathcal{G}^{\perp}$ by the case (6) of Theorem 2.30. □

3. Spectra of Pure Subjordan Operators.

3a. Introduction

The goal of this section is to characterize those compact subsets of the real line which arise as spectra of pure (real) subjordan operators. We first define all the terms.

Definition 3.1. Let $\mathcal{K}$ be a Hilbert space. Then a (bounded, linear) operator J on $\mathcal{K}$ will be said to be (real) *jordan* if J has the form $J = S + N$ where

$$S = S^* \tag{3.1.1}$$

$$N^2 = 0 \tag{3.1.2}$$

$$SN = NS \tag{3.1.3}$$

Definition 3.2. An operator T on the Hilbert space $\mathcal{H}$ is said to be *subjordan* if there is a larger Hilbert space $\mathcal{K} \supset \mathcal{H}$ and a jordan operator J on $\mathcal{K}$ such that $\mathcal{H}$ is invariant for J and $T = J|\mathcal{H}$. If

$J = S + N$ as in Definition 3.1 and if $\mathcal{K}$ is the smallest subspace of $\mathcal{K}$ containing $\mathcal{H}$ which is invariant for both S and N, we say that J is the *minimal jordan extension* of T. Finally, if $\mathcal{H}_0 = \{0\}$ is the only subspace of $\mathcal{H}$ such that (i) $\mathcal{H}_0$ is invariant for T and (ii) $T|\mathcal{H}_0$ is jordan, then we say that T is pure.

The following two propositions give equivalent descriptions of purity.

PROPOSITION 3.3. *T is a pure subjordan operator on $\mathcal{H}$ if and only if there are no nonzero invariant subspaces for T on which T is self-adjoint.*

PROOF: Let $\mathcal{H}_0 \subseteq \mathcal{H}$ with $T(\mathcal{H}_0) \subseteq \mathcal{H}_0$. First, if $T|\mathcal{H}_0$ is self-adjoint then it is trivially jordan on $\mathcal{H}_0$. Conversely, let $T|\mathcal{H}_0$ have jordan decomposition $T|\mathcal{H}_0 = S_0 + N_0$ with $S_0^* = S_0, N_0^2 = 0$, and $S_0N_0 = N_0S_0$. If $N_0 = 0$ then $T|\mathcal{H}_0$ is self-adjoint and we are done. Otherwise, let $\mathcal{H}_1 = Ran\ N_0 \neq 0$, and we consider $T|\mathcal{H}_1$. Let $h_0 \in \mathcal{H}_0$ and set $h_1 = Nh_0 \in \mathcal{H}_1$.

Then

$$\begin{aligned} Th_1 = (S_0 + N_0)Nh_0 &= S_0N_0h_0 + N_0^2h_0 \\ &= S_0h_1 \end{aligned}$$

and

$$S_0h_1 = S_0N_0h_0 = N_0(S_0h_0) \subseteq \mathcal{H}_1.$$

So we see $\mathcal{H}_1$ is a nonzero invariant subspace for T for which the restriction is self-adjoint. □

We next show that the only manner in which the subjordan operator $T = J|\mathcal{H} = S_0 + N_0$ can be jordan is in the obvious way where $J = S + N$, $\mathcal{H}$ is invariant for S and N, $S_0 = S|\mathcal{H}$ and $N_0 = N|\mathcal{H}$. The proof relies on some preliminary notions from [BH], and a technical lemma. This then will lead to a second equivalent form of purity.

Definition 3.4. The *symbol expansion* of an operator T is the function

$$Q_T(s) = e^{-isT^*}e^{isT} = \sum_{k=0}^{\infty} B_k s^k$$

for some sequence of operators $\{B_k\}_{k=1}^{\infty}$. An operator T is said to be *coadjoint* (of order 2) if there are operators B_1 and B_2 so that

$$Q_T(s) = I + B_1 s + B_2 s^2.$$

An operator A is said to be a *nilpart* for a coadjoint operator T, if

$$A^*A \leq B_2 \tag{3.4.1}$$

$$A - A^* = T - T^* \tag{3.4.2}$$

and

$$T^*(A + A^*) = (A + A^*)T. \tag{3.4.3}$$

As a point of reference, nilparts are the basic building blocks used by Ball and Helton in constructing jordan extensions (again, see [BH]). In [H3], Helton proved that an operator is jordan if and only if both it and its adjoint are co-adjoint. Using that we present this technical result.

LEMMA 3.5. *Let $J = S + N$ be a jordan operator on $\mathcal{K}$. Then the nilpart for J is unique and equal to N.*

PROOF: Let B_1 and B_2 be operators so that

$$Q_J(s) = I + sB_1 + s^2B_2,$$

and suppose A is a nilpart for J. By Lemma 11.5 in [BH], $B_2 = N^*N$ and since $J = S + N$ is jordan, $J - J^* = N - N^*$. Therefore 3.4.1 and 3.4.2 become

$$A^*A \leq N^*N \tag{3.5.1}$$

and

$$A - A^* = N - N^*. \tag{3.5.2}$$

Now write $\mathcal{K} = (RanN)^{\perp} \oplus (RanN)$ and for some $\Gamma : (RanN)^{\perp} \to RanN$ we have

$$N = \begin{bmatrix} 0 & 0 \\ \Gamma & 0 \end{bmatrix}.$$

By (3.5.1), we have $RanA^* \subseteq RanN^* \subseteq kerN^* = (RanN)^{\perp}$. Thus, if with respect to $\mathcal{K} = (RanN)^{\perp} \oplus RanN$,

$$A = \begin{bmatrix} A_{11} & A_{12} \\ A_{21} & A_{22} \end{bmatrix}$$

it follows that

$$A^* = \begin{bmatrix} A_{11}^* & A_{21}^* \\ 0 & 0 \end{bmatrix}.$$

Hence $A_{12} = A_{22} = 0$. Using (3.5.2), we have

$$\begin{bmatrix} A_{11} - A_{11}^* & -A_{11}^* \\ A_{21} & 0 \end{bmatrix} = \begin{bmatrix} 0 & -\Gamma^* \\ \Gamma & 0 \end{bmatrix}.$$

Thus $A_{11} = A_{11}^*$ and $A_{21} = \Gamma$. Returning to (3.5.1) and computing products yields

$$\begin{bmatrix} A_{11}^2 + \Gamma^*\Gamma & 0 \\ 0 & 0 \end{bmatrix} \leq \begin{bmatrix} \Gamma^*\Gamma & 0 \\ 0 & 0 \end{bmatrix},$$

or $A_{11}^2 + \Gamma^*\Gamma \leq \Gamma^*\Gamma$. Therefore $A_{11} = A_{12} = A_{22} = 0$ and $A_{21} = \Gamma$ from which we conclude $A = N$.

□

PROPOSITION 3.6. *Let $J = S + N$ be a jordan operator on $\mathcal{K}$. Let $\mathcal{H} \subseteq \mathcal{K}$ be invariant for J with $T = J|\mathcal{H}$. Then T is jordan if and only if $\mathcal{H}$ is invariant for both S and N.*

PROOF: Let $J = S + N$ on $\mathcal{K}$, let $\mathcal{H} \subseteq K$ be invariant for J, let $T = J|\mathcal{H} = S_0 + N_0$ and assume both J and T are jordan. Further let

$$Q_J(s) = 1 + B_1 s + B_2 s^2$$

and

$$Q_T(s) = 1 + B_{10} s + B_{20} s^2$$

be the symbol expansions for J and T. We claim $B_{20} = P_{\mathcal{H}} B_2 |\mathcal{H}$. To see this consider

$$\begin{aligned} P_{\mathcal{H}} Q_J(s)|\mathcal{H} &= P_{\mathcal{H}} e^{-isJ^*} e^{isJ} |\mathcal{H} \\ &= P_{\mathcal{H}} e^{-isT^*} P_{\mathcal{H}} e^{isT} |\mathcal{H} \\ &= e^{-isT^*} e^{isT} = Q_T(s). \end{aligned}$$

Now by Lemma 1.5 of [BH], we have $B_2 = N^*N$ and $B_{20} = N_0^* N_0$. Thus

$$N_0^* N_0 = P_{\mathcal{H}} N^* N |\mathcal{H}.$$

But from Theorem 1.1 of [BH], we have $P_{\mathcal{H}}N|\mathcal{H}$ is a nilpart for T, which, by applying Lemma 3.5, gives

$$(P_{\mathcal{H}}N|\mathcal{H})^* P_{\mathcal{H}}N|\mathcal{H} = P_{\mathcal{H}}N^*N|\mathcal{H},$$

or

$$P_{\mathcal{H}}N^* P_{\mathcal{H}}N|\mathcal{H} = P_{\mathcal{H}}N^*N|\mathcal{H}.$$

This gives

$$P_{\mathcal{H}}N^*(I - P_{\mathcal{H}})N|\mathcal{H} = 0,$$

or

$$(I - P_{\mathcal{H}})N|\mathcal{H} = 0.$$

Hence $\mathcal{H}$ is invariant for N and hence also invariant for S.

Conversely, if $\mathcal{H}$ is invariant for both S and N, then $T = (S|\mathcal{H}) + (N|\mathcal{H})$, where $(S|\mathcal{H})^* = S|\mathcal{H}, (N|\mathcal{H})^2 = 0$ and $(S|\mathcal{H})(N|\mathcal{H}) = SN|\mathcal{H} = NS|\mathcal{H} = (N|\mathcal{H})(S|\mathcal{H})$. Hence T is jordan on $\mathcal{H}$. □

We conclude this introductory section with an elementary result concerning the spectrum of a jordan operator. We state it in a more general form than what will be needed.

PROPOSITION 3.7. *Let A and N be bounded operators on a Hilbert space $\mathcal{K}$ so that $AN = NA$ and $N^2 = 0$. Then $\sigma(A + N) = \sigma(A)$.*

PROOF: Since A and N commute, we have

$$[(\lambda I - A)^{-1}N]^2 = (\lambda I - A)^{-2}N^2 = 0$$

for all λ in the resolvent set of A.

Thus $(\lambda I - A)^{-1}N$ is nilpotent from which it follows that $I - (\lambda I - A)^{-1}N$ is invertible. So considering the formal equation

$$(\lambda I - (A + N))^{-1} = (\lambda I - A)^{-1}[I - (\lambda I - A)^{-1}N]^{-1},$$

we see $\lambda \in \rho(A + N)$ if and only if $\lambda \in \rho(A)$. Hence $\sigma(A + N) = \sigma(A)$. □

3b. The Model

We now introduce a concrete construction of a subjordan operator $T = T(\mu, \nu, \theta)$ which we will later show serves as a model for general pure subjordan operators. Earlier models for co-adjoint (or 2-symmetric) operators ([H1], [H2], [H3], [A]) were more complicated and involved the theory of distributions. The model introduced in [BH] assumes that the operator is already known to be subjordan; the former models were built to yield properties under which a jordan extension existed. The model from [BH] uses only measure theory while avoiding distributions. Our contribution here is to specialize it to characterize pure subjordan operators.

Definition 3.8. Let μ and ν be two real measures whose supports are compact and contained in some finite closed real interval, satisfying the following:

$$spp t\mu = sppt\nu \equiv E \tag{3.8.1}$$

$$\nu << \mu \tag{3.8.2}$$

$$\textit{The inclusion map } \Gamma : L^2(\mu) \to L^2(\nu) \textit{ is continuous} \tag{3.8.3}$$

$$\nu << m, \textit{ where } m \textit{ as before denotes Lebesgue measure (restricted to } E) \tag{3.8.4}$$

$$\nu \textit{ is carried by the union of all MILI's for } 1/w, \textit{ where } w = d\nu/dm \in L^1(m). \tag{3.8.5}$$

Set $\mathcal{K}_1 = L^2(\mu), \mathcal{K}_2 = L^2(\nu)$, $\mathcal{K} = \mathcal{K}_1 \oplus \mathcal{K}_2$, let S_i be multiplication by x on $\mathcal{K}_i (i = 1, 2)$. Let $\Gamma : \mathcal{K}_1 \to \mathcal{K}_2$ be the inclusion map and finally let $J = S + N : \mathcal{K} \to \mathcal{K}$, where

$$S = \begin{bmatrix} S_1 & 0 \\ 0 & S_2 \end{bmatrix}, \; N = \begin{bmatrix} 0 & 0 \\ \Gamma & 0 \end{bmatrix}$$

is thought of as acting on an element f of $\mathcal{K}$ written as $f = \begin{bmatrix} f_1 \\ f_2 \end{bmatrix}$ with $f_i \in \mathcal{K}_i (i = 1, 2)$. Further, let $\theta \in L^2(\nu)$, and define the unbounded operator $D : L^2(\mu) \to L^2(\nu)$, densely on all polynomials p by

$$Dp = \theta p + p'.$$

Let $\mathcal{H}$ denote the subspace of $\mathcal{K}$ obtained by closing the graph $\mathcal{G} \equiv \left\{ \begin{bmatrix} p \\ Dp \end{bmatrix} : p \text{ a polynomial} \right\}$ of D. Then we define the opeator $T = T(\mu, \nu, \theta)$ on $\mathcal{H}$ by

$$T(\mu, \nu, \theta) = J|\mathcal{H}.$$

Later we will show that a cyclic pure subjordan operator is unitarily equivalent to $T(\mu,\nu,\theta)$ for some μ,ν, and θ. The following theorem states that a given $T = T(\mu,\nu,\theta)$ is subjordan, and with an additional hypothesis (which we think is probably unnecessary) is in fact, pure.

THEOREM 3.9. *The operator $T = T(\mu,\nu,\theta)$, as defined by Definition 3.8, with μ and ν satisfying 3.8.1 through 3.8.5, is subjordan on $\mathcal{H}$ with minimal jordan extension J. Moreover if the following is also satisfied*

$$\text{The measure } \mu \text{ is carried by the union of the } MILI\text{'s for } 1/w, \tag{3.9.1}$$

then $T = T(\mu,\nu,\theta)$ is pure.

PROOF: We first show J is jordan. Clearly, $S = S_1 \oplus S_2$ is self-adjoint and N is nilpotent. Thus J is jordan, if $SN = NS$. Now

$$SN = \begin{bmatrix} S_1 & 0 \\ 0 & S_2 \end{bmatrix} \begin{bmatrix} 0 & 0 \\ \Gamma & 0 \end{bmatrix} = \begin{bmatrix} 0 & 0 \\ S_2\Gamma & 0 \end{bmatrix}$$

and

$$NS = \begin{bmatrix} 0 & 0 \\ \Gamma & 0 \end{bmatrix} \begin{bmatrix} S_1 & 0 \\ 0 & S_2 \end{bmatrix} = \begin{bmatrix} 0 & 0 \\ \Gamma S_1 & 0 \end{bmatrix}.$$

Since by (3.8.1), $sppt\mu = sppt\nu$ for all $f \in \mathcal{K}_1$,

$$S_1\Gamma(f(x)) = S_2(f(x)) = xf(x)$$

and

$$\Gamma S_1(f(x)) = \Gamma(xf(x)) = xf(x),$$

it follows that J is jordan on $\mathcal{K}$.

Next we note that D is closable. This follows from Theorem 2.24, since $\nu << m$ by (3.8.4), $\nu(K) = 0$ by (3.8.5), and the complement of $sppt\mu$ contains no intervals in a MILI for $1/w$ by (3.8.1).

To see that $\mathcal{H} = \mathcal{G}^-$ is invariant for J, let p be a polynomial and denote by J the operator of multiplication by $\begin{bmatrix} x & 0 \\ 1 & x \end{bmatrix}$. Then

$$\begin{bmatrix} x & 0 \\ 1 & x \end{bmatrix} \begin{bmatrix} p \\ Dp \end{bmatrix} = \begin{bmatrix} xp \\ p + xDp \end{bmatrix} = \begin{bmatrix} xp \\ p + x\theta p + xp' \end{bmatrix} = \begin{bmatrix} (xp) \\ D(xp) \end{bmatrix} \in \mathcal{G}$$

Thus $J(\mathcal{H}) \subseteq \mathcal{H}$ since $\mathcal{H} = \mathcal{G}^-$.

Now let $\hat{\mathcal{K}}$ be a subspace so that $\mathcal{H} \subseteq \hat{\mathcal{K}} \subseteq \mathcal{K}$ and which is invariant for both S and N. Then

$$N \begin{bmatrix} p \\ Dp \end{bmatrix} = \begin{bmatrix} 0 & 0 \\ 1 & 0 \end{bmatrix} \begin{bmatrix} p \\ Dp \end{bmatrix} = \begin{bmatrix} 0 \\ p \end{bmatrix} \in \hat{\mathcal{K}}$$

for all polynomials p. Thus, $[N(\mathcal{H})]^- = [N(\mathcal{G}^-)]^- = \left\{ \begin{bmatrix} 0 \\ p \end{bmatrix} : p \text{ a polynomial} \right\}^- = \begin{bmatrix} 0 \\ L^2(\nu) \end{bmatrix} \subseteq \hat{\mathcal{K}}$. Moreover since $Dp \in L^2(\nu)$ for all polynomials, $\begin{bmatrix} 0 \\ Dp \end{bmatrix} \in \hat{\mathcal{K}}$ for all polynomials. Finally since $\mathcal{H} \subseteq \hat{\mathcal{K}}$, we have $\begin{bmatrix} p \\ Dp \end{bmatrix} \in \hat{\mathcal{K}}$ for all polynomials, and it follows that

$$\begin{bmatrix} p \\ Dp \end{bmatrix} - \begin{bmatrix} 0 \\ Dp \end{bmatrix} = \begin{bmatrix} p \\ 0 \end{bmatrix} \in \hat{\mathcal{K}}.$$

Taking the closure over all polynomials p, we have $\begin{bmatrix} L^2(\mu) \\ 0 \end{bmatrix} \subseteq \hat{\mathcal{K}}$. Thus

$$\hat{\mathcal{K}} \supseteq \begin{bmatrix} L^2(\mu) \\ 0 \end{bmatrix} \oplus \begin{bmatrix} 0 \\ L^2(\nu) \end{bmatrix} = \mathcal{K}$$

which proves that J is a minimal jordan extension of T.

It is left to show that if in addition μ satisfies 3.9.1, then T is in fact pure. To do this we show that every nonzero invariant subspace for T contains an invariant subspace on which T is not self-adjoint. Since the restriction of a self-adjoint operator to an invariant subspace is also self-adjoint, it follows that T has no invariant subspaces on which it is self-adjoint, hence by Proposition 3.3, T is pure.

Let $\mathcal{H}_0$ be a nonzero invariant subspace for T. Since $\mathcal{H}_0 \subseteq \mathcal{H} = \begin{bmatrix} I \\ D \end{bmatrix} DomD$, there is a $h_0 \in DomD \subseteq L^2(\mu), h_0 \neq 0$, so that $\begin{bmatrix} h_0 \\ Dh_0 \end{bmatrix} \in \mathcal{H}_0$. Using the definition of $\hat{\mathcal{G}}$ (Definition 2.6) and the fact that $\hat{\mathcal{G}} \subseteq \mathcal{G}^-$ (Theorem 2.10), we may assume $h_0 \in AC_{loc}(I)$ for each MILI I for $1/w$. Moreover, by 3.9.1, μ is carried by the union of the MILI's for $1/w$, and hence is not identically zero on some MILI.

Consider the cyclic subspace $\hat{\mathcal{H}}_0$ of $\mathcal{H}_0$ generated by $\begin{bmatrix} h_0 \\ Dh_0 \end{bmatrix}$. Precisely, let

$$\hat{\mathcal{H}}_0 = [span\{T^n \begin{bmatrix} h_0 \\ Dh_0 \end{bmatrix} : n = 0, 1, 2, \dots\}]^-.$$

By an inductive argument one can show in general that

$$T^n\begin{bmatrix} h \\ Dh \end{bmatrix} = \begin{bmatrix} x^n h \\ D(x^n h) \end{bmatrix},$$

for $h \in DomD, n = 1,2,\dots$; hence $\hat{\mathcal{H}}_0$ can also be identified as

$$\hat{\mathcal{H}}_0 = \left\{ \begin{bmatrix} ph_0 \\ D(ph_0) \end{bmatrix} : p \text{ a polynomial} \right\}^- .$$

Now let $d\hat{\mu} = |h_0|^2 d\mu$, $d\hat{\nu} = |h_0|^2 d\nu$, and $\hat{\theta} = \theta + h_0'/h_0$, and consider the transformation

$$U : \begin{bmatrix} L^2(d\mu|supph_0) \\ L^2(d\nu|supph_0) \end{bmatrix} \to \begin{bmatrix} L^2(d\hat{\mu}) \\ L^2(d\hat{\nu}) \end{bmatrix}$$

defined by

$$U\begin{bmatrix} h \\ k \end{bmatrix} = \begin{bmatrix} h/h_0 \\ k/h_0 \end{bmatrix}.$$

Since for polynomials p,

$$\left\| \begin{bmatrix} ph_0 \\ D(ph_0) \end{bmatrix} \right\|_{\mathcal{G}}^2 = \|ph_0\|^2_{L^2(\mu)} + \|\theta ph_0 + ph_0' + p'h_0\|^2_{L^2(\nu)} = \|p\|^2_{L^2(\hat{\mu})} + \|\hat{\theta}p + p'\|^2_{L^2(\hat{\nu})},$$

(where $\|\cdot\|_{\mathcal{G}}$ denotes the graph norm on $\mathcal{H}$) we see that U is a unitary map from $\hat{\mathcal{H}}_0$ to $\hat{\mathcal{H}}_1$, where $\hat{\mathcal{H}}_1$ is the closure of the graph of $\hat{D} : Dom\hat{D} \subseteq L^2(\hat{\mu}) \to L^2(\hat{\nu})$, defined densely on the polynomials as

$$\hat{D}p = \hat{\theta}p + p'.$$

Noting that $\hat{\theta} \in L^2(\hat{\nu})$, since $\|\hat{\theta}\|_{L^2(\hat{\nu})} = \|Dh_0\|^2_{L^2(\nu)}$, and letting $\hat{T} : \begin{bmatrix} I \\ \hat{D} \end{bmatrix} Dom\hat{D} \to \hat{\mathcal{H}}_1$ be defined densely on all polynomials p by

$$\hat{T}\begin{bmatrix} p \\ \hat{D}p \end{bmatrix} = T\begin{bmatrix} ph_0 \\ D(ph_0) \end{bmatrix},$$

we see that $T|\hat{\mathcal{H}}_0$ is unitarily equivalent to $\hat{T}$.

We next argue that $\hat{T} = T(\hat{\mu}, \hat{\nu}, \hat{\theta})$ is an operator of the type defined in 3.8, i.e. $\hat{\mu}$ and $\hat{\nu}$ satisfy 3.8.1 through 3.8.5. First note that

$$spp t\hat{\mu} = (sppt|h_0|^2) \cap E = (sppt|h_0|^2) \cap sppt\nu = sppt\hat{\nu}.$$

Further we have

$$d\hat{\nu} = |h_0|^2 d\nu = |h_0|^2 w dm,$$

and

$$d\hat{\nu} = |h_0|^2 d\nu = |h_0|^2 (d\nu/d\mu)d\mu = (d\nu/d\mu)d\hat{\mu}.$$

Thus $\hat{\mu}$ and $\hat{\nu}$ satisfy 3.8.1-3.8.4.

Let $G = \cup\{Int\ I : I \text{ a MILI for } 1/w\}$. By 3.9.1, the previously noted fact that h_0 is $\mu - a.e.$ locally absolutely continuous, and assuming, since $h_0 \in L^2(\nu)$, $h_0 = 0$ when $w = 0$, we may conclude $|h_0|^2$ is carried by the open subset $\hat{G}$ of G where

$$\hat{G} = E \setminus \{x : h_0(x) = 0\}.$$

Thus $d\hat{\nu} = |h_0|^2 d\nu$ is carried by $\hat{G}$ which is a countable union of open intervals contained in G. Furthermore, $(|h_0|^2 w)^{-1}$ is locally integrable on each of these intervals and not integrable on $E \setminus \hat{G}$. Hence the union of the MILI's for $(|h_0|^2 w)^{-1}$ must be contained in $(\hat{G})^-$. We conclude $d\hat{\nu} = |h_0|^2 w d\mu$ is carried by the union of the MILI's for $(|h_0|^2 w)^{-1}$. Thus $\hat{\nu}$ satisfies 3.8.5, and it follows that $\hat{T} = T(\hat{\mu}, \hat{\nu}, \hat{\theta})$ is subjordan on $\hat{\mathcal{H}}_1$. The proof of Theorem 3.9 is completed by the following.

PROPOSITION 3.10. *If $T = T(\mu,\nu,\theta)$ is defined as in Definition 3.8, and T is self-adjoint, then $\mu = \nu = 0$.*

PROOF: Let p and q be polynomials. We compare graph inner products

$$< T\begin{bmatrix} p \\ Dp \end{bmatrix}, \begin{bmatrix} q \\ Dq \end{bmatrix} >_{\mathcal{G}} \qquad (3.10.1)$$

and

$$< \begin{bmatrix} p \\ Dp \end{bmatrix}, T\begin{bmatrix} q \\ Dq \end{bmatrix} >_{\mathcal{G}}, \qquad (3.10.2)$$

where T is multiplication by $\begin{bmatrix} x & 0 \\ 1 & x \end{bmatrix}$. Expanding (3.10.1) gives

$$\int xp\bar{q}d\mu + \int [(xp)' + \theta xp][(\bar{q}' + \overline{\theta q})]d\nu,$$

while (3.10.2) yields

$$\int xp\bar{q}d\mu + \int [p' + \theta p][(x\bar{q})' + x\bar{\theta}\bar{q}]d\nu.$$

Canceling the first term of each and then considering just the integrands of the second terms, we have

$$xp'\bar{q}' + p\bar{q}' + xp'\bar{\theta}\bar{q} + p\bar{\theta}\bar{q} + \theta xp\bar{q}' + |\theta|^2 xp\bar{q}$$

and

$$xp'\bar{q}' + p'\bar{q} + p'x\bar{\theta}\bar{q} + x\theta p\bar{q}' + \theta p\bar{q} + |\theta|^2 xp\bar{q}.$$

Dropping like terms, we see that self-adjointness of T requires

$$p\bar{q}' + p\bar{\theta}\bar{q} = p'\bar{q} + \theta p\bar{q} \qquad \nu - a.e. \tag{3.10.3}$$

for all polynomials p and q. Letting $p = x, q = 1$, (3.10.3) yields

$$x\bar{\theta} = 1 + \theta x$$

or

$$x(\frac{1}{2}Im(\theta)) = i,$$

for $\nu - a.e.x$. This forces $\nu = 0$ whence by (3.8.1) $\mu = 0$ as well. □

3c. The Cyclic Case

In the following we show that an arbitrary subjordan operator T with cyclic vector is unitarily equivalent to an operator of the form presented in Definition 3.8. This is done through a series of lemmas. The results of Section 2 are then cited to conclude that $\sigma(T)$ must be regularly closed.

LEMMA 3.11. *Let J be a Jordan operator on $\mathcal{K}$. Let $\mathcal{H} \subseteq \mathcal{K}$ be invariant for J and let $T \equiv J|\mathcal{H}$. Then J can be represented on $\mathcal{K} = \mathcal{K}_1 \oplus \mathcal{K}_2$ by*

$$J = \begin{bmatrix} S_1 & 0 \\ \Gamma & S_2 \end{bmatrix}$$

where

$$S_i : \mathcal{K}_i \to \mathcal{K}_i \text{ is self-adjoint } (i = 1, 2). \tag{3.11.1}$$

$$\Gamma S_1 = S_2 \Gamma, \tag{3.11.2}$$

$$RanΓ \text{ is dense in } \mathcal{K}_2, \tag{3.11.3}$$

and

$$\text{There is a closed operator } D : \mathcal{K}_1 \to \mathcal{K}_2 \tag{3.11.4}$$

so that $\mathcal{H} = \begin{bmatrix} I \\ D \end{bmatrix} Dom\ D + \begin{bmatrix} 0 \\ \mathcal{N} \end{bmatrix}$, $S_1(Dom\ D) \subseteq Dom\ D, \mathcal{N} \subseteq (RanD)^\perp, S_2\mathcal{N} \subseteq \mathcal{N}$ *and* $DS_1 = S_2D + \Gamma$.

PROOF: Let $J = S + N$ where $S = S^*, N^2 = 0$ and $SN = NS$. Since S is self-adjoint, it can be represented as multiplication by x on a direct integral space (see [Cl])

$$\mathcal{K} = \int \oplus \mathcal{K}(x) d\alpha(x)$$

for some measure α. Further since $SN = NS$, N can be represented as a diagonal operator of multiplication by the measurable operator function $N(x)$ on $\mathcal{K}$. Now for each x, put $\mathcal{K}(x) = \mathcal{K}_1(x) \oplus \mathcal{K}_2(x)$, where $\mathcal{K} = [Ran(N(x))]^\perp$ and $\mathcal{K}_2(x) = RanN(x)$. Then we can write $N(x)$ as

$$N(x) = \begin{bmatrix} 0 & 0 \\ \Gamma(x) & 0 \end{bmatrix}$$

for some operator $\Gamma(x) : \mathcal{K}_1(x) \to \mathcal{K}_2(x)$ which has dense range for $\alpha - a.e.\ x$ (3.11.3). With respect to $\mathcal{K}(x) = \mathcal{K}_1(x) \oplus \mathcal{K}_2(x)$, write

$$S(x) = \begin{bmatrix} S_1(x) & 0 \\ 0 & S_2(x) \end{bmatrix}$$

where for $i = 1, 2$, $S_i(x)$ is multiplication by x on $\mathcal{K}_i(x)$. Then

$$J = \begin{bmatrix} S_1 & 0 \\ \Gamma & S_2 \end{bmatrix}$$

on $\mathcal{K}_1 \oplus \mathcal{K}_2 = (\int \oplus \mathcal{K}_1(x) d\alpha(x)) \oplus (\int \oplus \mathcal{K}_2(x) d\alpha(x))$. Further, since $SN = NS$, a quick calculation shows $\Gamma(x)S_1(x) = S_2(x)\Gamma(x)$, for each x.

It is left to prove (3.11.4). Towards this, define a subset, $Dom\ D$, of $\mathcal{K}$ via

$$DomD = \{h \in \mathcal{K}_1 \ :\ \exists k \in \mathcal{K}_2 \text{ with } \begin{bmatrix} h \\ k \end{bmatrix} \in \mathcal{H}\}.$$

Let D be the linear operator on $DomD$ defined by: for $h \in DomD$, let $Dh = k$ where k is the element of minimal norm in the closed subspace $\{\ell \in \mathcal{K}_2 \ :\ \begin{bmatrix} h \\ \ell \end{bmatrix} \in \mathcal{H}\}$.

We first show D is closed. Let $\{f_n\}$ be a sequence of functions in $DomD$ so that $f_n \to f$ and $Df_n \to g$. Now, $\begin{bmatrix} f_n \\ Df_n \end{bmatrix} \in \mathcal{H}$ for all n, and since $\mathcal{H}$ is closed and $\begin{bmatrix} f_n \\ Df_n \end{bmatrix} \to \begin{bmatrix} f \\ g \end{bmatrix}$, $\begin{bmatrix} f \\ g \end{bmatrix} \in \mathcal{H}$. Thus $f \in DomD$. Let

$$\hat{g}_n = D(f - f_n) \text{ and } g_n = Df_n.$$

Then for each n,

$$\|\hat{g}_n\| = inf\{\|k\| \mid \begin{pmatrix} f - f_n \\ k \end{pmatrix} \in \mathcal{H}\}.$$

Further, since $\begin{pmatrix} f - f_n \\ g - g_n \end{pmatrix} \in \mathcal{H}$, we have

$$\|\hat{g}_n\| \leq \|g - g_n\|$$

for all n. But $\lim_{n\to\infty} \|g - g_n\| = 0$. So

$$\begin{aligned} \|Df - g\| &\leq \|Df - Df_n\| + \|Df_n - g\| \\ &= \|\hat{g}_n\| + \|g_n - g\| \underset{n\to\infty}{\to} 0. \end{aligned}$$

Thus $Df = g$.

Let $h \in DomD$ and choose $k \in \mathcal{K}_2$ so that $\begin{bmatrix} h \\ k \end{bmatrix} \in \mathcal{H}$. Thus, by the invariance of $\mathcal{H}$ for J, we get

$$\begin{bmatrix} S_1 & 0 \\ \Gamma & S_2 \end{bmatrix} \begin{bmatrix} h \\ k \end{bmatrix} = \begin{bmatrix} S_1 h \\ \Gamma h + S_2 k \end{bmatrix} \in \mathcal{H}.$$

So $S_1 h \in DomD$ and we conclude that $DomD$ is invariant for S_1. Further, using $\begin{bmatrix} S_1 h \\ DS_1 h \end{bmatrix}$ in the previous calculation gives $DS_1 = \Gamma + S_2 D$.

Finally, define $\mathcal{N}$ to be the subspace of $\mathcal{K}_2$ given by

$$\mathcal{N} = \{k \in \mathcal{K}_2 \ : \ \begin{bmatrix} 0 \\ k \end{bmatrix} \in \mathcal{H}\}.$$

Then $\mathcal{N} \subseteq (RanD)^{\perp}$ and

$$\mathcal{H} = \begin{bmatrix} I \\ D \end{bmatrix} DomD + \begin{bmatrix} 0 \\ \mathcal{N} \end{bmatrix}.$$

Moreover, if $k \in \mathcal{N}$, then $\begin{bmatrix} 0 \\ k \end{bmatrix} \in \mathcal{H}$ and

$$J \begin{bmatrix} 0 \\ k \end{bmatrix} = \begin{bmatrix} 0 \\ S_2 k \end{bmatrix} \in \mathcal{H}.$$

Thus $S_2 \mathcal{N} \subseteq \mathcal{N}$, and the proof is complete. □

LEMMA 3.12. *Let T be subjordan on $\mathcal{H}$ with cyclic vector. Let J be a minimal Jordan extension of T on $\mathcal{K} = \mathcal{K}_1 \oplus \mathcal{K}_2$ of the form $J = \begin{bmatrix} S_1 & 0 \\ \Gamma & S_2 \end{bmatrix}$ as in Lemma 3.11 with S_1, S_2, Γ satisfying (3.11.1), (3.11.2), and (3.11.3). Then there are compactly supported real measures μ and ν, so that up to unitary equivalence we may take*

$$\mathcal{K}_1 = L^2(\mu), \mathcal{K}_2 = L^2(\nu) \tag{3.12.1}$$

and

$$S_i \text{ is multiplication by } x \text{ on } \mathcal{K}_i \ (i = 1, 2). \tag{3.12.2}$$

PROOF: By the proof of Lemma 3.11, for $i = 1, 2$

$$\mathcal{K}_i = (\int \oplus \mathcal{K}_i(x) d\alpha(x)).$$

Thus, to show (3.12.1) and (3.12.2), we argue that the fiber of $\mathcal{K}_i$ over x, for each x, is one-dimensional. Let $\hat{\xi} = \begin{bmatrix} \xi_1 \\ \xi_2 \end{bmatrix}$ be a cyclic vector for T. We look for an invariant subspace for both S and N (and thus for J) which contains

$$\mathcal{H} = \left\{ p(T) \begin{bmatrix} \xi_1 \\ \xi_2 \end{bmatrix} \ : \ p \text{ a polynomial} \right\}^{-}.$$

Using the easily verified identity

$$T^n = \begin{bmatrix} x & 0 \\ \Gamma(x) & x \end{bmatrix}^n = \begin{bmatrix} x^n & 0 \\ n\Gamma(x)x^{n-1} & x^n \end{bmatrix}$$

we see that $\mathcal{H}$ is also given by

$$= \{ \begin{bmatrix} p(x)\xi_1(x) \\ p'(x)\Gamma(x)\xi_1(x) + p(x)\xi_2(x) \end{bmatrix} \ : \ p \text{ a polynomial}\}^{-}.$$

Let $\hat{\mathcal{K}} = \hat{\mathcal{K}}_1 \oplus \hat{\mathcal{K}}_2$ where $\hat{\mathcal{K}}_i = \int \oplus \hat{\mathcal{K}}_i(x) d\alpha(x)$ and

$$\hat{\mathcal{K}}_1(x) = \vee\{\xi_1(x)\}$$

and

$$\hat{\mathcal{K}}_2(x) = \vee\{\Gamma(x)\xi_1(x),\ \xi_2(x)\}.$$

Notice that $\mathcal{H} \subset \hat{\mathcal{K}}$ and since, if $\begin{bmatrix} f \\ g \end{bmatrix} \in \hat{\mathcal{K}}$ then

$$N \begin{bmatrix} f \\ g \end{bmatrix} = \begin{bmatrix} 0 \\ \Gamma(x) f(x) \end{bmatrix} \in \hat{\mathcal{K}}$$

and

$$S \begin{bmatrix} f \\ g \end{bmatrix} = \begin{bmatrix} x f(x) \\ x g(x) \end{bmatrix} \in \hat{\mathcal{K}},$$

we have $\hat{\mathcal{K}}$ is invariant for both S and N.

Hence, by minimality of $J, \mathcal{K} \subseteq \hat{\mathcal{K}}$, and in particular $\mathcal{K}_1 \subseteq \hat{\mathcal{K}}_1$ which has a one dimensional fiber. Thus $\mathcal{K}_1$ also has a one dimensional fiber, for $\alpha - a.e.\ x$. To see that $\mathcal{K}_2$ has a one dimensional fiber, note that $\mathcal{K}_2 = RanN$ and

$$N \begin{bmatrix} \xi_1(x) \\ \xi_2(x) \end{bmatrix} = \begin{bmatrix} 0 \\ \Gamma(x) \xi_1(x) \end{bmatrix}.$$

That is, the fiber of $RanN$ over x is spanned by $\begin{bmatrix} 0 \\ \Gamma(x) \xi_1(x) \end{bmatrix}$ and therefore is one-dimensional.

We conclude by the spectral theorem that there exists real measures μ and ν so that up to unitary equivalence $\mathcal{K}_1 = L^2(\mu)$ and $\mathcal{K}_2 = L^2(\nu)$, and that S_i is multiplication by x on $\mathcal{K}_i$ $(i = 1, 2)$. □

LEMMA 3.13. *Under the assumptions of Lemma 3.12, the following hold:*

$$\text{There is a function } \Gamma(x) \in L^2(\nu) \text{ so that for all } h \in L^2(\mu), (\Gamma h)(x) = \Gamma(x) h(x) \tag{3.13.1}$$

$$\nu << \mu \tag{3.13.2}$$

$$\Gamma(x) \neq 0 \quad \nu - a.e. \tag{3.13.3}$$

PROOF: By (3.11.2), (3.12.1), and (3.12.2), the operator $\Gamma : L^2(\mu) \to L^2(\nu)$ intertwines multiplication by x on $L^2(\mu)$ and $L^2(\nu)$. Citing Abrahamse [A], we thus have there exists a function $\Gamma(x)$ so that for all $h \in L^2(\mu)$, $(\Gamma h)(x) = \Gamma(x) h(x)$ and which satisfies μ-a.e. on the set $E = \{x | (d\mu/d\nu)(x) > 0\}$

$$(\Gamma(x)) \leq c (d\mu/d\nu)^{1/2}, \text{ (some } c > 0). \tag{3.13.4}$$

Further

$$\Gamma(x) = 0, \qquad x \notin E. \tag{3.13.5}$$

To show $\nu << \mu$, we argue by contradiction as follows. Let $\nu = \nu_a + \nu_s$ where $\nu_a << \mu$ and $\nu_s \perp \mu$. Suppose there is a set A so that $\nu_s(A) > 0$ while $\nu_a(A) = \mu(A) = 0$. Consider the function $\chi_A(x)$, a non-zero element of $L^2(\nu)$. Pick $h \in Ran\Gamma$. Then $h(x) = \Gamma(x)g(x)$ $\nu-$ a.e., for some $g \in L^2(\mu)$. Since $\mu(A) = 0$, we may assume $g = 0$ on A. Then

$$\begin{aligned} < h, \chi_A >_{\mathcal{K}_2} &= \int \Gamma(x)g(x)\overline{\chi_A(x)}d\nu \\ &= \int_A \Gamma(x)g(x)d\nu = 0. \end{aligned}$$

Thus χ_A is orthogonal to $Ran\Gamma$ in $L^2(\nu)$. But by (3.11.3), $Ran\Gamma$ is dense in $\mathcal{K}_2$, hence $\chi_A = 0$ $\nu - a.e.$, a contradiction. We conclude $\nu_s = 0$, and hence $\nu << \mu$.

Now note that since $\nu << \mu$, the set

$$\begin{aligned} E &= \{x | (d\mu/d\nu)(x) > 0\} \\ &= \{x | (d\nu/d\mu)(x) > 0\} \end{aligned}$$

is a carrier for ν. Thus, by (3.13.4)

$$\begin{aligned} \int |\Gamma|^2 d\nu &= \int_E |\Gamma|^2 d\nu \leq \int_E c|d\mu/d\nu| d\nu \\ &= \int_E c d\mu < \infty, \end{aligned}$$

and it follows that $\Gamma(x) \in L^2(\nu)$.

Finally, using again that E is a carrier for ν and (3.13.4), we have

$$\frac{1}{\Gamma(x)} \geq \frac{1}{c}(d\nu/d\mu)^{1/2} > 0$$

ν-a.e., and (3.13.3) follows. □

LEMMA 3.14. *Suppose T is a cyclic subjordan operator on $\mathcal{H}$. Then, up to unitary equivalence, there exist compactly supported measures μ and ν on $\mathbb{R}$ and a function $\theta \in L^2(\nu)$ such that:*

$$\nu << \mu \text{ and the inclusion map } \Gamma : L^2(\mu) \to L^2(\nu) \text{ is continuous.} \tag{3.14.1}$$

A minimal jordan extension J of T is given by multiplication by $\begin{bmatrix} x & 0 \\ 1 & x \end{bmatrix}$ *on* $\mathcal{K} = L^2(\mu) \oplus L^2(\nu)$.
(3.14.2)

$$\text{The space } \mathcal{H} \text{ is given by } \mathcal{H} = \{ \begin{bmatrix} p \\ Dp \end{bmatrix} : p \text{ a polynomial} \}^- \tag{3.14.3}$$

where D is a closable operator from $L^2(\mu)$ into $L^2(\nu)$ given by $D = M_\theta + d/dx$ and M_θ=multiplication by θ.

PROOF: By Lemmas 3.12 and 3.13, up to unitary equivalence we may assume that a minimal jordan extension J for T is multiplication by $\begin{bmatrix} x & 0 \\ \Gamma(x) & x \end{bmatrix}$ on $L^2(\mu) \oplus L^2(\nu)$ for some compactly supported measures μ and ν on $\mathbb{R}$ with $\nu << \mu$ and $\Gamma(x) \neq 0$ $\nu - a.e.$ As in (3.11.4) we have that $\mathcal{H}$ in this representation for J has the form

$$\mathcal{H} = \begin{bmatrix} I \\ D \end{bmatrix} DomD + \begin{bmatrix} 0 \\ \mathcal{N} \end{bmatrix}$$

where $D : L^2(\mu) \to L^2(\nu)$ is a closed operator satisfying

$$DM_x = M_x D + M_\Gamma. \tag{3.14.4}$$

We first argue that purity of T forces $\mathcal{N} = 0$. Indeed, the identity

$$\begin{bmatrix} 0 \\ \mathcal{N} \end{bmatrix} = \mathcal{H} \cap \begin{bmatrix} 0 \\ L^2(\nu) \end{bmatrix}$$

shows that $\begin{bmatrix} 0 \\ \mathcal{N} \end{bmatrix}$ is the intersection of subspaces invariant for J, and hence is itself invariant for J. Moreover, since the restriction of T to $\begin{bmatrix} 0 \\ \mathcal{N} \end{bmatrix}$ is clearly selfadjoint, it follows from Proposition 3.3 that $\mathcal{N}$ must be 0 if T is pure. We conclude that $\mathcal{N} = 0$, and hence

$$\mathcal{H} = \begin{bmatrix} I \\ D \end{bmatrix} DomD \tag{3.14.5}$$

Since $\Gamma(x) \neq 0$ $\nu - a.e.$, we may define a unitary transformation U: $L^2(\nu) \to L^2(|\Gamma|^2 d\nu)$ via

$$(Uf)(x) = (f/\Gamma)(x)$$

for all $f \in L^2(\nu)$. Then $(I \oplus U)J(I \oplus U^*) = \tilde{J}$ is multiplication by $\begin{bmatrix} x & 0 \\ 1 & x \end{bmatrix}$ on $L^2(\mu) \oplus L^2(|\Gamma|^2 d\nu)$. Replacing $d\nu$ by $|\Gamma|^2 d\nu$, we may assume that without loss of generality that $\Gamma = 1$ and M_Γ is simply the (bounded) inclusion map of $L^2(\mu)$ into $L^2(\nu)$.

Next let $\xi = \begin{bmatrix} I \\ D \end{bmatrix} \xi_1 = \begin{bmatrix} \xi_1 \\ \xi_2 \end{bmatrix}$ be a cyclic vector for T. We claim that $\xi_1 \neq 0$ $\mu - a.e.$ Otherwise there would be a set A with $\mu(A) > 0$ and $\xi_1 \equiv 0$ on A. By considering

$$\hat{\mathcal{K}} = \chi_A L^2(\mu) \oplus L^2(\nu)$$

we see that $\hat{\mathcal{K}}$ is invariant for both $S = M_x$ and $N = M_{\begin{bmatrix} 0 & 0 \\ 1 & 0 \end{bmatrix}}$, contains $\mathcal{H}$, and is strictly smaller than $\mathcal{K}$. This contradicts the assumption that J is a minimal Jordan extension.

Since $\xi_1 \neq 0$ $\mu - a.e.$, we may define a unitary transformation $V : L^2(d\mu) \to L^2(|\xi_1|^2 d\mu)$ by $Vf = \xi_1^{-1} f$. Via this transformation and replacing $d\mu$ by $|\xi_1|^2 d\mu$ if necessary, we may assume without loss of generality that $\xi_1 = 1$.

We conclude that

$$\begin{aligned} \mathcal{H} &= \{p(J) \begin{bmatrix} I \\ D \end{bmatrix} 1 : \ p \text{ a polynomial}\}^- \\ &= \{\begin{bmatrix} p(x) \\ p'(x) + p(x)(D1) \end{bmatrix} : \ p \text{ a polynomial}\}^- \\ &= \{\begin{bmatrix} I \\ D \end{bmatrix} p(x) : \ p \text{ a polynomial}\}^-, \end{aligned}$$

the closure being taken in $L^2(\mu) \oplus L^2(\nu)$. From (3.14.5) we see that D is closable. If we set $\theta = D(1) \in L^2(\nu)$, then all assertions of the Lemma follow. □

THEOREM 3.15. *If T is a pure subjordan operator with cyclic vector, then there are real measures μ and ν satisfying (3.8.1) through (3.8.5) and a $\theta \in L^2(\nu)$ so that $T \cong T(\mu, \nu, \theta)$ as defined in Definition 3.8.*

PROOF: Lemmas 3.11 through 3.14 give all of Definition 3.8, but (3.8.1), (3.8.4), and (3.8.5). Citing Theorem 2.24 gives (3.8.4) and (3.8.5).

It is left to show 3.8.1, that is $spp t\mu = spp t\nu$. Now since $\nu << \mu, spp t\nu \subseteq spp t\mu$ so we need only show $spp t\mu \subseteq spp t\nu$. Suppose there is a set $A \subseteq (spp t\mu) \backslash (spp t\nu)$ with $\mu(A) > 0$. Let $\mathcal{A} = \chi_A L^2(\mu)$. Then $\mathcal{A} \subseteq ker D$ and it follows that the subspace $\mathcal{M} = \begin{bmatrix} \mathcal{A} \\ 0 \end{bmatrix}$ is contained in $\mathcal{H}$. Moreover, since $S_1 \mathcal{A} \subseteq \mathcal{A}$ and

$$T|\mathcal{M} = \begin{bmatrix} S_1 & 0 \\ 0 & 0 \end{bmatrix}$$

we have by Proposition 3.3, T is not pure, a contradiction. We conclude $spp t\mu = spp t\nu$. □

COROLLARY 3.16. *A set E is the spectrum of a pure subjordan operator with cyclic vector if and only if E is regularly closed.*

PROOF: Suppose T is a cyclic pure subjordan operator on $\mathcal{H}$ with minimal Jordan extension J on $\mathcal{K}$. By Theorem 3.15, we have T is unitarily equivalent to an operator of the form $T(\mu,\nu,\theta)$ as in Definition 3.8 on $\mathcal{H} = \left\{ \begin{bmatrix} p \\ Dp \end{bmatrix} | p \text{ a polynomial} \right\}^{-}$ where $D = \theta + d/dx : L^2(\mu) \to L^2(\nu)$. Let $E = spp t\mu = sppt\nu$. By Corollary 2.25, since D is closed, it follows that E is a regularly closed compact set. Thus we need only argue that $\sigma(T) = E$.

Since J is assumed minimal, by Lemma 6.7 of [BH], we know $\sigma(J) \subseteq \sigma(T)$. Moreover, by Theorem 3.8 of Agler [A2], we have $\sigma(T)\backslash\sigma(J)$ is either empty or a union of components of $\mathbb{C}\backslash\sigma(J)$. But since $\sigma(J) \subseteq \mathbb{R}, \mathbb{C} \setminus \sigma(J)$ has exactly one component, itself. Thus $\sigma(T) \setminus \sigma(J) = \mathbb{C} \setminus \sigma(J)$ or is empty. The former is impossible since $\sigma(T)$ is bounded, so we conclude $\sigma(J) = \sigma(T)$. (Note: This argument does not rely on T being cyclic.)

It is left to show $\sigma(J) = E$. Since

$$J = \begin{bmatrix} x & 0 \\ 1 & x \end{bmatrix} : L^2(\mu) \oplus L^2(\nu) \to L^2(\mu) \oplus L^2(\nu),$$

we have by Proposition 3.7 that $\sigma(J) = \sigma(S)$, where $S = \begin{bmatrix} x & 0 \\ 0 & x \end{bmatrix}$. That is, $S = S_1 \oplus S_2$ with $S_1 = M_x$ on $L^2(\mu)$ and $S_2 = M_x$ on $L^2(\nu)$. Thus

$$\begin{aligned} \sigma(J) &= \sigma(S) = \sigma(S_1) \cup \sigma(S_2) \\ &= (sppt\mu) \cup (sppt\nu) = E \cup E = E. \end{aligned}$$

Now let E be any regularly closed compact subset of $\mathbb{R}$. Let $\mu = \nu = m_E$, where m_E is Lebesgue measure restricted to E. Then μ and ν satisfy 3.8.1 through 3.8.5 of Definition 3.8, and 3.9.1 of Theorem 3.9. Constructing the operator $T = T(\mu,\nu,0)$ as in Definition 3.8, we have T is pure subjordan by Theorem 3.5. Finally, by the argument in the first part of the proof $\sigma(T) = E$.
□

3d. The General Case

Here we prove the main theorem of this chapter. To do this we model the general (i.e. possibly non-cyclic) pure subjordan in a slightly different manner than in the proof of Lemma 3.11.

THEOREM 3.17. *A compact subset E of the real line is the spectrum of a pure subjordan operator if and only if E is regularly closed.*

PROOF: One direction of the implication follows from Corollary 3.16. For the other direction, let T be pure subjordan on $\mathcal{H}$ with minimal Jordan extension $J = S+N$ on $\mathcal{K}$. Let $\mathcal{K}_1 = (RanN)^\perp, \mathcal{K}_2 = RanN$, and decompose

$$J = \begin{bmatrix} S_1 & 0 \\ 0 & S_2 \end{bmatrix} + \begin{bmatrix} 0 & 0 \\ \Gamma & 0 \end{bmatrix}$$

with respect to $\mathcal{K} = \mathcal{K}_1 \oplus \mathcal{K}_2$. Then (via Theorem 9.1, Chapter II, [C]) we represent S_1 as $\oplus\sum_j M_x$ on $\oplus \sum L^2(\mu_j)$ with $\mu_{j+1} << \mu_j$ for all j and $\sigma(S_1) = spp t\mu_1$. Similarly S_2 can be represented as $\oplus\sum_i M_x$ on $\oplus\sum_i L^2(\nu_i)$ with $\nu_{i+1} << \nu_i$ for all i and $\sigma(S_2) = sppt\nu_1$. Finally, Γ can be viewed as a matrix operator, $\Gamma = [\Gamma_{ij}]$, where

$$\Gamma_{ij} : L^2(\mu_j) \to L^2(\nu_i)$$

is a bounded multiplication operator for all i and j, via Abrahamse [A].

The idea of the proof is find a vector ξ so that T restricted to the cyclic subspace generated by ξ has spectrum the same as T and then cite Corollary 3.16.

We begin by showing $\sigma(T) = \sigma(J) = sppt\mu_1$. That $\sigma(T) = \sigma(J)$ follows in exactly the same manner as in the proof of Corollary 3.16. To see that $\sigma(J) = sppt\mu_1$, it suffices to show $\nu_1 << \mu_1$, since this implies $sppt\nu_1 \subseteq sppt\mu_1$, and thus

$$\begin{aligned} \sigma(J) &= \sigma(S) = \sigma(S_1) \cup \sigma(S_2) \\ &= sppt\mu_1 \cup sppt\nu_1 = sppt\mu_1. \end{aligned} \tag{3.17.1}$$

So suppose ν_1 is not absolutely continuous with respect to μ_1. Then, if we write $\nu_1 = \nu_a + \nu_s$ where $\nu_a << \mu_1$, and $\nu_s \perp \mu_1$, there is a set A with $\nu_s(A) > 0$ and $\nu_a(A) = \mu_1(A) = 0$. Let $f(x) = \oplus\sum_i f_i(x)$, where

$$f_i(x) = \begin{cases} \chi_A(x) & i = 1 \\ 0 & i \neq 1 \end{cases}.$$

Then $f(x)$ is nonzero as an element of $\mathcal{K}_2$. Let $h(x) = (\Gamma g)(x)$ for some $g \in \mathcal{K}_1$. Then $h \in Ran\Gamma$ which is dense in $\mathcal{K}_2$ by definition. Furthermore, since $\mu_j << \mu_1$ for all j and $\mu_1(A) = 0$, we may assume $g(x) = \oplus \sum g_j(x)$ satisfies $g_j(x) = 0$ on A for all j. It follows that

$$\begin{aligned} < h, f >_{\mathcal{K}_2} &= \sum_{i=1}^{\infty} \int (\sum_{j=1}^{\infty} \Gamma_{ij}(x) g_j(x)) \overline{f_i(x)} d\nu_i(x) \\ &= \int (\sum_{j=1}^{\infty} \Gamma_{1j}(x) g_j(x)) \overline{f_1(x)} d\nu_1(x) \\ &= \int_A (\sum_{j=1}^{\infty} \Gamma_{1j}(x) g_j(x)) d\nu_1(x) = 0. \end{aligned}$$

Thus f is orthogonal to $Ran\Gamma$, a contradiction. We next show that there is vector $\xi = \begin{bmatrix} \xi_1 \\ \xi_2 \end{bmatrix} \in \mathcal{H}$ so that if $\xi_1(x) = \oplus \sum_{j=1}^{\infty} \xi_{1j}$, then $\xi_{11} \neq 0$ μ_1-a.e. Let $\mathcal{C}$ be the nonempty linear submanifold of $L^2(\mu_1)$ defined by

$$\mathcal{C} = \{f_1 \in L^2(\mu_1) \ : \ \text{there is a } g \in \mathcal{K}_2,\ f_j \in L^2(\mu) \text{ for } j > 1 \text{ so that } f = \oplus \sum_{j=1}^{\infty} f_j \in \mathcal{K}_1, \begin{bmatrix} f \\ g \end{bmatrix} \in \mathcal{H}\}.$$

Define a norm, $\|\cdot\|_{\mathcal{C}}$, on $\mathcal{C}$ via

$$\|f_1\|_{\mathcal{C}} = inf\{\| \begin{bmatrix} f \\ g \end{bmatrix} \|_{\mathcal{K}} : f_j \in L^2(\mu_j)(j \geq 2), g \in \mathcal{K}_2$$

$$\text{such that } f = \oplus \sum_{j=1}^{\infty} f_j \in \mathcal{K}_1, g \in \mathcal{K}_2 \text{ and } \begin{bmatrix} f \\ g \end{bmatrix} \in \mathcal{H}\}.$$

Under this norm, $\mathcal{C}$ is closed. We now cite a result of Chaumat [Ch][2], to get the existence of ξ_{11} in $\mathcal{C} \subseteq L^2(\mu_1)$, so that for all $f \in \mathcal{C}, f d\mu_1 << \xi_{11} d\mu_1$.

Before proceeding to prove $\xi_{11} \neq 0$ $\mu_1 - a.e.$, we first argue that for any set A with $\mu_1(A) > 0$, there is a $\begin{bmatrix} f \\ g \end{bmatrix} \in \mathcal{H}$ so that $f_1 \neq 0$ on A where $f = \oplus \sum_j f_j$. If this were not the case, that is, if for

[2]Chaumat's Lemma 3 states that if $\mathcal{C}$ is a closed, nonempty convex subset of $L^1(\mu)$ for a finite measure μ, then there exists a $f_c \in \mathcal{C}$ so that for all $f \in \mathcal{C}, f d\mu << f_c d\mu$. But he actually proves a more general result: $\mathcal{C}$ need only be closed in some norm. Using this and noting that $L^1(\mu_1) \subseteq L^2(\mu_1)$, we have the result cited.

all $\begin{bmatrix} f \\ g \end{bmatrix} \in \mathcal{H}, f_1 \equiv 0$ on A, then letting $B = (spp t\mu_1) \setminus A$,

$$\hat{\mathcal{K}}_1 = \chi_B L^2(\mu_1) \oplus \sum_{j=2}^{\infty} L^2(\mu_j)$$

and $\hat{\mathcal{K}} = \hat{\mathcal{K}}_1 \oplus \mathcal{K}_2$, we have $\hat{\mathcal{K}}$ contains $\mathcal{H}$, is strictly smaller than $\mathcal{K}$, and is invariant for both S and N (since these are multiplication operators), and hence for J; this contradicts minimality.

Now suppose $\xi_{11} = 0$ on some set A with $\mu_1(A) > 0$. By the above there is a $\begin{bmatrix} f \\ g \end{bmatrix} \in \mathcal{H}$ so that $f_1 \neq 0$ on A. But then

$$\int_A |f_1| d\mu_1 > 0, \text{ while } \int_A \xi_{11} d\mu_1 = 0$$

contradicting that $f_1 d\mu << \xi_{11} d\mu_1$. It follows that $\xi_{11} \neq 0$ $\mu_1 - a.e.$, or equivalently $esssppt\xi_{11} = sppt\mu_1$.

Consider the closed subspace

$$\mathcal{M} = [\vee\{T^n \xi\}]^-.$$

By definition, $\mathcal{M}$ is an invariant cyclic subspace for T and if we let $\hat{T} = T|\mathcal{M}$, then $\hat{T}$ is subjordan. In fact, $\hat{T}$ is pure, for if there is an invariant subspace $\mathcal{M}_0$ of $\mathcal{M}$ on which $\hat{T}$ is self-adjoint, then $\mathcal{M}_0$ is an invariant subspace for T on which it is self-adjoint. Let $\hat{J} = \hat{S} + \hat{N}$ be a minimal Jordan extension. We claim

$$\sigma(\hat{J}) = \sigma(\hat{T}) = sppt\mu_1 = \sigma(T) = \sigma(J). \tag{3.17.2}$$

Since T is pure subjordan with cyclic vector, by Corollary 3.16, the spectrum is necessarily regularly closed. Thus once (3.17.2) is established it follows that $\sigma(T)$ is regularly closed and the proof is complete.

Now the first (and as already noted, the last) equality of (3.17.2) follows from a phenomenon previously mentioned: the spectrum of a (real) subjordan operator is equal to the spectrum of any minimal (real) Jordan extension (see proof of Corollary 3.16). So it is left to show $\sigma(\hat{T}) = sppt\mu_1$. Let $\lambda \in \mathbb{R} \setminus (sppt\mu_1)$. Since $\mu_j << \mu_1, \nu_1 << \mu_1$, and $\nu_i << \nu_1$, for all i, j, it follows that

$$\lambda \in [\mathbb{C} \setminus (sppt\mu_j)] \cap [\mathbb{C} \setminus (sppt\nu_i)]$$

for all i, j. Thus $g(x) = (x - \lambda)^{-1}$ is continuous on

$$\bigcup_{i,j}[(spp t\mu_j) \cup (sppt\nu_i)] \subseteq sppt\mu_1.$$

By adjusting the definition of $g(x)$ on a small neighborhood of λ if necessary, we may assume that there is a continuously differentiable function g_0 defined on the closed convex hull $[a, b]$ of $sppt\mu_1$ such that $g_0|sppt\mu_1 = g|sppt\mu_1$. Since g_0' is continuous on $[a, b]$, there is a sequence of polynomials $\{p_n'\}$ converging uniformly to g_0' on $[a, b]$. Define a sequence of polynomials p_n by

$$p_n(x) = \int_a^x p_n'(t)dt + g_0(a).$$

Since, for $x \in [a, b]$

$$g_0(x) = \int_a^x g'(t)dt + g_0(a),$$

it follows that, for all $x \in [a, b]$,

$$\begin{aligned} |p_n(x) - g(x)| &= |\int_a^x (p_n'(t) - g'(t))dt| \\ &\leq (b - a)sup\{|p_n'(x) - g_0'(x)| : x \in [a, b]\}. \end{aligned}$$

Thus the sequence $\{p_n\}$ and its derived sequence $\{p_n'\}$ converge to g and g_0', respectively uniformly on $[a, b]$, and hence also on $sppt\mu_1$. Using this, we have

$$\lim_{n\to\infty} p_n(\hat{T})\xi = \begin{bmatrix} g(x)\xi_1(x) \\ \Gamma(x)g_0'(x)\xi_1 + g(x)\xi_2(x) \end{bmatrix} \in \mathcal{M}.$$

Further, since $p_n \to g$ and $p_n' \to g_0'$ uniformly on $sppt\mu_1$, we have $p_n(\hat{T}) \to g(\hat{T}) = (\hat{T} - \lambda)^{-1}$ strongly. Therefore, $\lambda \in \rho(\hat{T})$. We conclude that $\sigma(\hat{T}) \subseteq sppt\mu_1$.

Now let $\lambda_0 \in \rho(\hat{T})$. Then, since $\sigma(\hat{T}) = \sigma(\hat{J}) = \sigma(\hat{S}_1)$ (see (3.17.1)), we have $\lambda_0 \in \rho(\hat{S}_1)$. Choose an open set G so that $\lambda_0 \in G \subseteq \rho(\hat{S}_1)$. Then, for all $\lambda \in G, (x - \lambda)^{-1}$ is a bounded multiplication operator on $L^2(\mu_1)$. Thus $(x - \lambda)^{-1}\xi_{11}(x) \in L^2(\mu_1) \subseteq L^1(\mu_1)$, for all $\lambda \in G$. Hence $G \cap esssppt\xi_{11} = \emptyset$. In particular, $\lambda_0 \notin esssppt\xi_{11} = sppt\mu_1$. Thus $sppt\mu_1 \subseteq \sigma(\hat{T})$, which completes the proof. $\square$

References

[A] M. B. Abrahamse, Some examples on lifting the commutant of a subnormal operator, Ann. Polon. Math. 37 (1980), No. 3, 289-298.

[Ag1] J. Agler, Subjordan operators, Thesis, Indiana University, (1980).

[Ag2] ———, Subjordan operators: Bishop's theorem, spectral inclusion and spectral sets, J. Operator Theory 7 (1982), 373-395.

[BH] J. A. Ball and J. W. Helton, Nonnormal dilations, disconjugacy and constrained spectral factorization, Integral Equations and Operator Theory 3 (1980), 216-309.

[C] J. Conway, *Subnormal Operators*, Pitman Advanced Publishing Program, Boston (1981).

[CP] K. F. Clancey and C. R. Putnam, The local spectral behavior of completely subnormal operators, Trans. Amer. Math. Soc. 163 (1973), 239-244.

[Ch] J. Chaumat, Adherence faible etoile d'algebras de fractions rationelles, A. Inst. Fourier 24 (1974), 93-120.

[Cl] K. F. Clancey, *Seminormal Operators*, Springer-Verlag, New York (1979).

[DF] J. Dixmier et C. Foias, Sur le spectre ponctuel d'un opérateur, Colloquia Mathematica Societas János Bolyai 5. Hilbert Space Operators, Tihany (Hungary), 1970.

[Fa] T. R. Fanney, Closability of differential operators and subjordan operators, Ph.D. dissertation, Virginia Polytechnic Institute and State University, 1989.

[Fu] Fukushima, *Dirichlet Forms and Markov Processes*, North Holland, Amsterdam–Oxford–New York (1980).

[Ham] M. Hamza, Determination des Formes de Dirichlet Sur $\mathbb{R}^n$, Thése 3e Cycle, Orsay (1975).

[H1] J. W. Helton, Operators with representation as multiplication by x on a Sobolev space, Colloquia Math. Soc. Danos Bolyai 5. Hilbert Space Operators, Tihany, Hungary (1970), 279-287.

[H2] ———, Jordan operators in infinite dimensions and Sturm-Liouville conjugate point theory, Bull. Amer. Math. Soc. 78 (1972), 57-62.

[H3] ———, Infinite dimensional Jordan operators and Sturm-Liouville conjugate point theory, Trans. Amer. Math. Soc. 170 (1972), 305-331.

Joseph A. Ball
Department of Mathematics
Virginia Tech
Blacksburg, VA 24061

Thomas R. Fanney
Department of Mathematics
Virginia Tech
Blacksburg, VA 24061

Operator Theory:
Advances and Applications, Vol. 48

INVERTIBILITY AND DICHOTOMY OF SINGULAR DIFFERENCE EQUATIONS

A. Ben-Artzi, I. Gohberg, M.A. Kaashoek

Dedicated to the memory of Ernst Hellinger

Unique solvability in vector-valued ℓ_2 spaces of a system of singular difference equations with initial values in a certain subspace is proved to be equivalent to the existence of a normal dichotomy for the equations. The inverse operator is described explicitly in terms of the dichotomy and conversely. For the inverse operator and the dichotomy the dependence on the space of initial values is analysed.

1. INTRODUCTION

In this paper we treat the following class of difference equations:

$$\begin{cases} B_{n+1}x_{n+1} = A_n x_n + f_n, & n = 0, 1, \dots, \\ x_0 \in L. \end{cases} \tag{1.1}$$

Here $(A_n)_{n=0}^{\infty}$ and $(B_{n+1})_{n=0}^{\infty}$ are given sequences of complex $r \times r$ matrices, and L is a subspace of $\mathbb{C}^r$. This notation will be preserved throughout this paper. If some of the matrices B_{n+1} are not invertible, then the difference equation (1.1) is called *singular*. In what follows ℓ_r^2 denotes the Hilbert spaces of all square summable sequences $(x_n)_{n=0}^{\infty}$ with entries in $\mathbb{C}^r$, and

$$\ell_{r,L}^2 = \{(x_n)_{n=0}^{\infty} \in \ell_r^2 \mid x_0 \in L\}.$$

Assume that the sequences $(A_n)_{n=0}^{\infty}$ and $(B_{n+1})_{n=0}^{\infty}$ are bounded in the usual matrix norm, and consider the operator

$$T = (-\delta_{ij}A_j + \delta_{i+1,j}B_j)_{i,j=0}^{\infty} : \ell_{r,L}^2 \to \ell_r^2.$$

By definition, the difference equation (1.1) is unique solvable in ℓ_r^2 if and only if $T : \ell_{r,L}^2 \to \ell_r^2$ is one to one and onto. By the open mapping theorem, this is equivalent to the bounded invertibility of T. We call T the *operator associated with* the difference equation (1.1). The following three theorems contain the main results of this paper. The results are stated in terms of dichotomy of systems. This notion is defined immediately after the first theorem.

THEOREM 1.1 *Let a difference equation (1.1) be given, where $(A_n)_{n=0}^{\infty}$ and $(B_{n+1})_{n=0}^{\infty}$ are bounded sequences of $r \times r$ matrices, and let $T : \ell_{r,L}^2 \to \ell_r^2$ be the associated operator. Then T is invertible if and only if the system*

$$B_{n+1}x_{n+1} = A_n x_n, \quad n = 0, 1, \ldots,$$

admits a unique normal dichotomy $(P_n)_{n=0}^{\infty}$ such that Ker $P_0 = L$.

Moreover, if the operator T is invertible, then the dichotomy $(P_n)_{n=0}^{\infty}$ is related to the inverse operator $T^{-1} = (\Gamma_{ij})_{i,j=0}^{\infty} : \ell_r^2 \to \ell_{r,L}^2$ in the following way. The entries Γ_{ij} of T^{-1} are given by

$$(1.2)\quad \Gamma_{ij} = \begin{cases} -(A_i|_{\mathrm{Ker}P_i})^{-1}B_{i+1}\cdots(A_{j-1}|_{\mathrm{Ker}P_{j-1}})^{-1}B_j(A_j|_{\mathrm{Ker}P_j})^{-1}(I_r - U_j), & i \le j, \\ (B_i|_{\mathrm{Im}P_i})^{-1}A_{i-1}\cdots(B_{j+2}|_{\mathrm{Im}P_{j+2}})^{-1}A_{j+1}(B_{j+1}|_{\mathrm{Im}Pj+1})^{-1}U_j, & i > j, \end{cases}$$

where U_n $(n = 0, 1, \ldots)$ is the projection onto $\mathrm{Im}(B_{n+1}P_{n+1})$ along $\mathrm{Im}(A_n(I_r - P_n))$, and $I_r = (\delta_{ij})_{ij=1}^{r}$. In addition, the projections $P_1, P_2 \ldots$ and $U_0, U_1, \ldots$ are determined by T^{-1} as follows

$$(1.3)\quad P_n = \Gamma_{n,n-1}B_n \quad (n = 1, 2, \ldots),$$

$$(1.4)\quad U_n = B_{n+1}\Gamma_{n+1,n} \quad (n = 0, 1, \ldots).$$

Finally, P_0 is given, in terms of its image and kernel, by

$$(1.5)\quad \mathrm{Ker}P_0 = L,$$

$$(1.6)\quad \mathrm{Im}P_0 = \{x_0 \in \mathbb{C}^r : \exists (x_n)_{n=1}^{\infty} \in \ell_r^2 \ \ \text{such that} \ \ B_{n+1}x_{n+1} = A_n x_n \ \ (n = 0, 1, \ldots)\}.$$

Formula (1.2) above appears in [GKvS] under certain additional conditions.

The dichotomy of regular systems is defined in [CS]; see also [GKvS]. The following extension of this notion to the case of general systems appears in [BG2]. Let a system

$$(1.7) \qquad B_{n+1}x_{n+1} = A_n x_n, \quad n = 0, 1, \ldots,$$

be given, where $(A_n)_{n=0}^{\infty}$, and $(B_n)_{n=0}^{\infty}$ are two sequences of $r \times r$ matrices. Note that the sequences are not required to be bounded. We consider bounded sequences of projections in $\mathbb{C}^r$, $(P_n)_{n=0}^{\infty}$, satisfying the following conditions:

$$(1.8) \qquad \text{Rank } P_n \text{ is constant}, \quad n = 0, 1, \ldots,$$

$$(1.9) \qquad \text{Im}(A_n P_n) \subset \text{Im}(B_{n+1} P_{n+1}), \text{Im}(B_{n+1}(I_r - P_{n+1})) \subset \text{Im}(A_n(I_r - P_n)), n \geq 0,$$

$$(1.10) \qquad \text{Im}(B_{n+1}P_{n+1}) + \text{Im}(A_n(I_r - P_n)) = \mathbb{C}^r, \ n \geq 0,$$

where $I_r = (\delta_{ij})_{ij=1}^{r}$. Since Rank P_n ($n = 0, 1, \ldots$) is constant, (1.10) shows that the restricted mappings

$$A_n\big|_{\text{Ker} P_n} : \text{Ker} P_n \to \text{Im}(A_n(I_r - P_n)), \quad n = 0, 1, \ldots,$$

$$B_{n+1}\big|_{\text{Im} P_{n+1}} : \text{Im} P_{n+1} \to \text{Im}(B_{n+1}P_{n+1}), \quad n = 0, 1, \ldots,$$

are invertible. We can now define a dichotomy for the system (1.7). A bounded sequence of projections $(P_n)_{n=0}^{\infty}$, with properties (1.8) - (1.10), is called a *dichotomy* for the system (1.7), if there exist positive numbers a and M, with $a < 1$, such that the following inequalities hold

$$(1.11) \qquad \|(B_{n+j}\big|_{\text{Im} P_{n+j}})^{-1} A_{n+j-i} \cdots (B_{n+1}\big|_{\text{Im} P_{n+1}})^{-1} A_n P_n\| \leq M a^j, \ n \geq 0, j \geq 1,$$

$$(1.12) \qquad \begin{aligned} &\|(A_{n-j}\big|_{\text{Ker} P_{n-j}})^{-1} B_{n-j+1} \cdots (A_{n-1}\big|_{\text{Ker} P_{n-1}})^{-1} B_n (I_r - P_n)\| \\ &\qquad \leq M a^j, \quad n \geq 1, 1 \leq j \leq n. \end{aligned}$$

The integer $p = \text{Rank } P_n$ $(n = 0, 1, \ldots)$ is called the *rank* of the dichotomy . The dichotomy $(P_n)_{n=0}^{\infty}$ is called a *normal dichotomy* if the following additional inequality holds

(1.13)
$$\sup_n \|(B_{n+1}|_{\text{Im}\,P_{n+1}}, A_n|_{\text{Ker}\,P_n})^{-1}\| < \infty.$$

Here, the mappings $(B_{n+1}|_{\text{Im}\,P_{n+1}}, A_n|_{\text{Ker}\,P_n}) : \text{Im}P_{n+1} \oplus \text{Ker}P_n \to \mathbb{C}^r$, are defined by

$$(B_{n+1}|_{\text{Im}\,P_{n+1}}, A_n|_{\text{Ker}\,P_n})\begin{pmatrix} u \\ v \end{pmatrix} = B_{n+1}u + A_n v,$$

where $u \in \text{Im}P_{n+1}$, and $v \in \text{Ker}P_n$ $(n = 0, 1, \ldots)$. Note that the invertibility of $(B_{n+1}|_{\text{Im}\,P_{n+1}}, A_n|_{\text{Ker}\,P_n})$ follows from (1.10). Condition (1.13) is clearly equivalent to the following three inequalities:

(1.14)
$$\inf_{n=0,1,\ldots} \{\text{dist}(u, \text{Im}(A_n(I_r - P_n))) \mid u \in \text{Im}(B_{n+1}P_{n+1}), \ \|u\| = 1\} > 0,$$

(1.15)
$$\sup_{n=0,1,\ldots} \|(B_{n+1}|_{\text{Im}\,P_{n+1}})^{-1}\| < \infty,$$

(1.16)
$$\sup_{n=0,1,\ldots} \|(A_n|_{\text{Ker}\,P_n})^{-1}\| < \infty.$$

For $n = 0, 1, 2, \ldots$ let U_n be the projection of $\mathbb{C}^r$ onto $\text{Im}(B_{n+1}P_{n+1})$ along $\text{Im}(A_n(I_r - P_n))$. From (1.14) it follows that the sequence $(U_n)_{n=0}^{\infty}$ is bounded whenever $(P_n)_{n=0}^{\infty}$ is a normal dichotomy.

Let us consider two special cases of Theorem 1.1.

(I) Assume that $L = \mathbb{C}^n$. Then $\ell^2_{r,L} = \ell^2_r$, and T is an upper triangular operator in ℓ^2_r. The condition $\text{Ker } P_0 = L$ implies that $P_0 = 0$. Since $\text{Rank } P_n$ $(n = 0, 1, \ldots)$ is constant, it follows that $P_n = 0$ $(n = 0, 1, \ldots)$ and therefore $U_n = 0$ $(n = 0, 1, \ldots)$. Hence, formula (1.2) merely expresses the fact that the inverse of an upper triangular operator is upper triangular, and gives a formula for its inverse. The dichotomy inequalities (1.12) and (1.16), and formula (1.2), imply a known property of the entries of the inverse of a band matrix, namely, exponential decay away from the main diagonal (see [BG 1]).

(II) Assume that the sequences $(A_n)_{n=0}^{\infty}$ and $(B_n)_{n=0}^{\infty}$ are constant. Let $A_n = A, B_{n+1} = B$ $(n = 0, 1, \ldots)$. If T is invertible, then the operator $\widetilde{T} = (-\delta_{ij}A + \delta_{i+1,j}B)_{ij}^{\infty}$ on

ℓ_r^2 is Fredholm. It is well known that this implies that the symbol, $A+\lambda B$, is invertible for $|\lambda|=1$. By Theorem 2.1 of [GK], it follows that the system $Bx_{n+1}=Ax_n$ $(n=0,1,\ldots)$, admits a dichotomy $P_n=P$ $(n=0,1,\ldots)$. In this case, Theorem 1.1 follows immediately from these considerations.

The next two results deal with the dependence of the dichotomy $(P_n)_{n=0}^{\infty}$ and the inverse T^{-1} upon the subspace L.

THEOREM 1.2 *Assume that the system*

$$B_{n+1}x_{n+1}=A_nx_n \quad (n=0,1,\ldots), \tag{1.17}$$

admits a dichotomy $(P_n)_{n=0}^{\infty}$. *Then, for each subspace* L *of* $\mathbb{C}^r$, *satisfying* $L\oplus \mathrm{Im}P_0=\mathbb{C}^r$, *there exists a unique dichotomy* $(\overline{P}_n)_{n=0}^{\infty}$ *of (1.17) such that* $\mathrm{Ker}\overline{P}_0=L$. *This dichotomy is given by the formula*

$$\begin{aligned}\overline{P}_n=P_n-(B_n|_{\mathrm{Im}P_n})^{-1}A_{n-1}\cdots(B_1|_{\mathrm{Im}P_1})^{-1}A_0\Phi_0\times\\ \times(A_0|_{\mathrm{Ker}P_0})^{-1}B_1\ldots(A_{n-1}|_{\mathrm{Ker}P_{n-1}})^{-1}B_n(I_r-P_n)\end{aligned} \tag{1.18}$$

for $n=0,1,\ldots,$ *where* $\Phi_0:\mathrm{Ker}P_0\to\mathrm{Im}P_0$ *is given by*

$$\Phi_0=P_0((I_r-P_0)|_L)^{-1}. \tag{1.19}$$

Furthermore, all the dichotomies of the system (1.17) can be obtained in this way by varying L, *and if one dichotomy of the system (1.17) is normal, then all its dichotomies are normal.*

This result appears in [GKvS] in the case when $B_{n+1}=I_r$ and A_n is invertible $(n=0,1,\ldots)$. In Theorem 1.3 below we will consider two difference equations.

$$\begin{cases}B_{n+1}x_{n+1}=A_nx_n \quad (n=0,1,\ldots),\\ x_0\in L_1\end{cases} \tag{1.20}$$

and

$$\begin{cases}B_{n+1}x_{n+1}=A_nx_n \quad (n=0,1,\ldots),\\ x_0\in L_2\end{cases} \tag{1.21}$$

which differ only in the boundary conditions. In the statement below, and in the rest of the paper, we denote by $Q_0 : \ell_r^2 \to \mathbb{C}^r$ the projection onto the first coordinate space, i.e., $Q_0((u_n)_{n=0}^\infty) = u_0$.

THEOREM 1.3 *Assume that the difference equations (1.20) and (1.21), where $(A_n)_{n=0}^\infty$ and $(B_{n+1})_{n=0}^\infty$ are bounded sequences of $r \times r$ matrices, are uniquely solvable in ℓ_r^2, and let T_1 and T_2 be their corresponding associate operators. Let $(P_n)_{n=0}^\infty$ be an arbitrary dichotomy of the system $B_{n+1}x_{n+1} = A_n x_n$ $(n = 0, 1, \ldots)$, and denote by $\overline{P}$ the projection onto Im P_0 along L_2. Then,*

$$T_2^{-1} = T_1^{-1} - ZQ_0T_1^{-1}. \tag{1.22}$$

Here, the operator $Z : \mathbb{C}^r \to \ell_r^2$ is given by

$$Zx = ((B_j|_{\operatorname{Im} P_j})^{-1}A_{j-1}\cdots(B_1|_{\operatorname{Im} P_1})^{-1}A_0\overline{P}x)_{j=0}^\infty \quad (x \in \mathbb{C}^r). \tag{1.23}$$

This paper consists of four sections. The first is the present introduction. The second section contains the proof of Theorem 1.1. The proof is based on an application of Theorem 2.1 in [GK]. The third section deals with dichotomy and contains the proof of Theorem 1.2. The last section contains results about perturbations of systems and the proof of Theorem 1.3.

2. INVERTIBILITY

In this section we prove Theorem 1.1. The proof will be based on the spectral decomposition theorem for operator pencils which appears in [GK]. The latter result, which we need for the unit circle only, may be stated as follows.

THEOREM 2.1. *Let X and Y be complex Banach spaces, and let $G, H : X \to Y$ be bounded linear operators. Assume that $\lambda G - H$ is invertible for $|\lambda| = 1$. Then there exist bounded projections V and U in X and Y, respectively, such that the following block partitioning holds:*

$$\lambda G - H = \begin{pmatrix} \lambda G' - H' & 0 \\ 0 & \lambda G'' - H'' \end{pmatrix} : \operatorname{Ker} V \oplus \operatorname{Im} V \to \operatorname{Ker} U \oplus \operatorname{Im} U, \tag{2.1}$$

the operators H' and G'' are invertible, and the spectral radii of the operators $(H')^{-1}G'$ and $(G'')^{-1}H''$ are strictly less than one. Furthermore, V and U are uniquely determined.

In fact

$$V=\frac{1}{2\pi i}\int_{\mathbb{T}}(\lambda G-H)^{-1}Gd\lambda, \tag{2.2}$$

and

$$U=\frac{1}{2\pi i}\int_{\mathbb{T}}G(\lambda G-H)^{-1}d\lambda, \tag{2.3}$$

where $\mathbb{T}=\{\lambda:|\lambda|=1\}$.

In the proof of Theorem 1.1 we will also make use of the following result, which is Corollary 6.5 of [BG2].

PROPOSITION 2.2. *If the system* $B_{n+1}x_{n+1}=A_nx_n \quad (n=0,1,\ldots)$ *admits a dichotomy* $(P_n)_{n=0}^{\infty}$, *then, for* $k=0,1,\ldots$

$$\operatorname{Im}P_k=\{x_k\in\mathbb{C}^r:\exists x_{k+1},x_{k+2}\ldots \text{in } \mathbb{C}^r \text{ such that}$$

$$B_{n+1}x_{n+1}=A_nx_n,\quad n\geq k,\ \text{and}\ \lim_{n\to\infty}x_n=0\}.$$

Furthermore, all the dichotomies of the system $B_{n+1}x_{n+1}=A_nx_n \quad (n=0,1,\ldots)$ *have the same rank.*

We now proceed with the proof of Theorem 1.1.

PROOF OF THEOREM 1.1. The proof is divided into parts (A), (B) and (C), and each part has a further subdivision.

(A) In this part we assume that the operator

$$T=(-\delta_{ij}A_j+\delta_{i+1,j}B_j)_{i,j=0}^{\infty}:\ell^2_{r,L}\to\ell^2_r \tag{2.4}$$

is invertible, and we prove the existence of a normal dichotomy.

(A.1) Let R be an arbitrary projection of $\mathbb{C}^r$ onto L. Define

$$H_0=R,\quad G_1=I_r-R, \tag{2.5a}$$

$$H_n=A_{n-1},\quad G_{n+1}=B_n,\quad n=1,2,\ldots. \tag{2.5b}$$

Consider the operator

$$\widehat{T}=(-\delta_{ij}H_j+\delta_{i+1,j}G_j)_{i,j=0}^{\infty}:\ell^2_{r,L}\to\ell^2_r. \tag{2.6}$$

We now show that $\widehat{T}$ is invertible. Assume $y = (y_n)_{n=0}^{\infty} \in \operatorname{Ker} \widehat{T}$. Then $y_0 \in L$ and $-H_0 y_0 + G_1 y_1 = 0$. Hence $-Ry_0 + (I_r - R)y_1 = 0$. Since R is a projection and Im $R = L$, we conclude that $y_0 = 0$ and $y_1 \in L$. Put $y' = (y_1, y_2, \ldots)$. Since $y_1 \in L$, we have $y' \in \ell^2_{r,L}$. Moreover, $y_0 = 0$ and $\widehat{T}y = 0$ lead to $y\prime \in \operatorname{Ker} T$. Since T is invertible, it follows that $y' = 0$, and hence, $y_0 = 0$ shows that $y = 0$. Thus Ker $\widehat{T} = \{0\}$. On the other hand, since T is invertible, the operator $\widehat{T}$ is Fredholm and index $\widehat{T} = 0$. Thus $\widehat{T}$ is invertible.

(A.2) Define two operators $H, G : \ell^2_{r,L} \to \ell^2_r$ via $H = (\delta_{ij} H_j)_{j=0}^{\infty}$ and $G = (\delta_{i+1,j} G_j)_{i,j=0}^{\infty}$. We shall prove that $\lambda G - H : \ell^2_{r,L} \to \ell^2_r$ is invertible for each $\lambda \in \mathbb{T}$.

For $\lambda \in \mathbb{T}$ let $U(\lambda)$ be the unitary operator on ℓ^2_r defined by

$$U(\lambda) = ((x_n)_{n=0}^{\infty}) = (\lambda^n x_n)_{n=0}^{\infty}.$$

Obviously, $\ell^2_{r,L}$ is invariant under $U(\lambda)$, and the restriction $U_0(\lambda)$ of $U(\lambda)$ to $\ell^2_{r,L}$ is a unitary operator on $\ell^2_{r,L}$. From the diagonal structure of H and G we see that

$$\lambda G - H = U(\lambda)(G - H)U_0(\lambda) \quad , \lambda \in \mathbb{T}. \tag{2.7}$$

Since $G - H = \widehat{T}$, we know from the result proved under (A.1) that $G - H$ is invertible, and hence $\lambda G - H$ is invertible for each $\lambda \in \mathbb{T}$.

(A.3) Since $\lambda G - H$ is invertible for each $\lambda \in \mathbb{T}$, we can apply Theorem 2.1. Let

$$V : \ell^2_{r,L} \to \ell^2_{r,L}, \qquad U : \ell^2_r \to \ell^2_r$$

be the projections defined by (2.2) and (2.3). We shall prove that V and U are block diagonal operators.

Consider the operator matrix representation for $\widehat{T} = (G - H)^{-1}$:

$$\widehat{T} = (\widehat{\Gamma}_{ij})_{i,j=0}^{\infty} : \ell^2_r \to \ell^2_{r,L}.$$

This, together with (2.7), leads to $(\lambda G - H)^{-1} = (\lambda^{j-i} \widehat{\Gamma}_{ij})_{i,j=0}^{\infty}$, and hence

$$\frac{1}{2\pi i} \int_{\mathbb{T}} (\lambda G - H)^{-1} d\lambda = (\delta_{i,j+1} \widehat{\Gamma}_{i,j})_{i,j=0}^{\infty}.$$

We can now compute V and U by (2.4) and (2.5). It follows that

$$V = \operatorname{diag}\,(0, \widehat{\Gamma}_{10} G_1, \widehat{\Gamma}_{21} G_2, \ldots),$$

$$U = \text{diag}\,(G_1\widehat{\Gamma}_{10}, G_2\widehat{\Gamma}_{21}, G_3\widehat{\Gamma}_{32}, \ldots).$$

Since V and U are projections in $\ell^2_{r,L}$ and ℓ^2_r, respectively, the operators $\widehat{\Gamma}_{n+1,n}G_{n+1}$ and $G_{n+1}\widehat{\Gamma}_{n+1,n}$ $(n = 0, 1, 2, \ldots)$ are projections of $\mathbb{C}^n$. Put

$$P_n = \widehat{\Gamma}_{n+1,n}G_{n+1}, \quad n = 0, 1, 2, \ldots, \tag{2.8a}$$

$$U_n = G_{n+1}\widehat{\Gamma}_{n+1,n}, \quad n = 0, 1, 2, \ldots. \tag{2.8b}$$

Since V and U are bounded, the sequences $(P_n)_{n=0}^{\infty}$ and $(U_n)_{n=0}^{\infty}$ are bounded.

(A.4) We now prove that $(P_n)_{n=0}^{\infty}$ is a normal dichotomy for the system $B_{n+1}x_{n+1} = A_n x_n$, $n = 0, 1, \ldots$. From (2.1) we know that $GV = UG$ and $HV = UH$, which implies that

$$\text{Im}\,(I - R)(I - P_0) \subset \text{Im}\,(I - U_0), \tag{2.9}$$

$$\text{Im}\,B_{n+1}(I - P_{n+1}) \subset \text{Im}\,(I - U_{n+1}), \quad n \geq 0, \tag{2.10}$$

$$\text{Im}\,A_n P_n \subset \text{Im}\,U_{n+1}, \quad n \geq 0. \tag{2.11}$$

Moreover, the invertibility of $G'' : \text{Im}\,V \to \text{Im}\,U$ and $H' : \text{Ker}\,V \to \text{Ker}\,K$ shows that the operators

$$B_{n+1}\big|_{\text{Im}\,P_{n+1}} : \text{Im}\,P_{n+1} \to \text{Im}\,U_{n+1}, \quad n \geq 0, \tag{2.12}$$

$$A_n\big|_{\text{Ker}\,P_n} : \text{Ker}\,P_n \to \text{Ker}\,U_{n+1}, \quad n \geq 0, \tag{2.13}$$

are invertible and that the following inequalities hold

$$\sup_{n=0,1,\ldots} \|(B_{n+1}\big|_{\text{Im}\,P_{n+1}})^{-1}\| < \infty, \quad n \geq 0, \tag{2.14}$$

$$\sup_{n=0,1,\ldots} \|(A_n\big|_{\text{Ker}\,P_n})^{-1}\| < \infty, \quad n \geq 0. \tag{2.15}$$

The invertibility of the maps G'' and H' also implies that the operators

$$I - R\big|_{\mathrm{Im}\ P_0} : \mathrm{Im}\ P_0 \to \mathrm{Im}\ U_0, \tag{2.16}$$

$$R\big|_L : L \to \mathrm{Ker}\ U_0, \tag{2.17}$$

are invertible. Since R is a projection of $\mathbb{C}^r$ onto L, formula (2.17) implies that $L = \mathrm{Ker}\ U_0$.

From the invertibility of the operators in (2.12), (2.13) and (2.16) it also follows that

$$\mathrm{rank}\ P_n = \mathrm{rank}\ U_n = \mathrm{rank}\ P_{n+1}, \quad n \geq 0. \tag{2.18}$$

Now use that $\mathrm{Ker}\ U_0 = L$. So we may conclude $(P_n)_{n=0}^{\infty}$ is a bounded sequence of projections of $\mathbb{C}^r$ of constant rank p, where $p = r - \dim L$.

The invertibility of the operators (2.12) and (2.13) also imply

$$\mathrm{Im}\ B_{n+1}P_{n+1} \oplus \mathrm{Im}\ A_n(I_r - P_n) = \mathbb{C}^r, \quad n \geq 0. \tag{2.19}$$

Moreover, the fact that the sequence $(U_n)_{n=0}^{\infty}$ is a bounded sequence of projection of $\mathbb{C}^r$ leads to

$$\inf_{n=0,1,\ldots} \{\mathrm{dist}\ (u, \mathrm{Im}\ A_n(I_r - P_n)) \big| u \in \mathrm{Im}\ B_{n+1}P_{n+1},\ \|u\| = 1\} > 0. \tag{2.20}$$

The equalities $GV = UG$ and $HV = UH$ also yield the following inequalities:

$$\mathrm{Im}\ A_nP_n \subset \mathrm{Im}\ U_{n+1}, \quad \mathrm{Im}\ B_{n+1}(I_r - P_{n+1}) \subset \mathrm{Im}\ (I_r - U_{n+1}), \quad n \geq 0.$$

This, together with the invertibility of the operators (2.12) and (2.13), shows that

$$\mathrm{Im}\ A_nP_n \subset \mathrm{Im}\ B_{n+1}P_{n+1}, \quad \mathrm{Im}\ B_{n+1}(I_r - P_{n+1}) \subset \mathrm{Im}\ A_n(I_r - P_n), \quad n \geq 0. \tag{2.21}$$

To prove that $(P_n)_{n=0}^{\infty}$ is a normal dichotomy, it remains to establish the inequalities (1.11) and (1.12). By Theorem 2.1, the spectral radii of $(H')^{-1}G'$ and $(G'')^{-1}H''$ are strictly less than 1. Hence there exist positive numbers M_0 and a, $a < 1$, such that

$$\|((H')^{-1}G')^j\| \leq M_0 a^j, \quad \|((G'')^{-1}H'')^j\| \leq M_0 a^j, \quad j \geq 1.$$

Now use (2.1), the block diagonal structure of V and U, the special structure of the operators G and H, and the relations (2.9), (2.12), (2.13) and (2.21). It follows that

$$\|(B_{n+j}|_{\text{Im } P_{n+j}})^{-1}A_{n+j-1}\cdots(B_{n+1}|_{\text{Im } P_{n+1}})^{-1}A_n x\|$$
$$\le M_0 a^j\|x\|, \quad x\in \text{Im } P_n \quad (n\ge 0, j\ge 1),$$

and

$$\|(A_{n-j}|_{\text{Ker } P_{n-j}})^{-1}B_{n-j+1}\cdots(A_{n-1}|_{\text{Ker } P_{n-1}})^{-1}B_n x\|$$
$$\le M_0 a^j\|x\|, \quad x\in \text{Ker } P_n \quad (n\ge 1, 1\le j\le n-1).$$

The sequence $(P_n)_{n=0}^\infty$ is a bounded sequence of projections of $\mathbb{C}^r$. So we see that (1.11) and (1.12) hold with $M = M_0(1+\sup_{n=0,1,\dots}\|P_n\|)$.

(A.5) Let us show that for the dichotomy $(P_n)_{n=0}^\infty$ constructed above Ker P_0 $= L$. The invertibility of the operator (2.16) implies that Ker $(I-R)$ has a trivial intersection with Im P_0. But R is a projection with Im $R = L$. Thus Im $P_0\cap L = \{0\}$. On the other hand, formula (2.8a) shows that $L\subset$ Ker P_0. Since rank $P_0 = r - \dim L$, we have $L =$ Ker P_0.

(B) In this part we assume that the system

$$B_{n+1}x_{n+1} = A_n x_n, \quad n = 0,1,\dots, \tag{2.22}$$

admits a normal dichotomy $(P_n)_{n=0}^\infty$ satisfying

$$\text{Ker } P_0 = L. \tag{2.23}$$

We shall prove that T is invertible and derive the formula for its inverse.

(B.1) Let us first show that Ker $T = \{0\}$. Assume that $x = (x_0, x_1,\dots)\in$ Ker T. Now $x\in\ell_r^2$ leads to $x_n\to 0$ $(n\to\infty)$. Since $x\in$ Ker T, we have $B_{n+1}x_{n+1} = A_n x_n$, $(n = 0,1,\dots)$. Combining this with the limit above, it follows from Proposition 2.2 that $x_n\in$ Im P_n $(n = 0,1,\dots)$. But $B_{n+1}|_{\text{Im } P_{n+1}}$ is invertible for $n = 0,1,\dots$, and hence

$$x_{n+1} = (B_{n+1}|_{\text{Im } P_{n+1}})^{-1}A_n x_n, \quad n = 0,1,\dots. \tag{2.24}$$

Now $x \in \ell^2_{r,L}$, and hence $x_0 \in L = \text{Ker } P_0$. But $x_0 \in \text{Im } P_0$. So we obtain $x_0 = 0$. The equality (2.24) now shows that $x_n = 0$ for $n \geq 0$. Hence $x = 0$. Since $x \in \text{Ker } T$ is arbitrary, we proved

$$\text{(2.25)} \qquad \text{Ker } T = 0.$$

(B.2) Define Γ_{ij} by (1.2). Note that $\text{Im } \Gamma_{0j} \subset \text{Ker } P_0 = L, (j = 0, 1, \ldots)$. Hence we can define $\Gamma = (\Gamma_{ij})_{ij=0}^{\infty} : \ell^2_r \to \ell^2_{r,L}$. It follows from $\sup_n \|U_n\| < \infty$, the inequalities (1.15) - (1.16) and (1.11) - (1.12), that the operators Γ_{ij} satisfy a bound of type $\|\Gamma_{ij}\| \leq M_1 a^{(i-j)}$ for a suitable $M_1 > 0$, and $1 > a > 0$. Hence Γ is a bounded operator in ℓ^2_r. Note that

$$\text{(2.26)} \quad \Gamma_{nn} = -(A_n|_{\text{Ker } P_n})^{-1}(I_r - U_n); \Gamma_{n+1,n} = (B_{n+1}|_{\text{Im } P_{n+1}})^{-1}U_n \quad (n = 0, 1, \ldots).$$

Therefore,

$$-A_n\Gamma_{nn} + B_{n+1}\Gamma_{n+1,n} = I_r, \quad n = 0, 1, \ldots.$$

In addition, it is clear that $-A_k\Gamma_{kn} + B_{k+1}\Gamma_{k+1,n} = 0, \quad k \neq n$. It follows from these equalities that

$$\text{(2.27)} \qquad T\Gamma = I_{l^2_r},$$

where $I_{l^2_r}$ is the identity on l^2_r. In particular, T is onto. By (2.25) this shows that T is invertible. Hence (2.29) leads to

$$\text{(2.28)} \qquad T^{-1} = \Gamma.$$

(B.3) Equality (1.6) follows from Proposition 2.2. Equality (1.5) holds by definition, and (1.4) follows immediately from the second part of (2.26).

(C) In this part we prove the uniqueness of the dichotomy. So, let us assume that $(P_n)_{n=0}^{\infty}$ is a normal dichotomy of the system

$$\text{(2.29)} \qquad B_{n+1}x_{n+1} = A_n x_n, \quad n = 0, 1, 2, \ldots.$$

such that $\text{Ker } P_0 = L$. We have to show that the sequence $(P_n)_{n=0}^{\infty}$ is uniquely determined by T.

By Proposition 2.2, the range of the projection P_0 is uniquely determined by the system (2.29). Since Ker $P_0 = L$, we conclude that P_0 is uniquely determined by T. To prove the analogous statement for the other P_n, it suffices to prove (1.3). Note that, by part (B) of the proof, we know that T is invertible.

Let $n = 1, 2, \ldots$. Since

$$\Gamma_{n,n-1} = (B_n|_{\text{Im } P_n})^{-1} U_{n-1}, \quad \text{Im } U_{n-1} = \text{Im } B_n P_n,$$

formula (1.3) is equivalent to

$$B_n P_n = U_{n-1} B_n. \tag{2.30}$$

To prove the latter equality, note that

$$U_{n-1} B_n = U_{n-1} B_n P_n + U_{n-1} B_n (I_r - P_n). \tag{2.31}$$

However U_{n-1} is the projection onto Im $B_n P_n$ along Im $A_{n-1}(I_r - P_{n-1})$. Hence,

$$U_{n-1} B_n P_n = B_n P_n, \tag{2.32}$$

$$U_{n-1} A_{n-1} (I_r - P_{n-1}) = 0. \tag{2.33}$$

On the other hand, it follows from (1.9) that Im $B_n(I_r - P_n) \subset$ Im $A_{n-1}(I_r - P_{n-1})$. By this inclusion and (2.33), we have $U_{n-1} B_n (I_r - P_n) = 0$. This equality and (2.31), (2.32) imply that $U_{n-1} B_n = B_n P_n$. Hence (2.30) holds $\diamondsuit$

Let us note here the following corollary of Theorems 1.1 and 2.1, which gives an alternative description of the projections $P_1, P_2, \ldots, U_0, U_1, \ldots$.

PROPOSITION 2.3. *Let $(A_n)_{n=0}^{\infty}$ and $(B_{n+1})_{n=0}^{\infty}$ be bounded sequences of $r \times r$ matrices, and let L be a subspace of $\mathbb{C}^r$. Assume that the system $B_{n+1} x_{n+1} = A_n x_n$, $n = 0, 1, 2, \ldots$, admits a normal dichotomy $(P_n)_{n=0}^{\infty}$ such that Ker $P_0 = L$. Introduce two operators $A, B : \ell^2_{r,L} \to \ell^2_r$ by setting $A = (\delta_{ij} A_j)_{ij=0}^{\infty}$ and $B = (\delta_{i+1,j} B_j)_{ij=0}^{\infty}$. Then*

$$\begin{pmatrix} 0 & 0 & 0 & \cdots \\ 0 & P_1 & 0 & \cdots \\ 0 & 0 & P_2 & \cdots \\ \vdots & \vdots & \vdots & \ddots \end{pmatrix} = \frac{1}{2\pi i} \int_{\mathbb{T}} (\lambda B - A)^{-1} B \, d\lambda, \tag{2.34}$$

and

$$(2.35),\qquad \begin{pmatrix} U_0 & 0 & 0 & \dots \\ 0 & U_1 & 0 & \dots \\ 0 & 0 & U_2 & \dots \\ \vdots & \vdots & \vdots & \ddots \end{pmatrix} = \frac{1}{2\pi i}\int_{\mathbb{T}} B(\lambda B - A)^{-1} d\lambda,$$

where U_n is the projection onto Im $B_{n+1}P_{n+1}$ *along* Im $A_n(I_r - P_n)$, $n = 0, 1.\dots.$ *The $(i,0)$-entry in the matrix in (2.34) is the zero operator from L into $\mathbb{C}^r$.*

PROOF. By Theorem 1.1, the operator $T := B - A$ is invertible. The arguments employed in part (A.2) of the proof of Theorem 1.1 show that $\lambda B - A$ is invertible for $|\lambda| = 1$. Thus the right hand sides of (2.34) and (2.35) are well-defined. As in part (A.3) of the proof of Theorem 1.1, one shows that

$$\frac{1}{2\pi i}\int_{\mathbb{T}}(\lambda B - A) d\lambda = (\delta_{i,j+1}\Gamma_{ij})_{i,j=0}^{\infty},$$

where $T^{-1} = (\Gamma_{ij})_{i,j=0}^{\infty}$. It follows that

$$\frac{1}{2\pi i}\int_{\mathbb{T}}(\lambda B - A)B d\lambda = \text{diag}\,(0, \Gamma_{10}B_1, \Gamma_{21}B_2, \dots),$$

$$\frac{1}{2\pi i}\int_{\mathbb{T}}(B(\lambda B - A) d\lambda = \text{diag}\,(B_1\Gamma_{10}, B_2\Gamma_{20}, \dots).$$

By (1.3) and (1.4), this yields the desired formulas (2.34) and (2.35). $\diamond$

3. DICHOTOMY

The purpose of this section is to prove Theorem 1.2. Let us first show how for normal dichotomies and bounded sequences $(A_n)_{n=0}^{\infty}$ and $(B_{n+1})_{n=0}^{\infty}$ the first statement in this theorem follows from the results of the preceding section. We use the notation of Theorem 1.2, and assume that $(P_n)_{n=0}^{\infty}$ us a normal dichotomy for (1.17). Consider the operator

$$\widetilde{T} = (-\delta_{ij}A_j + \delta_{i+1,j}B_j)_{i,j=0}^{\infty} : \ell_r^2 \to \ell_r^2$$

Put $N = \text{Ker } P_0$, and define

$$T_N : \ell_{r,N}^2 \to \ell_r^2 \quad , \quad T_N = \widetilde{T}|_{\ell_{r,N}^2}$$

$$T_L : \ell_{r,L}^2 \to \ell_r^2 \quad , \quad T_L = \widetilde{T}|_{\ell_{r,L}^2}$$

By Theorem 1.1, the operator T_N is invertible. Since $L \oplus \text{Im } P_0 = \mathbb{C}^r$, we have $\dim N = \dim L$, and hence $\ell^2_{r,N}$ and $\ell^2_{r,L}$ are subspaces of ℓ^2_r of equal codimension. It follows that T_L is a Fredholm operator of index zero. Assume $\mathbf{x} = (x_n)_{n=0}^\infty \in \text{Ker } T_L$. Then $B_{n+1}x_{n+1} = A_n x_n$ for $n = 0, 1, 2, \ldots$ and $x_n \to 0$ $(n \to \infty)$. By Proposition 2.2, this leads to $x \in \text{Im } P_0$. However, $x_0 \in L$, and hence $x_0 \in L \cap \text{Im } P_0 = \{0\}$. Thus $x_0 = 0$. But then we may conclude that $\mathbf{x} \in \text{Ker } T_N$, and therefore $\mathbf{x} = 0$. This proves that $\text{Ker } T_L = \{0\}$. Since index $T_L = 0$, we see that T_L is invertible, and so, by Theorem 1.1, there exists a unique normal dichotomy $(\widetilde{P}_n)_{n=0}^\infty$ of (1.17) satisfying $\text{Ker } \widetilde{P}_0 = L$.

We now return to the general case. Throughout this section, $(A_n)_{n=0}^\infty$ and $(B_{n+1})_{n=0}^\infty$ are sequences of complex $r \times r$ matrices.

PROOF OF THEOREM 1.2. The proof is divided into several parts

Part (A). We first prove the uniqueness part of the statement. Assume that $(\overline{P}_n)_{n=0}^\infty$ and $(\widetilde{P}_n)_{n=0}^\infty$ are dichotomies for the system (1.17) such that $\text{Ker } \overline{P}_0 = \text{Ker } \widetilde{P}_0$. We shall show that they are equal. First, it follows from Proposition 2.2 that always

$$\text{Im } \overline{P}_k = \text{Im } \widetilde{P}_k, \quad k = 0, 1, \ldots, \tag{3.1}$$

and hence,

$$\overline{P}_0 = \widetilde{P}_0. \tag{3.2}$$

Let k be a natural number and x_k be an arbitrary vector in $\text{Ker } \overline{P}_k$. Set

$$z_k = (I_r - \widetilde{P}_k)x_k, \tag{3.3}$$

and define vectors $x_0, \ldots, x_{k-1}, z_0, \ldots, z_{k-1}$ via

$$z_j = (A_j|_{\text{Ker } \widetilde{P}_j})^{-1} B_{j+1} \cdots (A_{k-1}|_{\text{Ker } \widetilde{P}_{k-1}})^{-1} B_k z_k, \tag{3.4}$$

$$x_j = (A_j|_{\text{Ker } \overline{P}_j})^{-1} B_{j+1} \cdots (A_{k-1}|_{\text{Ker } \overline{P}_{k-1}})^{-1} B_k x_k, \tag{3.5}$$

for $j = 0, \ldots, k-1$. These definitions are legitimate because $z_k \in \text{Ker } \widetilde{P}_k$ and $x_k \in \text{Ker } \overline{P}_k$. Let

$$y_j = x_j - z_j, \quad j = 0, \ldots, k. \tag{3.6}$$

Then definitions (3.4) and (3.5) immediately lead to

$$A_j y_j = A_j(x_j - z_j) = B_{j+1}(x_{j+1} - z_{j+1}) = B_{j+1} y_{j+1}, \quad j = 0, \dots, k-1. \tag{3.7}$$

Now note that, by (3.3) and (3.6), we have

$$y_k = \widetilde{P}_k x_k \in \operatorname{Im} \widetilde{P}_k. \tag{3.8}$$

Hence, it follows from Proposition 2.2 that there exists a sequence of vectors $y_{k+1}, y_{k+2}, \dots$ such that

$$A_j y_j = B_{j+1} y_{j+1}, \quad j = k, k+1, \dots, \tag{3.9a}$$

$$\lim_{j \to +\infty} y_j = 0. \tag{3.9b}$$

Equalities (3.7) and (3.9a) lead to

$$B_{j+1} y_{j+1} = A_j y_j, \quad j = 0, 1, \dots \tag{3.10}$$

Again apply Proposition 2.2. It follows from these equalities and the limit in (3.9b) that

$$y_j \in \operatorname{Im} \widetilde{P}_j, \quad j = 0, 1, \dots. \tag{3.11}$$

Therefore, by (3.10), we have

$$y_j = (B_j|_{\operatorname{Im} \widetilde{P}_j})^{-1} A_{j-1} \cdots (B_1|_{\operatorname{Im} \widetilde{P}_1})^{-1} A_0 y_0, \quad j = 0, 1, \dots. \tag{3.12}$$

Finally, note that $y_0 = x_0 - z_0$, while $x_0 \in \operatorname{Ker} \overline{P}_0$, and $z_0 \in \operatorname{Ker} \widetilde{P}_0$. Since $\overline{P}_0 = \widetilde{P}_0$, by (3.2), it follows that $y_0 \in \operatorname{Ker} \widetilde{P}_0$. Combining this with (3.11) yields $y_0 = 0$. By (3.12), with $j = k$, we then have $y_k = 0$, and the equality in (3.8) leads to $x_k \in \operatorname{Ker} \widetilde{P}_k$. But x_k is an arbitrary vector in $\operatorname{Ker} \overline{P}_k$, and k is an arbitrary natural number. Hence,

$$\operatorname{Ker} \overline{P}_k \subset \operatorname{Ker} \widetilde{P}_k, \quad k = 0, 1, \dots.$$

Combining these inclusions with (3.1), we obtain that $\overline{P}_k = \widetilde{P}_k, \ k = 0, 1, \dots$, which concludes the proof of the uniqueness part.

Part (B). We now turn to the existence part of the proof. First we deal with the map Φ_0 defined in (1.19). Let $(P_n)_{n=0}^{\infty}$ be a dichotomy for the system (1.17). Since $L \oplus \text{Im } P_0 = \mathbb{C}^r$, we have

$$L \cap \text{Ker } (I_r - P_0) = L \cap \text{Im } P_0 = \{0\}, \quad \dim L = \dim \text{Ker } (P_0).$$

Therefore, the restricted mapping $(I_r - P_0)\big|_L : L \to \text{Ker } P_0$, is invertible. It follows that

$$\Phi_0 = P_0((I_r - P_0)\big|_L)^{-1} : \text{Ker } P_0 \to \text{Im } P_0, \tag{3.13}$$

is well defined, and

$$\text{Dom } (\Phi_0) = \text{Ker } P_0, \quad \text{Im } (\Phi_0) \subset \text{Im } P_0. \tag{3.14}$$

Moreover, for $x \in L$, we have $(I_r - P_0)x = ((I_r - P_0)\big|_L)x$. Hence

$$x = ((I_r - P_0)\big|_L)^{-1}(I_r - P_0)x,$$

and therefore $P_0 x = \Phi_0(I_r - P_0)x$. This leads to

$$x = P_0 x + (I_r - P_0)x = (I_r + \Phi_0)(I_r - P_0)x.$$

Consequently, $L \subset \text{Im } (I_r + \Phi_0)(I_r - P_0)$. But,

$$\dim L = \dim \text{Ker } P_0 = \dim \text{Im } (I_r - P_0) \geq \dim \text{Im } (I_r + \Phi_0)(I_r - P_0).$$

This implies that

$$L = \text{Im } (I_r + \Phi_0)(I_r - P_0). \tag{3.15}$$

Define a sequence of mappings

$$\Phi_n : \text{Ker } P_n \to \text{Im } P_n \quad (n = 1, 2, \ldots),$$

by setting

$$\begin{aligned} \Phi_n =& (B_n\big|_{\text{Im } P_n})^{-1} A_{n-1} \cdots (B_1\big|_{\text{Im } P_1})^{-1} A_0 \Phi_0 \times \\ & \times (A_0\big|_{\text{Ker } P_0})^{-1} B_1 \cdots (A_{n-1}\big|_{\text{Ker } P_{n-1}})^{-1} B_n. \end{aligned} \tag{3.16}$$

Note that Φ_n is well defined on Ker P_n by the properties (1.8) - (1.10) of the dichotomy and by (3.14). In addition, the dichotomy inequalities (1.11) and (1.12) lead to $\|\Phi_n\| \leq M^2 a^{2n} \|\Phi_0\| \quad (n = 1, 2, \ldots)$; in particular,

$$K := \sup_n \|\Phi_n\| < \infty. \tag{3.17}$$

Part (C). Let $(\overline{P}_n)_{n=0}^{\infty}$ be given by (1.18). We now show that $(\overline{P}_n)_{n=0}^{\infty}$ satisfies the requirements of the theorem. The proof is divided into four steps.

(C.I) Here we show that the sequence $(\overline{P}_n)_{n=0}^{\infty}$ is a bounded sequence of projections in $\mathbb{C}^r$ of constant rank p, where p is the rank of the dichotomy $(P_n)_{n=0}^{\infty}$, and that

$$L = \text{Ker } \overline{P}_0. \tag{3.18}$$

By (1.18), we have

$$\overline{P}_n = P_n - \Phi_n(I_r - P_n) \quad (n = 0, 1, \ldots). \tag{3.19}$$

From this formula, the boundedness of the dichotomy $(\overline{P}_n)_{n=0}^{\infty}$, and (3.17) it follows that

$$\sup_n \|\overline{P}_n\| < \infty. \tag{3.20}$$

Moreover, the inclusion Im $\Phi_n \subset$ Im $P_n \quad (n = 0, 1, \ldots)$, leads to

$$\begin{aligned} \overline{P}_n^2 &= P_n(P_n - \Phi_n(I_r - P_n)) - \Phi_n(I_r - P_n)(P_n - \Phi_n(I_r - P_n)) \\ &= P_n - \Phi_n(I_r - P_n) = \overline{P}_n. \end{aligned}$$

Hence $(\overline{P}_n)_{n=0}^{\infty}$ is a bounded sequence of projections. By (3.19), we have

$$I_r - \overline{P}_n = (I_r + \Phi_n)(I_r - P_n), \quad n = 0, 1, \ldots. \tag{3.21}$$

Equality (3.18) follows from this equality (with $n = 0$) and (3.15). Finally, it follows from (3.19) that $\overline{P}_n P_n = P_n \quad (n = 0, 1, \ldots)$. Hence, Im $P_n \subset$ Im $\overline{P}_n \quad (n = 0, 1, \ldots)$. On the other hand, (3.19) and the inclusion Im $\Phi_n \subset$ Im P_n, lead to Im $\overline{P}_n \subset$ Im $P_n \quad (n = 0, 1, \ldots)$. Thus,

$$\text{Im } \overline{P}_n = \text{Im } P_n, \quad n = 0, 1, \ldots. \tag{3.22}$$

Consequently Rank $\overline{P}_n = \text{Rank } P_n = p \quad (n = 0, 1, \ldots)$.

(C.II) Here we prove (1.9) for the sequence $(\overline{P}_n)_{n=0}^{\infty}$. It follows immediately from equality (3.22) and the first part of (1.9) that

$$\text{(3.23)} \qquad \text{Im } A_n\overline{P}_n \subset \text{Im } B_{n+1}\overline{P}_{n+1}, \quad n = 0, 1, \ldots .$$

We now prove that

$$\text{(3.24)} \qquad \text{Im } B_{n+1}(I_r - \overline{P}_{n+1}) \subset \text{Im } A_n(I_r - \overline{P}_n), \quad n = 0, 1, \ldots .$$

Taking into account (3.21) and (3.16), we have, for $n = 0, 1, \ldots$,

$$\begin{aligned} B_{n+1}(I_r - \overline{P}_{n+1}) &= B_{n+1}(I_r - P_{n+1}) + B_{n+1}\Phi_{n+1}(I_r - P_{n+1}) \\ &= B_{n+1}(I_r - P_{n+1}) + A_n\Phi_n(A_n|_{\text{Ker } P_n})^{-1}B_{n+1}(I_r - P_{n+1}). \end{aligned}$$

By the second part of (1.9), the last term is equal to

$$A_n(I_r + \Phi_n)(A_n|_{\text{Ker } P_n})^{-1}B_{n+1}(I_r - P_{n+1}).$$

However, $(A_n|_{\text{Ker } P_n})^{-1} = (I_r - P_n)(A_n|_{\text{Ker } P_n})^{-1}$. Hence, we finally obtain that

$$\begin{aligned} B_{n+1}(I_r - \overline{P}_{n+1}) &= A_n(I_r + \Phi_n)(I_r - P_n)(A_n|_{\text{Ker } P_n})^{-1}B_{n+1}(I_r - P_{n+1}) \\ &= A_n(I_r - \overline{P}_n)(A|_{\text{Ker } P_n})^{-1}B_{n+1}(I_r - P_{n+1}), \end{aligned}$$

where we made use of (3.21) again. The inclusion (3.24) now follows immediately.

(C.III) We now show that (1.10) holds for $(\overline{P}_n)_{n=0}^{\infty}$, and that, if $(P_n)_{n=0}^{\infty}$ is a normal dichotomy, then $(\overline{P}_n)_{n=0}^{\infty}$ satisfies condition (1.13). Let n be an integer, and x be an arbitrary vector in $\mathbb{C}^r$. By (1.10), there exists vectors $u \in \text{Im } P_{n+1}$ and $v \in \text{Ker } P_n$ such that

$$\text{(3.25)} \qquad x = B_{n+1}u + A_nv.$$

Since $\text{Im } \Phi_n \subset \text{Im } P_n$,we have

$$A_n\Phi_nv \in \text{Im } A_nP_n \subset \text{Im } B_{n+1}P_{n+1}.$$

Hence, $A_n\Phi_n v = B_{n+1}(B_{n+1}|_{\text{Im } P_{n+1}})^{-1}A_n\Phi_n v$. It follows from this equality and (3.25) that

$$x = B_{n+1}(u - (B_{n+1}|_{\text{Im } P_{n+1}})^{-1}A_n\Phi_n v) + A_n(I_r + \Phi_n)v. \tag{3.26}$$

Now note that, by (3.22) and $u = \text{Im } P_{n+1}$, we have

$$u - (B_{n+1}|_{\text{Im } P_{n+1}})^{-1}A_n\Phi_n v \in \text{Im } P_{n+1} = \text{Im } \overline{P}_{n+1},$$

while (3.21) and $v \in \text{Ker } P_n$ imply that

$$(I_r + \Phi_n)v = (I_r + \Phi_n)(I_r - P_n)v \in \text{Ker } \overline{P}_n.$$

Hence, (3.26) and the fact that x is arbitrary imply that

$$\text{Im } (B_{n+1}\overline{P}_{n+1}) + \text{Im } (A_n(I_r - \overline{P}_n)) = \mathbb{C}^r.$$

Assume now that $(P_n)_{n=0}^{\infty}$ is a normal dichotomy, and let

$$\gamma = \sup_{n=0,1,\ldots} \|(B_{n+1}|_{\text{Im } P_{n+1}}, A_n|_{\text{Ker } P_n})^{-1}\|.$$

Note that, by (3.25) and (3.26),

$$(B_{n+1}|_{\text{Im } P_{n+1}}, A_n|_{\text{Ker } P_n})^{-1}x = \begin{pmatrix} u \\ v \end{pmatrix},$$

$$(B_{n+1}|_{\text{Im } \overline{P}_{n+1}}, A_n|_{\text{Ker } \overline{P}_n})^{-1}x = \begin{pmatrix} u - (B_{n+1}|_{\text{Im } P_{n+1}})^{-1}A_n\Phi_n v \\ (I_r + \Phi_n)v \end{pmatrix},$$

where we used the notation of the previous paragraph. It follows from the dichotomy inequality (1.11), with $j = 1$, the inequality (3.17), and the inclusion $\text{Im } \Phi_n \subset \text{Im } P_n$ that

$$\|(B_{n+1}|_{\text{Im } P_{n+1}})^{-1}A_n\Phi_n v\| = \|(B_{n+1}|_{\text{Im } P_{n+1}})^{-1}A_n P_n\Phi_n v\| \leq MaK\|v\|,$$

as well as $\|(I_r + \Phi_n)v\| \leq (1 + k)\|v\|$. Therefore,

$$\begin{aligned}\|(B_{n+1}|_{\text{Im } \overline{P}_{n+1}}, A_n|_{\text{Ker } \overline{P}_n})^{-1}x\| &\leq \|u\| + (1 + K + MaK)\|v\| \\ &\leq (2 + K + MaK)\|\begin{pmatrix} u \\ v \end{pmatrix}\|.\end{aligned}$$

However,

$$\left\|\begin{pmatrix} u \\ v \end{pmatrix}\right\| = \|(B_{n+1}|_{\text{Im } P_{n+1}}, A_n|_{\text{Ker } P_n})^{-1}x\| \le \gamma\|x\|,$$

and thus,

$$\|(B_{n+1}|_{\text{Im } \overline{P}_{n+1}}, A_n|_{\text{Ker } \overline{P}_n})^{-1}x\| \le (2 + K + MaK)\gamma\|x\|.$$

Hence $(\overline{P}_n)_{n=0}^{\infty}$ satisfies (1.13).

(C.IV) We prove that the inequalities (1.11) and (1.12) hold for $(\overline{P}_n)_{n=0}^{\infty}$. First note that (3.22) implies that $\overline{P}_n = P_n\overline{P}_n$, and $(B_n|_{\text{Im } P_n})^{-1} = (B_n|_{\text{Im } \overline{P}_n})^{-1}$. Hence, (1.11) and the inequality (3.20) show that the inequalities (1.11) hold for $(\overline{P}_n)_{n=0}^{\infty}$ too.

We now turn to (1.12). Let n be a natural number, and $x \in \mathbb{C}^r$ be arbitrary. Put $y_n = (I_r - \overline{P}_n)x$ and

$$y_{n-k} = (A_{n-k}|_{\text{Ker } \overline{P}_{n-k}})^{-1}B_{n-k+1}\cdots(A_{n-1}|_{\text{Ker } \overline{P}_{n-1}})^{-1}B_n(I_r - \overline{P}_n)x, \tag{3.27}$$

for $k = 1, 2, \ldots, n$. Then $y_{n-k} \in \text{Ker } \overline{P}_{n-k}$ for $k = 0, \ldots, n$. Hence, by (3.21), there exist vectors $z_{n-k} \in \text{Ker } \overline{P}_{n-k}, k = 0, \ldots, n$, such that

$$y_{n-k} = (I_r + \Phi_{n-k})z_{n-k}, \quad k = 0, \ldots, n. \tag{3.28}$$

By (3.27), $A_{n-k}y_{n-k} = B_{n-k+1}y_{n-k+1}$, and therefore

$$A_{n-k}z_{n-k} - B_{n-k+1}z_{n-k+1} = B_{n-k+1}\Phi_{n-k+1}z_{n-k+1} - A_{n-k}\Phi_{n-k}z_{n-k}, \tag{3.29}$$

for $k = 1, \ldots, n$. However, $z_{n-k} \in \text{Ker } P_{n-k}$, and $\Phi_{n-k}z_{n-k} \in \text{Im } P_{n-k}$ for $k = 0, 1, \ldots, n$. Hence, (1.9) shows that the right hand side of (3.29) belongs to Im $B_{n-k+1}P_{n-k+1}$, while the left hand side belongs to Im $A_{n-k}(I_r - P_{n-k})$. From (1.8) and (1.10) it follows that both sides of (3.29) must vanish. In particular

$$A_{n-k}z_{n-k} = B_{n-k+1}z_{n-k+1}, \quad k = 1, \ldots, n.$$

Since $z_{n-k} \in \text{Ker } P_{n-k}$, this leads to

$$z_{n-k} = (A_{n-k}|_{\text{Ker } P_{n-k}})^{-1}B_{n-k+1}z_{n-k+1}, \quad k = 1, \ldots, n.$$

Iterating this equality yields

$$z_{n-j} = (A_{n-j}|_{\text{Ker } P_{n-j}})^{-1} B_{n-j+1} \cdots (A_{n-1}|_{\text{Ker } P_{n-1}})^{-1} B_n z_n, \quad j = 0, \dots, n.$$

Since $z_n \in \text{Ker } P_n$, it follows from this equality and (1.12) that

$$\|z_{n-j}\| \leq M a^j \|z_n\|, \quad j = 0, \dots, n. \tag{3.30}$$

On the other hand, $z_n \in \text{Ker } P_n$, and by (3.18), $y_n = (I_r + \Phi_n) z_n$. Since $\text{Im } \Phi_n \subset \text{Im } P_n$, this leads to $(I_r - P_n) y_n = (I_r - P_n)(I_r + \Phi_n) z_n = z_n$. However, $y_n = (I_r - \overline{P}_n) x$, and thus $z_n = (I_r - P_n)(I_r - \overline{P}_n) x$. Therefore,

$$\|z_n\| \leq M_1 \|x\|, \tag{3.31}$$

where $M_1 = \sup_n \|I_r - P_n\| \|(I_r - \overline{P}_n\| < \infty$. Finally, it follows from (3.17) and (3.28), that

$$\|y_{n-j}\| \leq (1+K) \|z_{n-j}\|, \quad j = 0, \dots, n. \tag{3.32}$$

By combining (3.27) with the inequalities (3.30) - (3.32), we obtain

$$\|(A_{n-j}|_{\text{Ker } \overline{P}_{n-j}})^{-1} B_{n-j+1} \cdots (A_{n-1}|_{\text{Ker } \overline{P}_{n-1}})^{-1} B_n (I_r - \overline{P}_n) x\| \leq$$

$$\leq (1+K) M a^j M_1 \|x\|, \quad j = 0, \dots, n.$$

Since n and x are arbitrary, this proves that (1.12) holds for $(\overline{P}_n)_{n=0}^{\infty}$, with M being replaced by $(1+K)MM_1$. It follows from (C.I) - (C.IV) above that $(\overline{P}_n)_{n=0}^{\infty}$ is a dichotomy for (1.17). This proves the first part of the theorem.

Part (D). We prove the remaining statements of the theorem. Let $(\widetilde{P}_n)_{n=0}^{\infty}$ be an arbitrary dichotomy of the system (1.17). By Proposition 2.2, $\text{Im } \widetilde{P}_0 = \text{Im } P_0$. Put $L = \text{Ker } \widetilde{P}_0$. Then $L \oplus \text{Im } P_0 = \mathbb{C}^r$, because $\text{Ker } \widetilde{P}_0 \oplus \text{Im } \widetilde{P}_0 = \mathbb{C}^r$. Let $(\overline{P}_n)_{n=0}^{\infty}$ be constructed as above with this subspace L. By construction, $\text{Ker } \overline{P}_0 = L = \text{Ker } \widetilde{P}_0$. It now follows from the uniqueness part of the proof that $(\overline{P}_n)_{n=0}^{\infty} = (\widetilde{P}_n)_{n=0}^{\infty}$. Hence, each dichotomy can be obtained via (1.18).

Finally, assume that (1.17) admits a normal dichotomy, $(P_n)_{n=0}^{\infty}$ say. By the previous paragraph, each dichotomy $(\overline{P}_n)_{n=0}^{\infty}$ of the system (1.17), can be obtained via the formula (1.18), with a suitable L. By part (C.III) of the proof, the dichotomy $(\overline{P}_n)_{n=0}^{\infty}$ is normal. Hence all the dichotomies of the system (1.17) are normal. $\diamond$

4. PERTURBATION OF SYSTEMS

In this section we prove Theorem 1.3 and present some additional results of independent interest. The following lemma will be used in the proof of Theorem 1.3.

LEMMA 4.1. *Let* $G = (-\delta_{ij}A_j + \delta_{i+1,j}B_j)_{ij=0}^{\infty}$ *be a bounded operator on* ℓ_r^2, *where* $(A_n)_{n=0}^{\infty}$ *and* $(B_{n+1})_{n=0}^{\infty}$ *are two bounded sequences of* $r \times r$ *matrices. If the system* $B_{n+1}x_{n+1} = A_n x_n$, $n = 0, 1, \ldots$, *admits a dichotomy* $(P_n)_{n=0}^{\infty}$, *then* Ker G = Im Y, *where* $Y : \mathbb{C}^r \to \ell_r^2$ *is the bounded operator given by*

$$Yx = ((B_j|_{\mathrm{Im}\ P_j})^{-1}A_{j-1}\cdots(B_1|_{\mathrm{Im}\ P_1})^{-1}A_0P_0x)_{j=0}^{\infty}, \quad x \in \mathbb{C}^r.$$

Note that $Y|_{\mathrm{Ker}\ P_0} = 0$. Hence, Im Y = Im $(Y|_{\mathrm{Im}\ P_0})$. Furthermore, Proposition 2.2 shows that Im P_0 is independent of the dichotomy. Thus, also $Y|_{\mathrm{Im}\ P_0}$ is independent of the dichotomy. In fact, if $x \in \mathrm{Im}\ P_0$, then

$$Yx = ((B_j|_{\mathrm{Im}\ P_j})^{-1}A_{j-1}\cdots(B_1|_{\mathrm{Im}\ P_1})^{-1}A_0x)_{j=0}^{\infty}.$$

Furthermore, note that the first entry of Yx is P_0x.

PROOF OF LEMMA 4.1. Assume that $y = (y_j)_{j=0}^{\infty} \in$ Ker G. Then $B_{n+1}y_{n+1} = A_n y_n$, $n = 0, 1, \ldots$, and, in addition, $\sum_{n=0}^{\infty} \|y_n\|^2 < \infty$ which implies that $y_n \to 0$ $(n \to \infty)$. By Proposition 2.2, we have $y_n \in \mathrm{Im}\ P_n$, $n = 0, 1, \ldots$. Hence, the equalities $B_{n+1}y_{n+1} = A_n y_n$, $n = 0, 1 \ldots$, lead to $Y_{n+1} = (B_{n+1}|_{\mathrm{Im}\ P_{n+1}})^{-1}A_n y_n$, $n = 0, 1, \ldots$. Iterating these equalities, we obtain that

$$y_n = (B_n|_{\mathrm{Im}\ P_n})^{-1}A_{n-1}\cdots(B_1|_{\mathrm{Im}\ P_1})^{-1}A_0P_0y_0, \ n = 0, 1, \ldots,$$

where we used $y_0 = P_0y_0$. Thus, $y = (y_n)_{n=0}^{\infty} = Yy_0 \in \mathrm{Im}\ Y$.

Conversely, the dichotomy inequalities in (1.11) show that Y is well defined and bounded, and it is clear that $Yx \in$ Ker G $(x \in \mathbb{C}^n)$. $\diamond$

In [GKvS] a formula for the dichotomy of operators with semi-separable kernels is proved. Here we prove a perturbation formula for the dichotomy, which is connected to the latter. The setting is the following. Let $(A_n)_{n=0}^{\infty}$, $(B_{n+1})_{n=0}^{\infty}$, $(\overline{A}_n)_{n=0}^{\infty}$ and $(\overline{B}_{n+1})_{n=0}^{\infty}$ be bounded sequences of complex $r\times r$ matrices. Assume that the difference equations

$$(4.1)\qquad \begin{cases} B_{n+1}x_{n+1} = A_n x_n + f_n, \; n = 0, 1, \ldots, \\ x_0 \in L, \end{cases}$$

and

$$(4.2)\qquad \begin{cases} \overline{B}_{n+1}x_{n+1} = \overline{A}_n x_n + f_n, \; n = 0, 1, \ldots, \\ x_0 \in L, \end{cases}$$

where L is a subspace of $\mathbb{C}^r$, are uniquely solvable. Consider the respective associate operators $T = (-\delta_{ij}A_j + \delta_{i+1,j}B_j)_{i,j=0}^{\infty}$ and $\overline{T} = (-\delta_{ij}\overline{A}_j + \delta_{i+1,j}\overline{B}_j)_{i,j=0}^{\infty}$, which act from $\ell_{r,L}^2$ into ℓ_r^2. Then T and $\overline{T}$ are invertible. By Theorem 1.1, the systems

$$(4.3)\qquad B_{n+1}x_{n+1} = A_n x_n, \quad n = 0, 1, \ldots,$$

$$(4.4)\qquad \overline{B}_{n+1}x_{n+1} = \overline{A}_n x_n, \quad n = 0, 1, \ldots,$$

admit normal dichotomies $(P_n)_{n=0}^{\infty}$ and $(\overline{P}_n)_{n=0}^{\infty}$, respectively, with $\operatorname{Ker} P_0 = \operatorname{Ker} \overline{P}_0 = L$. In the statement below, and in the sequel, we denote by $Q_0 : \ell^2 \to \mathbb{C}^r$ the first coordinate projection, i.e., $Q_0((x_n)_{n=0}^{\infty}) = x_0$.

THEOREM 4.2. *With the above notation, the following equality holds*

$$(4.5)\qquad P_0 - \overline{P}_0 = Q_0 \overline{T}^{-1} D Y.$$

Here $Y : \mathbb{C}^r \to \ell_r^2$ *is given by*

$$(4.6)\qquad Yx = ((B_j|_{\operatorname{Im} P_j})^{-1}A_{j-1}\cdots(B_1|_{\operatorname{Im} P_1})^{-1}A_0P_0x)_{j=0}^{\infty}, \; x \in \mathbb{C}^r,$$

and D *is the following operator*

$$(4.7)\qquad D = (-\delta_{ij}(\overline{A}_j - A_j) + \delta_{i+1,j}(\overline{B}_j - B_j))_{i,j=0}^{\infty} : \ell_r^2 \to \ell_r^2.$$

Moreover, if $D = BC$, where B and C are bounded operators in ℓ_r^2, then

$$P_0 - \overline{P}_0 = Q_0 T^{-1} B(I + CT^{-1}B)^{-1} CY, \tag{4.8}$$

where I is the identity operator in ℓ_r^2.

PROOF Define $\overline{Y} : \mathbb{C}^r \to \ell_r^2$ via

$$\overline{Y}x = ((\overline{B}_j|_{\text{Im } \overline{P}_j})^{-1}\overline{A}_{j-1} \cdots (\overline{B}_1|_{\text{Im } \overline{P}_1})^{-1}\overline{A}_0\overline{P}_0 x)_{n=0}^{\infty}, \ x \in \mathbb{C}^r,$$

and introduce operators

$$G = (-\delta_{ij}A_j + \delta_{i+1,j}B_j)_{i,j=0}^{\infty} : \ell_r^2 \to \ell_r^2,$$

$$\overline{G} = (-\delta_{ij}\overline{A}_j + \delta_{i+1,j}\overline{B}_j)_{i,j=0}^{\infty} : \ell_r^2 \to \ell_r^2.$$

By Lemma 4.1 we have

$$GY = \overline{G}\overline{Y} = 0. \tag{4.9}$$

Moreover, from the definitions of Y and $\overline{Y}$ it is clear that

$$Q_0 Y = P_0, \quad Q_0 \overline{Y} = \overline{P}_0. \tag{4.10}$$

Since Ker $P_0 = L =$ Ker $\overline{P}_0$, we have $\overline{P}_0(I - P_0) = 0$, and hence $P_0 - \overline{P}_0 = (I_r - \overline{P}_0)P_0$. The latter equality and (4.10) imply that

$$\text{Im } Q_0(Y - \overline{Y}) = \text{Im } (P_0 - \overline{P}_0) \subset \text{Ker } \overline{P}_0 = L.$$

Therefore, Im $(Y - \overline{Y}) \subset \ell_{r,L}^2$. Now, note that $D = \overline{G} - G$, hence, by using (4.9), we obtain

$$\overline{T}(Y - \overline{Y}) = \overline{G}(Y - \overline{Y}) = \overline{G}Y = (G + D)Y = DY.$$

But $\overline{T}$ is invertible. So $Y - \overline{Y} = \overline{T}^{-1}DY$, which, by (4.10), yields

$$P - \overline{P}_0 = Q_0(Y - \overline{Y}) = Q_0\overline{T}^{-1}DY,$$

and thus (4.5) is proved.

We now prove (4.8). Write C_0 for the restriction of C to $\ell^2_{r,L}$. Note that

$$\overline{T} = T + BC_0 = (I + BC_0T^{-1})T,$$

where I is the identity operator on ℓ^2_r. The operators B and C_0T^{-1} act on ℓ^2_r. Since $\overline{T}$ and T are invertible, the same holds true for $I + BC_0T^{-1}$. A standard computation shows that

$$(I + BC_0T^{-1})^{-1} = I - B(I + C_0T^{-1}B)^{-1}C_0T^{-1},$$

and hence

$$\begin{aligned}\overline{T}^{-1}D &= (T + BC_0)^{-1}BC = [(I + BC_0T^{-1})T]^{-1}BC \\ &= T^{-1}\{I - B(I + C_0T^{-1}B)^{-1}C_0T^{-1}\}BC \\ &= T^{-1}\{B - B(I + C_0T^{-1}B)^{-1}C_0T^{-1}B\}C \\ &= T^{-1}B(I + C_0T^{-1}B)^{-1}C.\end{aligned}$$

As $C_0T^{-1} = CT^{-1}$, this calculation and (4.5) yield the expression (4.8). ◇

In another form, for the case when the matrices A_n and B_{n+1} are invertible for all $n \geq 0$, formula (4.8) also appears in [GKvS].

We conclude this section with the proof of Theorem 1.3.

PROOF OF THEOREM 1.3. Since the difference equations (1.20) and (1.21) are uniquely solvable, the operators T_1 and T_2 are invertible. By Theorem 1.1, the system $B_{n+1}x_{n+1} = A_nx_n$, $n = 0, 1, \ldots$, admit a unique normal dichotomy $(\widetilde{P}_n)_{n=0}^{\infty}$ such that $L_2 = \text{Ker } \widetilde{P}_0$. Hence $L_2 \oplus \text{Im } \widetilde{P}_0 = \mathbb{C}^r$. According to Proposition 2.2, the equality $\text{Im } \widetilde{P}_0 = \text{Im } P_0$ holds. Thus $L_2 \oplus \text{Im } P_0 = \mathbb{C}^r$. It follows that $\overline{P}$ is well defined.

Let Y be the operator defined in Lemma 4.1. Since $\text{Im } \overline{P}_0 = \text{Im } P_0$, we have $\overline{P} = P_0\overline{P}$. Hence $Z = Y\overline{P}$, where Z is given by (1.23).

Let G be the operator on ℓ^2_r given by $G = (-\delta_{ij}A_j + \delta_{i+1,j}B_j)_{ij=0}^{\infty}$. By Lemma 4.1, we have

$$\text{Ker } G = \text{Im } Y. \tag{4.11}$$

Furthermore, it is clear that

$$Q_0Y = P_0, \quad YP_0 = Y. \tag{4.12}$$

Hence,

$$YQ_0Y = Y. \tag{4.13}$$

Let $x \in \ell_r^2$ be arbitrary. Since $G|_{\ell^2_{r,L_1}} = T_1$ and $G|_{\ell^2_{r,L_2}} = T_2$, we have

$$G(T_2^{-1} - T_1^{-1})x = T_2T_2^{-1}x - T_1T_1^{-1}x = 0.$$

By (4.11), this implies that $(T_2^{-1} - T_1^{-1})x = Yu$, for some $u \in \mathbb{C}^r$. Taking into account (4.13) it follows that

$$(T_2^{-1} - T_1^{-1})x = YQ_0Yu = YQ_0(T_2^{-1} - T_1^{-1})x. \tag{4.14}$$

Now, note that $Q_0(T_2^{-1} - T_1^{-1})x = Q_0Yu \in \text{Im } P_0$, by (4.12). But Im P_0 = Im $\overline{P}$ and hence,

$$Q_0(T_2^{-1} - T_1^{-1})x = \overline{P}Q_0(T_2^{-1} - T_1^{-1})x. \tag{4.15}$$

Moreover, $T_2^{-1}x \in \ell^2_{r,L_2}$. Hence, $Q_0T_2^{-1}x \in L_2 = \text{Ker } \overline{P}$, and therefore, $\overline{P}Q_0T_2^{-1}x = 0$. This equality and (4.15) lead to $Q_0(T_2^{-1} - T_1^{-1})x = -\overline{P}Q_0T_1^{-1}x$. By inserting this in (4.14), we obtain

$$(T_2^{-1} - T_1^{-1})x = -Y\overline{P}Q_0T_1^{-1}x = -ZQ_0T_1^{-1}x,$$

where the second equality follows from $Y\overline{P} = Z$. Since $x \in \ell_r^2$ is arbitrary, this proves (1.22). ◇

REFERENCES

[BG1] A. Ben-Artzi and I. Gohberg, Fredholm properties of band matrices and dichotomy, Topics in Operator Theory, Constantin Apostal Memorial Issue, OT 32, Birkhäuser Verlag, 1988; pp. 37-52.

[BG2] A. Ben-Artzi and I. Gohberg, Band matrices and dichotomy, Operator Theory and its Applications, Proceedings Rotterdam Workshop, to appear.

[CS] C.V. Coffman and J.J. Schäffer, Dichotomies for linear difference equations, Math. Ann. 172 (1967), 139-166.

[GF] I. Gohberg and I.A. Feldman, Convolution equations and projection methods for their solutions, Transl. Math. Monographs, Vol.41, Amer. Math. Soc., Providence, R.I., 1974.

[GK] I. Gohberg and M.A. Kaashoek, Block Toeplitz operators with rational symbols, Operator Theory and its Applications, Vol. 35, Birkhäuser Verlag, 1988; pp. 385-440.

[GKvS] I. Gohberg, M.A. Kaashoek and F. van Schagen, Non-compact integral operators with semi-separable kernels and their discrete analogous: inversion and Fredholm properties, Integral Equations and Operator Theory, 7 (1984), 642-703.

A. Ben-Artzi
Department of Mathematics
University of California at San Diego
La Jolla, Calif. 92093, USA

I. Gohberg
Raymond and Beverly Sackler Faculty of Exact Science
School of Mathematical Sciences
Tel-Aviv University, Ramat-Aviv, Israel

M.A. Kaashoek
Department of Mathematics and Computer Science
Vrije Universiteit

Amsterdam, The Netherlands

Operator Theory:
Advances and Applications, Vol. 48

ON THE ITERATES OF A COMPLETELY NONUNITARY CONTRACTION

Hari Bercovici*

Dedicated to the memory of Professor Ernst Hellinger

Let T be a contraction, i.e., $\|T\| \leq 1$, on a Hilbert space, and let A denote the disc algebra. Thus A consists of all continuous functions on $\overline{\mathbf{D}} = \{\lambda : |\lambda| \leq 1\}$ which are analytic in the open unit disc $\overline{\mathbf{D}}$. The von Neumann inequality implies that one can define a functional calculus $f \to f(T), f \in A$, such that $\|f(T)\| \leq \|f\|_\infty$.

Esterle, Strouse and Zouakia proved in [2] that the following assertions are equivalent for $f \in A$:

(1) $\lim_{n\to\infty} \|f(T)T^n\| = 0$; and

(2) $f|\sigma(T) \cap \mathbf{T} = 0$,

where $\mathbf{T} = \partial\mathbf{D}$ denotes the unit circle. In this note we will extend this result to the Sz.-Nagy—Foiaş functional calculus for the case in which T is completely nonunitary. Recall that T is completely nonunitary if there is no invariant subspace on which T acts like a unitary operator. If T is a completely nonunitary contraction then the Sz.-Nagy—Foiaş functional calculus (cf. Chapter III of [6]) associates an operator $f(T)$ with each f in the algebra H^∞ of bounded analytic functions on $\mathbf{D}$. Unlike the result of [2], our theorem does not give equivalent conditions, and we will see below

* The author was supported in part by grants from the National Science Foundation.

that it is very difficult to find equivalent conditions for (1) in our context.

For the remainder of this article T will be a fixed, completely nonunitary, contraction on the Hilbert space H. We will denote by M_∞ the maximal ideal space of H^∞; thus an element $\varphi \in M_\infty$ is a nonzero homomorphism $\varphi : H^\infty \to \mathbf{C}$. The extended spectrum of T is a closed subset $\sigma_{\text{ext}}(T) \subset M_\infty$ defined by Foiaş and Mlak [4] by

$$\sigma_{\text{ext}}(T) = \{\varphi \in M_\infty : \varphi(f) \in \sigma(f(T)) \text{ for all } f \in H^\infty\}.$$

It was shown in [4] that in fact

$$\sigma(f(T)) = \{\varphi(f) : \varphi \in \sigma_{\text{ext}}(T)\}$$

for every $f \in H^\infty$. Let us recall that $\mathbf{D}$ can be regarded as a subset of M_∞ (a point λ in $\mathbf{D}$ corresponds with point evaluation at λ in M_∞). We have $\sigma(T) \cap \mathbf{D} = \sigma_{\text{ext}}(T) \cap \mathbf{D}$. If $f \in H^\infty$ and $\zeta \in \mathbf{T}$ we denote by $f(\zeta)$ the radial limit of f at ζ; $f(\zeta)$ exists almost everywhere and $\|f\|_\infty = \operatorname{ess\,sup}\{|f(\zeta)| : \zeta \in \mathbf{T}\}$ (cf. [3]).

(3) THEOREM. *Let T be a completely nonunitary contraction, let $f \in H^\infty$, and consider the following assertions.*

(a) $\lim_{\lambda \to \zeta, \lambda \in \mathbf{D}} f(\lambda) = 0$ for every $\zeta \in \sigma(T) \setminus \mathbf{D}$.

(b) $\lim_{n \to \infty} \|f(T)T^n\| = 0$.

(c) $\varphi(f) = 0$ for every $\varphi \in \sigma_{\text{ext}}(T) \setminus \mathbf{D}$.

Then (a) implies (b) and (b) implies (c), but the converse implications are not true.

Proof. Assume first that (a) holds. As in [2], we introduce a new norm on H^∞ by setting

$$\|u\|' = \lim_{n \to \infty} \|u(T)T^n\|, \; u \in H^\infty.$$

Let us denote by B the completion of H^∞ in this norm, and let $\Phi : H^\infty \to B$ be the canonical 'inclusion' (which is not generally one-to-one). If $\chi \in H^\infty$ denotes the identity function of $\mathbf{D}$, then clearly $\|\chi u\|' = \|u\|'$, and it follows from [2] that $\Phi(\chi)$ is invertible in B. Let us denote by C the algebra of continuous functions on $\mathbf{T}$. We claim that Φ can be extended to a continuous homomorphism $\Phi' : H^\infty + C \to B$. Indeed, if $n \geq 1, u \in H^\infty$, and $v(\zeta) = \zeta^{-n}u(\zeta), \zeta \in \mathbf{T}$, we can set $\Phi'(v) = \Phi(\chi)^{-n}\Phi(u)$, and we have

$$\|\Phi'(v)\|' = \|\Phi(u)\|' \leq \|u\|_\infty = \|v\|_\infty.$$

Since $H^\infty + C$ is the uniform closure of all functions v of this form (cf. [5] or [3]), the existence of Φ' is established.

We are now ready to prove (b). What we have to do is show that $f \in \ker(\Phi)$. It follows from [2] that $\ker(\Phi')$ contains every function $g \in C$ such that $g|\sigma(T) \setminus \mathbf{D} = 0$. Fix $\varepsilon > 0$ and note that (a) implies that the set

$$\sigma_\varepsilon = \{\zeta \in \mathbf{T} : \limsup_{\lambda \to \zeta, \lambda \in \mathbf{D}} |f(\lambda)| < \varepsilon\}$$

is a neighborhood (in $\mathbf{T}$) of $\sigma(T) \setminus \mathbf{D}$. Thus there exists a function $g \in C$ such that $g|\sigma(T) \setminus \mathbf{D} = 0$ and $g|\mathbf{T} \setminus \sigma_\varepsilon = 1$. Clearly for such a g we have $\|f - gf\|_\infty \leq \varepsilon$, and since $\Phi'(g) = 0$,

$$\begin{aligned}\|\Phi(f)\|' = \|\Phi'(f)\|' &= \|\Phi'(f) - \Phi'(g)\Phi'(f)\|' \\ &= \|\Phi'(f - gf)\|' \leq \|f - gf\|_\infty \leq \varepsilon.\end{aligned}$$

Since $\varepsilon > 0$ is arbitrary, we have $\Phi(f) = 0$, as desired.

Assume now that (b) holds and $\varphi \in \sigma_{\text{ext}}(T) \setminus \mathbf{D}$. Then we have $|\varphi(\chi)| = 1$ and $\varphi(f\chi^n) \in \sigma(f(T)T^n)$ so that

$$\begin{aligned}|\varphi(f)| = |\varphi(f\chi^n)| &= \lim_{n \to \infty} |\varphi(f\chi^n)| \\ &\leq \lim_{n \to \infty} \|f(T)T^n\| = 0,\end{aligned}$$

and (c) follows at once.

In order to show that the converse implications do not hold, consider the singular inner function $\theta(\lambda) = \exp((\lambda + 1)/(\lambda - 1))$ and the operator T defined as follows. Let H^2 denote the Hardy space of square summable Taylor series, and let S denote the shift on H^2:

$$(Sf)(\lambda) = \lambda f(\lambda), \lambda \in \mathbf{D}, f \in H^2.$$

Then set $H = H^2 \ominus \theta H^2$ and $T = P_H S|H$. The following facts about T are well known. The spectrum of T reduces to $\{1\}$, $\theta(T) = 0$, and $\theta^{1/2}(T)$ is a partial isometry, where $\theta^{1/2}(\lambda) = \exp((\lambda + 1)/(2(\lambda - 1)))$ (see [6] and [1]). It is clear that θ does not extend continuously to $\zeta = 1$, but we have $\theta(T)T^n = 0$. Thus (b) does not imply (a). On the other hand, consider the function $f = \theta^{1/2}$. If $\varphi \in \sigma_{\text{ext}}(T)$ then $\varphi(\theta) \in \sigma(\theta(T)) = \{0\}$ so that

$\varphi(\theta) = \varphi(f^2) = 0$. Therefore $\varphi(f) = 0$ and we see that f satisfies (c). However, because $f(T)$ is a partial isometry, we have

$$\begin{aligned}\|f(T)T^n\| &= \|T^n|\text{ran}(f(T))\| \\ &= \|(T|\text{ran}(f(T)))^n\| = 1\end{aligned}$$

because $\sigma(T|\text{ran}(f(T))) = \{1\}$. Thus (c) does not imply (b). Q.E.D.

(4) REMARK. Note that (a) means that the cluster set of f at every $\zeta \in \sigma(T) \setminus \mathbf{D}$ is $\{0\}$, while (c) implies only that the cluster set at such points ζ contains zero. If $f \in A$ then these two conditions are obviously equivalent. The example given above concerning θ and $\theta^{1/2}$ shows that a characterization in terms of boundary values of functions satisfying (b) is probably impossible.

REFERENCES

1. H. Bercovici, Operator theory and arithmetic in H^∞, Math. Monographs and Surveys, No. 26, Amer. Math. Soc., Providence, Rhode Island, 1988.

2. J. Esterle, E. Strouse, and F. Zouakia, Theorems of Katznelson-Tzafriri type for contractions, preprint.

3. R.G. Douglas, Banach algebra techniques in operator theory, Academic press, New York, 1972.

4. C. Foiaş and W. Mlak, The extended spectrum of completely nonunitary contractions and the spectral mapping theorem, Studia Math. **26**(1966), 239-245.

5. D. Sarason, Generalized interpolation in H^∞, Trans. Amer. Math. Soc. **127**(1967), 179-203.

6. B. Sz.-Nagy and C. Foiaş, Harmonic analysis of operators on Hilbert space, North Holland, Amsterdam, 1970.

Department of Mathematics
Indiana University
Bloomington, IN 47405
U.S.A.

Operator Theory:
Advances and Applications, Vol. 48

A GOHBERG-KRUPNIK-SARASON SYMBOL CALCULUS FOR ALGEBRAS OF TOEPLITZ, HANKEL, CAUCHY, AND CARLEMAN OPERATORS

A. Böttcher, S. Roch, B. Silbermann, I.M. Spitkovskii

To the memory of Ernst Hellinger

The subject of this paper is operator algebras on weighted L^p spaces generated by PQC multiplications, the Cauchy singular integral operator, and a Carleman shift. Such algebras contain copies of the algebra generated by all Toeplitz operators and also of the algebra which is generated by all Toeplitz and Hankel operators with PQC symbols. We provide a symbol calculus for the Calkin and Allan-Douglas images of the aforementioned algebras using extensions of the Two Projections Theorem, and vice versa, we show how the operator algebras studied here motivate and manufacture several versions of the Two Projections Theorem.

1. INTRODUCTION

This paper unites the results of [21] and [6] and extends those of [25]. The latter paper is devoted to the description of the C^*-algebra generated by all Toeplitz and Hankel operators on H^2 with piecewise quasicontinuous symbols. The article [21] contains a new approach to various extensions of the symbol calculus of Gohberg and Krupnik [9],[10], Karapetyants and Samko [12], Vasilevskii and M.V. Shapiro [26], and Costabel [7] for operator algebras on weighted L^p spaces generated by singular integral operators with piecewise continuous coefficients and a Carleman shift. In [6] the Fredholm criterion of Sarason [24] for Toeplitz operators on H^2 with piecewise quasicontinuous symbols was generalized to operators acting on H^p.

The approach of [21] is based upon consistent application of the local principle of Allan and Douglas and upon

Banach algebra versions of the Two Projections Theorem. A crucial point of this construction is the identification of the spectrum of pqp, where p and q are certain well-chosen idempotents. In the context of singular integral operators one may interprete pqp as a local Toeplitz operator. The spectra of local Toeplitz operators with piecewise continuous symbols are known to be certain circular arcs (which may degenerate to single points). Only in [6] it was shown that local Toeplitz operators generated by piecewise quasicontinuous functions have spectra of the same kind. Once the latter result is available, the machinery developed in [21] can be used to establish a symbol calculus for singular integral operators with piecewise quasicontinuous coefficients and a Carleman shift on weighted L^p spaces. To carry out this program is the purpose of the present paper.

2. PRELIMINARIES

2.1. Weighted L^p spaces. The spaces we let the operators act on are certain weighted L^p spaces $L^p(\varrho)$ on the complex unit circle $\mathbb{T}$. Throughout what follows we assume that $1 < p < \infty$ and that q is defined by $1/p + 1/q = 1$. The weight ϱ is specified to be what we call a Khvedelidze weight, i.e. we always suppose that

$$\varrho(t) = \prod_{j=1}^{m} |t - t_j|^{\mu_j} \qquad (t \in \mathbb{T}),$$

where $t_1, \ldots, t_m$ are pairwise distinct points on $\mathbb{T}$ and $\mu_1, \ldots, \mu_m$ are real numbers subject to the condition $-1/p < \mu_j < 1/q$. The norm in $L^p(\varrho)$ is given by

$$\|f\| = \left(\int_{\mathbb{T}} |f(t)\, \varrho(t)|^p \, dm \right)^{1/p},$$

dm referring to Lebesgue measure on $\mathbb{T}$. In case $\varrho(t) \equiv 1$, we abbreviate $L^p(\varrho)$ to L^p.

The algebra of all bounded linear operators on $L^p(\varrho)$ will be denoted by $\mathcal{L}(L^p(\varrho))$. We let $\mathcal{K}(L^p(\varrho))$ stand for the ideal of all compact operators on $L^p(\varrho)$. The Calkin algebra $\mathcal{L}^\pi(L^p(\varrho))$ is defined as the quotient algebra

$$\mathcal{L}(L^p(\varrho))/\mathcal{K}(L^p(\varrho)).$$

2.2. The Cauchy singular integral. If $L^p(\varrho)$ is as in the previous subsection, then the singular integral operator

$$(S\varphi)(t) = \frac{1}{\pi i}\int_{\mathbb{T}} \frac{\varphi(\tau)}{\tau - t}\, d\tau \qquad (t \in \mathbb{T}),$$

the integral understood in the principal value sense, is bounded on $L^p(\varrho)$. Because $S^2 = I$, the operators defined by $P = (I+S)/2$ and $Q = (I-S)/2$ are (complementary) projections. The operator P is usually referred to as the Riesz projection on $L^p(\varrho)$.

2.3. Carleman shifts. A homeomorphism α of $\mathbb{T}$ onto itself will be called a Lyapunov homeomorphism if it is not the identity map and if it possesses a Hölder continuous derivative α' on $\mathbb{T}$ such that $\alpha'(t) \neq 0$ for all $t \in \mathbb{T}$. If $\alpha: \mathbb{T} \to \mathbb{T}$ is a Lyapunov homeomorphism, then

$$\int_{\mathbb{T}} |f(\alpha(t))\varrho(t)|^p\, dm = \int_{\mathbb{T}} |f(t)\varrho(\beta(t))|^p\, |\beta'(t)|\, dm,$$

β denoting the inverse of α, and hence the composition operator C_α defined by

$$(C_\alpha f)(t) = f(\alpha(t)) \qquad (t \in \mathbb{T})$$

is bounded on $L^p(\varrho)$ whenever there exists a constant M such that $|\varrho(t)| \leqq M\,|\varrho(\alpha(t))|$ for all $t \in \mathbb{T}$. Note that C_α is always bounded on L^p without weight.

Let $\alpha: \mathbb{T} \to \mathbb{T}$ be a Lyapunov homeomorphism. If $C_\alpha^n = I$ for some integer $n \geqq 2$ (equivalently, if the n-th iterate α_n of α is the identity map) then C_α is called a Carleman shift. The least integer $n \geqq 2$ for which $C_\alpha^n = I$ is referred to as the order of the Carleman shift C_α. The homeomorphism α either preserves or changes the orientation of $\mathbb{T}$. In the former case C_α is said to be a forward shift, in the latter case C_α is called a backward shift. There exist forward Carleman shifts of arbitrary orders: for instance, if $\alpha(t) = e^{2\pi i/n}t$ then the order of C_α is just n. On the other hand, backward Carleman shifts are always of order 2. The archetypal example of a backward Carleman shift is the reflection operator R defined by $(R\varphi)(t) = \varphi(\bar{t})$ for $t \in \mathbb{T}$; here $\bar{t} = 1/t$ is the complex conjugate of t. One can show (see e.g. [14, Lemma 19.3]) that if C_α is any backward Carleman shift, then there exists a Lyapunov homeomorphism $\gamma: \mathbb{T} \to \mathbb{T}$ preserving the orientation of $\mathbb{T}$ such that $C_\alpha = C_\gamma^{-1} R C_\gamma$. The operator J resulting from R by post-multiplication by $\bar{t}$,

$$(Jf)(t) = \bar{t}\, f(\bar{t}) \qquad (t \in \mathbb{T}),$$

is frequently referred to as the flip operator.

We remark that if C_α is a forward Carleman shift, then α has no fixed points on $\mathbb{T}$, while in the case where C_α is a backward Carleman shift the mapping α perpetually owns exactly two fixed points on $\mathbb{T}$ (see [16, §2]).

2.4. Multiplication operators. Let L^∞ be the Banach algebra of all measurable and essentially bounded functions on $\mathbb{T}$. The multiplication operator M_a induced by a function $a \in L^\infty$ is the bounded operator acting on $L^p(\varrho)$ by the rule

$$(M_a f)(t) = a(t)\, f(t) \qquad (t \in \mathbb{T}).$$

Sometimes it will be convenient to write simply a in place of M_a.

2.5. PQC. We denote by C and PC the C^*-subalgebras of L^∞ constituted by all continuous and piecewise continuous functions, respectively. Let H^∞ (resp. $\overline{H^\infty}$) stand for the algebra of all functions in L^∞ whose negative (resp. positive) Fourier coefficients vanish. The C^*-algebra QC of all quasicontinuous functions is defined as the intersection $(C+H^\infty) \cap (C+\overline{H^\infty})$. Note that $QC = VMO \cap L^\infty$ (see [23],[24], or [4]). Having recourse to the latter equality it is easy to produce discontinuous functions in QC. Notice that QC is actually much bigger than C. Finally, PQC, the C^*-algebra of all piecewise quasicontinuous functions, is the smallest closed subalgebra of L^∞ containing QC and PC.

Given a C^*-subalgebra A of L^∞ which contains the constants, we denote by M(A) the maximal ideal space of A and we freely identify functions in A with their Gelfand transform on M(A). The maximal ideal spaces of QC and PQC will be denoted by Y and Z, respectively. We so may think of functions in QC or PQC as continuous functions on Y or Z.

If A and B are C^*-subalgebras of L^∞ such that $C \subset B \subset A$, then for each point (functional) $\beta \in M(B)$ the so-called fiber

$$M_\beta(A) := \left\{ \alpha \in M(A) : \alpha(b) = \beta(b) \quad \forall\, b \in B \right\}$$

is a nonempty compact subset of M(A), and

$$M(A) = \bigcup_{\beta \in M(B)} M_\beta(A)$$

is a partition of M(A) into pairwise disjoint subsets.

The maximal ideal spaces M(C) and M(PC) may be naturally identified with $\mathbb{T}$ (endowed with the usual topology) and $\mathbb{T} \times \{0,1\}$ (equipped with an exotic topology), respectively. For τ in $\mathbb{T}$, we have $M_\tau(PC) = \{(\tau,0),\ (\tau,1)\}$ and $a(\tau,0) = a(\tau-0)$, $a(\tau,1) = a(\tau+0)$ for all $a \in PC$. The (by no means trivial) quasicontinuous counterpart of this fibration was found by Sarason [24], who showed the following.

For each $\tau \in \mathbb{T}$, the fiber $Y_\tau := M_\tau(QC)$ splits into three nonempty pairwise disjoint subsets Y_τ^-, Y_τ^0, Y_τ^+ with the following properties. If $y \in Y_\tau^-$, then $Z_y = M_y(PQC)$ is a singleton, denoted by $\{(y,0)\}$, and $a(y,0) = a(\tau-0)$ for every $a \in PC$. Similarly, if $y \in Y_\tau^+$, then $Z_y = M_y(PQC)$ is a singleton, $\{(y,1)\}$, and $a(y,1) = a(\tau+0)$ for every $a \in PC$. Finally, if $y \in Y_\tau^0$, then $Z_y = M_y(PQC)$ is a doubleton, $\{(y,0),\ (y,1)\}$, and we have $a(y,0) = a(\tau-0)$, $a(y,1) = a(\tau+0)$ for all $a \in PC$. Moreover, if $y \in Y_\tau^0$, where $\tau = e^{i\theta_0}$, then there exists a sequence $\{\lambda_n\}_{n=1}^\infty$ of numbers such that $\lambda_n \to +\infty$ and

$$a(y,0) = \lim_{n\to\infty} (\lambda_n/\pi) \int_{\theta_0-\pi/\lambda_n}^{\theta_0} a(e^{i\theta})\, d\theta ,$$

$$a(y,1) = \lim_{n\to\infty} (\lambda_n/\pi) \int_{\theta_0}^{\theta_0+\pi/\lambda_n} a(e^{i\theta})\, d\theta$$

for every a in PC. We remark that proofs of the results cited in this paragraph are also in [4].

<u>2.6. Symbol maps, matrix and Fredholm symbols</u>. Let $\mathfrak{A}$ be a Banach algebra with identity element. A mapping Sym of $\mathfrak{A}$ into an algebra $\mathfrak{B}$ (which need not be a Banach algebra but is required to have an identity element) is called a <u>symbol map</u> if

(i) Sym is an algebra homomorphism,

(ii) an element $a \in \mathfrak{A}$ is invertible in $\mathfrak{A}$ if and only if Sym a is invertible in $\mathfrak{B}$.

To have an example, we remark that if $\mathfrak{A}$ is a commutative Banach algebra with identity element, then the Gelfand map $\Gamma : \mathfrak{A} \to C(M(\mathfrak{A}))$ is a symbol map.

Given any set M, we denote by $C_{nxn}(M)$ the algebra of all $\mathbb{C}_{nxn}$-valued matrix-functions on M. Notice that M need not be a topological space and that, although the letter C is used, the members of $C_{nxn}(M)$ need not possess any continuity properties. A matrix-function $f \in C_{nxn}(M)$ is clearly invertible in $C_{nxn}(M)$ if and only if $(\det f)(x) \neq 0$ for all x in M. An algebra $\mathfrak{B}$ is referred to as an algebra of matrix-functions if it can be represented as a (finite) direct sum

$$\mathfrak{B} = \bigoplus_{l=1}^{n} C_{lxl}(M_l) ,$$

where some of the sets $M_1, \ldots, M_n$ are allowed to be empty.

A symbol map Sym : $\mathfrak{A} \to \mathfrak{B}$ will be called a <u>matrix symbol</u> if (in addition to (i) and (ii))

(iii) $\mathfrak{B} = \bigoplus_{l=1}^{n} C_{lxl}(M_l)$ is an algebra of matrix functions, and for each $l \in \{1, \ldots, n\}$ and each $x \in M_l$ the mapping $\mathfrak{A} \to \mathbb{C}_{lxl}$, $a \mapsto (\mathrm{Sym}\ a)(x)$ is continuous;

here $\mathbb{C}_{lxl}$ is thought of as being endowed with its usual topology.

Finally suppose $\mathfrak{A}$ is a closed subalgebra of $\mathcal{L}(L^p(\varrho))$ containing the identity operator and the collection $\mathcal{K}(L^p(\varrho))$ of all compact operators. Then the quotient algebra $\mathfrak{A}^\pi := \mathfrak{A}/\mathcal{K}(L^p(\varrho))$ is a closed subalgebra of the Calkin algebra $\mathcal{L}^\pi(L^p(\varrho))$. We denote the coset $A + \mathcal{K}(L^p(\varrho))$ by $\Phi^\pi(A)$. As usual, an operator $A \in \mathcal{L}(L^p(\varrho))$ is said to be Fredholm if $\Phi^\pi(A)$ is invertible in $\mathcal{L}^\pi(L^p(\varrho))$.

We call a mapping $\widetilde{\mathrm{Sym}} : \mathfrak{A}^\pi \to \mathcal{B}$ a <u>Fredholm</u> <u>symbol</u> if

(iv) $\mathcal{B} = \bigoplus_{l=1}^{n} C_{l\times l}(M_l)$ is an algebra of matrix-functions, and for each $l \in \{1,\dots, n\}$ and each $x \in M_l$ the mapping

$$\mathfrak{A}^\pi \to \mathbb{C}_{l\times l}\ , \quad \Phi^\pi(A) \mapsto (\widetilde{\mathrm{Sym}}\ \Phi^\pi(A))(x)$$

is continuous,

(v) $\widetilde{\mathrm{Sym}}$ is an algebra homomorphism,

(vi) an operator $A \in \mathfrak{A}$ is Fredholm on $L^p(\varrho)$ if and only if $\widetilde{\mathrm{Sym}}\ \Phi^\pi(A)$ is invertible in $\mathcal{B}$.

We emphasize that there is a subtle difference between matrix and Fredholm symbols for a subalgebra $\mathfrak{A}^\pi$ of $\mathcal{L}^\pi(L^p(\varrho))$: the former answer for the invertibility of $\Phi^\pi(A)$ in $\mathfrak{A}^\pi$, whereas the latter do so for the invertibility of $\Phi^\pi(A)$ in $\mathcal{L}^\pi(L^p(\varrho))$.

<u>2.7. Algebras generated by multiplications, the singular integral, and a Carleman shift</u>. Let $B \subset L^\infty$ be a C^*-subalgebra containing C and let C_α be a Carleman shift such that $C_\alpha \in \mathcal{L}(L^p(\varrho))$. Recall that every Carleman shift is bounded on L^p without weight. We denote by $\mathrm{alg}(B,P,C_\alpha)$ the smallest closed subalgebra of $\mathcal{L}(L^p(\varrho))$ containing the operators P (or S) and C_α and the set $\{M_a : a \in B\}$. One can show (see e.g. [4, 4.52(d)]) that $\mathcal{K}(L^p(\varrho))$ is contained in $\mathrm{alg}(C,P)$, the smallest closed subalgebra of $\mathcal{L}(L^p(\varrho))$ which contains P and the set $\{M_a : a \in C\}$. Thus $\mathcal{K}(L^p(\varrho))$ is all the more a subset of $\mathrm{alg}(B,P,C_\alpha)$.

The quotient algebra $\mathrm{alg}(B,P,C_\alpha)/\mathcal{K}(L^p(\varrho))$ is denoted by $\mathrm{alg}^\pi(B,P,C_\alpha)$, and we let Φ^π stand for the canonical homomorphism of $\mathrm{alg}(B,P,C_\alpha)$ onto $\mathrm{alg}^\pi(B,P,C_\alpha)$. Our objective is to find matrix and Fredholm symbols for $\mathrm{alg}^\pi(B,P,C_\alpha)$.

The algebra $\mathrm{alg}^\pi(PQC,P,C_\alpha)$ contains several subalgebras for which matrix and Fredholm symbols are well known. Let, for example, C_α be a Carleman shift of order 2. Then there exists a mapping

$$\mathrm{Sym} : \mathrm{alg}^\pi(C,P,C_\alpha) \to C_{4\times 4}(\mathbb{T})$$

which is simultaneously a matrix and a Fredholm symbol and which has the following property: if $t \in \mathbb{T}$, then

$$\operatorname{Sym} \Phi^{\pi}(M_a)(t) = \begin{bmatrix} a(t) & 0 & 0 & 0 \\ 0 & a(t) & 0 & 0 \\ 0 & 0 & a(\alpha(t)) & 0 \\ 0 & 0 & 0 & a(\alpha(t)) \end{bmatrix} \quad (a \in C) ,$$

$$\operatorname{Sym} \Phi^{\pi}(P)(t) = \begin{bmatrix} 1 & 0 & 0 & 0 \\ 0 & 0 & 0 & 0 \\ 0 & 0 & 1 & 0 \\ 0 & 0 & 0 & 0 \end{bmatrix} ,$$

$$\operatorname{Sym} \Phi^{\pi}(C_\alpha)(t) = \begin{bmatrix} 0 & 0 & 1 & 0 \\ 0 & 0 & 0 & 1 \\ 1 & 0 & 0 & 0 \\ 0 & 1 & 0 & 0 \end{bmatrix}$$ (for forward shifts), $$\operatorname{Sym} \Phi^{\pi}(C_\alpha)(t) = \begin{bmatrix} 0 & 0 & 0 & 1 \\ 0 & 0 & 1 & 0 \\ 0 & 1 & 0 & 0 \\ 1 & 0 & 0 & 0 \end{bmatrix}$$ (for backward shifts)

In a similar way one can construct a matrix and Fredholm symbol of $\operatorname{alg}^{\pi}(C,P,C_\alpha)$ into $C_{2n\times 2n}(\mathbb{T})$ in case C_α is a forward Carleman shift of order n. See the monographs by Litvinchuk [16] and Krupnik [14] for the material of this paragraph.

Matrix and Fredholm symbols for $\operatorname{alg}^{\pi}(PC,P,C_\alpha)$ were given in [9],[26],[12] for the case where C_α is a forward Carleman shift and in [9],[10],[7],[19],[20],[21] for backward Carleman shifts. For more about the PC case see also the recent book [13]. In the $p = 2$ and $\varrho \equiv 1$ case a matrix and Fredholm symbol for $\operatorname{alg}^{\pi}(PQC,P,C_\alpha)$ can be derived from the results of the paper [25].

We remark that there is an ingenious trick which reduces the study of the algebras $\operatorname{alg}^{\pi}(B,P,C_\alpha)$ to the algebra of all (nxn)-matrix-functions with entries in $\operatorname{alg}^{\pi}(B,P)$ in case C_α is a forward Carleman shift of the order n (see [9], [16, § 35.4], [14, §§ 19 and 23]). Since backward Carleman shifts are equivalent to the flip J in the sense of 2.3, it would therefore suffice to consider the matrix-valued edition of $\operatorname{alg}^{\pi}(B,P,J)$ only. However, the prime purpose of this paper is not to obtain matrix and Fredholm symbols as quickly as possible but rather to demonstrate how extensions of the Two Projections Theorem produce such mappings and to show how singular integral operators quite naturally create generalizations of the Two Projections Theorem (experts will of course realize that the trick alluded to above is in the following used "locally").

Finally, to get a first idea of the structure of the algebras $\mathrm{alg}^{\pi}(B,P,C_\alpha)$ we mention that if C_α has the order n and if $A \in \mathrm{alg}(B,P,C_\alpha)$, then

$$\Phi^{\pi}(A) = \sum_{j=0}^{n-1} \Phi^{\pi}(A_j)\left[\Phi^{\pi}(C_\alpha)\right]^j$$

with A_j in alg(B,P), and the operators A_j are determined uniquely up to compact operators. The proof of this fact given for n = 2 in [14, § 19] can be extended to general n without difficulty.

<u>2.8. Why do we restrict ourselves to Carleman shifts ?</u> Answer: because there do not exist Fredholm symbols for non-trivial algebras containing non-Carleman shifts. This was pointed out by Krupnik and Feldman [15] (also see [14, § 27]). The precise result is as follows.

Let $\alpha: \mathbb{T} \to \mathbb{T}$ be a Lyapunov homeomorphism and let B be any closed subalgebra of L^∞ containing the disk algebra $C \cap H^\infty$ or its conjugate, $C \cap \overline{H^\infty}$. If there exist a set M, a positive integer m, and a Fredholm symbol $\widetilde{\mathrm{Sym}} : \mathrm{alg}^{\pi}(B,C_\alpha) \to C_{m\times m}(M)$, then C_α must be a Carleman shift, i.e. we necessarily have $C_\alpha^n = I$ for some positive integer n.

<u>2.9. Toeplitz and Hankel operators</u>. Operators of the form $T_a := PM_aP + Q$ and $H_a := PM_aQJ + Q$ $(= PM_aJP + Q)$ are called Toeplitz and Hankel operators, respectively. Here J is the flip operator introduced in 2.3. If $a \in L^\infty$, then T_a is bounded on $L^p(\varrho)$ The Hankel operator H_a is bounded on $L^p(\varrho)$ for every $a \in L^\infty$ if and only if the weight ϱ is invariant under complex conjugation, i.e. if and only if $\varrho(t) = \varrho(\bar{t})$ for all $t \in \mathbb{T}$. The matrix representations of T_a and H_a with respect to the standard basis $\{t^n\}_{n=-\infty}^{\infty}$ of $L^p(\varrho)$ are

$$\begin{pmatrix} (a_{i-j})_{i,j=0}^{\infty} & 0 \\ 0 & I \end{pmatrix} \quad \text{and} \quad \begin{pmatrix} (a_{i+j+1})_{i,j=0}^{\infty} & 0 \\ 0 & I \end{pmatrix}$$

respectively, $\{a_n\}_{n=-\infty}^{\infty}$ denoting the sequence of the Fourier coefficients of a.

Given a C^*-algebra B between C and L^∞, we let alg T(B) and alg TH(B) denote the smallest closed subalgebra of $\mathcal{L}(L^p(\varrho))$ containing $\{T_a : a \in B\}$ and $\{T_a : a \in B\} \cup \{H_a : a \in B\}$, respectively. Clearly, alg T(B) $\subset$ alg(B,P) and alg TH(B) $\subset$ alg(B,P,J) = alg(B,P,R), where R is the reflection operator defined in 2.3. Also notice that $\mathcal{K}(L^p(\varrho)) \subset$ alg T(B) (see [4, 4.5]).

It is well known (and easily seen) that H_a is compact whenever a is continuous. Hence alg T(C) = alg TH(C). If, however, B contains at least one function which has a jump discontinuity at some point of $\mathbb{T} \setminus \{-1,+1\}$, then alg TH(B) is strictly larger than alg T(B) (as far as we know, this was first noticed by Sheldon Axler; see [17, p. 368] and [4, 4.51]).

<u>2.10. Intertwining relations</u>. We finally list up some relations between multiplication operators, singular integrals and shifts which will be frequently used in the following.

(a) If C_α is a forward (resp. backward) Carleman shift then

$$\Phi^\pi(PC_\alpha) = \Phi^\pi(C_\alpha P) \quad (\text{resp. } \Phi^\pi(PC_\alpha) = \Phi^\pi(C_\alpha Q)\ )$$

(see e.g. [16, § 4] and also see Section 2.11). The reflection operator R does not intertwine P and Q, i.e. we have $PR \neq RQ$, but the flip operator J satisies the equality $PJ = JQ$; notice that it is just this identity which makes using the flip J preferable to working with the reflection operator R.

(b) If $a \in QC$ then $\Phi^\pi(M_aP) = \Phi^\pi(PM_a)$.

(c) If $a \in L^\infty$ and C_α is a Carleman shift, then it is easy to verify that

$$C_\alpha M_a = M_{C_\alpha a} C_\alpha .$$

Thus, intertwining Carleman shifts and multiplications leads to considering the action of the composition operator C_α on subalgebras of L^∞. It is clear that C_α maps C and PC into themselves. It is less obvious that C_α converts QC functions into QC functions, but using the equality $QC = VMO \cap L^\infty$ it is not too difficult to show that $C_\alpha a$ is in QC whenever α is a Lyapunov homeomorphism and $a \in QC$. In particular, if C_α is a Carleman shift, then C_α leaves PQC invariant.

2.11. A few words about notation. Our notation follows traditional lines and is not dictated by the ambition to consistently designate different things by different symbols. So, of course, in 2.1 the letters p and q are exponents of L^p and L^q satisfying $1/p + 1/q = 1$, while throughout Section 4 the letters p and q denote idempotents. However, there are more delicate situations: in 2.10(a), PC_α is the product of the operators P and C_α, whereas in 3.2 the symbol QC_α stands for a well-defined subalgebra of QC and not for the product of the projection $Q = I - P$ and the composition operator C_α.

Nevertheless, the willing reader will certainly always catch the meaning of a notation from the context (and will, for example, recognize that PC is the C^*-algebra of piecewise continuous functions and not the subset of VMO constituted by the image of C under the projection P).

Whenever possible, we tried to avoid any ambiguities, but we confess that some readers will occasionally be bothered by our notational practice. Notwithstanding the objections, we believe that the people who already know something about the topics discussed here will appreciate that our notation is conventional and not mystified by inventing new and unusual symbols for old and familiar objects.

3. LOCALIZATION

3.1. The local principle of Allan and Douglas. Let $\mathfrak{A}$ be a Banach algebra with identity element. The center Cen $\mathfrak{A}$ is the collection of all c in $\mathfrak{A}$ such that $ac = ca$ for all a in $\mathfrak{A}$. Let B be a closed subalgebra of Cen $\mathfrak{A}$ containing the identity element. Then B is a commutative Banach algebra. For $\beta \in M(B)$, denote by J the closure in $\mathfrak{A}$ of the collection of all finite sums of the form $\sum_i a_i b_i$, where $a_i \in \mathfrak{A}$, $b_i \in B$, $b_i(\beta) = 0$. The set J is clearly a closed two-sided ideal in $\mathfrak{A}$. The canonical homomorphism of $\mathfrak{A}$ onto $\mathfrak{A}/J_\beta$ is denoted by Φ_β. The local principle of Allan [2] and Douglas [8] (see also [4, 1.34] or the paper [3]) says that

$$\Phi : \mathfrak{A} \to \bigoplus_{\beta \in M(B)} \mathfrak{A}/J_\beta \ , \quad a \mapsto \bigoplus_{\beta \in M(B)} \Phi_\beta(a)$$

is a symbol map of $\mathfrak{A}$ into the algebra $\bigoplus \mathfrak{A}/J_\beta$ of "vector-valued" functions on $M(B)$. Hence, an element $a \in \mathfrak{A}$ is invertible in $\mathfrak{A}$ if and only if $\Phi_\beta(a)$ is invertible in $\mathfrak{A}/J_\beta$ for all β in $M(B)$. The algebras $\mathfrak{A}/J_\beta$ are usually referred to as <u>local algebras</u> and the spectrum of $\Phi_\beta(a)$ in $\mathfrak{A}/J_\beta$ is often called the <u>local spectrum</u> of the element a at (the point) $\beta \in M(B)$.

<u>3.2. Subalgebras of the center of alg (PQC,P,C_α)</u>. If $\varphi \in QC$, then M_φ commutes with all multiplication operators, and M_φ commutes modulo $\mathcal{K}(L^p(\varrho))$ with P. From 2.10(c) we infer that $\Phi^\pi(M_\varphi C_\alpha) = \Phi^\pi(C_\alpha M_\varphi)$ whenever $C_\alpha \varphi = \varphi$. Thus, the C^*-algebra

$$QC_\alpha := \left\{ \Phi^\pi(M_\varphi) \ : \ \varphi \in QC \ , \ C_\alpha \varphi = \varphi \right\}$$

is a subalgebra of the center of $\mathrm{alg}^\pi(PQC,P,C_\alpha)$.

The Lyapunov homeomorphism $\alpha : \mathbb{T} \to \mathbb{T}$ induces a mapping $\alpha : Y \to Y$, $y \mapsto \alpha(y)$ in a natural manner. More precisely, for y in Y we define $\alpha(y) \in Y$ by

$$\varphi(\alpha(y)) = (C_\alpha \varphi)(y) \quad \forall \varphi \in QC \ .$$

The <u>orbit</u> $\xi = \mathcal{O}_\alpha(y)$ of an element $y \in Y$ is the set $\{y, \alpha(y), \alpha_2(y), \alpha_3(y), \dots\}$, where $\alpha_2(y) := \alpha(\alpha(y))$, $\alpha_3(y) := \alpha(\alpha_2(y))$ etc. Since C_α is always assumed to be a Carleman shift, each orbit is finite and its cardinality does not exceed the order of C_α. A moment's thought reveals that the maximal ideal space $M(QC_\alpha)$ may be identified with the set of all orbits $\{\mathcal{O}_\alpha(y) : y \in Y\}$. Given $\xi \in M(QC_\alpha)$, define the ideal J_ξ of $\mathrm{alg}^\pi(PQC,P,C_\alpha)$ as in 3.1, i.e. let J_ξ be the smallest closed two-sided ideal of the algebra $\mathrm{alg}^\pi(PQC,P,C_\alpha)$ containing all operators M_φ with $\varphi \in QC$ and $\varphi(y) = 0$ for some (and thus all) y in the orbit ξ. Denote by Φ_ξ the canonical homomorphism of $\mathrm{alg}^\pi(PQC,P,C_\alpha)$ onto the local algebra

$$\mathrm{alg}^\pi_\xi(PQC,P,C_\alpha) := \mathrm{alg}^\pi(PQC,P,C_\alpha)/J_\xi$$

and put $\Phi^\pi_\xi := \Phi_\xi \circ \Phi^\pi$.

By virtue of 3.1 the mapping

$$\Phi : \mathrm{alg}^{\pi}(PQC,P,C_{\alpha}) \to \bigoplus_{\xi \in M(QC_{\alpha})} \mathrm{alg}^{\pi}_{\xi}(PQC,P,C_{\alpha})$$

$$\Phi^{\pi}(A) \mapsto \bigoplus_{\xi \in M(QC\ )} \Phi^{\pi}_{\xi}(A)$$

is a symbol map.

We now show that the algebras $\mathrm{alg}^{\pi}_{\xi}(PQC,P,C_{\alpha})$ are finitely generated, i.e. that there exist finitely many operators $T_1, T_2, \ldots$ (depending on both the homeomorphism α and the choice of $\xi \in M(QC_{\alpha})$) such that $\mathrm{alg}^{\pi}_{\xi}(PQC,P,C_{\alpha})$ coincides with its smallest closed subalgebra containing $\Phi^{\pi}_{\xi}(T_1), \Phi^{\pi}_{\xi}(T_2), \ldots$. It is clear that $\mathrm{alg}^{\pi}_{\xi}(PQC,P,C_{\alpha})$ is generated by $\Phi^{\pi}_{\xi}(P)$, $\Phi^{\pi}_{\xi}(C_{\alpha})$, and the set $\{\Phi^{\pi}_{\xi}(M_a) : a \in PQC\}$. It is also not difficult to see that if a,b are in PQC, then $\Phi^{\pi}_{\xi}(M_a) = \Phi^{\pi}_{\xi}(M_b)$ whenever $a|Z_y = b|Z_y$ for all y in the orbit ξ ; here Z_y is the fiber $M_y(PQC)$ and $c|Z_y$ denotes the restriction of $c \in PQC$ to Z_y.

We remark that Douglas [8] was probably the first to realize that the local algebras $\mathrm{alg}^{\pi}_{\xi}(PC,P)$ are generated by two idempotents, which offers application of the C^*-algebra version of the Two Projections Theorem.

<u>3.3. Forward Carleman shifts</u>. Let C_{α} be a forward Carleman shift of the order n. Then each orbit $\xi \in M(QC_{\alpha})$ consists of exactly n elements, $\xi = \{y, \alpha(y), \ldots, \alpha_{n-1}(y)\}$ with some $y \in Y$. Put $y_o = y$ and $y_j = \alpha_j(y)$ for $j = 1, \ldots, n-1$. For each j, there is a unique $\tau_j \in \mathbb{T}$ such that $y_j \in Y_{\tau_j}$; the points $\tau_o, \ldots, \tau_{n-1}$ are of course pairwise distinct, and we have $\tau_j = \alpha_j(\tau_o)$ for $j = 1, \ldots, n-1$. It is readily seen that all y_j ($j = 1, \ldots, n-1$) belong to $Y^{o}_{\tau_j}$ (resp. $Y^{\pm}_{\tau_j}$) if $y_o \in Y^{o}_{\tau_o}$ (resp. $Y^{\pm}_{\tau_o}$).

Let U_j ($j = 0, \ldots, n-1$) be sufficiently small subarcs of $\mathbb{T}$ such that one endpoint of U_j is τ_j and the other endpoint lies between τ_j and τ_{j+1} ($\tau_n := \tau_o$) and denote by χ_j the characteristic function of U_j. Let φ_j ($j = 0, \ldots, n-1$) be any continuous function such that $\varphi_j(\tau_j) = 1$ and $\varphi_j(\tau_i) = 0$ for $i \neq j$.

Now pick any $a \in PQC$. If $y_j \in Y^o_{\tau_j}$ for all j, then the function $\psi \in PQC$ defined by

$$\psi = a - \left(\sum_{j=0}^{n-1} [a(y_j,1) - a(y_j,0)] \chi_j + \sum_{j=0}^{n-1} a(y_j,0) \varphi_j \right)$$

vanishes on $Z_{y_0} \cup \ldots \cup Z_{y_{n-1}}$. Therefore

$$\Phi^\pi_\xi(a) = \sum_{j=0}^{n-1} [a(y_j,1) - a(y_j,0)] \Phi^\pi_\xi(\chi_j) + \sum_{j=0}^{n-1} a(y_j,0) \Phi^\pi_\xi(\varphi_j),$$

where $\Phi^\pi_\xi(a)$ etc. stands for $\Phi^\pi_\xi(M_a)$ etc. Hence, $alg^\pi_\xi(PQC,P,C_\alpha)$ is generated by the 2n+2 elements

$$\Phi^\pi_\xi(P),\ \Phi^\pi_\xi(C_\alpha),\ \Phi^\pi_\xi(\chi_0), \ldots, \Phi^\pi_\xi(\chi_{n-1}),\ \Phi^\pi_\xi(\varphi_0), \ldots, \Phi^\pi_\xi(\varphi_{n-1}).$$

There is, however, a strong interdependence between these 2n+2 elements. Namely, because obviously $C^j_\alpha M_f C^{n-j}_\alpha = M_{C^j_\alpha f}$ for every $f \in L^\infty$, we have

$$\Phi^\pi_\xi(\chi_j) = \left[\Phi^\pi_\xi(C_\alpha)\right]^j \Phi^\pi_\xi(\chi_0) \left[\Phi^\pi_\xi(C_\alpha)\right]^{n-j},$$

$$\Phi^\pi_\xi(\varphi_j) = \left[\Phi^\pi_\xi(C_\alpha)\right]^j \Phi^\pi_\xi(\varphi_0) \left[\Phi^\pi_\xi(C_\alpha)\right]^{n-j}.$$

Thus, in fact the algebra $alg^\pi_\xi(PQC,P,C_\alpha)$ is generated by the four elements

$$p := \Phi^\pi_\xi(P), \quad v := \Phi^\pi_\xi(C_\alpha), \quad q := \Phi^\pi_\xi(\chi_0), \quad r := \Phi^\pi_\xi(\varphi_0).$$

Note that $p^2 = p$, $q^2 = q$, $r^2 = r$, and $v^n = e$, where $e = \Phi^\pi_\xi(I)$ is the identity element.

If $y_j \in Y^-_{\tau_j}$, we put $\psi = a - \sum_{j=0}^{n-1} a(y_j,0) \varphi_j$. Clearly, ψ is identically zero on $Z_{y_0} \cup \ldots \cup Z_{y_{n-1}} = \{(y_0,0), \ldots, (y_{n-1},0)\}$ and so it follows that

$$\Phi^\pi_\xi(a) = \sum_{j=0}^{n-1} a(y_j,0) \Phi^\pi_\xi(\varphi_j).$$

One similarly obtains that

$$\Phi^\pi_\xi(a) = \sum_{j=0}^{n-1} a(y_j,1) \Phi^\pi_\xi(\varphi_j)$$

in case $y_j \in Y^+_{\tau_j}$. Hence, if $\xi = \{y, \alpha(y), \ldots, \alpha_{n-1}(y)\}$ with y in $Y^-_{\tau_0} \cup Y^+_{\tau_0}$ for some $\tau_0 \in \mathbb{T}$, then $\mathrm{alg}^{\pi}_{\xi}(PQC, P, C_\alpha)$ is generated by the three elements

$$p := \Phi^{\pi}_{\xi}(P), \quad v := \Phi^{\pi}_{\xi}(C_\alpha), \quad r := \Phi^{\pi}_{\xi}(\varphi_0).$$

3.4. Backward Carleman shifts. Now let C_α be a backward Carleman shift. Then $C^2_\alpha = I$, and α has exactly two fixed points. Write $C_\alpha = C^{-1}_\gamma R C_\gamma$ as in 2.3.

Let τ be either of the fixed points of α. If $y \in Y^{\pm}_\tau$ then $\alpha(y) \in Y^{\mp}_\tau$, and hence the orbit $\xi = \mathcal{O}_\alpha(y)$ is the doubleton $\{y, \alpha(y)\}$. We claim that $\alpha(y) = y$ for all y in Y^0_τ. To see this, let φ be any function in QC and notice first that if $y \in Y^0_\tau$, then

$$\varphi(\alpha(y)) = (C_\alpha\varphi)(y) = (C^{-1}_\gamma RC_\gamma\varphi)(y) = (RC_\gamma\varphi)(\gamma^{-1}(y)).$$

It is clear that $\gamma^{-1}(y)$ also belongs to Y^0_τ. From what was stated at the end of 2.5 we conclude that

$$(RC_\gamma\varphi)(\gamma^{-1}(y)) = \lim_{n\to\infty} \frac{\lambda_n}{\pi} \int_{\theta_0-\pi/\lambda_n}^{\theta_0} (RC_\gamma\varphi)(e^{i\theta})\, d\theta$$

$$= \lim_{n\to\infty} \frac{\lambda_n}{\pi} \int_{\theta_0}^{\theta_0+\pi/\lambda_n} (C_\gamma\varphi)(e^{i\theta})\, d\theta = (C_\gamma\varphi)(\gamma^{-1}(y)) = \varphi(y),$$

which proves our claim.

If τ is not a fixed point of α, then the orbit $\xi = \mathcal{O}_\alpha(y)$ is the doubleton $\{y, \alpha(y)\}$, and if $y \in Y^0_\tau$ (resp. $y \in Y^{\pm}_\tau$) then $\alpha(y) \in Y^0_{\alpha(\tau)}$ (resp. $\alpha(y) \in Y^{\mp}_{\alpha(\tau)}$). We so shall consider localization at a fixed point and at a movable point of the homeomorphism α separately.

3.5. Backward shifts at a fixed point. Let τ be a fixed point of α, fix y in Y^0_τ and let ξ stand for the orbit $\{y\}$. We denote by χ_0 the characteristic function of a sufficiently small subarc of $\mathbb{T}$ of the form $\{\tau e^{i\theta} : 0 < \theta < \varepsilon\}$ and by $\varphi_0 \in C$ any function which equals 1 at τ. Then if $a \in PQC$, we have

$$a = [a(y,1) - a(y,0)]\,\chi_0 + a(y,0)\,\varphi_0 + \psi,$$

where $\psi \in PQC$ vanishes on $Z_y = \{(y,0), (y,1)\}$. So

$$\Phi_\xi^\pi(a) = [a(y,1) - a(y,0)]\,\Phi_\xi^\pi(\chi_0) + a(y,0)\,\Phi_\xi^\pi(\varphi_0),$$

and because $\Phi_\xi^\pi(\varphi_0) = \Phi_\xi^\pi(I)$, it follows that $\mathrm{alg}_\xi^\pi(PQC,P,C_\alpha)$ is generated by the three elements

$$p := \Phi_\xi^\pi(P)\ ,\quad q := \Phi_\xi^\pi(\chi_0)\ ,\quad j := \Phi_\xi^\pi(C_\alpha).$$

Notice that $p^2 = p$, $q^2 = q$, $j^2 = e$, $jpj = e-p$, $jqj = e-q$.

Now let $y \in Y_\tau^-$. Then $\alpha(y) \in Y_\tau^+$. Let $\varphi_y \in QC$ denote any function which equals 1 at y and is zero at $\alpha(y)$. Then the function $\varphi_{\alpha(y)} := C_\alpha \varphi_y$ is 1 at $\alpha(y)$ and 0 at y. Every function $a \in PQC$ can be written in the form

$$a = a(y,0)\,\varphi_y + a(\alpha(y),1)\,\varphi_{\alpha(y)} + \psi ,$$

where ψ vanishes on $Z_y \cup Z_{\alpha(y)} = \{(y,0),(\alpha(y),1)\}$. Hence, if $y \in Y_\tau^-$, then $\mathrm{alg}_\xi^\pi(PQC,P,C_\alpha)$ is generated by the three elements

$$p := \Phi_\xi^\pi(P)\ ,\quad j := \Phi_\xi^\pi(C_\alpha)\ ,\quad r := \Phi_\xi^\pi(\varphi_y)$$

(note that $\Phi_\xi^\pi(\varphi_{\alpha(y)}) = e - \Phi_\xi^\pi(\varphi_y)$). We have $p^2 = p$, $r^2 = r$, $j^2 = e$, $pr = rp$, $jpj = e-p$, $jrj = e-r$.

The situation is analogous for $y \in Y_\tau^+$.

3.6. Backward shifts at a movable point. Now suppose C_α is a backward Carleman shift and τ is not a fixed point of α . Let $y \in Y_\tau$ and denote by $\xi = \{y, \alpha(y)\}$ the orbit of y. Choose sufficiently small arcs $U_0 = \{\tau e^{i\theta} : 0<\theta<\varepsilon\}$ and $U_1 = \{\alpha(\tau)e^{i\theta} : 0<\theta<\varepsilon\}$, denote the characteristic functions of U_0 and U_1 by χ_0 and χ_1 , and let φ_0 and φ_1 be any continuous functions on $\mathbb{T}$ such that $\varphi_0(\tau) = \varphi_1(\alpha(\tau)) = 1$ and $\varphi_0(\alpha(\tau)) = \varphi_1(\tau) = 0$.

Let first $y \in Y_\tau^0$, which implies that $\alpha(y) \in Y_{\alpha(\tau)}^0$. If $a \in PQC$, then

$$a = [a(y,1) - a(y,0)]\,\chi_0 + [a(\alpha(y),1) - a(\alpha(y),0)]\,\chi_1 + a(y,0)\varphi_0 + a(\alpha(y),0)\,\varphi_1 + \psi ,$$

ψ vanishing on $Z_y \cup Z_{\alpha(y)}$. It results that $\mathrm{alg}_\xi^\pi(PQC,P,C_\alpha)$ is generated by

$$\Phi_\xi^\pi(P),\ \Phi_\xi^\pi(C_\alpha),\ \Phi_\xi^\pi(\chi_0),\ \Phi_\xi^\pi(\chi_1),\ \Phi_\xi^\pi(\varphi_0),\ \Phi_\xi^\pi(\varphi_1).$$

Since $\Phi_\xi^\pi(\varphi_1) = \Phi_\xi^\pi(I-\varphi_0)$ and $\Phi_\xi^\pi(\chi_1) = \Phi_\xi^\pi((I-\varphi_0)(I-C_\alpha\chi_0 C_\alpha))$, the local algebra is actually merely generated by the four elements

$$p := \Phi_\xi^\pi(P)\ ,\ q := \Phi_\xi^\pi(\chi_0)\ ,\ r := \Phi_\xi^\pi(\varphi_0)\ ,\ j := \Phi_\xi^\pi(C_\alpha)\,.$$

Now assume $y \in Y_\tau^-$. Then $\alpha(y) \in Y^+_{\alpha(\tau)}$, and if $a \in PQC$ then

$$a = a(y,0)\,\varphi_0 + a(\alpha(y),1)\,\varphi_1 + \psi\,,$$

where ψ vanishes on $Z_y \cup Z_{\alpha(y)}$. Analogously, if $y \in Y_\tau^+$ and thus $\alpha(y) \in Y^-_{\alpha(\tau)}$, then

$$a = a(y,1)\,\varphi_0 + a(\alpha(y),0)\,\varphi_1 + \psi\,,$$

ψ vanishing on $Z_y \cup Z_{\alpha(y)}$. Consequently, in the latter two cases $\mathrm{alg}_\xi^\pi(PQC,P,C_\alpha)$ is generated by the three elements

$$p := \Phi_\xi^\pi(P)\ ,\ r := \Phi_\xi^\pi(\varphi_0)\ ,\ j := \Phi_\xi^\pi(C_\alpha)\ .$$

4. ABSTRACTION

4.1. Prologue. The preceding section motivates the investigation of certain Banach algebras generated by idempotents, flips, and shifts. As usual, we say that a Banach algebra $\mathfrak{A}$ with identity element e is generated by a finite collection $a_1,\dots,a_m$ of its elements and write $\mathfrak{A} = \mathrm{alg}(a_1,\dots,a_m)$ in this case if the smallest closed subalgebra of $\mathfrak{A}$ containing $e,a_1,\dots,a_m$ is all of $\mathfrak{A}$. The spectrum of an element a in $\mathfrak{A}$ will be denoted by sp a and if necessary, by $\mathrm{sp}_{\mathfrak{A}}\, a$. An idempotent is an element p such that $p^2 = p$. Elements v satisfying $v^n = e$ for some $n \geq 2$ will be called shifts. Finally, if we are given an idempotent p and a shift j such that $j^2 = e$ and $jpj = e-p$, then j is referred to as a flip.

Our next objective is to provide matrix symbols for algebras generated by idempotents and a shift or a flip. The "axioms" we impose on the generating elements are dictated by the concrete situations that arose in 3.3-3.6.

4.2. F_{2n}-algebras. A Banach algebra $\mathfrak{A}$ with identity element is said to be an F_{2n}-algebra (n being a positive integer) if $F(a_1,\dots,a_{2n}) = 0$ for all $a_1,\dots,a_{2n}$ in $\mathfrak{A}$, where F stands for the standard polynomial of the order 2n,

$$F(a_1,\dots,a_{2n}) := \sum_{p \in S_{2n}} (-1)^p \, a_{p(1)} \, a_{p(2)} \cdots a_{p(2n)} ,$$

S_{2n} being the symmetric group and $(-1)^p$ denoting the sign of the permutation $p \in S_{2n}$.

Given an F_{2n}-algebra $\mathfrak{A}$, we denote by $M(\mathfrak{A})$ the collection of all maximal two-sided ideals of $\mathfrak{A}$. It is well known (see [14]) that the mapping

$$\eta : \mathfrak{A} \to \bigoplus_{m \in M(\mathfrak{A})} \mathfrak{A}/m , \quad a \mapsto \eta_m(a) := a + m$$

is a symbol map.

4.3. Krupnik symbols. Let $\mathfrak{A}$ be an F_{2n}-algebra. Krupnik [14, § 21] showed that for each $m \in M(\mathfrak{A})$ the algebra $\mathfrak{A}/m$ is isomorphic to the algebra $\mathbb{C}_{l \times l}$ of all complex $(l \times l)$-matrices for some $l = l(m) \leqq n$. Let $M_l(\mathfrak{A})$ $(1 \leqq l \leqq n)$ denote the set of all $m \in M(\mathfrak{A})$ for which $\mathfrak{A}/m$ is isomorphic to $\mathbb{C}_{l \times l}$. Thus, on choosing any isomorphism $\xi_m : \mathfrak{A}/m \to \mathbb{C}_{l \times l}$ for each m in $M_l(\mathfrak{A})$ the symbol map provided in 4.2 can be converted into a symbol map

$$\xi : \mathfrak{A} \to \bigoplus_{l=1}^{n} C_{l \times l}(M_l(\mathfrak{A})) , \quad a \mapsto \xi_m(\eta_m(a)) ,$$

and from [14, § 25] it follows that this mapping is "pointwise continuous", i.e., more precisely, that it is a matrix symbol (satisfying 2.6(iii)).

A symbol map σ of a Banach algebra $\mathfrak{A}$ into an algebra $\mathcal{L}$ will be called a Krupnik symbol if

(i) $\mathfrak{A}$ is an F_{2n}-algebra for some n,

(ii) $\mathcal{L}$ is isomorphic to $\bigoplus_{l=1}^{n} C_{l \times l}(M_l(\mathfrak{A}))$.

In other words, the Krupnik symbols for an F_{2n}-algebra $\mathfrak{A}$ are those symbol maps which emerge from specifying the isomorphisms $\xi_m : \mathfrak{A}/m \to \mathbb{C}_{l \times l}$ $(m \in M_l(\mathfrak{A}))$ and from possibly re-interpreting $C_{l \times l}(M_l(\mathfrak{A}))$ as a space of matrix-functions on another set. Note that every Krupnik symbol is necessarily a matrix symbol (in the sense of 2.6).

4.4. Tensorable symbol maps. Let $\mathfrak{A}$ be a (Banach) algebra. The collection $\mathfrak{A}_{k\times k}$ of all (kxk)-matrices with entries in $\mathfrak{A}$ is a (Banach) algebra in a natural fashion (the concrete choice of a Banach algebra norm in $\mathfrak{A}_{k\times k}$ is irrelevant to our purposes).

Now assume $\mathfrak{A}$ is a Banach algebra and $\mathfrak{B}$ is any algebra. Suppose both $\mathfrak{A}$ and $\mathfrak{B}$ possess an identity element. A symbol map $\mathrm{Sym} : \mathfrak{A}\to\mathfrak{B}$ is said to be tensorable if the mapping

$$\mathrm{Sym}_{k\times k} : \mathfrak{A}_{k\times k} \to \mathfrak{B}_{k\times k} , \quad [a_{ij}]_{i,j=1}^{k} \mapsto [\mathrm{Sym}\ a_{ij}]_{i,j=1}^{k}$$

is a symbol map for every $k \geqq 2$. By a tensorable matrix (resp. Fredholm) symbol we understand a matrix (resp. Fredholm) symbol $\mathrm{Sym} : \mathfrak{A}\to\mathfrak{B}$ such that $\mathrm{Sym}_{k\times k}$ is a matrix (resp. Fredholm) symbol for all $k \geqq 2$.

4.5. THEOREM. Krupnik symbols are tensorable.

PROOF. What we must prove is that if $\mathfrak{A}$ is an F_{2n}-algebra and $k \geqq 2$ an integer, then the mapping

$$\sigma : \mathfrak{A}_{k\times k} \to \bigoplus_{l=1}^{n} C_{lk\times lk}(M_l(\mathfrak{A})) ,$$

$$(\sigma [a_{ij}]_{i,j=1}^{k})(m) := [(\xi_m \circ \eta_m)(a_{ij})]_{i,j=1}^{k}$$

is a symbol map.

It is clear that σ is a continuous homomorphism. We so infer from the Amitsur-Levitzki theorem that

$$\sigma(F_{2nk}(a_1,\dots,a_{2nk})) = F_{2nk}(\sigma a_1,\dots,\sigma a_{2nk}) = 0$$

for all $a_1,\dots,a_{2nk}$ in $\mathfrak{A}_{k\times k}$.

Suppose first that the radical of $\mathfrak{A}$ is trivial. Then so also is the radical of $\mathfrak{A}_{k\times k}$, and since the kernel of σ is a subset of the radical of the algebra $\mathfrak{A}_{k\times k}$, it follows that

$$F_{2nk}(a_1,\dots,a_{2nk}) = 0 \quad \forall\ a_1,\dots,a_{2nk} \in \mathfrak{A}_{k\times k} .$$

Hence, $\mathfrak{A}_{k\times k}$ is an F_{2nk}-algebra.

It can be easily verified that every two-sided ideal J of $\mathfrak{A}_{k\times k}$ is of the form $J = m_{k\times k}$, where m is a two-sided ideal of $\mathfrak{A}$ and that J is maximal in $\mathfrak{A}_{kxk}$ if and only if m is so in $\mathfrak{A}$. Therefore $M_{lk}(\mathfrak{A}_{kxk}) = \{m_{kxk} : m \in M_l(\mathfrak{A})\}$ for l = 1,...,n and $M_j(\mathfrak{A}_{kxk}) = \emptyset$ for all j = 1,...,nk which are not divisible by k.

Because for $m \in M_l(\mathfrak{A})$ the algebra $\mathfrak{A}_{kxk}/m_{kxk}$ is isomorphic to $(\mathfrak{A}/m)_{kxk}$ and the mapping

$$(\mathfrak{A}/m)_{kxk} \to \mathbb{C}_{lkxlk}, \quad [a_{ij} + m]_{i,j=1}^{k} \mapsto [(\xi_m \circ \eta_m)a_{ij}]_{i,j=1}^{k}$$

is an isomorphism, the assertion is now immediate from 4.3.

Now let $\mathfrak{A}$ have a nontrivial radical $\mathcal{R}$. From what has just been proved we know that

$$\mu : (\mathfrak{A}/\mathcal{R})_{kxk} \to \bigoplus_{l=1}^{n} C_{lkxlk}(M_l(\mathfrak{A}/\mathcal{R})),$$

$$(\mu[a_{ij} + \mathcal{R}]_{i,j=1}^{k})(m) := [(\xi_m \circ \eta_m)(a_{ij} + \mathcal{R})]_{i,j=1}^{k}$$

is a symbol map. Since $\mathcal{R}$ is contained in all maximal two-sided ideals of $\mathfrak{A}$ and $(\mathfrak{A}/\mathcal{R})/(m/\mathcal{R})$ is isomorphic to $\mathfrak{A}/m$ for every $m \in M(\mathfrak{A})$, we may identify $M_l(\mathfrak{A}/\mathcal{R})$ and $M_l(\mathfrak{A})$, obtaining that

$$\nu : (\mathfrak{A}/\mathcal{R})_{kxk} \to \bigoplus_{l=1}^{n} C_{lkxlk}(M_l(\mathfrak{A})),$$

$$(\nu[a_{ij} + \mathcal{R}]_{i,j=1}^{k})(m) := [(\xi_m \circ \eta_m)(a_{ij})]_{i,j=1}^{k}$$

is a symbol map. Finally, as $(\mathfrak{A}/\mathcal{R})_{kxk}$ is isomorphic to the algebra $\mathfrak{A}_{kxk}/\mathcal{R}_{kxk}$ and $\mathcal{R}_{kxk}$ is the radical of $\mathfrak{A}_{kxk}$, it follows that a is invertible in $\mathfrak{A}_{kxk}$ whenever σa is invertible, which completes the proof. ■

4.6. Universal constructions. There are a few simple (and well known) tricks which produce symbol maps in an astonishing manner. Throughout this subsection let $\mathfrak{A}$ be a Banach algebra with identity element e.

(a) If $\mathfrak{A}$ contains idempotents $r_1,\dots,r_m$ such that $r_1 + \dots + r_m = e$ and $r_i r_j = 0$ for $i \neq j$, then the mapping

$$\beta : \mathfrak{A} \to \mathfrak{A}_{mxm}, \quad a \mapsto [r_i a r_j]_{i,j=1}^{m}$$

is a tensorable symbol map. To see this notice that if the

matrix $[b_{ij}]_{i,j=1}^m$ is the inverse of $\beta(a)$, then $\sum_{i,j=1}^{m} r_i b_{ij} r_j$ is the inverse of a.

(b) Putting m = 2 in (a) and replacing r_1 and r_2 by p and e-p, respectively, we deduce that if $\mathfrak{A}$ contains an idempotent p, then the mapping

$$\gamma : \mathfrak{A} \to \mathfrak{A}_{2\times 2} \quad , \quad a \mapsto \begin{bmatrix} pap & pa(e-p) \\ (e-p)ap & (e-p)a(e-p) \end{bmatrix}$$

is a tensorable symbol map.

(c) Let $\mathfrak{A}$ contain an idempotent p and a flip j (which means that $p^2 = p$, $j^2 = e$, and $jpj = e-p$). The set $p\mathfrak{A}p :=$ $:= \{pap : a \in \mathfrak{A}\}$ is a closed subalgebra of $\mathfrak{A}$, and we think of p as the identity element of $p\mathfrak{A}p$. The mapping

$$\delta : \mathfrak{A} \to (p\mathfrak{A}p)_{2\times 2} \quad , \quad a \mapsto \begin{bmatrix} pap & pajp \\ pjap & pjajp \end{bmatrix}$$

is a tensorable symbol map. This follows easily from the equality

$$(a) = \begin{bmatrix} e & 0 \\ 0 & j \end{bmatrix} \gamma(a) \begin{bmatrix} e & 0 \\ 0 & j \end{bmatrix} .$$

<u>4.7. Convention</u>. Throughout what follows we let w denote an arbitrarily fixed function $w : \mathbb{C} \to \mathbb{C}$ such that

$$[w(z)]^2 = z(1-z)$$

for all $z \in \mathbb{C}$. Notice that w is neither required to be analytic nor even to be continuous.

<u>4.8</u>. THEOREM.*) Let $\mathfrak{A} = \mathrm{alg}(p,q)$, where p and q are idempotents such that $\{0,1\} = \mathrm{sp}\, p = \mathrm{sp}\, q \subset \mathrm{sp}(pqp)$ and the set sp(pqp) is connected. Put

$$X := \mathrm{sp}\big(pqp + (e-p)(e-q)(e-p)\big) \cup \{0,1\} .$$

Then there exists a Krupnik symbol Sym of $\mathfrak{A}$ into $C_{2\times 2}(X)$ such that

$$(\mathrm{Sym}\ e)(x) = \begin{bmatrix} 1 & 0 \\ 0 & 1 \end{bmatrix}, \quad (\mathrm{Sym}\ p)(x) = \begin{bmatrix} 1 & 0 \\ 0 & 0 \end{bmatrix},$$

$$(\mathrm{Sym}\ q)(x) = \begin{bmatrix} x & w(x) \\ w(x) & 1-x \end{bmatrix} \quad \forall x \in X .$$

---

*) For some new aspects of the matter see also [22] and [27].

PROOF. It was shown in [21, Theorem 3] that $\mathfrak{A}$ is an F_4-algebra. Hence, by 4.3, there exists a Krupnik symbol

$$\xi : \mathfrak{A} \to C_{1\times 1}(M_1(\mathfrak{A})) \oplus C_{2\times 2}(M_2(\mathfrak{A})) .$$

From [21, Theorem 4] we infer that ξ can be specified so that the following requirements are met:

(i) $(\xi e)(m) = \begin{bmatrix} 1 & 0 \\ 0 & 1 \end{bmatrix} \quad \forall m \in M_2(\mathfrak{A})$;

(ii) $(\xi p)(m) = \begin{bmatrix} 1 & 0 \\ 0 & 0 \end{bmatrix} \quad \forall m \in M_2(\mathfrak{A})$;

(iii) for each $m \in M_2(\mathfrak{A})$ there exists an $x(m) \in \mathbb{C}$ such that

$$(\xi q)(m) = \begin{bmatrix} x(m) & w(x(m)) \\ w(x(m)) & 1-x(m) \end{bmatrix} .$$

It results that

$$\big(\xi(pqp + (e-p)(e-q)(e-p))\big)(m) = \begin{bmatrix} x(m) & 0 \\ 0 & x(m) \end{bmatrix}$$

and thus that the set $\{x(m) : m \in M_2(\mathfrak{A})\} \cup \{0,1\}$ coincides with the set X.

Furthermore, Theorem 4 of [21] also says that $M_1(\mathfrak{A})$ consists of exactly four ideals m_{00}, m_{01}, m_{10}, m_{11}. One can choose ξ so that if we define

$$(\psi a)(0) = \operatorname{diag}\,((\xi a)(m_{00}) , (\xi a)(m_{11})),$$

$$(\psi a)(1) = \operatorname{diag}\,((\xi a)(m_{01}) , (\xi a)(m_{10})),$$

then

$$(\psi e)(0) = (\psi e)(1) = \begin{bmatrix} 1 & 0 \\ 0 & 1 \end{bmatrix} , (\psi p)(0) = (\psi p)(1) = \begin{bmatrix} 1 & 0 \\ 0 & 0 \end{bmatrix} ,$$

$$(\psi q)(0) = \begin{bmatrix} 0 & 0 \\ 0 & 1 \end{bmatrix} , \quad (\psi q)(1) = \begin{bmatrix} 1 & 0 \\ 0 & 0 \end{bmatrix} .$$

Now if $x \in \{0,1\}$ put $(\operatorname{Sym} a)(x) = (\psi a)(x)$, and if x is in $\{x(m) : m \in M_2(\mathfrak{A})\}$ define $(\operatorname{Sym} a)(x)$ as $(\xi a)(m)$, where m is any point in $M_2(\mathfrak{A})$ such that $x = x(m)$ (note that $(\xi a)(m_1) = (\xi a)(m_2)$ whenever $x(m_1) = x(m_2)$). The mapping Sym of $\mathfrak{A}$ into $C_{2\times 2}(X)$ is the desired Krupnik symbol. ∎

The following theorem shows that the things are simpler if a Banach algebra is generated by two idempotents <u>and</u> a flip (we need no restrictions on the spectra of the idempotents).

4.9. THEOREM. Let $\mathfrak{A} = \mathrm{alg}(p,q,j)$, where $p^2=p$, $q^2=q$, $j^2=e$, $jpj = e-p$, $jqj = e-q$, and put

$$X := \mathrm{sp}\left(pqp + (e-p)(e-q)(e-p)\right) .$$

Then there exists a Krupnik symbol Sym of $\mathfrak{A}$ into $C_{2\times 2}(X)$ such that

$$(\mathrm{Sym}\ e)(x) = \begin{bmatrix} 1 & 0 \\ 0 & 1 \end{bmatrix}, \quad (\mathrm{Sym}\ p)(x) = \begin{bmatrix} 1 & 0 \\ 0 & 0 \end{bmatrix},$$

$$(\mathrm{Sym}\ q)(x) = \begin{bmatrix} x & w(x) \\ w(x) & 1-x \end{bmatrix},$$

and

$$\text{either} \quad (\mathrm{Sym}\ j)(x) = \begin{bmatrix} 0 & i \\ -i & 0 \end{bmatrix} \quad \text{or} \quad (\mathrm{Sym}\ j)(x) = \begin{bmatrix} 0 & -i \\ i & 0 \end{bmatrix}$$

for all $x \in X$.

PROOF. From [21, Theorems 6 and 7] we deduce that $\mathfrak{A}$ is an F_4-algebra, that $M_1(\mathfrak{A}) = \emptyset$, and that there is a Krupnik symbol ξ of $\mathfrak{A}$ into $C_{2\times 2}(M_2(\mathfrak{A}))$ such that

(i) $(\xi e)(m) = \begin{bmatrix} 1 & 0 \\ 0 & 1 \end{bmatrix} \ \forall m \in M_2(\mathfrak{A})$;

(ii) $(\xi p)(m) = \begin{bmatrix} 1 & 0 \\ 0 & 0 \end{bmatrix} \ \forall m \in M_2(\mathfrak{A})$;

(iii) either $(\xi j)(m) = \begin{bmatrix} 0 & i \\ -i & 0 \end{bmatrix} \ \forall m \in M_2(\mathfrak{A})$ or $(\xi j)(m) = \begin{bmatrix} 0 & -i \\ i & 0 \end{bmatrix} \ \forall m \in M_2(\mathfrak{A})$;

(iv) for each $m \in M_2(\mathfrak{A})$ there is an $x(m) \in \mathbb{C}$ such that

$$(\xi q)(m) = \begin{bmatrix} x(m) & w(x(m)) \\ w(x(m)) & 1-x(m) \end{bmatrix} .$$

Since

$$\xi(pqp + (e-p)(e-q)(e-p)) = \begin{bmatrix} x(m) & 0 \\ 0 & x(m) \end{bmatrix},$$

the assertion now follows at once. ∎

REMARKS. Notice that one can also apply the construction 4.6(c) to prove that there exists a tensorable matrix symbol $\mathrm{Sym} : \mathfrak{A} \to C_{2\times 2}(X)$ with the above properties. The point is that $p\mathfrak{A}p$ is commutative in the present case.

In order to find out whether

$$(\mathrm{Sym}\ j)(x) = \begin{bmatrix} 0 & i \\ -i & 0 \end{bmatrix} \ \forall x \in X \quad \text{or} \quad (\mathrm{Sym}\ j)(x) = \begin{bmatrix} 0 & -i \\ i & 0 \end{bmatrix} \ \forall x \in X$$

note that if $(\mathrm{Sym}\ j)(x) = \sigma\begin{bmatrix} 0 & i \\ -i & 0 \end{bmatrix}$, then clearly

$$(\mathrm{Sym}\ ipqjp)(x) = \sigma\begin{bmatrix} w(x) & 0 \\ 0 & 0 \end{bmatrix} \quad \forall x \in X,$$

and hence sp(ipqjp), the spectrum of the "abstract Hankel" ipqjp, equals $\sigma w(X)$. Thus, to decide which of the two signs $\sigma \in \{-1,+1\}$ has to be chosen it suffices to know a single point of $sp(ipqjp) \setminus \{0,1\}$.

4.10. THEOREM. Let $\mathfrak{A} = alg(p,q,r,j)$, where

$$p^2 = p,\ q^2 = q,\ r^2 = r,\ j^2 = e,\ jpj = e-p,\ jrj = e-r,$$

$$rp = pr,\ rq = qr = q,\ pr \neq 0,\ pr \neq r,\ q \neq 0,\ q \neq r.$$

Suppose sp(pqp) is a connected set containing $\{0,1\}$ and put

$$X := sp\big(r(pqp + (e-p)(e-q)(e-p))r\big) \cup \{0,1\}.$$

Then there exists a tensorable matrix symbol Sym of $\mathfrak{A}$ into $C_{4\times 4}(X)$ such that

$$(\mathrm{Sym}\ e)(x) = \begin{bmatrix} 1&0&0&0 \\ 0&1&0&0 \\ 0&0&1&0 \\ 0&0&0&1 \end{bmatrix}, \quad (\mathrm{Sym}\ j)(x) = \begin{bmatrix} 0&0&1&0 \\ 0&0&0&1 \\ 1&0&0&0 \\ 0&1&0&0 \end{bmatrix},$$

$$(\mathrm{Sym}\ r)(x) = \begin{bmatrix} 1&0&0&0 \\ 0&1&0&0 \\ 0&0&0&0 \\ 0&0&0&0 \end{bmatrix}, \quad (\mathrm{Sym}\ p)(x) = \begin{bmatrix} 1&0&0&0 \\ 0&0&0&0 \\ 0&0&0&0 \\ 0&0&0&1 \end{bmatrix},$$

$$(\mathrm{Sym}\ q)(x) = \begin{bmatrix} x & w(x) & 0 & 0 \\ w(x) & 1-x & 0 & 0 \\ 0 & 0 & 0 & 0 \\ 0 & 0 & 0 & 0 \end{bmatrix}$$

for all $x \in X$.

PROOF. From 4.6(c) we know that

$$\delta : \mathfrak{A} \to (r\mathfrak{A}r)_{2\times 2}, \quad a \mapsto \begin{bmatrix} rar & rajr \\ rjar & rjajr \end{bmatrix}$$

is a tensorable symbol map. Every element $a \in \mathfrak{A}$ is of the form $a = b+cj$ with $b,c \in alg(p,q,r,jqj)$. We so have $rar = rbr+rcjr$, and because $rp = pr = rpr$, $rq = qr = rqr$, and

$$rjqj = j(e-r)qj = jqj-jrqj = jqj-jqj = 0,$$

$$rjr = j(e-r)r = 0,$$

it follows that $r\mathfrak{A}r$ is generated by the two idempotents rpr and rqr (and its identity r), i.e. $r\mathfrak{A}r = \mathrm{alg}(rpr,rqr)$. Since rpr and rqr do not belong to the set $\{0,r\}$, we have

$$\mathrm{sp}_{r\mathfrak{A}r}(rpr) = \mathrm{sp}_{r\mathfrak{A}r}(rqr) = \{0,1\}.$$

If $a \in \mathfrak{A}$ is arbitrary, then

$$\mathrm{sp}_{r\mathfrak{A}r}(rar) \cup \{0\} = \mathrm{sp}_{\mathfrak{A}}(rar) \cup \{0\}.$$

This implies that the spectrum of $rpr\,rqr\,rpr = pqp$ in $r\mathfrak{A}r$ is connected and (due to its closedness) contains $\{0,1\}$, and we also conclude that the unions of $\{0,1\}$ and the spectra of

$$rpr\,rqr\,rpr + (r-rpr)(r-rqr)(r-rpr)$$
$$= r(pqp + (e-p)(e-q)(e-p))r$$

in $r\mathfrak{A}r$ and $\mathfrak{A}$ coincide. So Theorems 4.8 and 4.5 give the existence of a tensorable matrix symbol

$$\sigma_{2\times2} : (r\mathfrak{A}r)_{2\times2} \to C_{4\times4}(X), \quad [a_{ij}]_{i,j=1}^2 \mapsto [\sigma a_{ij}]_{i,j=1}^2$$

such that $\sigma(r)$, $\sigma(rpr)$, $\sigma(rqr)$ take the values

$$\begin{bmatrix} 1 & 0 \\ 0 & 1 \end{bmatrix}, \begin{bmatrix} 1 & 0 \\ 0 & 0 \end{bmatrix}, \begin{bmatrix} x & w(x) \\ w(x) & 1-x \end{bmatrix},$$

respectively, at $x \in X$. Finally, since

$$\delta(e) = \begin{bmatrix} r & 0 \\ 0 & r \end{bmatrix}, \quad \delta(r) = \begin{bmatrix} r & 0 \\ 0 & 0 \end{bmatrix}, \quad \delta(j) = \begin{bmatrix} 0 & r \\ r & 0 \end{bmatrix},$$
$$\delta(p) = \begin{bmatrix} rpr & 0 \\ 0 & r-rpr \end{bmatrix}, \quad \delta(q) = \begin{bmatrix} rqr & 0 \\ 0 & 0 \end{bmatrix},$$

we arrive at the assertion. ■

<u>4.11</u>. THEOREM. Let $\mathfrak{A} = \mathrm{alg}(p,q,r,v)$, where

$$p^2=p,\ q^2=q,\ r^2=r,\ v^n=e,\ rp = pr,\ rq = qr,\ pv = vp,$$
$$rp \neq 0,\ rp \neq r,\ rq \neq 0,\ rq \neq r.$$

Define

$$r_j := v^j\, r\, v^{n-j}, \quad q_j := v^j\, q\, v^{n-j}$$

for $j = 1,\dots,n$ and assume that

$$r_1 + \dots + r_n = e, \quad r_i r_j = 0 \text{ for } i \neq j,$$
$$r_i q_j = q_j r_i = 0 \text{ for } i \neq j .$$

Suppose sp(pqp) is connected and contains $\{0,1\}$. Put

$$X := \mathrm{sp}\Big(r(pqp + (e-p)(e-q)(e-p))r\Big) .$$

Then there exists a tensorable matrix symbol Sym of $\mathfrak{A}$ into $C_{2n\times 2n}(X)$ such that

$$(\mathrm{Sym}\ e)(x) = \mathrm{diag}\ (I,\dots,I) ,$$
$$(\mathrm{Sym}\ r)(x) = \mathrm{diag}\ (0,\dots,0,I) ,$$
$$(\mathrm{Sym}\ p)(x) = \mathrm{diag}\ (E,\dots,E) ,$$
$$(\mathrm{Sym}\ q)(x) = \mathrm{diag}\ (0,\dots,0,W(x)) ,$$

$$(\mathrm{Sym}\ v)(x) = \begin{bmatrix} 0 & \dots & 0 & I \\ I & \dots & 0 & 0 \\ \vdots & \ddots & \vdots & \vdots \\ 0 & \dots & I & 0 \end{bmatrix}$$

for all $x \in X$, where

$$I = \begin{bmatrix} 1 & 0 \\ 0 & 1 \end{bmatrix} , \quad E = \begin{bmatrix} 1 & 0 \\ 0 & 0 \end{bmatrix} , \quad W(x) = \begin{bmatrix} x & w(x) \\ w(x) & 1-x \end{bmatrix} .$$

PROOF. To simplify notation we confine ourselves to the case $n = 3$. From 4.6(a) we deduce that the mapping

$$\beta : \mathfrak{A} \to \mathfrak{A}_{3\times 3} , \quad a \mapsto \left[r_i\ a\ r_j\right]_{i,j=1}^{3}$$

is a tensorable symbol map. Since $V := \mathrm{diag}\ (v^2, v, e)$ is invertible in $\mathfrak{A}_{3\times 3}$, the mapping

$$\eta : \mathfrak{A} \to \mathfrak{A}_{3\times 3} , \quad a \mapsto V\,\beta(a)\,V^{-1}$$

is also a tensorable symbol map. Note that

$$\eta(a) = \begin{bmatrix} rv^2avr & rv^2av^2r & rv^2ar \\ rvavr & rvav^2r & rvar \\ ravr & rav^2r & rar \end{bmatrix} ,$$

and consequently, η is a symbol map of $\mathfrak{A}$ into $(r\mathfrak{A}r)_{3\times 3}$.

Put $\mathcal{L} = \mathrm{alg}(p, r_1, r_2, r_3, q_1, q_2, q_3)$. We claim that $r\mathfrak{A}r = r\mathcal{L}r$. Indeed, every element a in $\mathfrak{A}$ can be written in the form $a = b + cv + dv^2$, where b, c, d belong to $\mathcal{L}$, and since $rg = rgr$ for all $g \in \mathcal{L}$, it results that

$$rar = rbr + rcvr + rdv^2r = rbr + rcrvr + rdrv^2r$$
$$= rbr + rcvr_2r + rdv^2r_1r = rbr \in r\mathcal{L}r .$$

The reverse inclusion $r\mathcal{B}r \subset r\mathcal{A}r$ is trivial.

The algebra $r\mathcal{B}r$ is generated by rpr, rr_jr, rq_jr $(j = 1,2,3)$ and thus by rpr and rqr (and its identity r). So Theorems 4.8 and 4.5 provide a tensorable matrix symbol

$$\sigma_{3\times3} : (r\mathcal{A}r)_{3\times3} = (r\mathcal{B}r)_{3\times3} \to C_{6\times6}(X)$$

such that $\sigma(r) = I$, $\sigma(rpr) = E$, $\sigma(rqr)(x) = W(x)$, and since

$$\eta(e) = \operatorname{diag}(r,r,r), \quad \eta(r) = \operatorname{diag}(0,0,r),$$
$$\eta(p) = \operatorname{diag}(rpr,rpr,rpr), \quad \eta(q) = \operatorname{diag}(0,0,rqr),$$
$$\eta(v) = \begin{bmatrix} 0 & 0 & r \\ r & 0 & 0 \\ 0 & r & 0 \end{bmatrix},$$

we immediately get the assertion. ■

<u>4.12. Baby versions</u>. The last three theorems have certain baby variants, which will be needed in the following.

(a) Let $\mathcal{A} = \operatorname{alg}(p,j)$, where $p^2=p$, $j^2=e$, $jpj=e-p$, and suppose that $\operatorname{sp} p = \{0,1\}$. Then there exists a tensorable matrix symbol Sym of $\mathcal{A}$ into $\mathbb{C}_{2\times2}$ such that

$$\operatorname{Sym} e = \begin{bmatrix} 1 & 0 \\ 0 & 1 \end{bmatrix}, \quad \operatorname{Sym} p = \begin{bmatrix} 1 & 0 \\ 0 & 0 \end{bmatrix}, \quad \operatorname{Sym} j = \begin{bmatrix} 0 & i \\ -i & 0 \end{bmatrix}.$$

PROOF. Define δ as in 4.6(c) and let

$$\eta : \mathcal{A} \to (p\mathcal{A}p)_{2\times2}, \quad a \mapsto \begin{bmatrix} 1 & 0 \\ 0 & -i \end{bmatrix} \delta(a) \begin{bmatrix} 1 & 0 \\ 0 & i \end{bmatrix}.$$

Since $p\mathcal{A}p$ is nothing but $\{\mu p : \mu \in \mathbb{C}\}$, we may define an isomorphism $\vartheta : p\mathcal{A}p \to \mathbb{C}$ by $\vartheta(\mu p) := \mu$. The desired matrix symbol is $\vartheta_{2\times2} \circ \eta$. ■

(b) Let $\mathcal{A} = \operatorname{alg}(p,r,j)$, where $p^2=p$, $r^2=r$, $j^2=e$, $pr=rp=0$, $jpj=e-p$, $jrj=e-r$, $\operatorname{sp} p = \{0,1\}$. Then there is a tensorable matrix symbol Sym of $\mathcal{A}$ into $C_{4\times4}$ such that

$$\operatorname{Sym} e = \begin{bmatrix} 1&0&0&0 \\ 0&1&0&0 \\ 0&0&1&0 \\ 0&0&0&1 \end{bmatrix}, \quad \operatorname{Sym} p = \begin{bmatrix} 1&0&0&0 \\ 0&0&0&0 \\ 0&0&0&0 \\ 0&0&0&1 \end{bmatrix},$$

$$\operatorname{Sym} r = \begin{bmatrix} 1&0&0&0 \\ 0&1&0&0 \\ 0&0&0&0 \\ 0&0&0&0 \end{bmatrix}, \quad \operatorname{Sym} j = \begin{bmatrix} 0&0&1&0 \\ 0&0&0&1 \\ 1&0&0&0 \\ 0&1&0&0 \end{bmatrix}.$$

PROOF. By 4.6(c), the mapping

$$\delta : \mathfrak{A} \to (r\mathfrak{A}r)_{2\times 2}\ , \quad a \mapsto \begin{bmatrix} rar & rajr \\ rjar & rjajr \end{bmatrix}$$

is a tensorable symbol map. The algebra $r\mathfrak{A}r$ is generated by rpr (the identity in $r\mathfrak{A}r$ being r). Hence $r\mathfrak{A}r$ is commutative and its maximal ideal space is $\{0,1\}$. The mapping

$$\vartheta : C_{2\times 2}(\{0,1\}) \to C_{4\times 4}\ , \quad \begin{bmatrix} a & b \\ c & d \end{bmatrix} \mapsto \begin{bmatrix} a(1) & 0 & b(1) & 0 \\ 0 & a(0) & 0 & b(0) \\ c(1) & 0 & d(1) & 0 \\ 0 & c(0) & 0 & d(0) \end{bmatrix}$$

is a tensorable matrix symbol. The asserted matrix symbol can be easily seen to be just $\vartheta \circ \delta$. ∎

(c) Let $\mathfrak{A} = \mathrm{alg}(p,r,v)$, where $p^2=p$, $r^2=r$, $v^n=e$, $rp=$ $=pr=0$, $pv=vp$. Define r_j as $v^j r v^{n-j}$ for $j=1,\dots,n$ and suppose that $r_1+\dots+r_n = e$ and $r_i r_j = 0$ for $i\neq j$. Finally assume that $\mathrm{sp}\ p = \{0,1\}$. Then there exists a tensorable matrix symbol Sym of $\mathfrak{A}$ into $C_{2n\times 2n}$ such that

$$\mathrm{Sym}\ e = \mathrm{diag}\ (I,\dots,I)\ , \quad \mathrm{Sym}\ p = \mathrm{diag}\ (E,\dots,E)\ ,$$

$$\mathrm{Sym}\ r = \mathrm{diag}\ (0,\dots,0,I)\ , \quad \mathrm{Sym}\ v = \begin{bmatrix} 0 & \dots & 0 & I \\ I & \dots & 0 & 0 \\ \vdots & \ddots & \vdots & \vdots \\ 0 & \dots & I & 0 \end{bmatrix}\ ,$$

where $I = \begin{bmatrix} 1 & 0 \\ 0 & 1 \end{bmatrix}$ and $E = \begin{bmatrix} 1 & 0 \\ 0 & 0 \end{bmatrix}$.

PROOF. Define η as in the proof of Theorem 4.11. The algebra $(r\mathfrak{A}r)_{3\times 3}$ is generated by rpr (and by its identity r), and the assertion can so be obtained by a reasoning similar to the one in the proof of part (b) of this subsection. ∎

5. LOCAL SPECTRA

5.1. Definitions. Henceforth let

$$\mathfrak{A} = \mathrm{alg}(PQC,P,C_\alpha)\ , \quad \mathfrak{A}^\pi = \mathrm{alg}^\pi(PQC,P,C_\alpha)\ ,$$

$$\mathfrak{A}^\pi_\xi = \mathrm{alg}^\pi_\xi(PQC,P,C_\alpha)\ , \quad \mathfrak{L} = \mathrm{alg}(PQC,P)\ , \quad \mathfrak{L}^\pi = \mathrm{alg}^\pi(PQC,P).$$

Note that $\mathfrak{L}$ and $\mathfrak{L}^\pi$ are closed subalgebras of $\mathfrak{A}$ and $\mathfrak{A}^\pi$, respectively. The algebra $QC = \{\Phi^\pi(M_\varphi) : \varphi \in QC\}$ is contained in the center of $\mathcal{L}^\pi$. For $y \in Y = M(QC)$, denote by K_y the smallest closed two-sided ideal of $\mathcal{L}^\pi$ containing the set $\{\Phi^\pi(M_\varphi) : \varphi \in QC,\ \varphi(y) = 0\}$ and put $\mathcal{L}^\pi_y = \mathcal{L}^\pi / K_y$. The local principle of Allan and Douglas (see 3.1) tells us that

$$\Psi : \mathcal{L}^\pi \to \bigoplus_{y \in Y} \mathcal{L}^\pi_y , \quad a \mapsto \bigoplus_{y \in Y} \Psi^\pi_y(A) ,$$

where $\Psi^\pi_y(A) := \Phi^\pi(A) + K_y$, is a symbol map.

Recall the definition of Y^-_τ, Y^0_τ, Y^+_τ given in 2.5. Put

$$Y^\pm = \bigcup_{\tau \in \mathbb{T}} Y^\pm_\tau , \quad Y^0 = \bigcup_{\tau \in \mathbb{T}} Y^0_\tau .$$

Given a number $\lambda \in (1, \infty)$ and two points $z_1, z_2 \in \mathbb{C}$, we denote by $\mathcal{A}_\lambda[z_1, z_2]$ the circular arc (including its endpoints) which joins z_1 to z_2 and passes through the point

$$\frac{1}{2}(z_1+z_2) + \frac{i}{2}(z_2-z_1) \cot \frac{\pi}{\lambda} .$$

Let ϱ be a weight as in 2.1. We associate with each point $\tau \in \mathbb{T}$ a number p_τ by $p_\tau = p$ if $\tau \notin \{t_1, \ldots, t_m\}$ and $p_\tau = p/(1+\mu_j p)$ if $\tau = t_j$.

5.2. THEOREM. For $a \in PQC$ the following are equivalent:

(i) the Toeplitz operator $T_a = PaP + Q$ is Fredholm on $L^p(\varrho)$;

(ii) the coset $\Phi^\pi(T_a)$ is invertible in $\mathcal{L}^\pi$;

(iii) $a(y,0) \neq 0$ for all $y \in Y^-$, $a(y,1) \neq 0$ for all $y \in Y^+$, and $0 \notin \mathcal{A}_{p_\tau}[a(y,0), a(y,1)]$ for all $y \in Y^0_\tau$, $\tau \in \mathbb{T}$.

The implications (iii) $\Rightarrow$ (ii) $\Rightarrow$ (i) were established in [4, 5.41] ; an independent proof of the implication (iii) $\Rightarrow$ $\Rightarrow$ (i) is also in [5]. The difficult part of the theorem is the implication (i) $\Rightarrow$ (iii); it was proved in [6]. We remark that this theorem is Sarason's [24] for $p = 2$ and $\varrho \equiv 1$. ■

Our next objective is to use the preceding theorem in order to describe the local spectra $sp \Psi^\pi_y(T_a)$ for a in PQC. The following theorem disposes of this question; it was already announced in [6].

5.3. THEOREM. Let $\tau \in \mathbb{T}$, $y \in Y_\tau$, $a \in PQC$, $T_a = PaP+Q$. Then

$$\operatorname{sp} \Psi_y^\pi(T_a) = \{a(y,0)\} \quad \text{for } y \in Y_\tau^- ,$$
$$\operatorname{sp} \Psi_y^\pi(T_a) = \{a(y,1)\} \quad \text{for } y \in Y_\tau^+ ,$$
$$\operatorname{sp} \Psi_y^\pi(T_a) = \mathcal{A}_{p_\tau}[a(y,0),a(y,1)] \quad \text{for } y \in Y_\tau^0 .$$

PROOF. If $y \in Y_\tau^-$, then $\Psi_y^\pi(T_a) = a(y,0)\,\Psi_y^\pi(I)$, which implies the assertion. The case where $y \in Y_\tau^+$ can be settled in the same way. So let $y \in Y_\tau^0$. Because

$$\Psi_y^\pi(T_a) = a(y,0)\,\Psi_y^\pi(I) + [a(y,1) - a(y,0)]\,\Psi_y^\pi(\chi),$$

where $\chi \in PC$ is any function which is continuous on $\mathbb{T} \setminus \{\tau\}$ and satisfies $0 \leqq \chi \leqq 1$, $\chi(\tau-0) = 0$, $\chi(\tau+0) = 1$, we are left with proving that $\operatorname{sp} \Psi_y^\pi(T_\chi) = \mathcal{A}_{p_\tau}[0,1]$.

From Theorem 5.2 and the local principle quoted in 5.1 we infer that $\operatorname{sp} \Psi_y^\pi(T_\chi)$ is contained in $\mathcal{A}_{p_\tau}[0,1]$. To show the reverse inclusion assume the contrary, i.e. assume that there is a $\lambda \in \mathcal{A}_{p_\tau}[0,1]$ such that $\lambda \notin \operatorname{sp} \Psi_y^\pi(T_\chi)$.

Thus, $\Psi_y^\pi(T_\chi - \lambda I)$ is invertible. A standard upper semi-continuity argument (see [4, 1.34(b)] or [3, Theorem 3.2(b)]) shows that $\Psi_\eta^\pi(T_\chi - \lambda I)$ is invertible for all η in some open neighborhood $U \subset Y$ of y.

There exist a function $\varphi \in QC$ and a closed set $F \subset Y$ such that $y \in F \subset U$, $0 \leqq \varphi \leqq 1$, $\varphi(\eta) = 1$ for $\eta \in F$ and $|\varphi(\eta)| < 1$ for $\eta \in Y \setminus U$. This follows immediately from the fact that Y as a compact Hausdorff space is normal, but can be also seen as follows. The set $\{y\}$, being a singleton, is a weak peak set for QC and so there are closed sets $F_\alpha \subset Y$ and functions $\varphi_\alpha \in QC$ satisfying

$$\{y\} = \bigcap_\alpha F_\alpha, \quad \varphi_\alpha | F_\alpha = 1, \quad |\varphi_\alpha| < 1 \text{ on } Y \setminus F .$$

By the compactness of Y, there are finitely many $\alpha_1, \dots, \alpha_n$ such that $F := F_{\alpha_1} \cap \dots \cap F_{\alpha_n} \subset U$, and $\varphi := |\varphi_{\alpha_1} \cdots \varphi_{\alpha_n}|$ is the desired function.

Consider the function $b := \varphi\chi - \lambda$. Theorem 5.2 shows that sp $\Phi^\pi(T_b)$ is contained in the set bounded by the line segment $[-\lambda,1-\lambda]$ and the arc $\mathcal{A}_{p_\tau}[-\lambda,1-\lambda]$. Note that 0 is located on the arc $\mathcal{A}_{p_\tau}[-\lambda,1-\lambda]$. Our assumption yields that 0 belongs to sp $\Phi^\pi(T_b)$.

If $\eta \notin Y_\tau^0$ then clearly $0 \notin \operatorname{sp} \Psi_\eta^\pi(T_b)$. If $\eta \in Y_\tau^0 \setminus F$ then sp $\Psi_\eta^\pi(T_b)$ is contained in some arc $\mathcal{A}_{p_\tau}[-\lambda,\mu]$, where μ is an inner point of the segment $[-\lambda,1-\lambda]$. Hence $0 \notin \operatorname{sp} \Psi_\eta^\pi(T_b)$ for y in $Y_\tau^0 \setminus F$. Finally, if $\eta \in Y_\tau^0 \cap F$ then $\Psi_\eta^\pi(T_b) = \Psi_\eta^\pi(T_\chi - \lambda I)$ and thus $0 \notin \operatorname{sp} \Psi_\eta^\pi(T_b)$ by assumption. It follows that $0 \notin \operatorname{sp} \Phi^\pi(T_b)$. This contradiction completes the proof. ∎

5.4. Notation. Given a subset $\xi \subset Y$, we denote by J_ξ (resp. K_ξ) the smallest closed two-sided ideal of $\mathfrak{A}^\pi$ (resp. $\mathcal{L}^\pi$) containing the set $\{\Phi^\pi(M_\varphi) : \varphi \in QC,\ \varphi(y) = 0 \quad \forall y \in \xi\}$. It is easily seen that if $\xi = \mathcal{O}_\alpha(y)$ is an orbit, then the J_ξ introduced here coincides with the J_ξ defined in 3.2. Furthermore, if $\xi = \{y\}$ is a singleton, then K_ξ is just the ideal K_y defined in 5.1.

5.5. LEMMA. $J_\xi \cap \mathcal{L}^\pi = K_\xi$.

PROOF. The inclusion $K_\xi \subset J_\xi \cap \mathcal{L}^\pi$ is obvious. To show the reverse inclusion, notice first that if C_α is of the order n, then

$$N_\xi := K_\xi + K_\xi \Phi^\pi(C_\alpha) + \dots + K_\xi \Phi^\pi(C_\alpha^{n-1})$$
$$:= \left\{\Phi^\pi\Big(\sum_{j=0}^{n-1} B_j C_\alpha^j\Big) : \Phi^\pi(B_j) \in K_\xi\right\}$$

is a closed two-sided ideal of $\mathfrak{A}^\pi$ containing $\{\Phi^\pi(M_\varphi) : \varphi \in QC, \varphi|\xi = 0\}$. Therefore the (minimal) ideal J_ξ is contained in N_ξ. This implies that every element $\Phi^\pi(A) \in J_\xi$ is of the form

$$\Phi^\pi(A) = \sum_{j=0}^{n-1} \Phi^\pi(B_j)\, \Phi^\pi(C_\alpha^j)$$

with $\Phi^\pi(B_j)$ in $\mathcal{L}^\pi$. Since such a representation is unique (recall the last paragraph of 2.7), we conclude that actually $\Phi^\pi(B_j) = 0$ for $j \ge 1$ whenever $\Phi^\pi(A) \in \mathcal{L}^\pi$, i.e. that then $\Phi^\pi(A) = \Phi^\pi(B_0) \in K_\xi$. ∎

We are now in a position to apply the theorems of Section 4 to the situations discussed in Section 3.

5.6. THEOREM. Let the situation be as in 3.5 and let $\xi = \{y\}$ be in Y_τ^o. Then

$$sp(pqp) = sp\ \Phi_\xi^\pi(P\chi_0 P) = \mathcal{A}_{p_\tau}[0,1],$$

$$sp(pqp+(e-p)(e-q)(e-p)) = sp\ \Phi_\xi^\pi(P\chi_0 P+Q(1-\chi_0)Q) = \mathcal{A}_{p_\tau}[0,1].$$

Moreover,

$$sp(ipqjp) = sp\ \Phi_\xi^\pi(iP\chi_0 JP) = w\ (\mathcal{A}_{p_\tau}[0,1]),$$

where w denotes the branch of $w(z) = \sqrt{z(1-z)}$ which is analytic in $\mathbb{C}\setminus((-\infty,0)\cup(1,\infty))$ and assumes positive values on (0,1).

PROOF. We have $\Phi_\xi^\pi(P\chi_0 P) = \Phi_\xi^\pi(P\chi P)$, where χ is as in the proof of Theorem 5.3 and so the reasoning of that proof (with only minor modifications) gives the assertion of the present theorem concerning sp(pqp). One similarly obtains that $sp\ \Phi_\xi^\pi(Q(1-\chi_0)Q) = \mathcal{A}_{p_\tau}[0,1]$. Since

$$\begin{aligned} &P\chi_0 P + Q(1-\chi_0)Q - \lambda I \\ &= (P(\chi_0-\lambda)P + Q)\ (P + Q(1-\chi_0-\lambda)Q) \\ &= (P + Q(1-\chi_0-\lambda)Q)\ (P(\chi_0-\lambda)P + Q), \end{aligned}$$

we see that sp(pqp+(e-p)(e-q)(e-p)) is also $\mathcal{A}_{p_\tau}[0,1]$.

The argument of the proof of Proposition 3(b) of [21] shows that the spectrum of $\Phi^\pi(iP\chi_0 JP)$ in $\mathfrak{A}^\pi$ equals $w(\mathcal{A}_{p_\tau}[0,1])$. So the second part of the assertion follows from Theorem 4.9 (and the remark made in the last paragraph of 4.9) along with the local principle of 3.1. ∎

5.7. THEOREM. Consider the situation of 3.6, let y be in Y_τ^o and let $\xi = \{y, \alpha(y)\}$. Then both

$$sp(pqp) = sp\ \Phi_\xi^\pi(P\chi_0 P)$$

and

$$\begin{aligned} &sp(rpqpr + r(e-p)(e-q)(e-p)r) \\ &= sp\ \Phi_\xi^\pi(\varphi_0 P\chi_0 P\varphi_0 + \varphi_0 Q(1-\chi_0)Q\varphi_0) \end{aligned}$$

equal $\mathcal{A}_{p_\tau}[0,1]$.

PROOF. By 4.6(c), the mapping

$$\delta : \mathfrak{A}_\xi^\pi \to (r\,\mathfrak{A}_\xi^\pi r)_{2\times 2}\ ,\quad a \mapsto \begin{pmatrix} rar & rajr \\ rjar & rjajr \end{pmatrix}$$

is a symbol map. We have

$$\delta(rpqpr - \lambda e) = \begin{bmatrix} rpqpr - \lambda r & 0 \\ 0 & -\lambda r \end{bmatrix},$$

$$\delta(rpqpr + r(e-p)(e-q)(e-p)r - \lambda e)$$

$$= \begin{bmatrix} r(e-p)(e-q)(e-p)r - \lambda r & 0 \\ 0 & -\lambda r \end{bmatrix},$$

and therefore the spectra we are interested in coincide with the spectra of some elements in the algebra $r\mathfrak{A}_\xi^\pi r$ (whose identity is r).

Since $r\mathfrak{A}_\xi^\pi r$ is generated by $\Phi_\xi^\pi(\varphi_0 \chi_0 \varphi_0)$ and $\Phi_\xi^\pi(\varphi_0 P \varphi_0)$ (see the proof of Theorem 4.10), every element of $r\mathfrak{A}_\xi^\pi r$ is of the form $\Phi_\xi^\pi(\varphi_0 A \varphi_0)$ with $A \in \mathcal{L}$. Define $\mu: r\mathfrak{A}_\xi^\pi r \to \mathcal{L}_y^\pi$ by

$$\mu(\Phi_\xi^\pi(\varphi_0 A \varphi_0)) = \Psi_y^\pi(A) \qquad (A \in \mathcal{L}).$$

This mapping is well-defined: if $\Phi_\xi^\pi(\varphi_0 A \varphi_0) = 0$, then $\Phi^\pi(\varphi_0 A \varphi_0)$ is in $J_\xi \cap \mathcal{L}^\pi$, and hence, by Lemma 5.5, $\Phi^\pi(\varphi_0 A \varphi_0) \in K_\xi$, which implies that

$$\Phi^\pi(A) = \Phi^\pi(\varphi_0 A \varphi_0 + \varphi_0 A(1-\varphi_0) + (1-\varphi_0)A\varphi_0 + (1-\varphi_0)A(1-\varphi_0))$$

belongs to K_y. Thus $\Psi_y^\pi(A) = 0$. It is easy to see that μ is a tensorable symbol map (we only remark that if $\Psi_y^\pi(B)$ is the inverse of $\Psi_y^\pi(A)$, then $\Phi_\xi^\pi(\varphi_0 B \varphi_0)$ is the inverse of $\Phi_\xi^\pi(\varphi_0 A \varphi_0)$).

Consequently, our spectra coincide with the spectra of the elements

$$\Psi_y^\pi(P\chi_0 P) \quad \text{and} \quad \Psi_y^\pi(P\chi_0 P + Q(1-\chi_0)Q)$$

in $\mathcal{L}_y^\pi$. The first of these two spectra was identified in Theorem 5.3. The arguments applied in 5.3 to show that the spectrum of $\Psi_y^\pi(P\chi_0 P + Q)$ is $\mathcal{A}_{p_\tau}[0,1]$ also give that the spectrum of $\Psi_y^\pi(P + Q(1-\chi_0)Q)$ equals $\mathcal{A}_{p_\tau}[0,1]$. Hence, we obtain as in the proof of Theorem 5.6 that $\operatorname{sp} \Psi_y^\pi(P\chi_0 P + Q(1-\chi_0)Q) = \mathcal{A}_{p_\tau}[0,1]$. ■

<u>5.8</u>. THEOREM. Suppose we are given the situation of 3.3. Then if $y \in Y_\tau^0$ and $\xi = \{y, \alpha(y), \ldots, \alpha_{n-1}(y)\}$, the spectra

$$\operatorname{sp}(pqp) = \operatorname{sp} \Phi_\xi^\pi(P\chi_0 P)$$

and

$$\text{sp}\ (rpqpr + r(e-p)(e-q)(e-p)r)$$
$$= \text{sp}\ \Phi_\xi^\pi(\varphi_0 P \chi_0 P \varphi_0 + \varphi_0 Q(1-\chi_0)Q\varphi_0)$$

are both equal to $\mathcal{A}_{p_\tau}[0,1]$.

PROOF. For simplicity, let n = 3. Define a symbol map $\eta : \mathfrak{A}_\xi^\pi \to (r\mathfrak{A}_\xi^\pi r)_{3\times3}$ as in the proof of Theorem 4.11. We have $r\mathfrak{A}_\xi^\pi r = r\mathcal{C}r$, where $\mathcal{C}$ is the closed subalgebra of $\mathfrak{A}_\xi^\pi$ generated by

$$\Phi_\xi^\pi(P),\ \Phi_\xi^\pi(\varphi_0),\ \Phi_\xi^\pi(\varphi_1),\ \Phi_\xi^\pi(\varphi_2),\ \Phi_\xi^\pi(\chi_0),\ \Phi_\xi^\pi(\chi_1),\ \Phi_\xi^\pi(\chi_2)$$

(see the proof of Theorem 4.11). So every element of $r\mathfrak{A}_\xi^\pi r$ is of the form $\Phi_\xi^\pi(\varphi_0 A \varphi_0)$ with $A \in \mathcal{L}$. As in the proof of the preceding theorem one can see that the mapping

$$\mu:\ r\mathfrak{A}_\xi^\pi r \to \mathcal{L}_y^\pi\ ,\quad \Phi_\xi^\pi(\varphi_0 A \varphi_0) \mapsto \Psi_y^\pi(A)$$

is well-defined and a tensorable symbol map. Computing $\eta(pqp)$ and $\eta(rpqpr + r(e-p)(e-q)(e-p)r)$ one obtains two diagonal matrices, and applying μ to the nontrivial diagonal elements one gets the assertion as in the previous proof. ■

6. SYMBOL CONSTRUCTION

6.1. THEOREM. Let C_α be a forward Carleman shift of the order n. For $\tau \in \mathbb{T}$, set

$$\mathcal{M}_\tau = Y_\tau^- \cup (Y_\tau^0 \times \mathcal{A}_{p_\tau}[0,1]) \cup Y_\tau^+$$

and put $\mathcal{M} = \bigcup_{\tau\in\mathbb{T}} \mathcal{M}_\tau$. There exists a tensorable matrix symbol

$$\text{Sym} : \text{alg}^\pi(PQC, P, C\ ) \to C_{2n\times2n}(\mathcal{M})$$

such that

(i) $\quad \text{Sym}\ \Phi^\pi(P)(\mu) = \begin{bmatrix} E & & 0 \\ & \ddots & \\ 0 & & E \end{bmatrix} \quad \forall \mu \in \mathcal{M}$, where $E = \begin{bmatrix} 1 & 0 \\ 0 & 0 \end{bmatrix}$;

(ii) $\quad \text{Sym}\ \Phi^\pi(C_\alpha)(\mu) = \begin{bmatrix} 0 & \dots & 0 & I \\ I & \dots & 0 & 0 \\ \vdots & \ddots & \vdots & \vdots \\ 0 & \dots & I & 0 \end{bmatrix} \quad \forall \mu \in \mathcal{M}$, where $I = \begin{bmatrix} 1 & 0 \\ 0 & 1 \end{bmatrix}$;

(iii) $\quad$ if $a \in PQC$, then

$$\text{Sym}\ \Phi^\pi(M_a)(\mu) = \text{diag}\ (A_1(\mu), \dots, A_n(\mu)),$$

where $A_j(\mu)$ is given by

$$\begin{bmatrix} a(\alpha_j(y),1)x + a(\alpha_j(y),0)(1-x) & [a(\alpha_j(y),1) - a(\alpha_j(y),0)]\, w(x) \\ [a(\alpha_j(y),1) - a(\alpha_j(y),0)]\, w(x) & a(\alpha_j(y),0)x + a(\alpha_j(y),1)(1-x) \end{bmatrix}$$

for $\mu = (y,x)$ in $Y_\tau^0 \times \mathcal{A}_{p_\tau}[0,1]$ ($w : \mathbb{C} \to \mathbb{C}$ being any function such that $[w(z)]^2 = z(1-z)$ for all z) and given by

$$\begin{bmatrix} a(\alpha_j(y),0) & 0 \\ 0 & a(\alpha_j(y),0) \end{bmatrix} \quad \text{resp.} \quad \begin{bmatrix} a(\alpha_j(y),1) & 0 \\ 0 & a(\alpha_j(y),1) \end{bmatrix}$$

for $\mu = y \in Y_\tau^-$ resp. $\mu = y \in Y_\tau^+$.

PROOF. Let ϕ be the tensorable symbol map

$$\phi : \mathrm{alg}^\pi(PQC, P, C_\alpha) \to \bigoplus_{\xi \in M(QC_\alpha)} \mathrm{alg}_\xi^\pi(PQC, P, C_\alpha)$$

introduced in 3.2.

We first consider the case where $y \in Y_\tau^0$ and $\xi = \{y, \alpha(y), \ldots, \alpha_{n-1}(y)\}$. Theorem 4.11 in conjunction with Theorem 5.8 yields a tensorable matrix symbol

$$\mathrm{Sym}_\xi : \mathrm{alg}_\xi^\pi(PQC, P, C_\alpha) \to C_{2n\times 2n}(\mathcal{A}_{p_\tau}[0,1])$$

such that $\mathrm{Sym}_\xi\, \phi_\xi^\pi(P)(x)$ and $\mathrm{Sym}_\xi\, \phi_\xi^\pi(C_\alpha)(x)$ equal

$$\begin{bmatrix} E & & 0 \\ & \ddots & \\ 0 & & E \end{bmatrix} \quad \text{and} \quad \begin{bmatrix} 0 & \cdots & 0 & I \\ I & \cdots & 0 & 0 \\ \vdots & \ddots & \vdots & \vdots \\ 0 & \cdots & I & 0 \end{bmatrix} \qquad \forall x \in \mathcal{A}_{p_\tau}[0,1] \, ,$$

respectively, and such that, with χ_j and φ_j as in 3.3,

$$\mathrm{Sym}_\xi\, \phi_\xi^\pi(\chi_j)(x) = \mathrm{diag}\,(0,\ldots,0,W(x),0,\ldots,0) \, ,$$

$$\mathrm{Sym}_\xi\, \phi_\xi^\pi(\varphi_j)(x) = \mathrm{diag}\,(0,\ldots,0,I,0,\ldots,0) \, ,$$

where $W(x)$ resp. I are at the j-th place ($j = 1,\ldots,n$) and $\chi_n := \chi_0$, $\varphi_n := \varphi_0$. If $a \in PQC$, then

$$\phi_\xi^\pi(a) = \sum_{j=1}^n [a(\alpha_j(y),1) - a(\alpha_j(y),0)]\, \phi_\xi^\pi(\chi_j) + \sum_{j=1}^n a(\alpha_j(y),0)\, \phi_\xi^\pi(\varphi_j)$$

and consequently,

$$\mathrm{Sym}_\xi \Phi_\xi^\pi(M_a)(x) = \mathrm{diag}\,(B_1(x),\dots,B_n(x)) ,$$

where

$$B_j(x) = \left[a(\alpha_j(y),1) - a(\alpha_j(y),0)\right] W(x) + a(\alpha_j(y),0)I = A_j(y,x).$$

Using the baby version 4.12(c) we similarly obtain a tensorable matrix symbol

$$\mathrm{Sym}_\xi : \mathrm{alg}_\xi^\pi (PQC,P,C_\alpha) \to \mathbb{C}_{2n\times 2n}$$

in case $\xi = \{y, \alpha(y),\dots,\alpha_{n-1}(y)\}$ with $y \in Y_\tau^- \cup Y_\tau^+$. Composing Sym_ξ and Φ we finally arrive at the assertion. ■

6.2. THEOREM. Let C_α be a backward Carleman shift. Denote the fixed points of α by τ_1, τ_2 and let $\mathring{\mathbb{T}}$ be either of the two connected components of $\mathbb{T}\setminus\{\tau_1,\tau_2\}$. For $\tau \in \mathring{\mathbb{T}}$, put

$$\mathcal{N}_\tau = Y_\tau^- \cup (Y_\tau^0 \times \mathcal{A}_{p_\tau}[0,1]) \cup Y_\tau^+$$

and

$$\mathcal{N} = \Big(\bigcup_{\tau\in\mathring{\mathbb{T}}} \mathcal{N}_\tau\Big) \cup Y_{\tau_1}^- \cup Y_{\tau_1}^+ \cup Y_{\tau_2}^- \cup Y_{\tau_2}^+ ,$$

$$\mathcal{R} = (Y_{\tau_1}^0 \times \mathcal{A}_{p_{\tau_1}}[0,1]) \cup (Y_{\tau_2}^0 \times \mathcal{A}_{p_{\tau_2}}[0,1]) .$$

Let $w : \mathbb{C}\to\mathbb{C}$ denote the branch of the function $w(z) = \sqrt{z(1-z)}$ which is analytic in $\mathbb{C} \setminus ((-\infty,0) \cup (1,\infty))$ and positive on $(0,1)$. There exists a tensorable matrix symbol

$$\mathrm{Sym} : \mathrm{alg}^\pi(PQC,P,C_\alpha) \to C_{4\times 4}(\mathcal{N}) \oplus C_{2\times 2}(\mathcal{R})$$

such that

(i) $\mathrm{Sym}\, \Phi^\pi(P)(\nu) = \begin{bmatrix} 1&0&0&0\\ 0&0&0&0\\ 0&0&0&0\\ 0&0&0&1 \end{bmatrix}$, $\mathrm{Sym}\, \Phi^\pi(P)(\sigma) = \begin{bmatrix} 1&0\\ 0&0 \end{bmatrix}$

for all $\nu\in\mathcal{N}$ and all $\sigma\in\mathcal{R}$;

(ii) $\mathrm{Sym}\, \Phi^\pi(C_\alpha)(\nu) = \begin{bmatrix} 0&0&1&0\\ 0&0&0&1\\ 1&0&0&0\\ 0&1&0&0 \end{bmatrix}$, $\mathrm{Sym}\, \Phi^\pi(C_\alpha)(\sigma) = \begin{bmatrix} 0&i\\ -i&0 \end{bmatrix}$

for all $\nu\in\mathcal{N}$ and all $\sigma\in\mathcal{R}$;

(iii) if $a \in PQC$, then $\mathrm{Sym}\, \Phi^\pi(M_a)(y,x) = \begin{bmatrix} A(y,x) & 0\\ 0 & B(y,x) \end{bmatrix}$, where $A(y,x)$ is

$$\begin{bmatrix} a(y,1)x + a(y,0)(1-x) & [a(y,1) - a(y,0)]w(x) \\ [a(y,1) - a(y,0)]w(x) & a(y,1)(1-x) + a(y,0)x \end{bmatrix}$$

and $B(y,x)$ is

$$\begin{bmatrix} a(\alpha(y),1)(1-x) + a(\alpha(y),0)x & [a(\alpha(y),0) - a(\alpha(y),1)]w(x) \\ [a(\alpha(y),0) - a(\alpha(y),1)]w(x) & a(\alpha(y),1)x + a(\alpha(y),0)(1-x) \end{bmatrix}$$

for $(y,x) \in \bigcup_{\tau \in \mathring{\mathbb{T}}} (Y^{o}_{\tau} \times \mathcal{A}_{p_\tau}[0,1])$;

$$\operatorname{Sym} \Phi^{\pi}(M_a)(y) = \begin{bmatrix} a(y,0)I & 0 \\ 0 & a(\alpha(y),1)I \end{bmatrix}, \quad I = \begin{bmatrix} 1 & 0 \\ 0 & 1 \end{bmatrix}$$

for $y \in \left(\bigcup_{\tau \in \mathring{\mathbb{T}}} Y^{-}_{\tau} \right) \cup Y^{-}_{\tau_1} \cup Y^{-}_{\tau_2}$;

$$\operatorname{Sym} \Phi^{\pi}(M_a)(y) = \begin{bmatrix} a(y,1)I & 0 \\ 0 & a(\alpha(y),0)I \end{bmatrix}, \quad I = \begin{bmatrix} 1 & 0 \\ 0 & 1 \end{bmatrix}$$

for $y \in \left(\bigcup_{\tau \in \mathring{\mathbb{T}}} Y^{+}_{\tau} \right) \cup Y^{+}_{\tau_1} \cup Y^{+}_{\tau_2}$;

$\operatorname{Sym} \Phi^{\pi}(M_a)(y,x)$ equals

$$\begin{bmatrix} a(y,1)x + a(y,0)(1-x) & [a(y,1) - a(y,0)]w(x) \\ [a(y,1) - a(y,0)]w(x) & a(y,1)(1-x) + a(y,0)x \end{bmatrix}$$

for $(y,x) \in \mathcal{R}$.

PROOF. Using Theorems 4.9, 4.10, 5.6, 5.7 and the results of 4.12 (a),(b) one can prove this theorem in the same fashion as Theorem 6.1. ∎

7. FREDHOLM CRITERIA

7.1. THEOREM. The matrix symbols given by Theorems 6.1 and 6.2 are tensorable Fredholm symbols.

PROOF. What we must show is that if $A \in \mathfrak{A} = \operatorname{alg}(PQC, P, C_\alpha)$ is Fredholm, then $\Phi^{\pi}(A)$ is invertible in the algebra $\mathfrak{A}^{\pi} = \operatorname{alg}(PQC, P, C_\alpha)$. It suffices to prove this for the case where A is a polynomial of P, C_α, and multiplications M_{φ_j} and M_{g_j}, the φ_j's belonging to QC and the g_j's being piecewise polynomial functions with at most finitely many jumps. The approximation argument used in the proof of Corollary 5 of [21] then extends the result to all operators $A \in \mathfrak{A}$. So assume A is a polynomial of the afore-mentioned kind and A is Fredholm.

Let B be any regularizer of A. Put

$$\tilde{\mathfrak{A}} = \text{alg}\,(PQC, P, C_\alpha, B) \text{ and } \tilde{\mathfrak{A}}^\pi = \tilde{\mathfrak{A}} / \mathcal{K}(L^p(\varrho)).$$

Then QC_α is a closed subalgebra of the centers of $\mathfrak{A}^\pi$ and $\tilde{\mathfrak{A}}^\pi$. For $\xi \in M(QC_\alpha)$, denote by $\mathfrak{J}_\xi$ and $\tilde{\mathfrak{J}}_\xi$, respectively, the smallest closed two-sided ideal of $\mathfrak{A}^\pi$ and $\tilde{\mathfrak{A}}^\pi$ containing the set $\{\Phi^\pi(M_\varphi) : \varphi \in QC_\alpha, \varphi(\xi) = 0\}$. Put $\mathfrak{A}^\pi_\xi = \mathfrak{A}^\pi / \mathfrak{J}_\xi$ and $\tilde{\mathfrak{A}}^\pi_\xi = \tilde{\mathfrak{A}}^\pi / \tilde{\mathfrak{J}}_\xi$ and denote by Φ^π_ξ and $\tilde{\Phi}^\pi_\xi$ the canonical homomorphisms of $\mathfrak{A}^\pi$ and $\tilde{\mathfrak{A}}^\pi$ onto $\mathfrak{A}^\pi_\xi$ and $\tilde{\mathfrak{A}}^\pi_\xi$, respectively.

The algebra $\mathfrak{A}^\pi_\xi$ is generated by $\Phi^\pi_\xi(P)$, $\Phi^\pi_\xi(C_\alpha)$, and a finite number of elements $\Phi^\pi_\xi(a_1), \ldots, \Phi^\pi_\xi(a_m)$, where $a_1, \ldots, a_m$ are PQC functions (see 3.3 - 3.6). The results of Sections 4 and 5 involve a matrix symbol Sym_ξ for the algebras $\mathfrak{A}^\pi_\xi$ (see the proof of Theorem 6.1).

Let $\hat{\mathfrak{A}}^\pi_\xi$ be the closed subalgebra of $\tilde{\mathfrak{A}}^\pi_\xi$ which is generated by $\tilde{\Phi}^\pi_\xi(P)$, $\tilde{\Phi}^\pi_\xi(C_\alpha)$, $\tilde{\Phi}^\pi_\xi(a_1), \ldots, \tilde{\Phi}^\pi_\xi(a_m)$. The arguments of Sections 4 and 5 yield a matrix symbol $\widehat{\text{Sym}}_\xi$ for the algebra $\hat{\mathfrak{A}}^\pi_\xi$ such that

$$\text{Sym}_\xi\, \Phi^\pi_\xi(A) = \widehat{\text{Sym}}\, \tilde{\Phi}^\pi_\xi(A)$$

for the operator A in question. It follows that the spectrum of $\Phi^\pi_\xi(A)$ in $\mathfrak{A}^\pi_\xi$ coincides with the spectrum of $\tilde{\Phi}^\pi_\xi(A)$ in $\hat{\mathfrak{A}}^\pi_\xi$ and that, due to the particular structure of A, these spectra are a finite union of curves which cannot be obtained from another set by "filling in holes". Consequently, the spectrum of $\tilde{\Phi}^\pi_\xi(A)$ in $\hat{\mathfrak{A}}^\pi_\xi$ equals the spectrum of $\tilde{\Phi}^\pi_\xi(A)$ in $\tilde{\mathfrak{A}}^\pi_\xi$. Since A is Fredholm, the origin does not belong to the spectrum of $\tilde{\Phi}^\pi_\xi(A)$ in $\tilde{\mathfrak{A}}^\pi_\xi$, and thus neither to the spectrum of $\tilde{\Phi}^\pi_\xi(A)$ in $\hat{\mathfrak{A}}^\pi_\xi$ nor to the spectrum of $\Phi^\pi_\xi(A)$ in $\mathfrak{A}^\pi_\xi$. From the local principle quoted in 3.1 we finally deduce that $\Phi^\pi(A)$ is invertible in $\mathfrak{A}^\pi$, as desired. ■

The previous theorem involves the well known Fredholm criteria for singular integral operators with piecewise continuous coefficients and a Carleman shift.

7.2. THEOREM. Let C_α be a forward Carleman shift of the order n and put

$$\mathcal{M}_0 = \bigcup_{\tau\in\mathbb{T}} (\{\tau\}\times \mathcal{A}_{p_\tau}[0,1]).$$

There exists a tensorable Fredholm symbol $\widetilde{\mathrm{Sym}}$ of $\mathrm{alg}^\pi(PC,P,C_\alpha)$ into $C_{2n\times 2n}(\mathcal{M}_0)$ such that

(i) $$\widetilde{\mathrm{Sym}}\,\Phi^\pi(P)(\tau,x) = \begin{bmatrix} E & & 0 \\ & \ddots & \\ 0 & & E \end{bmatrix} \quad \forall(\tau,x)\in\mathcal{M}_0 \ ; \ E = \begin{bmatrix}1 & 0\\ 0 & 0\end{bmatrix};$$

(ii) $$\widetilde{\mathrm{Sym}}\,\Phi^\pi(C_\alpha)(\tau,x) = \begin{bmatrix} 0 & \dots & 0 & I \\ I & \dots & 0 & 0 \\ \vdots & \ddots & \vdots & \vdots \\ 0 & \dots & I & 0 \end{bmatrix} \quad \forall(\tau,x)\in\mathcal{M}_0 ; \ I = \begin{bmatrix}1 & 0\\ 0 & 1\end{bmatrix};$$

(iii) if $a\in PC$, then

$$\widetilde{\mathrm{Sym}}\,\Phi^\pi(M_a)(\tau,x) = \begin{bmatrix} A_1(\tau,x) & & 0 \\ & \ddots & \\ 0 & & A_n(\tau,x) \end{bmatrix},$$

where, for $(\tau,x)\in\mathcal{M}_0$, $A_j(\tau,x)$ is given by

$$\begin{bmatrix} a(\tau_j+0)x + a(\tau_j-0)(1-x) & [a(\tau_j+0) - a(\tau_j-0)]w(x) \\ [a(\tau_j+0) - a(\tau_j-0)]w(x) & a(\tau_j-0)x + a(\tau_j+0)(1-x) \end{bmatrix},$$

$\tau_j := \alpha_j(\tau)$, and $w : \mathbb{C}\to\mathbb{C}$ is any function such that $[w(z)]^2 = z(1-z)$ for all $z\in\mathbb{C}$.

PROOF. If $A\in\mathrm{alg}\,(PC,P,C_\alpha)$, then the ranges of the functions $\mathrm{Sym}\,\Phi^\pi(A)$ and $\widetilde{\mathrm{Sym}}\,\Phi^\pi(A)$ on $Y_\tau^0\times\mathcal{A}_{p_\tau}[0,1]$ and $\{\tau\}\times\mathcal{A}_{p_\tau}[0,1]$, respectively, coincide. The question we must settle is how the values of $\mathrm{Sym}\,\Phi^\pi(A)$ on $Y_\tau^-\cup Y_\tau^+$ disappear. The answer is : if $\mathrm{Sym}\,\Phi^\pi(A)$ is invertible on $Y_\tau^0\times\{0,1\}$, then so it is also on $Y_\tau^-\cup Y_\tau^+$. Let us prove the last claim.

For the sake of definiteness assume n = 3. It is clear that the mapping

$$\sigma : \Phi^\pi(A)\mapsto \begin{bmatrix} 1&0&0&0&0&0\\ 0&0&1&0&0&0\\ 0&0&0&0&1&0\\ 0&1&0&0&0&0\\ 0&0&0&1&0&0\\ 0&0&0&0&0&1 \end{bmatrix} \mathrm{Sym}\,\Phi^\pi(A) \begin{bmatrix} 1&0&0&0&0&0\\ 0&0&1&0&0&0\\ 0&0&0&0&1&0\\ 0&1&0&0&0&0\\ 0&0&0&1&0&0\\ 0&0&0&0&0&1 \end{bmatrix}$$

is also a Fredholm symbol. For all $\mu \in \mathfrak{M}$ we have

$$\sigma\Phi^{\pi}(P)(\mu) = \begin{bmatrix} I & 0 \\ 0 & 0 \end{bmatrix}, \quad \sigma\Phi^{\pi}(C_\alpha)(\mu) = \begin{bmatrix} V & 0 \\ 0 & V \end{bmatrix},$$

where

$$I = \begin{bmatrix} 1 & 0 & 0 \\ 0 & 1 & 0 \\ 0 & 0 & 1 \end{bmatrix}, \quad V = \begin{bmatrix} 0 & 0 & 1 \\ 1 & 0 & 0 \\ 0 & 1 & 0 \end{bmatrix}.$$

If $a \in PC$ and $y \in Y_\tau^0$, then

$$\sigma\Phi^{\pi}(M_a)(y,0) = \begin{bmatrix} D_- & 0 \\ 0 & D_- \end{bmatrix}, \quad \sigma\Phi^{\pi}(M_a)(y,1) = \begin{bmatrix} D_+ & 0 \\ 0 & D_+ \end{bmatrix},$$

where

$$D_\pm = \mathrm{diag}\,(a(\tau_1 \pm 0), a(\tau_2 \pm 0), a(\tau_3 \pm 0)),$$

while if $y^\pm \in Y_\tau^\pm$,

$$\sigma\Phi^{\pi}(M_a)(y^\pm) = \begin{bmatrix} D_\pm & 0 \\ 0 & D_\pm \end{bmatrix}.$$

Now it is easily seen that $\sigma\Phi^{\pi}(A)(y^\pm)$ are invertible whenever $\sigma\Phi^{\pi}(A)(y,0)$ and $\sigma\Phi^{\pi}(A)(y,1)$ are so. ■

7.3. THEOREM. Let C_α be a backward Carleman shift, let τ_1, τ_2 be the fixed points of α, and denote by $\mathring{\mathbb{T}}$ either of the connected components of $\mathbb{T} \setminus \{\alpha_1, \alpha_2\}$. Put

$$\mathfrak{N}_0 = \bigcup_{\tau \in \mathring{\mathbb{T}}} (\{\tau\} \times \mathcal{A}_{p_\tau}[0,1]), \quad \mathfrak{R}_0 = \bigcup_{i=1}^{2} (\{\tau_i\} \times \mathcal{A}_{p_{\tau_i}}[0,1]),$$

and let $w : \mathbb{C} \to \mathbb{C}$ denote the branch of $w(z) = \sqrt{z(1-z)}$ which is analytic for z in $\mathbb{C} \setminus ((-\infty,0) \cup (1,\infty))$ and assumes positive values for z in $(0,1)$.

There exists a tensorable Fredholm symbol $\widetilde{\mathrm{Sym}}$ of the algebra $\mathrm{alg}^{\pi}(PC, P, C_\alpha)$ into $C_{4\times4}(\mathfrak{N}_0) \oplus C_{2\times2}(\mathfrak{R}_0)$ such that

(i) $$\widetilde{\mathrm{Sym}}\, \Phi^{\pi}(P)(\nu) = \begin{bmatrix} 1 & 0 & 0 & 0 \\ 0 & 0 & 0 & 0 \\ 0 & 0 & 0 & 0 \\ 0 & 0 & 0 & 1 \end{bmatrix}, \quad \widetilde{\mathrm{Sym}}\, \Phi^{\pi}(P)(\sigma) = \begin{bmatrix} 1 & 0 \\ 0 & 0 \end{bmatrix}$$

for $\nu \in \mathfrak{N}_0$ and $\sigma \in \mathfrak{R}_0$;

(ii) $$\widetilde{\mathrm{Sym}}\, \Phi^{\pi}(C_\alpha)(\nu) = \begin{bmatrix} 0 & 0 & 1 & 0 \\ 0 & 0 & 0 & 1 \\ 1 & 0 & 0 & 0 \\ 0 & 1 & 0 & 0 \end{bmatrix}, \quad \widetilde{\mathrm{Sym}}\, \Phi^{\pi}(C_\alpha)(\sigma) = \begin{bmatrix} 0 & i \\ -i & 0 \end{bmatrix}$$

for $\nu \in \mathfrak{N}_0$ and $\sigma \in \mathfrak{R}_0$;

(iii) if $a \in PC$ and $(\tau,x) \in \mathfrak{M}_0$, then

$$\widetilde{\mathrm{Sym}}\, \Phi^{\pi}(M_a)(\tau,x) = \begin{bmatrix} A(\tau,x) & 0 \\ 0 & B(\tau,x) \end{bmatrix},$$

where $A(\tau,x)$ is given by

$$\begin{bmatrix} a(\tau+0)x + a(\tau-0)(1-x) & [a(\tau+0) - a(\tau-0)]w(x) \\ [a(\tau+0) - a(\tau-0)]w(x) & a(\tau+0)(1-x) + a(\tau-0)x \end{bmatrix}$$

and $B(\tau,x)$ equals

$$\begin{bmatrix} a(\alpha(\tau)+0)(1-x) + a(\alpha(\tau)-0)x & [a(\alpha(\tau)-0) - a(\alpha(\tau)+0)]w(x) \\ [a(\alpha(\tau)-0) - a(\alpha(\tau)+0)]w(x) & a(\alpha(\tau)+0)x + a(\alpha(\tau)-0)(1-x) \end{bmatrix},$$

and if $a \in PC$ and $(\tau,x) \in \mathfrak{R}_0$, then $\widetilde{\mathrm{Sym}}\, \Phi^{\pi}(M_a)(\tau,x)$ is

$$\begin{bmatrix} a(\tau+0)x + a(\tau-0)(1-x) & [a(\tau+0) - a(\tau-0)]w(x) \\ [a(\tau+0) - a(\tau-0)]w(x) & a(\tau+0)(1-x) + a(\tau-0)x \end{bmatrix}.$$

PROOF. Proceed as in the proof of the preceding theorem. ■

We finally show what Theorem 7.1 reduces to for some subalgebras of $\mathrm{alg}(PQC,P,C_\alpha)$. Because the following results are easy consequences of Theorems 7.1, 6.1 and 6.2, proofs will be omitted.

7.4. <u>Singular integral operators</u>. Let $\mathfrak{M}$ and w be as in 6.1. There exists a tensorable Fredholm symbol $\widetilde{\mathrm{Sym}}$ of the algebra $\mathrm{alg}\,(PQC,P)$ into $C_{2\times 2}(\mathfrak{M})$ such that

(i) $\widetilde{\mathrm{Sym}}\, \Phi^{\pi}(P)(\mu) = \begin{bmatrix} 1 & 0 \\ 0 & 0 \end{bmatrix}$ for all $\mu \in \mathfrak{M}$;

(ii) if $a \in PQC$, then $\widetilde{\mathrm{Sym}}\, \Phi^{\pi}(M_a)(y,x)$ is equal to

$$\begin{bmatrix} a(y,1)x + a(y,0)(1-x) & [a(y,1) - a(y,0)]w(x) \\ [a(y,1) - a(y,0)]w(x) & a(y,1)(1-x) + a(y,0)x \end{bmatrix}$$

for $(y,x) \in \bigcup_{\tau\in\mathbb{T}} (Y_\tau^0 \times \mathcal{A}_{p_\tau}[0,1])$;

$$\widetilde{\mathrm{Sym}}\, \Phi^{\pi}(M_a)(y) = \begin{bmatrix} a(y,0) & 0 \\ 0 & a(y,0) \end{bmatrix} \quad \text{for } y \in \bigcup_{\tau\in\mathbb{T}} Y_\tau^- ;$$

$$\widetilde{\mathrm{Sym}}\, \Phi^{\pi}(M_a)(y) = \begin{bmatrix} a(y,1) & 0 \\ 0 & a(y,1) \end{bmatrix} \quad \text{for } y \in \bigcup_{\tau\in\mathbb{T}} Y_\tau^+ .$$

We remark that the PC version of this symbol is well known.

<u>7.5. Toeplitz operators</u>. Let $\mathfrak{M}$ and w be as in 6.1. There exists a tensorable Fredholm symbol $\widetilde{\mathrm{Sym}}$ of the algebra alg $T(PQC)\,/\,\mathcal{K}\,(L^p(\varrho))$ into $C(\mathfrak{M})$ such that if $a\in PQC$, then

$$\widetilde{\mathrm{Sym}}\;\Phi^{\pi}(T_a)(y,x) = a(y,1)x + a(y,0)(1-x)$$

for $(y,x)\in \bigcup_{\tau\in\mathbb{T}} (Y_\tau^0 \times \mathcal{A}_{p_\tau}[0,1])$,

$$\widetilde{\mathrm{Sym}}\;\Phi^{\pi}(T_a)(y) = a(y,0) \quad \text{resp.} \quad \widetilde{\mathrm{Sym}}\;\Phi^{\pi}(T_a)(y) = a(y,1)$$

for $y\in \bigcup_{\tau\in\mathbb{T}} Y_\tau^-$ resp. $y\in\bigcup_{\tau\in\mathbb{T}} Y_\tau^+$. The algebra alg $T(PQC)\,/\,\mathcal{K}(L^p(\varrho))$ is commutative.

For $p = 2$ and $\varrho\equiv 1$ this result was established by Sarason [24], in the general case it was announced in [6].

<u>7.6. Toeplitz plus Hankel operators</u>. Let $\mathring{\mathbb{T}} = \{\tau\in\mathbb{T}:\ \mathrm{Im}\,\tau > 0\}$, and for $\tau\in\mathring{\mathbb{T}}$ put

$$\mathfrak{N}_\tau = Y_\tau^- \cup (Y_\tau^0 \times \mathcal{A}_{p_\tau}[0,1]) \cup Y_\tau^+ ,$$

$$\mathfrak{N} = \Big(\bigcup_{\tau\in\mathring{\mathbb{T}}} \mathfrak{N}_\tau\Big) \cup Y_{-1}^- \cup Y_{-1}^+ \cup Y_1^- \cup Y_1^+ ,$$

$$\mathfrak{R} = (Y_{-1}^0 \times \mathcal{A}_{p_{-1}}[0,1]) \cup (Y_1^0 \times \mathcal{A}_{p_1}[0,1]) ,$$

and let w be as in 6.2. Also assume that $\varrho(\tau) = \varrho(\bar\tau)$ for all $\tau\in\mathbb{T}$; this condition ensures the boundedness of the flip operator J and thus of all Hankel operators H_a generated by functions $a\in L^\infty$ on $L^p(\varrho)$.

As in 3.2, given $y\in Y$ we denote by $\bar y\in Y$ the uniquely determined point for which $\varphi(\bar y) = (R\varphi)(y)$ for all $\varphi\in QC$, R denoting the reflection operator.

There exists a tensorable Fredholm symbol $\widetilde{\mathrm{Sym}}$ of the algebra alg $TH(PQC)\,/\,\mathcal{K}\,(L^p(\varrho))$ into $C_{2\times 2}(\mathfrak{N}) \oplus C(\mathfrak{R})$ such that for all $a\in PQC$,

(i) $\widetilde{\mathrm{Sym}}\;\Phi^{\pi}(T_a)(y,x)$ equals

$$\begin{bmatrix} a(y,1)x + a(y,0)(1-x) & 0 \\ 0 & a(\bar y,1)x + a(\bar y,0)(1-x)\end{bmatrix}$$

for $(y,x)\in \bigcup_{\tau\in\mathbb{T}} (Y_\tau^0 \times \mathcal{A}_{p_\tau}[0,1])$;

$$\widetilde{\mathrm{Sym}}\ \Phi^{\pi}(T_a)(y) = \begin{bmatrix} a(y,0) & 0 \\ 0 & a(\bar{y},1) \end{bmatrix}$$

for $y \in \left(\bigcup_{\tau \in \overset{\circ}{\mathbb{T}}} Y_\tau^- \right) \cup Y_{-1}^- \cup Y_1^-$;

$$\widetilde{\mathrm{Sym}}\ \Phi^{\pi}(T_a)(y) = \begin{bmatrix} a(y,1) & 0 \\ 0 & a(\bar{y},0) \end{bmatrix}$$

for $y \in \left(\bigcup_{\tau \in \overset{\circ}{\mathbb{T}}} Y_\tau^+ \right) \cup Y_{-1}^+ \cup Y_1^+$;

$$\widetilde{\mathrm{Sym}}\ \Phi^{\pi}(T_a)(y,x) = a(y,1)x + a(y,0)(1-x)$$

for $(y,x) \in \mathcal{R}$;

(ii) $\widetilde{\mathrm{Sym}}\ \Phi^{\pi}(H_a)(y,x)$ equals

$$\begin{bmatrix} 0 & a(y,1) - a(y,0)\, w(x) \\ a(\bar{y},0) - a(\bar{y},1)\, w(x) & 0 \end{bmatrix}$$

for $(y,x) \in \bigcup_{\tau \in \overset{\circ}{\mathbb{T}}} (Y_\tau^0 \times \mathcal{A}_{p_\tau}[0,1])$;

$$\widetilde{\mathrm{Sym}}\ \Phi^{\pi}(H_a)(y) = \begin{bmatrix} 0 & 0 \\ 0 & 0 \end{bmatrix}$$

for $y \in \bigcup_{\tau \in \overset{\circ}{\mathbb{T}}} (Y_\tau^- \cup Y_\tau^+) \cup Y_{-1}^- \cup Y_{-1}^+ \cup Y_1^- \cup Y_1^+$;

$$\widetilde{\mathrm{Sym}}\ \Phi^{\pi}(H_a)(y,x) = i\left[a(y,0) - a(y,1)\right] w(x)$$

for $(y,x) \in \mathcal{R}$.

In the $p = 2$ and $\varrho \equiv 1$ case the above result was first obtained in [25]; a Fredholm symbol for alg $TH(PC)/\mathcal{K}(L^2)$ had been previously given by Power [19], [20].

<u>7.7. Essential spectrum of Hankel operators</u>. Let $\overset{\circ}{\mathbb{T}}$, w, ϱ be as in 7.6. If $a \in PQC$, then the essential spectrum of H_a on $L^p(\varrho)$ is

$$S_{-1} \cup \left(\bigcup_{\tau \in \overset{\circ}{\mathbb{T}}} S_\tau^+ \right) \cup \left(\bigcup_{\tau \in \overset{\circ}{\mathbb{T}}} S_\tau^- \right) \cup S_1$$

where

$$S_{\pm 1} = \bigcup_{y \in Y_{\pm 1}^0} i\left[a(y,0) - a(y,1)\right] w(\mathcal{A}_{p_{\pm 1}}[0,1]) ,$$

$$S_\tau^{\pm} = \bigcup_{y \in Y_\tau^0} (\pm i)\sqrt{\left[a(y,1)-a(y,0)\right]\left[a(y,1)-a(y,0)\right]}\ w(\mathcal{A}_{p_{\pm 1}}[0,1]) .$$

Notice that the particular choice of the branch of the root in in the latter equality does not influence the union $S_\tau^- \cup S_\tau^+$.

For $p = 2$ and $\varrho \equiv 1$ the preceding result is Power's [18], [20], for $a \in PC$ and general p and ϱ it was established in [21].

We conclude with showing that semi-Fredholm operators in $\mathrm{alg}(PQC, P, C_\alpha)$ are automatically Fredholm.

7.8. THEOREM. Let C_α be a Carleman shift and let A be in $\mathrm{alg}(PQC, P, C_\alpha)$. Then A is semi-Fredholm on $L^p(\varrho)$ (i.e. the range of A is closed and A has a finite-dimensional kernel or cokernel) if and only if A is Fredholm on $L^p(\varrho)$.

PROOF. Once the local spectra $\mathrm{sp}\, \Phi_\xi^\pi(A)$ are available, this theorem follows almost immediately from the results of Albrecht and Mehta [1].

Let $\mathfrak{A}_0$ denote the collection of all finite product-sums $\sum_j \prod_k A_{jk}$ where each A_{jk} is either P or C_α or a multiplication operator M_a with $a \in PQC$. From the proofs of Theorems 6.1 and 6.2 it is readily seen that if $A \in \mathfrak{A}_0$, then the local spectrum $\mathrm{sp}\, \Phi_\xi^\pi(A)$ has no interior points (as a subset of $\mathbb{C}$) for each $\xi \in M(QC_\alpha)$. Because QC_α is a C^*-algebra, it is normal and spectrally closed in the sense of [1, Definition 3.1]. So Theorem 3.6 of [1] can be applied to deduce that every operator A in $\mathfrak{A}_0$ has the essential single valued extension property. Using Proposition 4.1 of [1] we then obtain that an operator $A \in \mathfrak{A}_0$ is Fredholm whenever it is semi-Fredholm.

To extend the result to all of $\mathrm{alg}(PQC, P, C_\alpha)$, assume there would exist an operator A in $\mathrm{alg}(PQC, P, C_\alpha)$ which is semi-Fredholm but not Fredholm. Since the set of all operators which are semi-Fredholm but not Fredholm is open in $\mathcal{L}(L^p(\varrho))$ (see e.g. [11, Theorem IV.15.4]) and since $\mathfrak{A}_0$ is dense in the algebra $\mathrm{alg}(PQC, P, C_\alpha)$, it would follow that there were an operator in $\mathfrak{A}_0$ which is semi-Fredholm but not Fredholm. However, we have seen in the preceding paragraph that such a situation cannot happen. ∎

Because all our symbol maps are tensorable, Theorem 7.8 is also true in the matrix case.

REFERENCES

1. E. Albrecht, R.D. Mehta: Some remarks on local spectral theory. J. Operator Theory 12 (1984), 285-317.

2. G.R. Allan: Ideals of vector-valued functions. Proc. London Math. Soc., 3rd ser., 18 (1968), 193-216.

3. A. Böttcher, N.Ya. Krupnik, B. Silbermann: A general look at local principles with special emphasis on the norm computation aspect. Integral Equations and Operator Theory 11 (1988), 455-479.

4. A. Böttcher, B. Silbermann: Analysis of Toeplitz Operators. Akademie-Verlag, Berlin and Springer-Verlag, Heidelberg 1989.

5. A. Böttcher, B. Silbermann, I.M. Spitkovskii: Toeplitz operators with piecewise quasisectorial symbols. Bull. London Math. Soc. (to appear).

6. A. Böttcher, I.M. Spitkovskii: Toeplitz operators with PQC symbols on weighted Hardy spaces. (to appear).

7. M. Costabel: A contribution to the theory of singular integral equations with Carleman shift. Integral Equations and Operator Theory 2 (1979), 11-24.

8. R.G. Douglas: Local Toeplitz operators. Proc. London Math. Soc., 3rd ser., 36 (1978), 234-276.

9. I. Gohberg, N.Ya. Krupnik: On one-dimensional singular integral operators with shift. Izv. Akad. Nauk Armen. SSR, Ser. Matem., 8:1 (1973), 3-12 [Russian].

10. I. Gohberg, N.Ya. Krupnik: On algebras of singular integral operators with shift. Matem. Issled. 8:2 (1973), 170-175 [Russian].

11. I. Gohberg, N.Ya. Krupnik: Einführung in die Theorie der eindimensionalen singulären Integralgleichungen. Birkhäuser Verlag, Basel 1979.

12. N.K. Karapetyants, S.G. Samko: Singular integral equations with a Carleman shift in the case of discontinuous coefficients and the investigation of the Noether property of a class of linear operators with an involution. Dokl. Akad. Nauk SSSR 211:2 (1973), 281-284 [Russian].

13. N.K. Karapetyants, S.G. Samko: On algebras generated by singular integral operators with piecewise continuous coefficients and a Carleman shift. Rostov-on-Don State Univ. Press 1989 [Russian].

14. N.Ya. Krupnik: Banach Algebras with a Symbol and Singular Integral Operators. Birkhäuser Verlag, Basel 1987.

15. N.Ya. Krupnik, I.A. Feldman: On the impossibility of introducing a matrix symbol on certain operator algebras. Matem. Issled. 61 (1981), 75-85 [Russian].

16. G.S. Litvinchuk: Boundary Value Problems and Singular Integral Equations with Shift. Nauka, Moscow 1977 [Russian].

17. N.K. Nikolskii: Treatise on the Shift Operator. Springer-Verlag, Berlin and Heidelberg 1986.

18. S.C. Power: The essential spectrum of a Hankel operator with piecewise continuous symbol. Michigan Math. J. 25 (1978), 117-121.

19. S.C. Power: C^*-algebras generated by Hankel operators and Toeplitz operators. J. Funct. Anal. 31 (1979), 52-68.

20. S.C. Power: Hankel Operators on Hilbert Space. Pitman, Boston, London, Melbourne 1981.

21. S. Roch, B. Silbermann: Algebras generated by idempotents and the symbol calculus for singular integral operators. Integral Equations and Operator Theory 11 (1988), 385-419.

22. S. Roch, B. Silbermann: On algebras with standard identities. (to appear).

23. D. Sarason: Algebras of functions on the unit circle. Bull. Amer. Math. Soc. 79 (1973), 286-299.

24. D. Sarason: Toeplitz operators with piecewise quasicontinuous symbols. Indiana Univ. Math. J. 26 (1977), 817-838.

25. B. Silbermann: The C^*-algebra generated by Toeplitz and Hankel operators with piecewise quasicontinuous symbols. Integral Equations and Operator Theory 10 (1987), 730-738.

26. N.L. Vasilevskii, M.V. Shapiro: On the algebra generated by singular integral operators with a Carleman shift and piecewise continuous coefficients. Ukrain. Matem. Zh. 27 (1975), 216-223 [Russian].

27. Y. Weiss: On algebras generated by two idempotents. Seminar Analysis 1987/88 of the Karl-Weierstrass-Inst., 139-145 [Russian].

A. Böttcher, S. Roch, B. Silbermann
Technische Universität Karl-Marx-Stadt
Sektion Mathematik
PSF 964
Karl-Marx-Stadt 9010
GDR

I.M. Spitkovskii
Academy of Sciences of the Ukrainian SSR
Institute of Hydrophysics
Ul. Sov. Armii 3
270 100 Odessa
USSR

Operator Theory:
Advances and Applications, Vol. 48

PERTURBATION AND SIMILARITY OF TOEPLITZ OPERATORS

Douglas N. Clark

A Toeplitz operator T_φ with symbol of the form $\varphi(t) = e^{iv(t)}$, where v is real valued and has a bounded harmonic conjugate, is known to be similar to a unitary operator U with absolutely continuous spectrum; Peller [10], Clark [2]. In this note, perturbation theory is used to generalize known results on the spectral multiplicity of U. For an even, unimodular, symbol φ, we generalize a selfadjoint perturbation problem of Putnam [12], [13] and Rosenblum [15].

1. Real, even symbols.

The first Toeplitz operators which were analysed in enough depth to determine absolute continuity and spectral multiplicity were probably those with symbols $\varphi(t) = \cos t$, [Hilbert, 6], and $\varphi(t) = \cos nt$, $n = 2, 3, \ldots$, [Hellinger, 5].

In the late fifties, when Putnam and Rosenblum were engaged in proving that selfadjoint Toeplitz operators were absolutely continuous [12], [13], [15], more general examples with even symbols yielded up their multiplicities. Let $\varphi(t)$ be even

$$\varphi(t) = \varphi(-t) \qquad -\pi \le t \le \pi$$

(and not necessarily realvalued or unimodular). Let M_φ denote the operator of multiplication by φ on $L^2(0,\pi)$ and let L be the operator from H^2 to $L^2(0,\pi)$ defined by

$$(Lf)(t) = (2i)^{-1}[e^{it}f(e^{it}) - e^{-it}f(e^{-it})] .$$

If $f(e^{it}) = \Sigma_{n=0}^{\infty} a_n e^{int}$ is the Fourier series of $f \in H^2$, we have

$$Lf = \sum_{n=1}^{\infty} a_{n-1} \sin nt$$

which shows that L is unitary from H^2 onto $L^2(0,\pi)$. Further,

$$M_\varphi Lf = (2i)^{-1}[e^{it}\varphi(t)f(e^{it}) - e^{-it}\varphi(-t)f(e^{-it})]$$

$$= (2i)^{-1}\{P_1[e^{it}\varphi(t)f(e^{it}) - e^{-it}\varphi(-t)f(e^{-it})]$$

$$-(I - P)[e^{-it}\varphi(-t)f(e^{-it}) - e^{it}\varphi(t)f(e^{it})]\} ,$$

where P and P_1 are the projections defined on $L^2(-\pi,\pi)$ Fourier series by

$$P \sum_{-\infty}^{\infty} a_n e^{int} = \sum_{n=0}^{\infty} a_n e^{int}$$

$$P_1 \sum_{-\infty}^{\infty} a_n e^{int} = \sum_{n=1}^{\infty} a_n e^{int} .$$

We have for the above

$$M_\varphi Lf = Le^{-it}P_1[e^{it}\varphi(t)f(e^{it}) - e^{-it}\varphi(-t)f(e^{-it})]$$

$$= L(T_\varphi f - H_\varphi f) ,$$

where $H_\varphi f = e^{-it}P_1 e^{-it}\varphi(-t)f(e^{-it})$ is (one way of defining) the Hankel operator with symbol φ . Thus

$$T_\varphi = L^{-1}M_\varphi L + H_\varphi . \tag{1}$$

Putnam and Rosenblum used (1) for selfadjoint T_φ to conclude that, whenever

(I) T_φ is absolutely continuous, and

(II) H_φ is trace class,

then, by the Kato-Rosenblum Theorem [7, Theorem X 4.4], T_φ is unitarily equivalent to the absolutely continuous part of M_φ. Condition (I) was shown to hold for general selfadjoint T_φ in Rosenblum [15]. Putnam and Rosenblum knew sufficient conditions for (II). The problem posed by their work of finding necessary and sufficient conditions for (II) stimulated work by several authors, culminating in Peller's beautiful paper [9].

In [14], Putnam generalized the unitary equivalence of T_φ with the absolutely continuous part of M_φ to classes of φ with evenness replaced other periodicity conditions.

2. Operators similar to unitary operators.

We want to carry out the analysis of the previous section for Toeplitz operators T_φ similar to unitary operators. Central to §1 was the Kato-Rosenblum Theorem [7, Theorem X 4.4], which states that *if* S *and* T *are selfadjoint and* $S-T$ *lies in trace class, then the absolutely continuous parts of* S *and* T *are unitarily equivalent.*

The Kato-Rosenblum Theorem, in the above form, continues to hold if S and T are replaced by unitary operators. This is wellknown and not hard to see by taking Cayley transforms. Indeed, let U and V be unitary with $U-V$ of trace class and let $S = i(U + I)(U - I)^{-1}$ and $T = i(V + I)(V - I)^{-1}$ be their Cayley transforms. Since

$$S - T = 2i(U - I)^{-1}[V - U](V - I)^{-1} ,$$

it is not immediate that $S - T$ lies in trace class if, for example, $U - I$ or $V - I$ is singular. But it is easy to see that $(S - \alpha I)^{-1} - (T - \alpha I)^{-1}$ lies in trace class, for nonreal α, and this implies that the absolutely continuous parts of S and T, and hence of U and V, are unitarily equivalent [7, Theorem X 4.12].

Since our perturbation problem concerns (1), where T_φ is *similar to* a unitary operator, we actually need the Kato-Rosenblum Theorem for operators similar to unitary operators. We have

LEMMA. *Suppose* A *and* B *are bounded operators similar to unitary operators* U *and* V, *respectively, and suppose* $A - B$ *lies in trace class. Then the absolutely continuous parts of* U *and* V *are unitarily equivalent.*

PROOF. First of all, we may assume that B is itself unitary, i.e. that $B = V$, since

$$A - B = L_1^{-1}UL_1 - L_2^{-1}VL_2 = K ,$$

where K is in trace class, implies

$$(L_1L_2^{-1})^{-1}U(L_1L_2^{-1}) - V = L_2KL_2^{-1} ,$$

and $L_2KL_2^{-1}$ again lies in trace class. So we may suppose $A = L^{-1}UL$ and $A = V + K$, where K lies in trace class.

From $LA = UL$, we get $A^*L^*LA = L^*L$. Let J be the positive square root of L^*L. We have

$$A^*J^2A = J^2 . \tag{2}$$

It is immediate that $\|JAf\| = \|Jf\|$, for every vector f, and hence that

$$JA = WJ ,$$

where W is an isometry. Clearly L invertible implies J invertible. Therefore W must be unitary and hence W and U, being similar, are unitarily equivalent.

We claim that (2) implies

$$A^*JA = J \qquad \text{(modulo trace class)}. \tag{3}$$

To prove this, note that

$$(A^*JA)^2 = A^*JAA^*JA = A^*J(V + K)(V^* + K^*)JA$$

$$= A^*J^2A + A^*J(VK^* + KV^* + KK^*)JA$$

$$= J^2 + A^*J(VK^* + KV^* + KK^*)JA ,$$

by (2), so that the (selfadjoint, positive) operators $(A^*JA)^2$ and J^2 differ by an operator lying in trace class. This implies

$$f((A^*JA)^2) - f(J^2)$$

lies in trace class if, for example, f is analytic on the union of the spectra of $(A^*JA)^2$ and J^2 ; [Langer, 8]. Since both of these spectra must lie on the positive real axis (excluding 0), we can choose $f(z) = z^{1/2}$ and we have proved (3).

Now (3) implies

$$J = (V^* + K^*)JA = V^*JA$$

or

$$V - W = V - JAJ^{-1} = 0$$

modulo trace class. Since W is unitarily equivalent to U , an application of the Kato-Rosenblum Theorem for unitary operators completes the proof of the lemma.

3. Unimodular even symbols.

Let $\varphi(t) = e^{iv(t)}$, where v is a realvalued measurable function on $[-\pi,\pi]$ having a bounded harmonic conjugate and satisfying $v(t) = v(-t)$. The Toeplitz operator T_φ is known to be similar to a unitary operator U ; Peller [10]. Since (1) holds, we can carry out the Putnam-Rosenblum perturbation program of §1 for T_φ , once we know the analogues of (I) (with U replacing T_φ) and (II).

Condition (I), absolute continuity of U , follows by Sz.-Nagy-Foias theory. Indeed, since T_φ is a completely nonunitary contraction [4], U must be a part of the minimal unitary dilation of T_φ [16, Theorem IX 1.2], which is known to be absolutely continuous [16, Theorem II 6.4].

For (II), we have Peller's necessary and sufficient condition that H_φ belong to trace class. Let A_1^1 denote the set of all $f(z)$, analytic in $|z| < 1$, with

$$\int_{-\pi}^{\pi} \int_0^1 |f''(re^{i\theta})| r dr \, d\theta < \infty .$$

Peller proved in [9] that H_φ belongs to trace class if and only if $P\bar{\varphi} \in A_1^1$, where P is the projection of $L^2(-\pi,\pi)$ on H^2 (as in §1).

Combining the above versions of (I) and (II) with (1) and the lemma of §2, we have

THEOREM 1. *If* v *is realvalued, even,* $\tilde{v} \in L^\infty$ *and* $\varphi = e^{iv}$ *satisfies* $P\bar{\varphi} \in A_1^1$, *then* T_φ *is unitarily equivalent to the absolutely continuous part of multiplication by* φ *on* $L^2(0,\pi)$.

Versions of Theorem 1 with evenness of v replaced by other periodicity conditions can be obtained from Putnam [14] or, more generally, from Cowen [3].

Our earlier papers [1], as applied to unimodular symbols φ such that T_φ has index 0, and [2], give characterizations of the unitary operators U similar to T_φ. Although free from the hypothesis that φ is even, the analysis in [1] and [2] is much more complicated than that presented here. Furthermore, restrictive bounds on v were assumed ($v([-\pi,\pi]) \subsetneq [-\pi,\pi]$ in [1]; $|v| \le \pi$ in [2]) that do not appear in Theorem 1 above. In the next section, we deal further with the problem of removing such bounds from the hypothesis.

4. Unimodular noneven symbols.

Let v be a real measurable function on $[-\pi,\pi]$ with $\tilde{v}$ bounded and let $\varphi = e^{iv}$. The proof quoted in §3 that T_φ is similar to an absolutely continuous unitary operator made no use of the evenness assumption of that section and so there is an absolutely continuous unitary operator U , similar to T_{φ} .

In [2], we proved

THEOREM 2. *If* $0 \leq v \leq 2\pi$, *then the multiplicity of* $\lambda = e^{i\theta}$ *in the spectrum of* U *is* n , *if the set*

$$\Gamma_{\theta} = \{e^{it}: \theta > v(t)\}$$

is, modulo zero sets, the union of n *proper arcs of the unit circle* $\mathbf{T}$ *and is* ∞ *otherwise.*

The bound $0 \leq v \leq 2\pi$ is a serious part of the hypotheses, since redefining the argument of φ may produce a v with $\tilde{v}$ unbounded. An even more restrictive assumption $(\varphi([-\pi,\pi]) \subsetneqq \mathbf{T})$ appears in our similarity theorem for rational, unimodular φ ; [1]. The lemma of §2 permits us to show that such bounds on v are not always essential.

THEOREM 3. *Suppose* $\varphi = e^{iv}$ *satisfies*

$$\varphi^{1/m} \in B_2^{1/2} , \tag{4}$$

where m *is the least integer not less than* $1/\pi\|v\|_{\infty}$. *Then the multiplicity of* $\lambda = e^{i\theta}$ *in the specturm of* U *is the number described in Theorem 2.*

By definition, a function with a Fourier series

$$f(t) \sim \sum_{-\infty}^{\infty} a_n e^{int} \qquad -\pi \leq t \leq \pi$$

lies in $B_2^{1/2}$ if $\Sigma_{-\infty}^{\infty} n|a_n|^2 < \infty$. Thus a sufficient condition for (4) is that v be absolutely continuous, with $v' \in L^2(-\pi,\pi)$. From this, we have

COROLLARY. *If* φ *is the quotient of two finite Blaschke products of equal degree, then the conclusion of Theorem 3 holds.*

PROOF OF THEOREM 3. Since $|v/m| \leq \pi$, Theorem 2 implies that if $\psi = \exp(iv/m)$, then T_ψ is similar to an absolutely continuous unitary operator V with the multiplicity of $\lambda = e^{i\theta}$ in the spectrum of V equal to n if the set

$$\Gamma_{\theta,m} = \{e^{it}: \theta > v(t)/m\}$$

is the union of n proper subarcs of $\mathbf{T}$, and equals ∞ otherwise. Clearly V^m is again unitary and absolutely continuous and has spectral multiplicity n at $e^{i\theta}$ if $\Gamma_\theta = \Gamma_{\theta,1}$ is the union of n proper subarcs of $\mathbf{T}$ (and ∞ otherwise). Also V^m is similar to T_ψ^m .

On the other hand, T_φ , $\varphi = e^{iv} = \psi^m$, is similar to an absolutely continuous unitary operator U and, by (4), $T_\psi^m - T_\varphi$ lies in trace class [Peller, 11]. By the lemma of §2, U must therefore be unitarily equivalent to V^m , and the theorem is proved.

REFERENCES

1. Clark, D. N.: On a similarity theory for rational Toeplitz operators, J. Reine Angew. Math. 320(1980), 6-31.

2. ______: On Toeplitz operators with unimodular symbols, in Operator Theory: Advances and Applications, vol. 24, Birkhäuser-Verlag, Basel (1987), 59-68.

3. Cowen, C. C.: On equivalence of Toeplitz operators, J. Operator Theory 7(1982), 167-172.

4. Goor, R.: On Toeplitz operators which are contractions, Proc. Amer. Math. Soc. 34(1972), 191-192.

5. Hellinger, E. D.: Spectra of quadratic forms in infinitely many variables, Northwestern University Studies Vol. IV (1941).

6. Hilbert, D.: Grundzüge einer allgemeinen theorie der linearen Integralgleichungen, Leipzig (1912).

7. Kato, T.: Perturbation theory for linear operators, Second Edition, Springer Verlag (1984).

8. Langer, H.: Eine Erweiterung der Spurformel der Störingstheorie, Math. Nachr. 30(1965), 123-135.

9. Peller, V. V.: Nuclearity of Hankel operators, LOMI Preprint E-I-79, Leningrad (1979).

10. ______: Some problems of spectral theory of Toeplitz operators (spectra, similarity, invariant subspaces), ibid. Preprint E-2-84 (1984).

11. ______: When is a function of a Toeplitz operator close to a Toeplitz operator, ibid. Preprint E-2-88 (1988).

12. Putnam, C. R.: Commutators and absolutely continuous operators, Trans. Amer. Math. Soc. 87 (1958), 513-525.

13. ______: On Toeplitz matrices, absolute continuity and unitary equivalence, Pacific J. Math. 9 (1959), 837-846.

14. ______: A note on Toeplitz matrices and unitary equivalence, Bull. Un. Mat. Ital. 15 (1960), 6-9.

15. Rosenblum, M.: The absolute continuity of Toeplitz's matrices, Pacific J. Math. 10 (1960), 987-996.

16. Sz.-Nagy, B. and Foiaş, C.: Harmonic analysis of operators on Hilbert space, North Holland (1970).

Department of Mathematics
University of Georgia
Athens, Georgia 30602
U.S.A.

Operator Theory:
Advances and Applications, Vol. 48

FACTORIZATION OF POSITIVE-DEFINITE KERNELS

Tiberiu Constantinescu

Dedicated to the memory of Ernst Hellinger

In this paper a factorization result for positive-definite kernels on the set of integers is obtained. The role played by Lowdenslager's idea in [12] is also emphasized in this context, by adapting the Wold-von Neumann decomposition to families of isometries. A few applications are indicated, especially computations of angles in nonstationary processes connected with the first Szegö limit theorem.

INTRODUCTION

Factorizations play a central role in several areas as prediction theory of stationary processes, modelling of operators, interpolation and system theory. We are interested here in a type of factorization originating in a classical result of Fejér and F. Riesz by means of which a positive trigonometric polynomial is written as the square of the modulus of an analytic trigonometric polynomial.

This result was further developed in many ways (see for instance [8], [4], [6], [9], [17], to quote only a few references - see [14] or [16] for many other comments and references). But, an important moment was the appearence of Lowdenslager's paper [12], where a geometrical method based on Wold-von Neumann decomposition of an isometry was proposed.

Using this method, general results were obtained in [5], [2], [16], [13], [15].

The purpose of this paper is to obtain a variant of the result in [16] (actually, of its equivalent version in [15]) for positive-definite kernels on the set of integers and, especially, to emphasize the fact that the idea in [12] can be also used in this context by an adaptation of the Wold-von Neumann decomposition to families of isometries.

The paper has the following content. In the next section several preliminaries concerning positive-definite kernels and Wold-von Neumann decompositions for families of isometries are presented.

Section 2 contains the main result of the paper and the last section contains applications to prediction of nonstationary processes (especially, some generalizations of Szego's first limit theorem, where Wold-von Neumann decompositions for families of isometries are also well suited.)

1. PRELIMINARIES

Throughout this paper $\mathbf{Z}$ denotes the set of integers; for two Hilbert spaces H and H', $L(H, H')$ denotes the set of linear bounded operators from H into H' and $L(H) = L(H,H)$. For a contraction $T \in L(H,H')$ (i.e. $\|T\| \leq 1$), $D_T = (I - T^*T)^{\frac{1}{2}}$ and $\mathcal{D}_T$ is the closure of the range $R(D_T)$ of D_T. The unitary operator

$$J(T) : H \oplus \mathcal{D}_{T^*} \to H' \oplus \mathcal{D}_T$$

$$J(T) = \begin{bmatrix} T & D_{T^*} \\ D_T & -T^* \end{bmatrix} \tag{1.1}$$

is the *elementary rotation* of T (see [16]). O_H (I_H) is the zero (identity) operator on the underlying space and for a closed subspace L of H, P^H_L denotes the orthogonal projection of H onto L.

1.1. KOLMOGOROV DECOMPOSITIONS

We consider a family of Hilbert spaces $\{H_n\}_{n \in \mathbf{Z}}$. A map S on $\mathbf{Z} \times \mathbf{Z}$ such that $S(i,j) = S_{ij} \in L(H_j, H_i)$ is called a *positive definite kernel* (on $\mathbf{Z}$) if the operators

$$M_{mn}(S) = M_{mn} : \bigoplus_{k=m}^{n} H_k \to \bigoplus_{k=m}^{n} H_k$$

$$M_{mn} = (S_{ij})_{m \leq i,j \leq n} \tag{1.2}$$

are positive for $m,n \in \mathbf{Z}$, $m \leq n$. We suppose, without loss of generality, that $S_{ii} = I_{H_i}$ for $i \in \mathbf{Z}$.

A *Kolmogorov decomposition* of S is a family of operators $V(S) = V = \{V(n)\}_{n \in \mathbf{Z}}$, $V(n) \in L(H_n, K_o)$, where K_o is a Hilbert space, $S_{ij} = V^*(i)V(j)$ for all $i,j \in \mathbf{Z}$ and $K_o = \bigvee_{n \in \mathbf{Z}} V(n)H_n$. It is a standard fact (see [16]) that for two Kolmogorov decompositions of S, V and V', there exists a unitary operator $\Omega : K_o \to K'_o$ such that $V'(n) = \Omega V(n)$ for all $n \in \mathbf{Z}$.

A special structure of the Kolmogorov decomposition of S can be pointed-out by an analysis of the elementary rotation of a row contraction of infinite length. That is, if $E = \bigoplus_{n=1}^{\infty} E_n$ is a Hilbert space, then any contraction $T \in L(E,E')$, E' being another

Hilbert space, has the structure

$$(1.3)\qquad T = (T_1, D_{T_1^*}T_2, \ldots, D_{T_1^*} \ldots D_{T_{n-1}^*} T_n, \ldots)$$

where T_1 is a contraction in $L(E_1, E')$ and for $n \geq 2$, T_n is a contraction in $L(E_n, D_{T_{n-1}^*})$.
Suitable identifications of the spaces D_T and D_{T^*} can be obtained in the following way. Define the operators: $D_1(T) = D_{T_1}$ and for $n \geq 2$,

$$D_n(T) : \bigoplus_{k=1}^{n} E_k (= E^{[k]}) \to \bigoplus_{k=1}^{n} D_{T_k} \subset \bigoplus_{k=1}^{\infty} D_{T_k} (= D(T))$$

$$(1.4)\qquad D_n(T) = \begin{bmatrix} D_{n-1}(T), & -Y_{n-1}T_n \\ 0, & D_{T_n} \end{bmatrix}$$

where

$$(1.5)\qquad Y_n = (T_1^* D_{T_2^*} \ldots D_{T_n^*}, T_2^* D_{T_3^*} \ldots D_{T_n^*}, \ldots, T_n^*)^t$$

and "t" stands for the matrix transpose. Define also:

$$(1.6)\qquad \begin{aligned} &H_n(T) : E' \to D_{T_n^*} \\ &H_n(T) = D_{T_n^*} \ldots D_{T_1^*} \end{aligned}$$

and, finally, the formulae:

$$(1.7)\qquad \alpha(T)D_T = D_\infty(T) = \text{s-}\lim_{n\to\infty} D_n(T) P^E_{E^{[n]}}$$

and

$$(1.8)\qquad \beta(T)D_{T^*} = H_\infty(T) = (\text{s-}\lim_{n\to\infty} H_n^*(T)H_n(T))^{\frac{1}{2}}$$

determine unitary operators from D_T onto $D(T)$ and, respectively, from D_{T^*} onto $D_*(T) = \overline{R(H_\infty(T))}$.

Returning to the given positive-definite kernel S, note by Theorem 1.3 in [3] that it is uniquely determined by a family of contractions $G = \{G_{ij} / i,j \in \mathbf{Z}, i \leq j\}$, where $G_{ii} = O_{H_i}$ for $i \in \mathbf{Z}$ and for $i < j$, $G_{ij} \in L(D_{G_{i+1,j}}, D_{G^*_{i,j-1}})$. For every $i \in \mathbf{Z}$, the family $\{G_{ik}\}_{i<k<\infty}$ determines a row contraction of infinite length $R_i : \bigoplus_{k=i+1}^{\infty} D_{G_{i+1,k}} \longrightarrow H_i$ by means of formula (1.3) (with $T_n = G_{i,i+n}$). Consider the spaces

$$(1.9)\qquad K_i = \bigoplus_{j=-\infty}^{i-1} D_*(R_j) \oplus H_i \oplus D(R_i)$$

and the unitary operators:

$$W_i : K_{i+1} \to K_i$$

(1.10) $$W_i = I \oplus \begin{bmatrix} I & 0 \\ 0 & \alpha(R_i) \end{bmatrix} J(R_i) \begin{bmatrix} 0 & I \\ \beta^*(R_i) & 0 \end{bmatrix}$$

where the direct sum is written with respect to the decompositions $K_{i+1} = (\bigoplus_{j=-\infty}^{i-1} D_*(R_j) \oplus (D_*(R_i) \oplus H_{i+1} \oplus D(R_{i+1}))$ and $K_i = (\bigoplus_{j=-\infty}^{i-1} D_*(R_j)) \oplus (H_i \oplus D(R_i))$.

By Theorem 2.4 in [3], the Kolmogorov decomposition of S is given by $V = \{V(n)\}_{n \in \mathbf{Z}}$, where

$$V(n) : H_n \to K_o$$

(1.11) $$V(n) = \begin{cases} W^*_{-1} W^*_{-2} \cdots W^*_n / H_n, & n < 0 \\ P_{H_o}^{K_o} / H_o \quad , & n = 0 \\ W_o W_1 \cdots W_{n-1} / H_n , & n > 0 \end{cases} .$$

1.2. WOLD-VON NEUMANN DECOMPOSITIONS

Only for fixing the notation we describe here an obvious adaptation of the Wold-von Neumann decomposition of an isometry to the case of a family of isometries.

We fix $N \in \mathbf{Z}$ $\{\infty\}$, $M \in \mathbf{Z}$ $\{-\infty\}$, $M \leq N$ and define $\mathbf{I} = \{n \in \mathbf{Z} / M - 1 < n < N + 1\}$, $\mathbf{J} = \{n \in \mathbf{Z} / M - 1 < n < N\}$. Given a family of Hilbert spaces $\{E_n\}_{n \in \mathbf{I}}$ and a family of isometries $\{v_n\}_{n \in \mathbf{J}}$, $v_n \in L(E_{n+1}, E_n)$, we define for $n \in \mathbf{J}$,

(1.12) $$L_n = E_n \ominus v_n E_{n+1}$$

and discern two cases:

CASE 1: $N \in \mathbf{Z}$, then we also define $L_N = E_N$ and for any $n \in \mathbf{I}$,

(1.13) $$E_n = L_n \oplus v_n L_{n+1} \oplus \cdots \oplus v_n \cdots v_{N-1} L_N ,$$

which leads to the following model for the family $\{v_n\}_{n \in \mathbf{J}}$: define for $n \in \mathbf{J}$ the isometries,

(1.14) $$S_{+,n} : \bigoplus_{k=n+1}^{N} L_k \to \bigoplus_{k=n}^{N} L_k$$

$$S_{+,n}(e_{n+1}, e_{n+1}, \ldots, e_N) = (0, e_{n+1}, \ldots, e_N),$$

then

$$\Phi_n v_n = S_{+,n}\Phi_{n+1} \tag{1.15}$$

where Φ_n, $n \in \mathbf{I}$ are obvious unitary operators from E_n onto $\bigoplus_{k=n}^{N} L_k$.

CASE 2: $N = \infty$, then for any $n > M - 1$, $p \geq 0$,

$$E_n = (L_n \oplus v_n L_{n+1} \oplus \dots \oplus v_n \dots v_{n+p-1} L_{n+p}) \oplus v_n \dots v_{n+p} E_{n+p+1} \,. \tag{1.16}$$

Let for $n > M - 1$, $R_n = \bigcap_{p=0}^{\infty} v_n \dots v_{n+p} E_{n+p+1}$ be the residual spaces. In view of relation (1.16),

$$E_n = \bigoplus_{k=0}^{\infty} v_n \dots v_{n+k-1} L_{n+k} \oplus R_n \tag{1.17}$$

(of course, for $k = 0$, $v_n \dots v_{n+k-1} L_{n+k}$ is interpreted as L_n) and now we are led to the following model for the family $\{v_n\}_{n\in\mathbf{J}}$: define for $n > M - 1$ the isometries

$$\begin{gathered} S_{+,n} : \bigoplus_{k=n+1}^{\infty} L_k \to \bigoplus_{k=n}^{\infty} L_k \\ S_{+,n}(e_{n+1}, e_{n+2}, \dots) = (0, e_{n+1}, e_{n+2}, \dots), \end{gathered} \tag{1.18}$$

then, with respect to the decomposition (1.17), v_n can be written as

$$v_n = v_n^{(1)} \oplus v_n^{(2)} \tag{1.19}$$

where $v_n^{(2)} : R_{n+1} \to R_n$ are unitary operators and

$$\Phi_n v_n^{(1)} = S_{+,n}\Phi_{n+1} \tag{1.20}$$

for obvious unitary operators Φ_n from $\bigoplus_{k=0}^{\infty} v_n \dots v_{n+k-1} L_{n+k}$ onto $\bigoplus_{k=0}^{\infty} L_{n+k}$.

REMARK 1.1. The above considerations can be rephrased as a Wold-von Neumann decomposition for unital isometric representations of the algebra of lower triangular matrices.

□

2. SPECTRAL FACTORS

Consider a positive-definite kernel S acting on the family $H = \{H_n\}_{n\in\mathbf{Z}}$ of Hilbert spaces.

For an array F with elements $F_{ij} = F(i,j) \in L(H_j, L_i)$, where $L = \{L_n\}_{n\in\mathbf{Z}}$ is a certain family of Hilbert spaces, and having the properties $F_{ij} = 0$ for $i < j$ and $\mathrm{col}_i F$ are contractions for $i \in \mathbb{Z}$ ($\mathrm{col}_i F$ denoting the i-th column of the array F), we can define the positive-definite kernel S_F by

$$S_F(i,j) = (\mathrm{col}_i F)^*(\mathrm{col}_j F)\ . \tag{2.1}$$

An array with these properties will be called a *lower triangular array* and the set of all these objects is denoted by $H^2(H,L)$.

A lower triangular array F is called *outer* if

$$\bigvee_{i \geq k} (\mathrm{col}_i F) H_i = \bigoplus_{i \geq k} L_i\ , \qquad k \in \mathbb{Z} \tag{2.2}$$

and recall that there exists a natural order on the set of positive-definite kernels, $S_1 \leq S_2$ if $S_2 - S_1$ is positive-definite.

With these preliminaries we can obtain the main result concerning the factorization of positive-definite kernels, with a proof modelled after [16] and [15].

THEOREM 2.1. *Let S be a positive-definite kernel, then there exists an outer array F^o in $H^2(H,L^o)$, where $L^o = \{L_n^o\}_{n \in \mathbb{Z}}$ is another family of Hilbert spaces, having the properties:*

(i) $S_{F^o} \leq S$

(ii) *for any other lower triangular array $F' \in H^2(H,\{L_n'\}_{n \in \mathbb{Z}})$ satisfying $S_{F'} \leq S$, we have $S_{F'} \leq S_{F^o}$.*

PROOF. Fix S (we continue to suppose, without loss of generality, $S_i = I_{H_i}$) and let V be the Kolmogorov decomposition of S given by (1.11). Defining the spaces

$$K_i^+ = H_i \oplus D(R_i), \quad i \in \mathbb{Z} \tag{2.3}$$

then the operators

$$\begin{aligned} &W_i^+ : K_{i+1}^+ \to K_i^+ \\ &W_i^+ = W_i / K_{i+1}^+ \end{aligned} \tag{2.4}$$

are isometries for $i \in \mathbb{Z}$. For the family $\{W_n^+\}_{n \in \mathbb{Z}}$ of isometries we use the Wold-von Neumann decomposition (we are in the case $M = -\infty$, $N = \infty$) and define, according to (1.12),

(2.5) $$L_i^o = K_i^+ \ominus W_i^+ K_{i+1}^+ .$$

Taking into account the structure of the operators W_n, we get for $i \in \mathbb{Z}$,

(2.6) $$L_i^o = W_i(\ldots \oplus 0 \oplus D_*(R_i) \oplus 0_{H_{i+1}} \oplus \ldots).$$

Further on, the array F^o is defined by

(2.7) $$F_{ji}^o = \begin{cases} P_{L_j^o}^{K_j} W_{j-1}^* \ldots W_i^* / H_i, & i < j \\ P_{L_i^o}^{K_i} / H_i, & i = j \\ 0, & i > j \end{cases}$$

and it is obvious that F^o is outer. We prove now the other required properties of F^o. By the Wold-von Neumann decomposition,

(2.8) $$K_i = (\ldots \oplus W_{i-1}^* L_{i-1}^o \oplus L_i^o \oplus W_i L_{i+1}^o \oplus W_i W_{i+1} L_{i+2}^o \oplus \ldots) \oplus R_i$$

and denote by P_i the projection onto the space $(\ldots \oplus W_{i-1}^* L_{i-1}^o \oplus L_i^o \oplus W_i L_{i+1}^o \oplus \ldots)$. We remark that for $i \in \mathbb{Z}$,

(2.9) $$W_{i-1} P_i = P_{i-1} W_{i-1}$$

and from the structure of W_n (and (2.6)),

$$P_i / H_i = (P_{L_i^o}^{K_i} + W_i P_{L_{i+1}^o}^{K_{i+1}} W_i^* + \ldots) / H_i ,$$

consequently, for $i \leq j$,

(2.10) $$(\mathrm{col}_i F^o)^* (\mathrm{col}_j F^o) = P_{H_i}^{K_i} W_i \ldots W_{j-1} P_j / H_j = P_{H_i}^{K_i} P_i W_i \ldots W_{j-1} / H_j .$$

So that, $\mathrm{col}_i F^o$ are contractions for $i \in \mathbb{Z}$ and then, we get for arbitrary integers m,n, $m \leq n$ and vectors $h_i \in H_i$, $m \leq i \leq n$, that

$$\sum_{i,j=m}^{n} (S_{F^o}(i,j)h_j,h_i) = \sum_{\substack{i,j=m \\ i\leq j}}^{n} (P_{H_i}^{K_i} P_i W_i \dots W_{j-1}h_j,h_i) +$$

$$+ \sum_{\substack{i,j=m \\ i>j}}^{n} (h_j,P_{H_j}^{K_j} P_j W_j \dots W_{i-1}h_i) \leq \sum_{\substack{i,j=m \\ i\leq j}}^{n} (P_{H_i}^{K_i} W_i \dots W_{j-1}h_j,h_i) +$$

$$+ \sum_{\substack{i,j=m \\ i>j}}^{n} (h_j,P_{H_j}^{K_j} W_j \dots W_{i-1}h_i) = \sum_{i,j=m}^{n} (S(i,j)h_j,h_i) .$$

Consequently, $S_{F^o} \leq S$ and (i) is proved.

Now, take $F' \in H^2(H,L')$ another array such that $S_{F'} \leq S$. As $K_i^+ = H_i \vee W_i H_{i+1} \vee \dots$ we define for $i \in \mathbf{Z}$ the maps $X_i : K_i^+ \to \bigoplus_{k\geq i} L'_k$, for finite sums by

$$X_i(\sum_{k=i}^{n} W_i \dots W_{k-1}h_k) = \sum_{k=i}^{n} (\mathrm{col}_k F')h_k \tag{2.11}$$

where $n \geq i$ is arbitrary. We have

$$\| X_i(\sum_{k=i}^{n} W_i \dots W_{k-1}h_k) \|^2 = \sum_{j,k=i}^{n} (S_{F'}(j,k)h_k,h_j) \leq$$

$$\leq \sum_{j,k=i}^{n} (S(j,k)h_k,h_j) = \| \sum_{k=i}^{n} W_i \dots W_{k-1}h_k \|^2$$

and then, X_i extends to a contraction from K_i^+ into $\bigoplus_{k\geq i} L'_k$ also denoted by X_i and which satisfies

$$X_i R_i = X_i(\bigcap_{p\geq i} W_i \dots W_{p-1}K_p^+) \subset \bigcap_{p\geq i} X_i W_i \dots W_{p-1}K_p^+ \subset$$

$$\subset \bigcap_{p\geq i} (\mathrm{col}_p F')(\bigoplus_{q\geq i} L'_q) = 0,$$

or

$$X_i P_i = X_i . \tag{2.12}$$

Now, for $m \leq n$ and $h_k \in H_k$, $m \leq k \leq n$, we get

$$\sum_{j,k=m}^{n} (S_{F'}(j,k)h_k,h_j) = \| \sum_{k=m}^{n} (\mathrm{col}_k F')h_k \|^2 =$$

$$= \|X_m(\sum_{k=m}^{n} W_m \dots W_{k-1}h_k)\|^2 = \|X_m P_m(\sum_{k=m}^{n} W_m \dots W_{k-1}h_k)\|^2 \leq$$

$$\leq \|P_m(\sum_{k=m}^{n} W_m \dots W_{k-1}h_k)\|^2 = \sum_{j,k=m}^{n} (S_{F^o}(j,k)h_k,h_j),$$

i.e. $S_{F'} \leq S_{F^o}$. □

Two questions remain to be elucidated. The first one regards the uniqueness of the array F^o and the second one regards conditions in order to have equality in the statement (i) of above theorem.

In order to describe the uniqueness property of F^o we need another factorization result. Consider F' and F" two lower triangular arrays, $F' \in H^2(H,L')$ and $F'' \in H^2(L',L)$ where $H = \{H_n\}_{n\in\mathbf{Z}}$, $L = \{L_n\}_{n\in\mathbf{Z}}$, $L' = \{L'_n\}_{n\in\mathbf{Z}}$ are three families of Hilbert spaces. Supposing that F" is actually a contraction between $\bigoplus_{n\in\mathbf{Z}} L'_n$ and $\bigoplus_{n\in\mathbf{Z}} L_n$, (we write this shortly $F'' \in H^\infty(L',L)$) we remark that the product $F'' \cdot F'$ makes sense in an obvious way as a lower triangular array. Now, we have the result.

2.2. PROPOSITION. *Let* F *and* F' *be two lower triangular arrays,* $F \in H^2(H,L)$ *and* $F' \in H^2(H,L')$, *the second one is supposed to be outer and*

$$S_F \leq S_{F'}.$$

Then, there exists a lower triangular array $F'' \in H^\infty(L',L)$ *such that*

$$F = F'' \cdot F'.$$

When $S_F = S_{F'}$, F" *is an isometry and if, in addition,* F *is outer, then* F" *is a unitary diagonal operator.*

PROOF. For fixed $i \in \mathbf{Z}$ and arbitrary $n \geq i$, we define

$$Y_i(\sum_{k=i}^{n} (\mathrm{col}_k F')h_k) = \sum_{k=i}^{n} (\mathrm{col}_k F)h_k$$

and

$$\| Y_i(\sum_{k=i}^{n} (\mathrm{col}_k F')h_k)\|^2 = \sum_{j,k=i}^{n} (\mathcal{S}_F(j,k)h_k,h_j) \leq$$

$$\leq \sum_{j,k=i}^{n} (\mathcal{S}_{F'}(j,k)h_k,h_j) = \| \sum_{k=i}^{n} (\mathrm{col}_k F')h_k \|^2 .$$

Consequently, Y_i extends to a contraction from $\bigoplus_{k\geq i} L'_k$ into $\bigoplus_{k\geq i} L_k$ in view of the outerness of F'. Moreover, we remark that for $i \in \mathbf{Z}$,

$$S_{+,i} Y_{i+1} = Y_i S'_{+,i}$$

where $S_{+,i}$ are the operators defined by (1.18) for the spaces $\{L_n\}_{n\in\mathbf{Z}}$ and $S'_{+,i}$ are the same operators with respect to the spaces $\{L'_n\}_{n\in\mathbf{Z}}$.

It readily follows that the family $\{Y_n\}_{n\in\mathbf{Z}}$ induces a lower triangular contraction F" between $\bigoplus_{n\in\mathbf{Z}} L'_n$ and $\bigoplus_{n\in\mathbf{Z}} L_n$ such that

$$F = F'' \cdot F' .$$

When $\mathcal{S}_F = \mathcal{S}_{F'}$, then Y_i are isometries for all $i \in \mathbf{Z}$ and, consequently, F" is an isometry.

Finally, if F is also outer, then Y_i are unitary operators and F" must be unitary diagonal. □

2.3. COROLLARY. *The array* F^o *in Theorem 2.1 is uniquely determined up to a left unitary diagonal factor.*

PROOF. Let F be another outer array satisfying (i) and (ii) in Theorem 2.1. It follows that $\mathcal{S}_{F'} = \mathcal{S}_{F^o}$ and we have only to use Proposition 2.2. □

In view of this corollary, F^o will be refered to as the *spectral factor* of *S*.

Another factorization result which can be easily obtained now, resembles the classical inner-outer factorization.

2.4. COROLLARY. *Any array* $F \in H^2(H,H')$, $H = \{H_n\}_{n\in\mathbf{Z}}$, $H' = \{H'_n\}_{n\in\mathbf{Z}}$, *admits a uniquely determined (up to unitary diagonal factors) factorization* $F = F_i F_o$, *with outer* $F_o \in H^2(H,H'')$, *for a certain family* $H'' = \{H''_n\}_{n\in\mathbf{Z}}$ *of Hilbert spaces and* F_i *an array in* $H^\infty(H'',H')$ *which is isometric.*

PROOF. Let $F_o \in H^2(H,H'')$ be the spectral factor of S_F. Consequently, $S_F = S_{F_o}$ and by Proposition 2.2, there exists isometric $F_i \in H^\infty(H'',H')$ such that $F = F_iF_o$.

For uniqueness, take $F = F_i'F_o'$ another factorization of the required type. Then it readily follows that $S_F = S_{F_o'} = S_{F_o}$ and again by Proposition 2.2, $F_o' = UF_o$, where U is a unitary diagonal array, and then, $F_i' = F_iU^*$. □

Now, we pass on to the question regarding conditions for equality holds in the statement (i) of Theorem 2.1 (in this case we say that S is *factorable*). From the proof of this theorem, we immediately get.

2.5. PROPOSITION. *S is factorable if and only if $R_i = \bigcap_{p\geq i} W_i \ldots W_{p-1}K_p^+ = 0$ for all $i \in \mathbf{Z}$.*

Of course, this criterion is quite intricate and it is desirable to know other criteria, or, at least, to determine some classes of factorable positive-definite kernels.

2.6. COROLLARY. *Let S be a positive-definite kernel on $\mathbf{N}$, the set of positive integers* (i.e., $H_n = 0$ *for* $n > 0$), *then S is factorable.*

PROOF. This is an immediate consequence of Proposition 2.5 and of the fact that the family $\{W_i^+\}$ to which is applied the Wold-von Neumann decomposition in the proof of Theorem 2.1 is indexed on $\{\ldots -2,-1,0\}$ (and so, $N = 0$), consequently, no residual spaces appear. □

2.7. REMARK. In particular, in case S is a positive block-matrix (i.e. only a finite number of spaces H_n are different from zero), the spectral factor of S coincides (in the sense of Corollary 2.3) with the so-called Cholesky factor of S (see for instance [11]). □

3. APPLICATIONS

3.1. A first application is about angle computations in nonstationary discrete processes. By a nonstationary process we mean here a family V of operators, $V = \{V(n)\}_{n\in\mathbf{Z}}$, $V(n) \in L(H_n,K_o)$, where $\{H_n\}_{n\in\mathbf{Z}}$ is a given family of Hilbert spaces and K_o is also a Hilbert space. The covariance kernel $S = \{S_{ij}\}_{i,j\in\mathbf{Z}}$ of V is defined by $S_{ij} = V^*(i)V(j)$, and it is obvious that S is positive-definite.

In other words, the elements of the process are exactly given by the Kolmogorov decomposition of its covariance kernel, and we can suppose that $K_o = \bigvee_{n\in\mathbb{Z}} V(n)H_n$ and $S_{ii} = I_{H_i}$.

Define the spaces $K_+^{(n)} = \bigvee_{k=n}^{\infty} V(k)H_k$, $K_-^{(n)} = \bigvee_{k=-\infty}^{-n} V(k)H_k$ and, for $p < q$, $K_p^{(q)} = \bigvee_{k=p}^{q} V(k)H_k$. For two subspaces G and F of K_o, the operator angle between G and F is defined by $B(G,F) = P_G^{K_o} P_F^{K_o} P_G^{K_o}$ and some of the problems in the study of processes refer exactly to the operator angles between the spaces $K_+^{(n)}$, $K_-^{(n)}$ or $K_p^{(q)}$.

Thus, an analogue of the Wiener-Kolmogorov prediction problem is the computation of the operator $\Delta_1^{(o)} = I_{H_o} - B(H_o, K_+^{(1)})$. We have the following result.

3.1. THEOREM. *The following formulas for computing $\Delta_1^{(o)}$ hold:*

$$\Delta_1^{(o)} = H_\infty^2(R_o)$$

$$\Delta_1^{(o)} = F_{oo}^{o*} F_{oo}^{o}$$

(according to the notation in Section 1 and Section 2).

PROOF. We immediately get that

$$K_+^{(1)} = W_o K_1^+$$

where K_1^+ was defined by (2.3), consequently,

$$P_{K_+^{(1)}}^{K_o} = W_o P_{K_1^+}^{K_1} W_o^*$$

and then

$$\Delta_1^{(o)} = I_{H_o} - P_{H_o}^{K_o} W_o P_{K_1^+}^{K_1} W_o^* P_{H_o}^{K_o} =$$

$$= P_{H_o}^{K_o} W_o (I_{K_1} - P_{K_1^+}^{K_1}) W_o^* P_{H_o}^{K_o} .$$

Having the structure of the operator W_o given by (1.10) we obtain for $h_o \in H_o$

$$\Delta_1^{(o)} h_o = P_{H_o}^{K_o} W_o (I_{K_1} - P_{K_1^+}^{K_1}) W_o^* h_o =$$

$$= P_{H_o}^{K_o} W_o (I_{K_1} - P_{K_1^+}^{K_1})(\ldots, H_\infty(R_o) h_o, *, * \ldots) =$$

$$= P_{H_o}^{K_o} W_o (\ldots, 0, H_\infty(R_o) h_o, 0_{H_o}, 0 \ldots) = H_o^2(R_o) h_o$$

which is exactly the first required formula (the entries marked by "*" do not enter into computations).

For the second formula, we have by (2.7)

$$F_{oo}^o = P_{L_o}^{K_o} | H_o$$

and by (2.6),

$$P_{L_o}^{K_o} = W_o P^{K_1} (\ldots \oplus 0 \oplus D_*(R_o) \oplus 0_{H_o} \oplus \ldots) W_o^*$$

so that

$$F_{oo}^{o*} F_{oo}^o = P_{H_o}^{K_o} W_o P^{K_1} (\ldots \oplus 0 \oplus D_*(R_o) \oplus 0_{H_o} \oplus \ldots) W_o^* P_{H_o}^{K_o} = H_\infty^2(R_o) = \Delta_1^{(o)}.$$

□

3.2. REMARK. The first formula in the above Theorem is an analogue of a formula of Verblunsky in the scalar ($H_n = \mathbf{C}$, $n \in \mathbf{Z}$), stationary (S is Toeplitz) case. The second part is an analogue of the Szegö-Kolmogorov-Krein formula which received several generalizations to multivariate stationary case in [16], [13] or [15]. □

3.2. In connection with Theorem 3.1, we can mention some asymptotic properties of the considered process. Thus, it is quite obvious that

$$\text{s-}\lim_{n \to \infty} (I - B(H_o, K_1^{(n)})) = \Delta_1^{(o)}$$

and this actually reflects the first limit theorem of Szegö (see [8]). Indeed, we remark that in case $\dim H_n < \infty$ for all $n \in \mathbf{Z}$,

$$\frac{\det M_{on}}{\det M_{1n}} = \det (I - B(H_o, K_1^{(n)})$$

(we can use for instance Theorem 4.1 in [1]) and then, by Theorem 3.1

$$\lim_{n\to\infty} \frac{\det M_{on}}{\det M_{1n}} = \det \Delta_1^{(o)} = \prod_{n=1}^{\infty} \det D^2_{G_{on}} = |\det F^o_{oo}|^2 \tag{3.1}$$

which is a nonstationary variant (pointed out in [3]) of the first limit theorem of Szegö.

Further on, let us mention that, also in connection with this classical result of Szegö, we can require (see also [7]) to compute:

$$\text{s-}\lim_{n\to\infty} [M^{-1}_{on}]_{oo} .$$

(of course, when M_{on} are invertible operators for $n > 0$; $[M^{-1}_{on}]_{oo}$ denotes the element in position $(0,0)$ in M^{-1}_{on}).

Using the formula for the Cholesky factor of M_{on} which follows from Theorem 2.1 (i.e. formula (1.11) in [3]), we get (by supposing all $D_{G^*_{on}}$, $n \geq 1$, to be invertible)

$$\begin{aligned}[M^{-1}_{on}]_{oo} &= I + R_{o,n} D_n(R_o)^{-2} R^*_{o,n} = \\ &= I + D^{-1}_{R^*_{o,n}} R_{o,n} R^*_{o,n} D^{-1}_{R^*_{o,n}} = \\ &= D^{-2}_{R^*_{o,n}} = H^{-1}_n(R_o) H^{*-1}_n(R_o)\end{aligned}$$

where $R_{o,n}$ is the compresion of R_o to the first n components (i.e. $R_{o,n} = (G_{o1}, D_{G^*_{o1}} G_{o2}, \ldots, D_{G^*_{o1}} \ldots D_{G^*_{o,n-1}} G_{on})$), $D_n(R_o)$ and $H_n(R_o)$ are defined by (1.4) and respectively (1.6). Using Theorem 3.1 we get that if $H_\infty(R_o)$ is an invertible operator, then

$$\text{s-}\lim_{n\to\infty} [M^{-1}_{on}]_{oo} = H^{-2}_\infty(R_o) = (F^o_{oo})^{-1} (F^{o*}_{oo})^{-1} . \tag{3.2}$$

Now, return to the Wold-von Neumann decomposition of K^+_o. That is, by (1.16), we have

$$K^+_o = (L^o_o \oplus W_o L^o_1 \oplus \ldots \oplus W_o \ldots W_{n-2} L^o_{n-1}) \oplus W_o \ldots W_{n-1} K^+_n .$$

But, it is easy to see that

$$W_o \ldots W_{n-1} K^+_n = K^{(n)}_+ ,$$

consequently,

$$P_{K^{(n)}_+} = P_{K^+_o} - (P_{L^o_o} + P_{W_o L^o_1} + \ldots + P_{W_o \ldots W_{n-2} L^o_{n-1}}) .$$

Taking (2.7) into account, we obtain an extension of the second formula in Theorem 3.1,

$$B(H_o, K_+^{(n)}) = I_{H_o} - \sum_{k=0}^{n-1} F_{ok}^{o*} F_{ok}^{o} . \tag{3.3}$$

Moreover, let R_o be the residual space, $R_o = \bigcap_{p \geq 0} W_o \ldots W_{p-1} K_p^+$, then, by the formula (1.17),

$$\text{s-}\lim_{n \to \infty} B(H_o, K_+^{(n)}) = B(H_o, R_o)$$

and

$$B(H_o, R_o) = I_{H_o} - \sum_{k=0}^{\infty} F_{ok}^{o*} F_{ok}^{o} . \tag{3.4}$$

Following the idea in [10], we can see that (3.3) and (3.4) reflect some generalizations of the first limit theorem of Szegö (of its nonstationary variant (3.1)). Indeed, for $p \geq 0$, consider $S^{[p]}$ the positive-definite kernel obtained from S by deleating the rows and columns indexed by $\{1,2,\ldots,p\}$, i.e. for $i \leq j$

$$S^{[p]}(i,j) = \begin{cases} S(i,j), & i < 0 \\ S(0,j+p), & i = 0 \\ S(i+p,j+p), & i > 0 \end{cases}$$

(and of course, $S^{[p]}(i,j) = S^{[p]}(j,i)^*$, elsewhere), and let $V^{[p]}$ be the process having $S^{[p]}$ as covariance kernel.

We remark that for $n \geq p$,

$$K_p^{(n)}(V) = K_1^{(n)}(V^{[p]})$$

and using again Theorem 4.1 in [1], we obtain that (when $\dim H_n < \infty$ for all $n \in \mathbf{Z}$),

$$\frac{\det M_{on}^{[p]}}{\det M_{p+1,n}} = \frac{\det M_{on}^{[p]}}{\det M_{1n}^{[p]}} = \det (I - B(H_o, K_1^{(n)}(V^{[p]})) =$$

$$= \det (I - B(H_o, K_p^{(n)}(V)) .$$

But, $\text{s-}\lim_{n \to \infty} B(H_o, K_p^{(n)}) = B(H_o, K_+^{(p)})$, consequently

$$\lim_{n \to \infty} \frac{\det M_{on}^{[p]}}{\det M_{p+1,n}} = \det\Big(\sum_{k=0}^{p-1} F_{ok}^{o*} F_{ok}^{o}\Big) \tag{3.5}$$

which represents a nonstationary variant of a result in [10].

REFERENCES

1. **Arsene, Gr.; Constantinescu, T.** : Structure of positive block-matrices and nonstationary prediction, *J. Funct. Anal.* **70** (1987), 402-425.

2. **de Branges, L.** : Appendix on square summable power series, The expansion theorem for Hilbert spaces of entire functions, Entire Functions and Related Parts of Analysis, Proc. Symp. Pure Math., vol. XI, Amer Math. Soc., Providence, RI, 1968.

3. **Constantinescu, T.** : Schur analysis of positive block-matrices, in "*I. Schur Methods in Operator Theory and Signal Processing*", pp. 191-206, (OT Series, vol. 18), Birkhauser, Basel, 1986.

4. **Devinatz, A.** : The factorization of operator valued functions, *Ann. of Math.* (2) **73** (1961), 458-495.

5. **Douglas, R.G.** : On factoring positive operator functions, *J. Math. Mech.* **16** (1966), 119-126.

6. **Gohberg, I.; Krein, M.G.** : *Theory and Applications of Volterra Operators in Hilbert Spaces (Russian)*, Moscow, 1967.

7. **Gohberg, I.; Levin, S.** : On an open problem for block-Toeplitz matrices, *Integral Equation Operator Theory*, 12 (1979), 121-129.

8. **Grenander, U.; Szego, G.** : *Toeplitz forms and their Applications*, Univ. California Press, Berkeley, 1958.

9. **Helson, H.; Lowdenslager, D.** : Prediction theory and Fourier series in several variables. I, II, *Acta Math.* 99 (1958), 165-202; 106 (1961), 175-213.

10. **Krein, M.G.; Spitkovskii, I.M.** : On some generalizations of Szego first limit theorem (Russian), *Anal. Math.* 9 (1983), 23-41.

11. **Lev-Ari, H.; Kailath, T.** : Lattice filter parametrization of nonstationary processes, IEEE Trans. Info. Theory, IT-30:2-16 (1984).

12. **Lowdenslager, D.** : On factoring matrix valued functions, *Ann. of Math.* (2) **78** (1963), 450-454.

13. **Rosenblum, M.; Rovnyak, J.** : The factorization problem for nonnegative operator valued functions, *Bull. Amer. Math. Soc.* **77** (1971), 287-318.

14. **Rosenblum, M.; Rovnyak, J.** : *Hardy Classes and Operator Theory*, Clarendon Press, Oxford, 1985.

15. **Suciu, I.; Valusescu, I.**: Factorization theorems and prediction theory, *Rev. Roumaine Math. Pures Appl.* **23** (1978), 1393-1424.

16. **Sz.Nagy, B.; Foias, C.** : *Harmonic Analysis of Operators on Hilbert space*, North-Holland/Akad. Kiado, Amsterdam/Budapest, 1970.

17. **Wiener, N.; Masani, P.** : The prediction theory of multivariate stochastic processes. I, II, *Acta Math.* **98** (1957), 111-150; **99** (1958), 93-139.

Department of Mathematics, INCREST
Bdul Pacii 220, 79622 Bucharest, Romania.

Operator Theory:
Advances and Applications, Vol. 48

GENERALIZED CORESOLVENTS OF STANDARD ISOMETRIC OPERATORS AND GENERALIZED RESOLVENTS OF STANDARD SYMMETRIC RELATIONS IN KREIN SPACES

Aad Dijksma, Heinz Langer, Henk de Snoo

All unitary (selfadjoint) Kreĭn space extensions of a given closed isometric operator V (symmetric relation S) as well as all generalized coresolvents of V (generalized resolvents of S) are parametrized over operator functions, holomorhic in a neighborhood of 0. The main assumption is that the defect spaces of V (S, respectively) are orthocomplemented. The principal tools are the representation of operator functions, holomorphic in a neighborhood of 0 as characteristic functions of unitary colligations and the Cayley transform.

0. Introduction

In a Kreĭn space $\mathfrak{H}$ let V be a closed isometric operator and assume that it is standard, which in this note means that its domain $\mathfrak{D}(V)$ and range $\mathfrak{R}(V)$ are orthocomplemented subspaces of $\mathfrak{H}$. Let W in a Kreĭn space $\mathfrak{K}$ be a unitary extension of V. By definition then $\mathfrak{K}$ is such that (*i*) $\mathfrak{H}\subset\mathfrak{K}$ and the inner product on $\mathfrak{K}$ restricted to $\mathfrak{H}$ coincides with the one on $\mathfrak{H}$, and (*ii*) W is a unitary (i.e., bijective and isometric) operator in $\mathfrak{K}$ and $V\subset W$. The compression $K(z)=P_{\mathfrak{H}}(I-zW)^{-1}|_{\mathfrak{H}}$ of the coresolvent $(I-zW)^{-1}$ of W to $\mathfrak{H}$ is called a generalized coresolvent of V, where by $P_{\mathfrak{L}}$ we denote the orthogonal projection from $\mathfrak{K}$ onto $\mathfrak{L}$. Generalized coresolvents $K(z)$ of V are holomorphic in $\mathfrak{d}$, the generic notation for a neighborhood of 0 contained in the open unit disc $\mathbb{D}\subset\mathbb{C}$, have values in $\mathbf{L}(\mathfrak{H})$ and satisfy $(I-zV)K(z)=I$ on $\mathfrak{d}$. Here, for Kreĭn spaces $\mathfrak{F}$ and $\mathfrak{G}$, $\mathbf{L}(\mathfrak{F},\mathfrak{G})$ is the linear space of bounded linear operators from $\mathfrak{F}$ to $\mathfrak{G}$ and we write $\mathbf{L}(\mathfrak{F})$ for $\mathbf{L}(\mathfrak{F},\mathfrak{F})$. We denote by $\mathbf{S}(\mathfrak{F},\mathfrak{G})$ the generalized Schur class, i.e., the collection of all mappings $\Theta(z)$ which are defined and holomorphic on some $\mathfrak{d}$ and have values in $\mathbf{L}(\mathfrak{F},\mathfrak{G})$. In Section 1 of this note we prove parametrizations of the sets of unitary extensions and of generalized coresolvents of V over the class $\mathbf{S}(\mathfrak{D}(V)^{\perp},\mathfrak{R}(V)^{\perp})$. Our description is based on a representation theorem of the mappings in $\mathbf{S}(\mathfrak{F},\mathfrak{G})$ as characteristic functions of unitary colligations, see, e.g., [AI] and [DLS]. It also involves the characteristic function and the so called Q function of V. Some results in this section complement those in the book of Azizov and Iohvidov [AI], where an isometry is called standard if only it is closed, bounded and boundedly invertible. In [DLS] we discussed the case where $\mathfrak{H}$ is a Hilbert space and $\mathfrak{K}$ is a Kreĭn space. The case where $\mathfrak{H}$ and $\mathfrak{K}$ are Pontryagin spaces (with possibly different numbers of negative squares) was discussed in [L], [LS], [S] and, under the extra condition that the subspaces $\mathfrak{D}(V)^{\perp}$ and $\mathfrak{R}(V)^{\perp}$ are isomorphic, in [KL2]. Sorjonen [S] even discusses the situation when V is not standard. If both $\mathfrak{H}$ and $\mathfrak{K}$ are Hilbert spaces some of these results can be found in [Č] and [G].

In Section 2 we consider the case of a closed symmetric relation S in a Kreĭn space $\mathfrak{H}$, which is standard, by which we mean that S has the property that for some $\mu \in \mathbb{C}\backslash\mathbb{R}$ the ranges $\mathfrak{R}(S-\mu)$ and $\mathfrak{R}(S-\bar{\mu})$ are orthocomplemented subspaces of $\mathfrak{H}$. If A in a Kreĭn space $\mathfrak{K}$ is a selfadjoint extension of S, i.e., A is a selfadjoint relation in $\mathfrak{K}$ such that (i) above is valid and $S \subset A$, and if $\mu \in \rho(A)$, then the compression $R(\ell) = P_{\mathfrak{H}}(A-\ell)^{-1}|_{\mathfrak{H}}$ of the resolvent $(A-\ell)^{-1}$ of A to $\mathfrak{H}$ is called a generalized resolvent of S. Generalized resolvents of S are holomorphic in $\mathcal{D}(\mu,\bar{\mu})$, have values in $\mathbf{L}(\mathfrak{H})$ and satisfy the relations $R(\ell)^{+}=R(\bar{\ell})$ and $(S-\ell)R(\ell) \subset I$ for $\ell \in \mathcal{D}(\mu,\bar{\mu})$, where $\mathcal{D}(\mu,\bar{\mu})$ is a neigborhood in $\mathbb{C}\backslash\mathbb{R}$ of the points μ and $\bar{\mu}$ which is symmetric with respect to the real axis. Using the Cayley transform we apply the results of Section 1 and obtain descriptions of all generalized resolvents of S. For particular cases we refer to [DLS], [LS], [KL1] and [LT].

Our presentation connects the work of M.A. Naĭmark, M.G. Kreĭn and A.V. Štraus and is related to the paper [AG] by D.Z. Arov and L.Z. Grossman. Applications of the results in this paper to boundary value problems will be given elsewhere.

1. Unitary extensions of standard isometric operators

In the following when not stated otherwise $\mathfrak{K}$, $\mathfrak{H}$ will stand for separable Kreĭn spaces. If $\mathfrak{L}$ is an orthocomplemented subspace of $\mathfrak{K}$ and $\mathfrak{L} = \mathfrak{L}^{+} + \mathfrak{L}^{-}$ is a fundamental decomposition of $\mathfrak{L}$ we define the signature $\operatorname{sign}\mathfrak{L}$ as the ordered pair of numbers $\{\dim\mathfrak{L}^{+}, \dim\mathfrak{L}^{-}\}$, which is actually independent of the chosen fundamental decomposition. V will always be a closed standard isometric operator in $\mathfrak{H}$. The orthogonal complements $\mathfrak{D}(V)^{\perp}$ and $\mathfrak{R}(V)^{\perp}$ are called the defect spaces and their signatures will be called the defect signatures of V. In this section we want to characterize all unitary extensions W of V. If T is an operator or relation from one Kreĭn space to another T^{+} stands for the adjoint of T. Interpreting $V^{-+} = (V^{+})^{-1} = (V^{-1})^{+}$ in terms of linear relations we have

$$V^{-+} = V \dot{+} \{\{0,\varphi\} \mid \varphi \in \mathfrak{R}(V)^{\perp}\} \dot{+} \{\{\psi,0\} \mid \psi \in \mathfrak{D}(V)^{\perp}\}, \text{ direct sums in } \mathfrak{H}^2,$$

which is an analog of Von Neumann's formula for symmetric relations, see Section 2 below. A unitary extension W in $\mathfrak{K}$ of V is called canonical if $\mathfrak{K} = \mathfrak{H}$ and then $V \subset W = W^{-+} \subset V^{-+}$. Such extensions exist if and only if $\operatorname{sign} \mathfrak{D}(V)^{\perp} = \operatorname{sign} \mathfrak{R}(V)^{\perp}$. If these defect signatures of V are equal, then the formula

$$(1.1) \qquad W = \Theta \oplus V = \begin{pmatrix} \Theta & O \\ O & V \end{pmatrix} : \begin{pmatrix} \mathfrak{D}(V)^{\perp} \\ \mathfrak{D}(V) \end{pmatrix} \rightarrow \begin{pmatrix} \mathfrak{R}(V)^{\perp} \\ \mathfrak{R}(V) \end{pmatrix}$$

gives a one to one correspondence between all canonical unitary extensions W of V and all unitary $\Theta \in \mathbf{L}(\mathfrak{D}(V)^{\perp}, \mathfrak{R}(V)^{\perp})$. We shall generalize this formula to one which characterizes all unitary extensions of a closed standard isometric operator with arbitrary defect signatures. If W is a unitary operator in $\mathfrak{K}$ and $\mathfrak{L}$ a subset of $\mathfrak{K}$, then W and (or) $\mathfrak{K}$ are (is) called $\mathfrak{L}$ minimal if

$$\mathfrak{K} = \text{c.l.s.}\{ (W-z)^{-1} f \mid f \in \mathfrak{L},\ z \in \rho(W) \}.$$

Here, if T is a relation or an operator in a Kreĭn space $\mathfrak{K}$, we have used the notation $\rho(T) = \{ z \in \mathbb{C} \mid (T-z)^{-1} \in \mathbf{L}(\mathfrak{K}) \}$ for the resolvent set of T and we recall that $(T-z)^{-1}$ on $\rho(T)$ is called the resolvent operator.

Now let V be a closed standard isometric operator (not necessarily satisfying $\operatorname{sign} \mathfrak{D}(V)^{\perp} = \operatorname{sign} \mathfrak{R}(V)^{\perp}$), assume that W in $\mathfrak{K}$ is a unitary extension of V and put $\hat{\mathfrak{H}} = \mathfrak{K} \ominus \mathfrak{H}$. Then W has the matrix representation

$$(1.2) \qquad W = \begin{pmatrix} T & F & O \\ G & H & O \\ O & O & V \end{pmatrix} : \begin{pmatrix} \hat{\mathfrak{H}} \\ \mathfrak{D}(V)^{\perp} \\ \mathfrak{D}(V) \end{pmatrix} \rightarrow \begin{pmatrix} \hat{\mathfrak{H}} \\ \mathfrak{R}(V)^{\perp} \\ \mathfrak{R}(V) \end{pmatrix},$$

where $T \in \mathbf{L}(\hat{\mathfrak{H}})$, $F \in \mathbf{L}(\mathfrak{D}(V)^{\perp}, \hat{\mathfrak{H}})$, $G \in \mathbf{L}(\hat{\mathfrak{H}}, \mathfrak{R}(V)^{\perp})$, $H \in \mathbf{L}(\mathfrak{D}(V)^{\perp}, \mathfrak{R}(V)^{\perp})$. The colligation

$$(1.3) \qquad \Delta = \left(\hat{\mathfrak{H}}, \mathfrak{D}(V)^{\perp}, \mathfrak{R}(V)^{\perp}; \left(\begin{smallmatrix} T & F \\ G & H \end{smallmatrix}\right) \right)$$

is unitary and will be called the unitary colligation associated with the extension W. Its characteristic function $\Theta(z) = \Theta_{\Delta}(z) = H + zG(I - zT)^{-1}F$, $z=0$ or $(1/z) \in \rho(T)$, will be called the characteristic function associated with W. We put

$$(1.4) \qquad \tilde{\Theta}(z) = \Theta(z) \oplus V = \begin{pmatrix} \Theta(z) & O \\ O & V \end{pmatrix} : \begin{pmatrix} \mathfrak{D}(V)^{\perp} \\ \mathfrak{D}(V) \end{pmatrix} \rightarrow \begin{pmatrix} \mathfrak{R}(V)^{\perp} \\ \mathfrak{R}(V) \end{pmatrix}$$

and find after some straightforward calculations that

$$(1.5) \qquad \{\{P_{\mathfrak{H}}\tilde{h}, P_{\mathfrak{H}}W\tilde{h}\} \mid \tilde{h} \in \mathfrak{K},\ (I - zW)\tilde{h} \in \mathfrak{H}\} = \{\{h, \tilde{\Theta}(z)h\} \mid h \in \mathfrak{H}\}.$$

The set on the lefthand side is defined for all $z \in \mathbb{C}$ and will be called the Štraus relation in $\mathfrak{H}$ associated with the unitary extension W.

We recall the notion of a weak isomorphism between two Kreĭn spaces. A linear map $Z: \mathfrak{D}(Z) \subset \mathfrak{K} \to \mathfrak{H}$ is called a weak isomorphism from $\mathfrak{K}$ to $\mathfrak{H}$ if $\mathfrak{D}(Z)$ is dense in $\mathfrak{K}$, $\mathfrak{R}(Z)$ is dense in $\mathfrak{H}$ and $[Zx,Zy]=[x,y]$ for all $x,y \in \mathfrak{D}(Z)$, where $[.,.]$ stands for any of the inner products on $\mathfrak{K}$ and $\mathfrak{H}$. If such a map exists $\mathfrak{K}$ and $\mathfrak{H}$ are called weakly isomorphic. A weak isomorphism between two Pontryagin spaces is bounded and can be extended by continuity to an isomorphism. In general, this is not true for Kreĭn spaces. Two unitary extensions W_j in $\mathfrak{K}_j$ of V, $j=1,2$, are called weakly isomorphic, if there exists a weak isomorphism Z from $\mathfrak{K}_1$ to $\mathfrak{K}_2$ with the properties that $Z|_{\mathfrak{H}}=I$, the identity on $\mathfrak{H}$, and that $ZW_1x=W_2Zx$ for all $x \in \mathfrak{D}(Z)$, so that, in particular, W_1 maps $\mathfrak{D}(Z)$ to itself. For a function $\Theta \in \mathbf{S}(\mathfrak{K},\mathfrak{H})$ we introduce the matrix kernel $\mathsf{S}_\Theta(z,w)$ defined by

$$\mathsf{S}_\Theta(z,w)=\begin{pmatrix} \dfrac{I-\Theta(w)^+\Theta(z)}{1-z\bar{w}} & \dfrac{\Theta(\bar{z})^+-\Theta(w)^+}{z-\bar{w}} \\ \dfrac{\Theta(z)-\Theta(\bar{w})}{z-\bar{w}} & \dfrac{I-\Theta(\bar{w})\Theta(\bar{z})^+}{1-z\bar{w}} \end{pmatrix}, \quad z,w \in \mathcal{D},\ z \neq \bar{w},$$

for some $\mathcal{D}$ which is symmetric with respect to the real axis. The next theorem provides the characterization of all unitary extensions which generalizes the one for all canonical unitary extensions based on formula (1.1). We do not assume that V has equal defect signatures.

THEOREM 1.1. *Let V be a closed standard isometric operator in a Kreĭn space $\mathfrak{H}$. Then the characteristic function Θ associated with a unitary extension W of V belongs to the class $\mathbf{S}(\mathfrak{D}(V)^\perp,\mathfrak{R}(V)^\perp)$. Conversely, if $\Theta \in \mathbf{S}(\mathfrak{D}(V)^\perp,\mathfrak{R}(V)^\perp)$ then there exists a unitary extension W of V in a Kreĭn space $\mathfrak{K}$ such that $\Theta(z)=\Theta_\Delta(z)$, $z \in \mathcal{D}$, where Δ is the unitary colligation associated with W, via (1.2) defined by (1.3). The extension W can be chosen $\mathfrak{H}$ minimal, in which case it is uniquely determined up to weak isomorphisms. In that case also, the number of positive (negative) squares of the kernel $\mathsf{S}_\Theta(z,w)$ equals the number of positive (negative) squares of the extending space $\hat{\mathfrak{H}}=\mathfrak{K}\ominus\mathfrak{H}$.*

As to the proof of this theorem we note that the first part is a summary of the preceding discussion. Its converse follows from a general result which states that each mapping $\Theta \in \mathbf{S}(\mathfrak{K},\mathfrak{H})$ coincides on some $\mathcal{D}$ with the characteristic function Θ_Δ of a unitary colligation $\Delta=(\tilde{\mathfrak{K}},\mathfrak{K},\mathfrak{H};\left(\begin{smallmatrix} T & F \\ G & H\end{smallmatrix}\right))$. The last statement also follows from this representation and the formula (in selfexplanatory notation)

$$[\mathsf{S}_{\Theta_\Delta}(z,w)\begin{pmatrix}k_1\\h_1\end{pmatrix},\begin{pmatrix}k_2\\h_2\end{pmatrix}]=[(I-zT)^{-1}Fk_1+(I-zT^+)^{-1}G^+h_1,(I-wT)^{-1}Fk_2+(I-wT^+)^{-1}G^+h_2].$$

For further details we refer to [DLS].

The coresolvent operator for a relation or an operator T in a Kreĭn space $\mathfrak{K}$ is defined to be the operator $K_T(z)=(I-zT)^{-1}$. It is locally holomorphic, has values in $\mathbf{L}(\mathfrak{K})$ for all $z\in\hat{\rho}(T)$, where $\hat{\rho}(T)=\{z\in\mathbb{C}\,|\,z=0 \text{ or } (1/z)\in\rho(T)\}$, and satisfies the relation

$$zK_T(z)-wK_T(w)=(z-w)K_T(z)K_T(w), \quad z,w\in\hat{\rho}(T).$$

If W is unitary, then the set $\hat{\rho}(W)\backslash\{0\}$ is symmetric with respect to the unit circle, $K_W(1/\bar{z})=I-K_W(z)^+$, $z\in\hat{\rho}(W)\backslash\{0\}$ and hence,

$$K_W(z)+K_W(w)^+-I-(1-z\bar{w})K_W(w)^+K_W(z)=0, \qquad z,w\in\hat{\rho}(W). \tag{1.6}$$

Let W in $\mathfrak{K}$ be a unitary operator and let $\mathfrak{H}$ be an orthocomplemented subspace of $\mathfrak{K}$. Then the compressed coresolvent $K(z)=P_{\mathfrak{H}}(I-zW)^{-1}|_{\mathfrak{H}}$ of W to $\mathfrak{H}$ is locally holomorphic on $\hat{\rho}(W)$, has values in $\mathbf{L}(\mathfrak{H})$ and, on account of (1.6), satisfies

$$\frac{K(z)+K(w)^+-I}{1-z\bar{w}}-K(w)^+K(z)=P_{\mathfrak{H}}K_W(w)^+(I-P_{\mathfrak{H}})K_W(z)|_{\mathfrak{H}}, \qquad z,w\in\hat{\rho}(W),\ z\bar{w}\neq 1. \tag{1.7}$$

We denote the kernel on the lefthand side of this equality by $\mathsf{R}_K(z,w)$,

$$\mathsf{R}_K(z,w)=\frac{K(z)+K(w)^+-I}{1-z\bar{w}}-K(w)^+K(z), \qquad z,w\in\hat{\rho}(W).$$

The link between the generalized coresolvent $K(z)$ and the characteristic function $\Theta(z)$ associated with an extension W of a closed standard isometric operator V is given by

$$K(z)=P_{\mathfrak{H}}(I-zW)^{-1}|_{\mathfrak{H}}=(I-z\tilde{\Theta}(z))^{-1}, \qquad z\in\mathfrak{D}, \tag{1.8}$$

which follows immediately from formulas (1.4) and (1.5).

With each unitary extension W of V we have associated two kernels, namely $\mathsf{R}_K(z,w)$ and $\mathsf{S}_\Theta(z,w)$. In order to express the relation between the two we define the operators $\Delta(z)\in\mathbf{L}(\mathfrak{D}(V)^\perp,\mathfrak{H})$, $\nabla(z)\in\mathbf{L}(\mathfrak{R}(V)^\perp,\mathfrak{H})$ by

$$\Delta(z)=P_{\mathfrak{H}}(I-zW^+)^{-1}|_{\mathfrak{D}(V)^\perp}, \quad \nabla(z)=zP_{\mathfrak{H}}(z-W)^{-1}|_{\mathfrak{R}(V)^\perp}, \qquad z\in\rho(W).$$

Note that by (1.8) for $z\in\mathfrak{D}$,

$$\Delta(z)^+=P_{\mathfrak{D}(V)^\perp}(I-\bar{z}\tilde{\Theta}(\bar{z}))^{-1}, \quad \nabla(1/z)=(I-z\tilde{\Theta}(z))^{-1}|_{\mathfrak{R}(V)^\perp}, \tag{1.9}$$

where in the last identity $z\neq 0$.

PROPOSITION 1.2. *(i) For each $z\in\mathfrak{D}$, $\Delta(z)$ maps $\mathfrak{D}(V)^\perp$ onto $\mathfrak{R}(I-\bar{z}V)^\perp$ and is boundedly invertible. (ii) For each z such that $1/z\in\mathfrak{D}$, $\nabla(z)$ maps $\mathfrak{R}(V)^\perp$ onto $\mathfrak{R}(I-\bar{z}V)^\perp$ and is boundedly invertible. (iii) The following relations hold*

$$\begin{aligned}
\mathsf{R}_K(z,w) &= \bar{w}\Delta(\bar{w})\,(1-z\bar{w})^{-1}(I-\Theta(w)^+\Theta(z))\,z\Delta(\bar{z})^+, && \\
\mathsf{R}_K(z,1/w) &= -\bar{w}\nabla(1/\bar{w})\,(z-\bar{w})^{-1}(\Theta(z)-\Theta(\bar{w}))\,z\Delta(\bar{z})^+, && z\neq\bar{w},\quad w\neq 0, \\
\mathsf{R}_K(1/z,w) &= -\bar{w}\Delta(\bar{w})\,(z-\bar{w})^{-1}(\Theta(\bar{z})^+-\Theta(w)^+)\,z\nabla(1/\bar{z})^+, && z\neq\bar{w},\quad z\neq 0,
\end{aligned}$$

$$\mathsf{R}_K(1/z,1/w) \;=\bar{w}\nabla(1/\bar{w})\,(1-z\bar{w})^{-1}(I-\Theta(\bar{w})\Theta(\bar{z})^+)\,z\nabla(1/\bar{z})^+, \qquad z,w\neq 0,$$

for z,w in some $\mathfrak{D}$ which is symmetric with respect to the real axis.

Proof. We briefly indicate the arguments leading to (*i*). Since, for $f\in\mathfrak{D}(V)^\perp$ and $g\in\mathfrak{D}(V)$

$$[\Delta(z)f,(I-\bar{z}V)g]=[(I-zW^+)^{-1}f,(I-\bar{z}V)g]=[f,g]=0,$$

$\Delta(z)$ maps $\mathfrak{D}(V)^\perp$ into $\mathfrak{R}(I-\bar{z}V)^\perp$. The operator $P_{\mathfrak{H}}(I-zW^+)^{-1}\in\mathbf{L}(\mathfrak{H})$ is boundedly invertible in some $\mathfrak{D}$ and hence for each $g\in\mathfrak{R}(I-\bar{z}V)^\perp$, $z\in\mathfrak{D}$, there exists an $f\in\mathfrak{H}$ so that $g=P_{\mathfrak{H}}(I-zW^+)^{-1}f$. Write $f=f_1+f_2$, $f_1\in\mathfrak{D}(V)^\perp$, $f_2\in\mathfrak{D}(V)$. Then

$$0=[g,(I-\bar{z}V)f_2]=[(I-zW^+)^{-1}f,(I-\bar{z}V)f_2]=[f,(I-\bar{z}W)^{-1}(I-\bar{z}V)f_2]=[f_2,f_2],$$

which implies that $f_2=0$ and hence, $g=\Delta(z)f_1\in\mathfrak{R}(\Delta(z))$. Part (*ii*) can be proved in a similar way. The first equality of (*iii*) follows from

$$\begin{aligned}\mathsf{R}_K(z,w)&=\bar{w}(I-w\tilde{\Theta}(w))^{-+}\;(1-z\bar{w})^{-1}(I-\tilde{\Theta}(w)^+\tilde{\Theta}(z))\;z(I-z\tilde{\Theta}(z))^{-1}\\&=\bar{w}(I-w\tilde{\Theta}(w))^{-+}P_{\mathfrak{D}(V)^\perp}\;(1-z\bar{w})^{-1}(I-\Theta(w)^+\Theta(z))\;P_{\mathfrak{D}(V)^\perp}z(I-z\tilde{\Theta}(z))^{-1}\\&=\bar{w}\Delta(\bar{w})\;(1-z\bar{w})^{-1}(I-\Theta(w)^+\Theta(z))\;z\Delta(\bar{z})^+,\end{aligned}$$

where we have used (1.9). The other equalities can be shown in the same way. We omit the details.

It follows from Theorem 1.1 and Proposition 1.2 that if the extending space $\mathfrak{K}$ is $\mathfrak{H}$ minimal, then the number of positive (negative) squares of $\mathfrak{K}\ominus\mathfrak{H}$ is equal to the number of positive (negative) squares of the kernel $\mathsf{R}_K(z,w)$. Clearly this also follows directly from (1.7).

Now we turn to a description of all generalized coresolvents $K(z)$ of V in terms of one fixed generalized coresolvent. For this we recall the definition of a characteristic function of a closed standard isometric operator V in a Kreĭn space $\mathfrak{H}$: it is the function

$$X(z)=zP_{\mathfrak{D}(V)^\perp}\,(I-z\tilde{V})^{-1}|_{\mathfrak{R}(V)^\perp},$$

where $\tilde{V}$ denotes the trivial linear extension of V to all of $\mathfrak{H}$, i.e., $\tilde{V}f=Vf$ if $f\in\mathfrak{D}(V)$, $\tilde{V}f=0$ if $f\in\mathfrak{D}(V)^\perp$. Note that $X(z)$ is the characteristic function of the unitary colligation $(\mathfrak{H},\mathfrak{R}(V)^\perp,\mathfrak{D}(V)^\perp;(\begin{smallmatrix}T&F\\G&0\end{smallmatrix}))$, where $T=\tilde{V}$, $F=I|_{\mathfrak{R}(V)^\perp}$ and $G=P_{\mathfrak{D}(V)^\perp}$. The operator $U=(\begin{smallmatrix}T&F\\G&0\end{smallmatrix})\oplus V^{-1}$ is a unitary extension in $\mathfrak{H}\oplus\mathfrak{H}$ of V^{-1} and we have

$$\{\{P_{\mathfrak{H}}\tilde{h},P_{\mathfrak{H}}U\tilde{h}\}\mid\tilde{h}\in\mathfrak{K},\;(I-zU)\tilde{h}\in\mathfrak{H}\}=\{\{h,\tilde{X}(z)h\}\mid h\in\mathfrak{H}\}$$

with $\tilde{X}(z)=X(z)\oplus V^{-1}$, cf. (1.4) and (1.5). It is well known that if $\mathfrak{H}$ is a Hilbert space then the characteristic function uniquely determines the simple part of the isometry up to isomorphisms. We need the following result, which is taken from [LS], where it appears in a Pontryagin space context, see also [DLS].

PROPOSITION 1.3. *Let V be a closed standard isometric operator in a Kreĭn space $\mathfrak{H}$ and let $X(z)$ stand for its characteristic function. Let W in $\mathfrak{K}$ be a unitary extension of V and let Θ and $\tilde{\Theta}$ be as in Theorem 1.1 and (1.4), respectively. Then for $z\in\mathfrak{D}$, we have*

(*i*) $P_{\mathfrak{D}(V)^\perp}(I-z\tilde{\Theta}(z))^{-1}|_{\mathfrak{D}(V)^\perp}=(I-X(z)\Theta(z))^{-1}$,

(*ii*) $P_{\mathfrak{R}(V)^\perp}(I-z\tilde{\Theta}(z))^{-1}|_{\mathfrak{R}(V)^\perp}=(I-\Theta(z)X(z))^{-1}$,

(*iii*) $P_{\mathfrak{D}(V)^\perp}(I-z\tilde{\Theta}(z))^{-1}|_{\mathfrak{R}(V)^\perp}=(1/z)(I-X(z)\Theta(z))^{-1}X(z)=(1/z)X(z)(I-\Theta(z)X(z))^{-1}$,

(*iv*) $P_{\mathfrak{R}(V)^\perp}((I-z\tilde{\Theta}(z))^{-1}\tilde{\Theta}(z))|_{\mathfrak{D}(V)^\perp}=(I-\Theta(z)X(z))^{-1}\Theta(z)=\Theta(z)(I-X(z)\Theta(z))^{-1}$.

Proof. From the identity $I-z\tilde{\Theta}(z)=I-z\tilde{V}-z\Theta(z)P_{\mathfrak{D}(V)^\perp}$, it follows that

$$(I-z\tilde{\Theta}(z))^{-1}(I-z\tilde{V})=(I-z(I-z\tilde{V})^{-1}\Theta(z)P_{\mathfrak{D}(V)^\perp})^{-1}$$

and

$$(I-z\tilde{V})(I-z\tilde{\Theta}(z))^{-1}=(I-z\Theta(z)P_{\mathfrak{D}(V)^\perp}(I-z\tilde{V})^{-1})^{-1}.$$

The last two equalities lead to (*i*), (*ii*) and (*iii*) by considering the power series expansions of their righthand sides. For example, the first of these then yields

$$P_{\mathfrak{D}(V)^\perp}(I-z\tilde{\Theta}(z))^{-1}|_{\mathfrak{D}(V)^\perp}=P_{\mathfrak{D}(V)^\perp}(I-z\tilde{\Theta}(z))^{-1}(I-z\tilde{V})|_{\mathfrak{D}(V)^\perp}=$$
$$=\sum_{n=0}^{\infty}(zP_{\mathfrak{D}(V)^\perp}(I-z\tilde{V})^{-1}\Theta(z))^n|_{\mathfrak{D}(V)^\perp}=\sum_{n=0}^{\infty}(X(z)\Theta(z))^n|_{\mathfrak{D}(V)^\perp}=(I-X(z)\Theta(z))^{-1},$$

which proves (*i*). The other two items can be obtained after similar calculations. Item (*iv*) follows directly from (*ii*).

Now we select and fix a (not necessarily canonical) unitary extension W_0 of V. We denote its generalized coresolvent by $K_0(z)$, and the corresponding characteristic function by $\Theta_0\in\mathbf{S}(\mathfrak{D}(V)^\perp,\mathfrak{R}(V)^\perp)$, see Theorem 1.1. As before we associate with W_0 the operators

$$\Delta_0(z)=P_{\mathfrak{H}}(I-zW_0^+)^{-1}|_{\mathfrak{D}(V)^\perp},\quad \nabla_0(z)=zP_{\mathfrak{H}}(z-W_0)^{-1}|_{\mathfrak{R}(V)^\perp},\qquad z\in\rho(W_0).$$

Their properties are as in Proposition 1.2 (*i*) and (*ii*).

THEOREM 1.4. *Let V be a closed standard isometric operator in a Kreĭn space $\mathfrak{H}$ with characteristic function $X(z)$. Let W_0 be a fixed unitary extension of V with characteristic function $\Theta_0\in\mathbf{S}(\mathfrak{D}(V)^\perp,\mathfrak{R}(V)^\perp)$. Then the collection of all generalized coresolvents of V is given by the relation*

(1.10) $K(z)=K_0(z)+\nabla_0(1/z)D(z)\Delta_0(\bar{z})^+,\qquad z,\bar{z}\in\mathfrak{D},$

with $D(z)\in\mathbf{L}(\mathfrak{D}(V)^\perp,\mathfrak{R}(V)^\perp)$ *defined by*

$$D(z)=z(I-\Theta_0(z)X(z))(I-\Theta(z)X(z))^{-1}(\Theta(z)-\Theta_0(z))$$
$$=z(\Theta(z)-\Theta_0(z))(I-X(z)\Theta(z))^{-1}(I-X(z)\Theta_0(z)),$$

where Θ *runs through the class* $\mathbf{S}(\mathfrak{D}(V)^\perp,\mathfrak{R}(V)^\perp)$.

It should be pointed out that the parameter Θ in this theorem is the same as in Theorem 1.1, that is, the parametrization is independent of the choice of W_0. So the formula (1.10) expresses, for arbitrary extensions W_0, W, the difference $K(z)-K_0(z)$ in terms of $\Theta(z)$ and $\Theta_0(z)$, which is "more" than a parametrization.

Proof. Let $K(z)$ and $\Theta\in\mathbf{S}(\mathfrak{D}(V)^\perp,\mathfrak{R}(V)^\perp)$ be the generalized resolvent and the characteristic function, respectively, of a unitary extension W of V and let $\tilde{\Theta}(z)$ be given by (1.4). Then on account of (1.8) and (1.9)

$$\begin{aligned}K(z)-K_0(z)&=(I-z\tilde{\Theta}(z))^{-1}-(I-z\tilde{\Theta}_0(z))^{-1}\\&=z(I-z\tilde{\Theta}(z))^{-1}(\tilde{\Theta}(z)-\tilde{\Theta}_0(z))(I-z\tilde{\Theta}_0(z))^{-1}\\&=z(I-z\tilde{\Theta}(z))^{-1}P_{\mathfrak{R}(V)^\perp}(\Theta(z)-\Theta_0(z))P_{\mathfrak{D}(V)^\perp}(I-z\tilde{\Theta}_0(z))^{-1}\\&=z(I-z\tilde{\Theta}_0(z))^{-1}(I-z\tilde{\Theta}_0(z))(I-z\tilde{\Theta}(z))^{-1}P_{\mathfrak{R}(V)^\perp}(\Theta(z)-\Theta_0(z))\Delta_0(\bar{z})^+\\&=\nabla_0(1/z)P_{\mathfrak{R}(V)^\perp}(I-z\tilde{\Theta}_0(z))(I-z\tilde{\Theta}(z))^{-1}P_{\mathfrak{R}(V)^\perp}(\Theta(z)-\Theta_0(z))\Delta_0(\bar{z})^+.\end{aligned}$$

Here we have used the fact that $P_{\mathfrak{R}(V)^\perp}$ commutes with $(I-z\tilde{\Theta}_0(z))(I-z\tilde{\Theta}(z))^{-1}$, since

$$(I-z\tilde{\Theta}_0(z))(I-z\tilde{\Theta}(z))^{-1}=I+z(\Theta(z)-\Theta_0(z))P_{\mathfrak{D}(V)^\perp}(I-z\tilde{\Theta}(z))^{-1}$$

and $\Theta(z)-\Theta_0(z)$ maps $\mathfrak{D}(V)^\perp$ into $\mathfrak{R}(V)^\perp$. This last relation and Proposition 1.3 readily yield that the above chain of equalities can be continued with

$$\begin{aligned}&=\nabla_0(1/z)\left(P_{\mathfrak{R}(V)^\perp}+z(\Theta(z)-\Theta_0(z))P_{\mathfrak{D}(V)^\perp}(I-z\tilde{\Theta}(z))^{-1}P_{\mathfrak{R}(V)^\perp}\right)(\Theta(z)-\Theta_0(z))\Delta_0(\bar{z})^+\\&=\nabla_0(1/z)\left(P_{\mathfrak{R}(V)^\perp}+z(\Theta(z)-\Theta_0(z))(1/z)X(z)(I-\Theta(z)X(z))^{-1}P_{\mathfrak{R}(V)^\perp}\right)(\Theta(z)-\Theta_0(z))\Delta_0(\bar{z})^+\\&=\nabla_0(1/z)(I-\Theta_0(z)X(z))(I-\Theta(z)X(z))^{-1}P_{\mathfrak{R}(V)^\perp}(\Theta(z)-\Theta_0(z))\Delta_0(\bar{z})^+.\end{aligned}$$

This proves the desired relation between $K(z)$ and $\Theta(z)$ with $D(z)$ as after the first equality sign. That the second equality also holds can be proved in a similar way. It is not difficult to verify that the correspondence between $K(z)$ and $\Theta(z)$ is one to one. Here ends the proof.

The parametrization in Theorem 1.4 takes a special form, when the defect signatures $\operatorname{sign}\mathfrak{D}(V)^\perp$ and $\operatorname{sign}\mathfrak{R}(V)^\perp$ are equal and we choose W_0 to be a canonical unitary extension of V. Then $\Theta_0=W_0|_{\mathfrak{D}(V)^\perp}$ is unitary, independent of z, and $\nabla_0(z)W_0|_{\mathfrak{D}(V)^\perp}=zP_{\mathfrak{H}}(z-W_0)^{-1}W_0|_{\mathfrak{D}(V)^\perp}=-z\Delta_0(z)$, $z\in\rho(W_0)$. This leads to the following result.

THEOREM 1.5. *Assume in addition to the hypotheses of Theorem 1.4 that V has equal defect signatures and that W_0 is canonical, and put for some $\mu \in \mathbb{C}\setminus\mathbb{R}$*

$$q(z) = -\tfrac{1}{2}(\mu-\bar{\mu})^{-1}(\mu+\bar{\mu}) + \tfrac{1}{2}(I-X(z)\Theta_0)^{-1}(I+X(z)\Theta_0).$$

Then the collection of all generalized coresolvents of V is given by the formula

$$K(z) = K_0(z) + \Delta_0(1/z)(q(z)+t(z))^{-1}\Delta_0(\bar{z})^+, \qquad z \in \mathfrak{D},$$

where the relation $t(z)$ is defined by

$$t(z) = \{\{(\mu-\bar{\mu})(\Theta_0^+\Theta(z)-I)f, \bar{\mu}(\Theta_0^+\Theta(z)-\mu)f\} \mid f \in \mathfrak{D}(V)^\perp\}$$

and Θ runs through the class $\mathbf{S}(\mathfrak{D}(V)^\perp, \mathfrak{R}(V)^\perp)$.

Proof. Since $(t(z)-\bar{\mu}(\mu-\bar{\mu})^{-1})^{-1} = -(\Theta_0^+\Theta(z)-I)$ we find that

$$\begin{aligned}(q(z)+t(z))^{-1} &= -(\Theta_0^+\Theta(z)-I)(I-(q(z)+\bar{\mu}(\mu-\bar{\mu})^{-1})(\Theta_0^+\Theta(z)-I)^{-1} = \\ &= -(\Theta_0^+\Theta(z)-I)(I-X(z)\Theta(z))^{-1}(I-X(z)\Theta_0).\end{aligned}$$

The theorem now follows from the relations

$$\begin{aligned}K(z)-K_0(z) &= \nabla_0(1/z)W_0W_0^+D(z)\Delta_0(\bar{z})^+ = -(1/z)\Delta_0(1/z)W_0^+D(z)\Delta_0(\bar{z})^+ \\ &= -\Delta_0(1/z)W_0^+(\Theta(z)-W_0)(I-X(z)\Theta(z))^{-1}(I-X(z)W_0)\Delta_0(\bar{z})^+ \\ &= \Delta_0(1/z)(q(z)+t(z))^{-1}\Delta_0(\bar{z})^+.\end{aligned}$$

The function $q(z)$ defined in Theorem 1.5 is called a Q function for the isometric operator V, see [LT]. It has values in $\mathbf{L}(\mathfrak{D}(V)^\perp)$, can be written as

$$q(z) = -\mu(\mu-\bar{\mu})^{-1} + P_{\mathfrak{D}(V)^\perp}(I-zW_0)^{-1}|_{\mathfrak{D}(V)^\perp}$$

and satisfies the equality

$$(1-z\bar{w})^{-1}(q(z)+q(w)^+) = \Delta_0(\bar{z})^+\Delta_0(\bar{w}), \qquad z,w \in \mathfrak{D}.$$

As to the relation $t(z)$ in Theorem 1.5 we note that for two elements of $t(z)$

$$\{\varphi,\psi\} = \{(\mu-\bar{\mu})(\Theta_0^+\Theta(z)-I)f, \bar{\mu}(\Theta_0^+\Theta(z)-\mu)f\}$$

and

$$\{\alpha,\beta\} = \{(\mu-\bar{\mu})(\Theta_0^+\Theta(z)-I)h, \bar{\mu}(\Theta_0^+\Theta(z)-\mu)h\}$$

with $f,h \in \mathfrak{D}(V)^\perp$, say, we have that $[\psi,\alpha]+[\varphi,\beta] = 4(\mathrm{Im}\,\mu)^2[(I-\Theta(z)^+\Theta(z))f,h]$.

2. SELFADJOINT EXTENSIONS OF STANDARD SYMMETRIC RELATIONS

In this section we let S be a closed standard symmetric linear relation in a Kreĭn space $\mathfrak{H}$ and $\mu \in \mathbb{C}\setminus\mathbb{R}$ be a point such that the ranges $\mathfrak{R}(S-\mu)$ and $\mathfrak{R}(S-\bar{\mu})$ are orthocomplemented subspaces of $\mathfrak{H}$. Then Von Neumann's formula

$$S^+ = S \dotplus \mathfrak{M}_\mu(S^+) \dotplus \mathfrak{M}_{\bar{\mu}}(S^+), \text{ direct sums in } \mathfrak{H}^2,$$

is valid, where, for $\ell \in \mathbb{C}$, $\mathfrak{M}_\ell(S^+) = \{\{f,g\} \in S^+ \mid g = \ell f\}$. We shall describe all selfadjoint extensions A of S for which $\mu \in \rho(A)$, so that also $\bar{\mu} \in \rho(A)$, since the resolvent set of a selfadjoint relation is symmetric with respect to $\mathbb{R}$. To that end we use the Cayley transform and apply the results of Section 1. Recall that if T is a relation in a Kreĭn space its Cayley transform is defined by

$$C_\tau(T) = \{\{g-\tau f, g-\bar{\tau} f\} \mid \{f,g\} \in T\}, \; \tau \in \mathbb{C}\backslash\mathbb{R}.$$

We put $V = C_\mu(S)$, then

$$\mathfrak{R}(S-\mu) = \mathfrak{D}(V), \; \mathfrak{R}(S-\bar{\mu}) = \mathfrak{R}(V), \; \mathfrak{D}(V)^\perp = \mathfrak{N}(S^+-\bar{\mu}), \; \mathfrak{R}(V)^\perp = \mathfrak{N}(S^+-\mu),$$

where $\mathfrak{N}(T)$ stands for the null space of T, and V is a closed standard isometric operator. Through the relation $W = C_\mu(A)$ the unitary extensions W of V correspond precisely to the selfadjoint extensions A of S with $\mu \in \rho(A)$. Each such extension A gives rise to a so called Štraus extension of S defined by

(2.1) $$T(\ell) = \{\{P_{\mathfrak{H}}\tilde{h}, P_{\mathfrak{H}}\tilde{k}\} \mid \{\tilde{h},\tilde{k}\} \in A, \; \tilde{k} - \ell\tilde{h} \in \mathfrak{H}\}, \quad \ell \in \mathbb{C}.$$

Clearly, $S \subset T(\ell) \subset T(\bar{\ell})^+ \subset S^+$ for all $\ell \in \mathbb{C}$ and $T(\ell) = T(\bar{\ell})^+$ if $\ell \in \rho(A)$. Furthermore, $C_\mu(T(\ell)) = \tilde{\Theta}(z(\ell))$, $\ell \in \mathfrak{D}(\mu)$, see (1.5), where for the remainder of this section $z(\ell) = (\ell-\mu)/(\ell-\bar{\mu})$, $z(\infty) = 1$ and $z(\bar{\mu}) = \infty$, and $\mathfrak{D}(\mu)$ stands for an arbitrary neighborhood of μ in $\mathbb{C}_\mu = \{\ell \in \mathbb{C}\backslash\mathbb{R} \mid \operatorname{Im}\mu . \operatorname{Im}\ell > 0\}$. Finally, the extension A in $\mathfrak{K}$ or $\mathfrak{K}$ itself is called $\mathfrak{H}$ minimal if

$$\mathfrak{K} = \text{c.l.s.}\{(I + (\ell-\mu)(A-\ell)^{-1})f \mid f \in \mathfrak{H}, \; \ell \in \rho(A)\},$$

where $\mu \in \rho(A)$, or, equivalently, if $W = C_\mu(A)$ is $\mathfrak{H}$ minimal. Theorem 1.1 now implies the following result.

THEOREM 2.1. *Let S be a closed standard symmetric relation in a Kreĭn space $\mathfrak{H}$ with $\mu \in \mathbb{C}\backslash\mathbb{R}$ as above. Then the Štraus extension $T(\ell)$ associated with a selfadjoint extension A of S in a Kreĭn space $\mathfrak{K}$ with $\mu \in \rho(A)$ is of the form*

(2.2) $$T(\ell) = S \dotplus \{\{(I - \Theta(z(\ell)))f, (\bar{\mu} - \mu\Theta(z(\ell)))f\} \mid f \in \mathfrak{N}(S^+ - \bar{\mu})\}, \text{ direct sum in } \mathfrak{H}^2,$$

with $\Theta \in \mathbf{S}(\mathfrak{N}(S^+-\bar{\mu}), \mathfrak{N}(S^+-\mu))$ and $\ell \in \mathfrak{D}(\mu)$. Conversely, if $\Theta \in \mathbf{S}(\mathfrak{N}(S^+-\bar{\mu}), \mathfrak{N}(S^+-\mu))$, then there exists a selfadjoint extension A of S in a Kreĭn space $\mathfrak{K}$ with $\mu \in \rho(A)$ such that its corresponding Štraus relation coincides with $T(\ell)$ given by (2.2) on some $\mathfrak{D}(\mu)$. The Kreĭn space $\mathfrak{K}$ can be chosen $\mathfrak{H}$ minimal, in which case A is uniquely determined up to weak isomorphisms, which when restricted to $\mathfrak{H}$ are equal to the identity operator on $\mathfrak{H}$.

In that case, the number of positive (negative) squares of the kernel $\mathsf{S}_\Theta(z(\ell),z(\lambda))$ *equals the number of positive (negative) squares of the extending space* $\mathfrak{K}\ominus\mathfrak{H}$.

The link between the resolvent $R_A(\ell)=(A-\ell)^{-1}$ of the selfadjoint extension A of S with $\mu\in\rho(A)$ and the coresolvent $K_W(z)=(I-zW)^{-1}$ of the corresponding $W=C_\mu(A)$ is given by the identity

$$(\mu-\bar{\mu})K_W(z(\ell))=(\ell-\bar{\mu})(I+(\ell-\mu)R_A(\ell)).$$

For the corresponding generalized coresolvent $K(z)$ of V and the generalized resolvent $R(\ell)$ of S this implies that

$$(\mu-\bar{\mu})K(z(\ell))=(\ell-\bar{\mu})(I+(\ell-\mu)R(\ell)).$$

From definition (2.1) it follows that $R(\ell)=(T(\ell)-\ell)^{-1}$ for $\ell\in\rho(A)$. On account of the identity $R_A(\ell)-R_A(\lambda)^+=(\ell-\bar{\lambda})R_A(\lambda)^+R_A(\ell)$, the generalized resolvent satisfies

$$\frac{R(\ell)-R(\lambda)^+}{\ell-\bar{\lambda}}-R(\lambda)^+R(\ell)=P_{\mathfrak{H}}R_A(\lambda)^+(I-P_{\mathfrak{H}})R_A(\ell)|_{\mathfrak{H}}. \tag{2.3}$$

The kernel on the lefthand side will be denoted by $\mathsf{K}_R(\ell,\lambda)$,

$$\mathsf{K}_R(\ell,\lambda)=\frac{R(\ell)-R(\lambda)^+}{\ell-\bar{\lambda}}-R(\lambda)^+R(\ell).$$

Its connection with the kernel $\mathsf{R}_K(z,w)$ is given by the formula

$$(\mu-\bar{\mu})(\bar{\mu}-\mu)\mathsf{R}_K(z(\ell),z(\lambda))=(\ell-\mu)(\ell-\bar{\mu})(\bar{\lambda}-\mu)(\bar{\lambda}-\bar{\mu})\mathsf{K}_R(\ell,\lambda).$$

We associate with each A the operator functions

$$\Gamma_\tau(\ell)=P_{\mathfrak{H}}(I+(\ell-\tau)R_A(\ell))|_{\mathfrak{R}(S^+-\tau)},\qquad \tau=\mu,\bar{\mu}.$$

It is easy to verify that

$$(\bar{\mu}-\mu)\Delta(\overline{z(\ell)})=(\bar{\ell}-\mu)\Gamma_{\bar{\mu}}(\bar{\ell}),\quad (\mu-\bar{\mu})\nabla(1/z(\ell))=(\ell-\bar{\mu})\Gamma_\mu(\ell),$$

where Δ and ∇ are defined just before Proposition 1.2. Note that $\mathfrak{R}(I-z(\ell)V)=\mathfrak{R}(S-\ell)$. The next result follows from Proposition 1.3 and the preceding relations.

PROPOSITION 2.2. *(i) For* $\tau\in\{\mu,\bar{\mu}\}$ *and* ℓ *in some* $\mathfrak{D}(\tau)$ *the mapping* $\Gamma_\tau(\ell)$ *maps* $\mathfrak{R}(S^+-\tau)$ *onto* $\mathfrak{R}(S^+-\ell)$ *and is boundedly invertible. (ii) The following formulas are valid for* $\ell,\lambda\in\mathfrak{D}(\mu)$

$$\mathsf{K}_R(\ell,\lambda)=(\bar{\lambda}-\mu)^{-1}\Gamma_{\bar{\mu}}(\bar{\lambda})\,(1-z(\ell)\overline{z(\lambda)})^{-1}(I-\Theta(z(\lambda))^+\Theta(z(\ell)))\,(\ell-\bar{\mu})^{-1}\Gamma_{\bar{\mu}}(\bar{\ell})^+,$$
$$\mathsf{K}_R(\ell,\bar{\lambda})=(\lambda-\bar{\mu})^{-1}\Gamma_\mu(\lambda)\,(z(\ell)-z(\lambda))^{-1}(\Theta(z(\ell))-\Theta(z(\lambda)))\,(\ell-\bar{\mu})^{-1}\Gamma_{\bar{\mu}}(\bar{\ell})^+,$$
$$\mathsf{K}_R(\bar{\ell},\lambda)=(\bar{\lambda}-\mu)^{-1}\Gamma_{\bar{\mu}}(\bar{\lambda})\,(\overline{z(\ell)}-\overline{z(\lambda)})^{-1}(\Theta(z(\ell))^+-\Theta(z(\lambda))^+)\,(\bar{\ell}-\mu)^{-1}\Gamma_\mu(\ell)^+,$$
$$\mathsf{K}_R(\bar{\ell},\bar{\lambda})=(\lambda-\bar{\mu})^{-1}\Gamma_\mu(\lambda)\,(1-\overline{z(\ell)}z(\lambda))^{-1}(I-\Theta(z(\lambda))\Theta(z(\ell))^+)\,(\bar{\ell}-\mu)^{-1}\Gamma_\mu(\ell)^+.$$

Part (*i*) of Proposition 2.2 can be proved directly by using the equalities

$$I+(\ell-\tau)R_A(\ell)=\{\{g-\ell f,g-\tau f\}\mid\{f,g\}\in A\},$$

$$[(I+(\ell-\tau)R_A(\ell))h,\psi-\bar{\ell}\varphi]=[h,\psi-\bar{\ell}\varphi]+[h,(\bar{\ell}-\bar{\tau})R_A(\bar{\ell})(\psi-\bar{\ell}\varphi)]=[h,\psi-\bar{\tau}\varphi],$$

where $h\in\mathfrak{H}$ and $\{\varphi,\psi\}\in S$, so that $R_A(\bar{\ell})(\psi-\bar{\ell}\varphi)=\varphi$. The first equality implies that, for ℓ in some $\mathfrak{D}(\tau)\subset\rho(A)$, $I+(\ell-\tau)R(\ell)\in\mathbf{L}(\mathfrak{H})$ and is boundly invertible and that $\Gamma_\tau(\ell)$ restricted to $\mathfrak{N}(S^+-\tau)$ is a bijection onto its range. That this is equal to $\mathfrak{N}(S^+-\ell)$ follows from the second equality, the proof is similar to the proof of Proposition 1.2(*i*).

It follows from Theorem 2.1 and the equalities in part (*ii*) of Proposition 2.2, but also directly from (2.3), that if the extending space $\mathfrak{K}$ is $\mathfrak{H}$ minimal, then the number of positive (negative) squares of $\mathfrak{K}\ominus\mathfrak{H}$ is equal to the number of positive (negative) squares of the kernel $\mathsf{K}_R(\ell,\lambda)$. The characteristic function of S is the function $X(z(\ell))$, where $X(z)$ is the characteristic funtion of $V=C_\mu(S)$, defined in Section 1. It can be shown that

$$X(z(\ell))=C_{\bar{\mu}}(S(\ell))|_{\mathfrak{N}(S^+-\mu)},\quad \ell\in\mathfrak{D}(\mu),$$

where $S(\ell)=S\dot{+}\mathfrak{M}_\ell(S^+)$, $\ell\in\mathbb{C}$, cf. [DLS]. Note that $S(\ell)$ is a Štraus extension of S, in fact,

$$S(\ell)=\{\{P_{\mathfrak{H}}\tilde{h},P_{\mathfrak{H}}\tilde{k}\}\mid\{\tilde{h},\tilde{k}\}\in A,\ \tilde{k}-\ell\tilde{h}\in\mathfrak{H}\},\qquad \ell\in\mathbb{C},$$

where A is the selfadjoint extension of S such that $C_{\bar{\mu}}(A)=U$ is the unitary extension of $V^{-1}=C_{\bar{\mu}}(S)$ defined just before Proposition 1.3.

Now we fix a (not necessarily canonical) selfadjoint extension A_0 of S in a Kreĭn space $\mathfrak{K}$ with $\mu,\bar{\mu}\in\rho(A_0)$. We put $R_0(\ell)=P_{\mathfrak{H}}(A_0-\ell)^{-1}|_{\mathfrak{H}}$,

$$\mathring{\Gamma}_\tau(\ell)=(I+(\ell-\tau)R_0(\ell))|_{\mathfrak{N}(S^+-\tau)},\quad \tau=\mu,\bar{\mu},$$

and let $\Theta_0\in\mathbf{S}(\mathfrak{N}(S^+-\bar{\mu}),\mathfrak{N}(S^+-\mu))$ correspond to A_0 in accordance with Theorem 2.1. Then Theorem 1.4 implies the following parametrization of the generalized resolvents of S.

THEOREM 2.3. *Let S be a closed standard symmetric relation in a Kreĭn space $\mathfrak{H}$ with $\mu\in\mathbb{C}\backslash\mathbb{R}$ as above and characteristic function $X(z(\ell))$. Let A_0 be a fixed selfadjoint extension of S. Then the collection of all generalized resolvents of S is given by the formula*

$$R(\ell)=R_0(\ell))+(\mu-\bar{\mu})^{-1}\mathring{\Gamma}_\mu(\ell)E(\ell))\mathring{\Gamma}_{\bar{\mu}}(\bar{\ell})^+,\qquad \ell\in\mathfrak{D}(\mu),$$

with $E(\ell)\in\mathbf{L}(\mathfrak{N}(S^+-\bar{\mu}),\mathfrak{N}(S^+-\mu))$ defined by

$$E(\ell)=(I-\Theta_0(z(\ell))X(z(\ell)))(I-\Theta(z(\ell))X(z(\ell)))^{-1}(\Theta(z(\ell))-\Theta_0(z(\ell)))$$

$$=(\Theta(z(\ell))-\Theta_0(z(\ell)))(I-X(z(\ell))\Theta(z(\ell)))^{-1}(I-X(z(\ell))\Theta_0(z(\ell))),$$

where Θ *runs through the class* $\mathbf{S}(\mathfrak{N}(S^+-\bar{\mu}),\mathfrak{N}(S^+-\mu))$.

This parametrization takes a special form, when the defect signatures $\operatorname{sign}\mathfrak{N}(S^+-\mu)$ and $\operatorname{sign}\mathfrak{N}(S^+-\bar{\mu})$ of S are equal and A_0 is a canonical selfadjoint extension of S.

THEOREM 2.4. *Assume in addition to the hypotheses of Theorem 2.3 that S has equal defect signatures and A_0 is canonical, and put*

$$Q(\ell)=-\tfrac{1}{2}(\mu+\bar{\mu})+\tfrac{1}{2}(\mu-\bar{\mu})(I-X(z(\ell))\Theta_0)^{-1}(I+X(z(\ell))\Theta_0).$$

Then the collection of all generalized resolvents of S is given by the formula

$$R(\ell)=R_0(\ell)-\mathring{\Gamma}_{\bar{\mu}}(\ell)(Q(\ell)+\mathcal{T}(\ell))^{-1}\mathring{\Gamma}_{\bar{\mu}}(\bar{\ell})^+,\qquad \ell\in\mathfrak{D}(\mu),$$

where the relation $\mathcal{T}(\ell)$ is defined by

$$\mathcal{T}(\ell)=\{\{(\Theta(z(\ell))-I)f,(\bar{\mu}\Theta(z(\ell))-\mu)f\}\mid f\in\mathfrak{N}(S^+-\bar{\mu})\}$$

and Θ *runs through the class* $\mathbf{S}(\mathfrak{N}(S^+-\bar{\mu}),\mathfrak{N}(S^+-\mu))$.

The function $Q(\ell)$ defined in Theorem 2.4 is a so called Q function for S and is discussed, in the particular case where $\mathfrak{H}$ is a Hilbert space, in [LT]. The remarks at the end of Section 1 easily yield that

$$Q(\ell)=-\mathrm{Re}\mu+P_{\mathfrak{N}(S^+-\bar{\mu})}\big((\ell-\mathrm{Re}\mu)+(\ell-\mu)(\ell-\bar{\mu})R_0(\ell)\big)\big|_{\mathfrak{N}(S^+-\bar{\mu})},$$

and hence, that

$$\frac{Q(\ell)-Q(\lambda)^+}{\ell-\bar{\lambda}}=\mathring{\Gamma}_{\bar{\mu}}(\lambda)^+\mathring{\Gamma}_{\bar{\mu}}(\ell).$$

As to the relation $\mathcal{T}(\ell)$ in Theorem 2.4 we note that for two elements of $\mathcal{T}(\ell)$

$$\{f,g\}=\{(\Theta(z(\ell))-I)\varphi,\bar{\mu}(\Theta(z(\ell))-\mu)\varphi\}$$

and

$$\{h,k\}=\{(\Theta(z(\ell))-I)\psi,\bar{\mu}(\Theta(z(\ell))-\mu)\psi\}$$

with $\varphi,\psi\in\mathfrak{D}(S^+-\bar{\mu})^{\perp}$, say, we have that

$$(\bar{\lambda}-\mu)\frac{[g,h]-[f,k]}{\ell-\bar{\lambda}}(\ell-\bar{\mu})=4(\mathrm{Im}\mu)^2\frac{[\varphi,\psi]-[\Theta(z(\ell))\varphi,\Theta(z(\lambda))\psi]}{1-z(\ell)\overline{z(\lambda)}}.$$

REFERENCES

[AG] D.Z. Arov, L.Z. Grossman, "Scattering matrices in the theory of dilations of isometric operators", Dokl. Akad. Nauk SSSR, 270 (1983), 17–20, (Russian) (English translation: Sov. Math. Dokl., 27 (1983), 518–522).

[AI] T.Ya. Azizov, I.S. Iohvidov, *Linear operators in spaces with an indefinite metric*, Nauka, Moscow, 1986 (Russian) (English translation: John Wiley, New York, 1989).

[Č] M.E. Čumakin, "Generalized resolvents of isometric operators", Sibirsk. Mat. Zh., 8 (1967), 876–892 (Russian) (English translation: Siberian Math. J., 8 (1967) 666–678).

[DLS] A. Dijksma, H. Langer, H.S.V. de Snoo, "Unitary colligations in Kreĭn spaces and their role in the extension theory of isometries and symmetric linear relations in Hilbert spaces", Functional Analysis II, Proceedings Dubrovnik 1985, Lecture Notes in Mathematics, 1242 (1986), 1–42.

[G] A.G. Gibson, "Triples of operator-valued functions related to the unit circle", Pacific J. Math. 28 (1969), 504–531.

[KL1] M.G. Kreĭn, H. Langer, "On defect subspaces and generalized resolvents of a Hermitian operator in the space Π_κ", Funktsional. Anal. i Prilozhen., 5 No. 2 (1971), 59–71; 5 No. 3 (1971) 54–69 (Russian) (English translation: Functional Anal. Appl., 5 (1971/1972), 136–146, 217–228).

[KL2] M.G. Kreĭn, H. Langer, "Über die verallgemeinerten Resolventen und die charakteristische Funktion eines isometrischen Operators im Raume Π_κ", Hilbert space operators and operator algebras (Proc. Int. Conf., Tihany, 1970) Colloq. Math. Soc. János Bolyai, no. 5, North-Holland, Amsterdam, 1972, 353–399.

[L] H. Langer, "Generalized coresolvents of a π-isometric operator with unequal defect numbers", Funktsional. Anal. i Prilozhen, 5 (1971), 73–75 (Russian) (English translation: Functional Anal. Appl., 5 (1970/1971), 329–331).

[LS] H. Langer, P. Sorjonen, "Verallgemeinerte Resolventen hermitescher und isometrischer Operatoren im Pontryaginraum", Ann. Acad. Sci. Fenn. Ser. A I Math., 561 (1974), 1–45.

[LT] H. Langer, B. Textorius, "On generalized resolvents and Q-functions of symmetric linear relations (subspaces) in Hilbert space", Pacific J. Math., 72 (1977), 135–165.

[S] P. Sorjonen, "Generalized resolvents of an isometric operator in a Pontrjagin space", Z. Anal. Anwendungen, 4 (69) (1985), 543–555.

A. Dijksma, H.S.V. de Snoo
Department of Mathematics
University of Groningen
Postbox 800
9700 AV Groningen
Nederland

H. Langer
Department of Mathematics
University of Dortmund
Postbox 500500
4600 Dortmund 50
B.R.D.

Operator Theory:
Advances and Applications, Vol. 48

ON THE STRUCTURE OF LINEAR CONTRACTIONS WITH RESPECT TO FUNDAMENTAL DECOMPOSITIONS

Aurelian Gheondea

Dedicated to the memory of Ernst Hellinger

The structure of (doubly) contractive operators with respect to fundamental decompositions is considered in connection with the problem of lifting of operators with prescribed negative signatures of defect.

§ 1. Introduction

Let $\mathcal{H}_1$, $\mathcal{H}_2$, $\mathcal{H}'_1$, and $\mathcal{H}'_2$ be Krein spaces. Consider the Krein spaces $\tilde{\mathcal{H}}_1 = \mathcal{H}_1[+]\mathcal{H}'_1$ and $\tilde{\mathcal{H}}_2 = \mathcal{H}_2[+]\mathcal{H}'_2$. For a given operator $T \in \mathcal{L}(\mathcal{H}_1, \mathcal{H}_2)$ and cardinal numbers $\varkappa_1$ and $\varkappa_2$ one considers the following problem of lifting of operators with prescribed negative signatures of defect (cf. [2] and [5]).

(*) <u>Determine all operators</u> $\tilde{T} \in \mathcal{L}(\tilde{\mathcal{H}}_1, \tilde{\mathcal{H}}_2)$ <u>such that</u>

$$P^{\tilde{\mathcal{H}}_2}_{\mathcal{H}_2} \tilde{T} \mid \mathcal{H}_1 = T, \quad \varkappa^-[I - \tilde{T}^{\#}\tilde{T}] = \varkappa_1, \quad \underline{\text{and}} \quad \varkappa^-[I - \tilde{T}\tilde{T}^{\#}] = \varkappa_2.$$

(For notations see Section 2.) In connection with this it is useful to consider the following problems, which are particular cases of the Problem (*).

$(*)_r$ <u>Determine all operators</u> $T_r \in \mathcal{L}(\tilde{\mathcal{H}}_1, \mathcal{H}_2)$ <u>such that</u>

$$T_r \mid \mathcal{H}_1 = T, \quad \varkappa^-[I - T_r^{\#} T_r] = \varkappa_1 \quad \underline{\text{and}} \quad \varkappa^-[I - T_r T_r^{\#}] = \varkappa_2.$$

$(*)_c$ <u>Determine all operators</u> $T_c \in \mathcal{L}(\mathcal{H}_1, \tilde{\mathcal{H}}_2)$ <u>such that</u>

$$P^{\tilde{\mathcal{H}}_2}_{\mathcal{H}_2} T_c = T, \quad \varkappa^-[I - T_c^{\#} T_c] = \varkappa_1, \quad \underline{\text{and}} \quad \varkappa^-[I - T_c T_c^{\#}] = \varkappa_2.$$

Let us come now to the main concern of this paper. For two Hilbert spaces $\mathcal{H}^-$ and $\mathcal{H}^+$ consider

$$\mathcal{H} = \mathcal{H}^- \oplus \mathcal{H}^+. \tag{1.1}$$

We regard $\mathcal{H}$ as a <u>Krein space</u> endowed with the indefinite inner product $[\cdot,\cdot]$ defined thus

$$[x^- + x^+, y^- + y^+] = -(x^-,y^-) + (x^+,y^+), \quad x^{\pm}, y^{\pm} \in \mathcal{H}^{\pm}. \tag{1.2}$$

Then (1.1) is called a fundamental decomposition of the Krein space $\mathcal{H}$. Denote by $J \in \mathcal{L}(\mathcal{H})$ the fundamental symmetry associated to (1.1), i.e.

$$J(x^- + x^+) = -x^- + x^+, \quad x^{\pm} \in \mathcal{H}^{\pm}. \tag{1.3}$$

Let us denote by $\#$ the involution, acting on the algebra $\mathcal{L}(\mathcal{H})$, which is determined by the inner product $[\cdot,\cdot]$. An operator $T \in \mathcal{L}(\mathcal{H})$ is called a contraction if

$$[Tx,Tx] \leq [x,x], \quad x \in \mathcal{H}, \tag{1.4}$$

or, equivalently,

$$[(I - T^{\#}T)x, x] \geq 0, \quad x \in \mathcal{H}. \tag{1.5}$$

T is called double contraction if both of T and $T^{\#}$ are contractions.

Our aim is to obtain a formula which describes the set of all (double) contractions, with respect to the decomposition (1.1), and such that the (operatorial) parameters are as simple as possible.

Let $\tilde{T} \in \mathcal{L}(\mathcal{H})$ be a double contraction and consider its 2x2 block-matrix representation, with respect to the decomposition (1.1),

$$\tilde{T} = \begin{bmatrix} T & X \\ Y & Z \end{bmatrix}. \tag{1.6}$$

Then it is easy to see that the only conditions that the operator $T \in \mathcal{L}(\mathcal{H}^-)$ must satisfy are $T^*T \geq I$ and $TT^* \geq I$, i.e. T must be a doubly expansive operator. On the other hand, the condition that $\tilde{T}$ is doubly contractive can be equivalently written

$$\kappa^-[I - \tilde{T}^{\#}\tilde{T}] = 0, \quad \kappa^-[I - \tilde{T}\tilde{T}^{\#}] = 0, \tag{1.7}$$

(see Section 2 for notations), hence our problem fits into the frame of Problem (*). Then, using the results from one of the papers [2], [10], and [5], it follows

1.1. PROPOSITION. With respect to the decomposition (1.1), the formula

$$\tilde{T} = \begin{bmatrix} T & (TT^* - I)^{\frac{1}{2}}\Gamma_1 \\ \Gamma_2(T^*T - I)^{\frac{1}{2}} & -\Gamma_2 T^*\Gamma_1 + (I - \Gamma_2\Gamma_2^*)^{\frac{1}{2}}\Gamma(I - \Gamma_1^*\Gamma_1)^{\frac{1}{2}} \end{bmatrix} \tag{1.8}$$

establishes a bijective correspondence between the set of all double contractions $\tilde{T} \in \mathcal{L}(\mathcal{H})$ and the set of all quadruples $\{T, \Gamma_1, \Gamma_2, \Gamma\}$ such that

$$(1.9)\quad \begin{cases} T \in \mathcal{L}(\mathcal{H}^-), \quad T^*T \geq I, \quad TT^* \geq I, \\ \Gamma_1 \in \mathcal{L}(\mathcal{H}^+, \overline{\mathcal{R}(I - TT^*)}), \quad \Gamma_1^* \Gamma_1 \leq I, \\ \Gamma_2 \in \mathcal{L}(\overline{\mathcal{R}(I - T^*T)}, \mathcal{H}^+), \quad \Gamma_2^* \Gamma_2 \leq I, \\ \Gamma \in \mathcal{L}(\overline{\mathcal{R}(I - \Gamma_1^* \Gamma_1)}, \overline{\mathcal{R}(I - \Gamma_2 \Gamma_2^*)}), \quad \Gamma^* \Gamma \leq I. \end{cases}$$

Moreover, there exist the unitary operators

$$(1.10)\quad U(T) : \overline{\mathcal{R}(J - T^*JT)} \rightarrow \overline{\mathcal{R}(I - \Gamma_2^* \Gamma_2)} \oplus \overline{\mathcal{R}(I - \Gamma^* \Gamma)},$$

$$(1.11)\quad U_*(T) : \overline{\mathcal{R}(J - TJT^*)} \rightarrow \overline{\mathcal{R}(I - \Gamma_1 \Gamma_1^*)} + \overline{\mathcal{R}(I - \Gamma\Gamma^*)}.$$

The unitary operators $U(T)$ and $U_*(T)$ from (1.10) and (1.11) can be written explicitly in terms of the parameters T, Γ_1, Γ_2, and Γ (e.g. see [2] or [5]).

Using these unitary operators, from Proposition 1.1 we obtain the following

1.2. COROLLARY. With respect to the decomposition (1.1), the formula (1.8) establishes a bijective correspondence between the set of all operators $\tilde{T} \in \mathcal{L}(\mathcal{H})$ which are isometric with contractive adjoint (unitary) and the set of all quadruples $\{T, \Gamma_1, \Gamma_2, \Gamma\}$ such that

$$(1.12)\quad \begin{cases} T \in \mathcal{L}(\mathcal{H}^-), \quad T^*T \geq I, \quad TT^* \geq I, \\ \Gamma_1 \in \mathcal{L}(\mathcal{H}^+, \overline{\mathcal{R}(I - TT^*)}), \quad \Gamma_1^* \Gamma_1 \leq I \ (\Gamma_1 \Gamma_1^* = I), \\ \Gamma_2 \in \mathcal{L}(\overline{\mathcal{R}(I - T^*T)}, \mathcal{H}^+), \quad \Gamma_2^* \Gamma_2 = I, \\ \Gamma \in \mathcal{L}(\overline{\mathcal{R}(I - \Gamma_1^* \Gamma_1)}, \overline{\mathcal{R}(I - \Gamma_2 \Gamma_2^*)}), \quad \Gamma^* \Gamma = I \ (\Gamma^* \Gamma = I \text{ and } \Gamma \Gamma^* = I). \end{cases}$$

For isometric and unitary operators, a description with respect to fundamental decomposition was obtained in [12], [14] (see also [13]). In those papers the parameters are two unitary operators (of Hilbert spaces) and a uniform contraction (i.e. an operator with norm strictly less than one).

Coming back to the description of contractive operators which was obtained in Proposition 1.1, we observe that the parameters Γ_1, Γ_2, and Γ are (Hilbert space) contractions. Since there exists a rich literature dealing with the structure of block--matrix contractions in Hilbert spaces (e.g. see [17], [8], [1]) we can consider these parameters as "simple".

The parameter T is doubly expansive. Considering $\mathcal{H}^-$ with the inner product inherited from $\mathcal{H}$ ($\mathcal{H}^-$ is negative definite), the conditions on T are equivalent with

$$\mathcal{H}^-[I - T^\# T] = \mathcal{H}^-[I - TT^\#] = 0.$$

Thus, the structure of these operators will be known if we can solve the Problem (*) for negative definite spaces $\mathcal{H}'_1$ and $\mathcal{H}'_2$. This is the main concern of this paper.

For the case of finite dimensional spaces $\tilde{\mathcal{H}}_1$ and $\tilde{\mathcal{H}}_2$ we obtained a solution of our problem in Theorem 5.3. The solution obtained there is only in form of an iterative scheme because some of the parameters are also controlled by their negative signature of defect.

In Section 3 we obtain a procedure to calculate the signatures of a hermitian 2x2 block-matrix and show by an example that this does not extend to the infinite dimensional case. The result contained in Lemma 3.2 is used in Section 5. Also as a preparatory fact, we investigate the Problems $(*)_r$ and $(*)_c$ in Section 4. One of the main technical tool obtained there is the identification of the defect space $\mathcal{D}_{T_r}$ from (4.7). Having this we can solve the Problem $(*)_r$ in the particular case which is needed during Section 5.

If any of the spaces $\tilde{\mathcal{H}}_1$ or $\tilde{\mathcal{H}}_2$ has infinite dimension the formulae obtained in Theorem 5.3 still produce a rich class of solutions, though in general these do not cover the whole set of solutions. This is one reason for which we presented the results from Section 4 in a setting which is a bit more general than we actually needed.

§ 2. Notations and terminology

2.1. The terminology from the theory of Krein spaces and their linear operators will be that used in [6] (as general references see the monographs [3] and [4]). Here we want to fix some notations.

If $\mathcal{H}$ is a Krein space then we always will denote its "indefinite" inner product by the symbol $[\cdot,\cdot]$. A fundamental decomposition of $\mathcal{H}$ (f.d. for short) will be denoted by $\mathcal{H} = \mathcal{H}^+ [+] \mathcal{H}^-$. In this case the operator J on $\mathcal{H}$ defined by

$$J(x^+ + x^-) = x^+ - x^-, \quad x^\pm \in \mathcal{H}^\pm,$$

is the corresponding fundamental symmetry (f.s. for short). Further on, the J-inner product $(\cdot,\cdot)_J$ defined by

$$(x,y)_J = [Jx,y], \quad x, y \in \mathcal{H},$$

turns $\mathcal{H}$ into a Hilbert space. The norm $\|.\|$ associated with the J-inner product is called a unitary norm and the strong topology on $\mathcal{K}$ is determined by an arbitrary unitary norm, thus, for $\mathcal{K}_1$ and $\mathcal{K}_2$ Krein spaces $\mathcal{L}(\mathcal{K}_1,\mathcal{K}_2)$ will denote the space of linear and strongly continuous operators defined on $\mathcal{K}_1$ and valued in $\mathcal{K}_2$.

The signature numbers $\varkappa^{\pm}[\mathcal{K}] = \mathbf{dim}\ \mathcal{K}^{\pm}$, where $\mathcal{K} = \mathcal{K}^{+}[+]\mathcal{K}^{-}$ is an arbitrary f.d. of $\mathcal{K}$, are independent of the f.d. The rank of indefiniteness of $\mathcal{K}$ is defined by $\varkappa[\mathcal{K}] = \mathbf{min}\{\varkappa^{+}[\mathcal{K}], \varkappa^{-}[\mathcal{K}]\}$. If $\varkappa[\mathcal{K}]$ is finite then $\mathcal{K}$ is called a Pontryagin space (in our context this will always mean that $\varkappa^{-}[\mathcal{K}]$ is finite).

Let $\mathcal{K}_1$ and $\mathcal{K}_2$ be Krein spaces. Then $\mathcal{K}_1[+]\mathcal{K}_2$ denotes the direct sum Krein space of $\mathcal{K}_1$ and $\mathcal{K}_2$ and we have

$$\varkappa^{\pm}[\mathcal{K}_1[+]\mathcal{K}_2] = \varkappa^{\pm}[\mathcal{K}_1] + \varkappa^{\pm}[\mathcal{K}_2].$$

For any operator $T \in \mathcal{L}(\mathcal{K}_1,\mathcal{K}_2)$ one defines the adjoint $T^{\#} \in \mathcal{L}(\mathcal{K}_2,\mathcal{K}_1)$ by

$$[Tx,y] = [x,T^{\#}y], \quad x \in \mathcal{K}_1,\ y \in \mathcal{K}_2.$$

Operators which are unitary, isometric or selfadjoint are understood with respect to the involution $\#$. The operator $T \in \mathcal{L}(\mathcal{K}_1,\mathcal{K}_2)$ is contractive if

$$[Tx, Tx] \leq [x,x], \quad x \in \mathcal{K}_1,$$

or, equivalently,

$$[(I - T^{\#}T)x,x] \geq 0, \quad x \in \mathcal{K}_1,$$

i.e. $I - T^{\#}T$ is positive. T is called doubly contractive if both of T and $T^{\#}$ are contractive.

If we fix two f.s.'s J_1 on $\mathcal{K}_1$ and J_2 on $\mathcal{K}_2$ and, for $T \in \mathcal{L}(\mathcal{K}_1,\mathcal{K}_2)$, let $T^{*} \in \mathcal{L}(\mathcal{K}_2,\mathcal{K}_1)$ denote the adjoint of T with respect to the Hilbert spaces $(\mathcal{K}_i, (\cdot,\cdot)_{J_i})$, $i = 1, 2$, then we have

$$T^{\#} = J_1 T^{*} J_2.$$

2.2. Let $\mathcal{H}$ be a Hilbert space and $A \in \mathcal{L}(\mathcal{H})$, $A = A^{*}$. Denote $S_A = \mathbf{sgn}(A)$, where **sgn** stands for the real function signum (by definition $\mathbf{sgn}(0) = 0$). Then S_A is the selfadjoint partial isometry which appears in the polar decomposition of A

$$A = S_A|A|, \quad \mathbf{ker}\, S_A = \mathbf{ker}\, A, \quad S_A\mathcal{H} = \overline{\mathcal{R}(A)},$$

where $\mathcal{R}(A)$ denotes as usually the range of the operator A and the upper bar is used for the closure. The signature numbers of A are defined as follows

$$\varkappa^{\pm}(A) = \mathbf{dim}\,\mathbf{ker}(I \mp S_A), \quad \varkappa^{0}(A) = \mathbf{dim}\,\mathbf{ker}\, S_A. \tag{2.1}$$

Denote by $\mathcal{H}_A$ the Krein space obtained from the space $S_A\mathcal{H}$ endowed with the inner product $[\cdot,\cdot]$

$$[x,y] = (S_A x,y), \quad x,y \in \mathcal{H},$$

where $(\cdot,\cdot)$ denotes the inner product of the Hilbert space $\mathcal{H}$. With this definition it is clear that $\varkappa^{\pm}(A) = \varkappa^{\pm}[\mathcal{H}_A]$.

2.3. Let $\mathcal{K}$ be a Krein space and J a f.s. on it. If $A \in \mathcal{L}(\mathcal{K})$, $A = A^{\#}$ then the operator $JA \in \mathcal{L}(\mathcal{K})$ is selfadjoint with respect to the inner product $(\cdot,\cdot)_J$ and

$$(JAx,y)_J = [Ax,y], \quad x,y \in \mathcal{K}.$$

This shows that the signature numbers $\varkappa^{\pm}[A]$ and $\varkappa^{0}[A]$ defined by

$$\varkappa^{\pm}[A] = \varkappa^{\pm}(JA), \quad \varkappa^{0}[A] = \varkappa^{0}(JA), \tag{2.2}$$

are not depending on the f.s. J.

Let $\mathcal{K}_1$ and $\mathcal{K}_2$ be Krein spaces and assume that J_1 and J_2 are fixed f.s. on $\mathcal{K}_1$ and, respectively, $\mathcal{K}_2$. For an arbitrary operator $T \in \mathcal{L}(\mathcal{K}_1,\mathcal{K}_2)$, the <u>signature numbers of defect</u> of T are defined by

$$\varkappa^{\pm}[I - T^{\#}T] = \varkappa^{\pm}(J_1 - T^* J_2 T), \tag{2.3}$$

$$\varkappa^{0}[I - T\ \ T] = \varkappa^{0}(J_1 - T^* J_2 T). \tag{2.4}$$

As a consequence of [2, Proposition 3.1] the following identities hold

$$\varkappa^{\pm}[I - T^{\#}T] + \varkappa^{\pm}[\mathcal{K}_2] = \varkappa^{\pm}[I - TT^{\#}] + \varkappa^{\pm}[\mathcal{K}_1], \tag{2.5}$$

$$\varkappa^{0}[I - T^{\#}T] = \varkappa^{0}[I - TT^{\#}]. \tag{2.6}$$

Considering the <u>sign operator</u> J_T and the <u>defect operator</u> D_T defined by

$$J_T = \mathbf{sgn}(J_1 - T^* J_2 T), \quad D_T = |J_1 - T^* J_2 T|^{\frac{1}{2}}, \tag{2.7}$$

one introduces the <u>defect space</u> $\mathcal{D}_T$ as the Krein space obtained from $\overline{\mathcal{R}(D_T)}$ with the inner product

$$[x,y] = (J_T x,y)_{J_1}, \quad x,y \in \overline{\mathcal{R}(D_T)},$$

i.e. the strong topology of $\mathcal{D}_T$ is those inherited from the strong topology of $\mathcal{K}_1$ and J_T is a f.s. of $\mathcal{D}_T$. Also by [2] there exist the so-called <u>link operators</u> $L_T \in \mathcal{L}(\mathcal{D}_T, \mathcal{D}_{T^*})$ and $L_{T^*} \in \mathcal{L}(\mathcal{D}_{T^*}, \mathcal{D}_T)$, uniquely determined such that

$$D_{T^*} L_T = TJ_1 D_T | \mathcal{D}_T, \quad D_T L_{T^*} = T^* J_2 D_{T^*} | \mathcal{D}_{T^*}, \tag{2.8}$$

which in addition satisfy the following identities

(2.9) $$L_{T^*} = J_T L_T^* J_{T^*} | \mathcal{D}_{T^*},$$

(2.10) $$(J_T - D_T J_1 D_T) | \mathcal{D}_T = L_T^* J_{T^*} L_T.$$

Recall that in case $TJ_1 = J_2 T$ (in particular, if $\mathcal{H}_1$ and $\mathcal{H}_2$ are Hilbert spaces, cf. [18], [7]) then $L_T = TJ_1$ and $L_{T^*} = T^* J_2$ (respectively $L_T = T$ and $L_{T^*} = T^*$).

§ 3. The signatures of hermitian 2 x 2 block-matrices

Let $\mathcal{H}_1$ and $\mathcal{H}_2$ be Hilbert spaces and consider the operators $A \in \mathcal{L}(\mathcal{H}_1)$, $A = A^*$, $C \in \mathcal{L}(\mathcal{H}_2)$, $C = C^*$ and $B \in \mathcal{L}(\mathcal{H}_2, \mathcal{H}_1)$. With these given operators define $H \in \mathcal{L}(\mathcal{H}_1 \oplus \mathcal{H}_2)$ by

(3.1) $$H = \begin{bmatrix} A & B \\ B^* & C \end{bmatrix}.$$

Our aim is to calculate the signature numbers $\varkappa^+(H)$ and $\varkappa^-(H)$ in terms of A,B and C.

First we recall the result for the case $\varkappa^-(H) = 0$ (cf. [16]).

3.1. LEMMA. The operator H is nonnegative if and only if A and C are nonnegative and $B = A^{\frac{1}{2}} \Gamma C^{\frac{1}{2}}$ where $\Gamma \in \mathcal{L}(\mathcal{H}_C, \mathcal{H}_A)$ is a uniquely determined contraction.

Moreover, in this case, there exist the unitary operator

(3.2) $$W : \mathcal{H}_H \to \mathcal{H}_A \oplus \mathcal{D}_\Gamma,$$
$$WH^{\frac{1}{2}} = \begin{bmatrix} A^{\frac{1}{2}} & C^{\frac{1}{2}} \\ 0 & D_\Gamma C^{\frac{1}{2}} \end{bmatrix}$$

in particular we have

(3.3) $$\varkappa^+(H) = \varkappa^+(A) + \varkappa^+(I - \Gamma^*\Gamma) = \varkappa^+(C) + \varkappa^+(I - \Gamma\Gamma^*).$$

If H is indefinite (i.e. both of $\varkappa^+(H)$ and $\varkappa^-(H)$ are nontrivial) then the corresponding result is much more complicated and it applies only in case $\mathcal{H}_1$ and $\mathcal{H}_2$ are finite dimensional.

We will first assume that $\mathcal{H}_1$ has dimension 1. Then, with respect to the decomposition

$$\mathcal{H} = \mathbf{ker}\, C \oplus \mathcal{H}_C,$$

there exists the representation

$$B = [B_1 \quad \vartheta |C|^{\frac{1}{2}}]$$

where $\vartheta : \mathcal{H}_C \to \mathbf{C}$ and $B_1 : \mathbf{ker}\, C \to \mathbf{C}$ are uniquely determined. Consequently, H has the representation

$$(3.4) \qquad H = \begin{bmatrix} a & B_1 & \vartheta|C|^{\frac{1}{2}} \\ B_1^* & 0 & 0 \\ |C|^{\frac{1}{2}}\vartheta^* & 0 & C \end{bmatrix} : \begin{matrix} \mathbf{C} \\ \oplus \\ \mathbf{ker}\, C \\ \oplus \\ \mathcal{H}_C \end{matrix} \longrightarrow \begin{matrix} \mathbf{C} \\ \oplus \\ \mathbf{ker}\, C \\ \oplus \\ \mathcal{H}_C \end{matrix}.$$

In the following, for a Hilbert space $\mathcal{G}$, we denote by $[\mathcal{G} \oplus \mathcal{G}]$ the Krein space obtained from $\mathcal{G} \oplus \mathcal{G}$ endowed with the symmetry

$$(3.5) \qquad J_o = \begin{bmatrix} 0 & I \\ I & 0 \end{bmatrix}.$$

3.2. LEMMA. For the hermitian matrix given in (3.4) we have

$$(3.6) \qquad \varkappa^-(H) = \varkappa^-(C) + \varkappa^-(a - \vartheta S_C \vartheta^*) + (1 - \varkappa^-(a - \vartheta S_C \vartheta^*)) \cdot \mathbf{rank}(B_1).$$

In addition,

(i) if $a - \vartheta S_C \vartheta^* = 0$, then there exists a unitary operator W, uniquely determined such that

$$(3.7) \qquad \begin{cases} W : \mathcal{H}_H \to [\mathcal{R}(B_1^*) \oplus \mathcal{R}(B_1^*)] [+] \mathcal{H}_C, \\ W|H|^{\frac{1}{2}} = \begin{bmatrix} B_1^* & 0 & 0 \\ 0 & P^{\mathbf{ker}\, C}_{\mathcal{R}(B_1^*)} & 0 \\ S_C \vartheta^* & 0 & |C|^{\frac{1}{2}} \end{bmatrix} : \begin{matrix} \mathbf{C} \\ \oplus \\ \mathbf{ker}\, C \\ \oplus \\ \mathcal{H}_C \end{matrix} \longrightarrow \begin{matrix} \mathcal{R}(B_1^*) \\ \oplus \\ \mathcal{R}(B_1^*) \\ \oplus \\ \mathcal{H}_C \end{matrix}. \end{cases}$$

(ii) if $a_1 = a - \vartheta S_C \vartheta^* \neq 0$ then, denoting $\beta_1 = |a_1|^{-\frac{1}{2}} B_1$, there exists a unitary operator W, uniquely determined such that

$$W : \mathcal{H}_H \to \mathcal{H}_{a_1} [+] \mathcal{H}_{-\beta_1^* a_1^{-1} \beta_1} [+] \mathcal{H}_C,$$

$$(3.8)\qquad \left\{ W|H|^{\frac{1}{2}} = \begin{bmatrix} |a_1|^{\frac{1}{2}} & S_a\beta_1 & 0 \\ 0 & |\beta_1| & 0 \\ S_C\vartheta^* & 0 & |C|^{\frac{1}{2}} \end{bmatrix} : \begin{matrix} \mathbf{C} \\ \oplus \\ \mathbf{ker}\, C \\ \oplus \\ \mathcal{H}_C \end{matrix} \longrightarrow \begin{matrix} \mathcal{H}_{a_1} \\ \oplus \\ \mathcal{H}_{-\beta_1^* a_1^{-1}\beta_1} \\ \oplus \\ \mathcal{H}_C \end{matrix} \right. .$$

Proof. In case $a - \vartheta S_C \vartheta^* = 0$, the hermitian matrix H given at (3.4) can be factored as follows

$$(3.9)\qquad H = \begin{bmatrix} B_1 & 0 & \vartheta S_C \\ 0 & P^{\ker C}_{\mathcal{R}(B_1^*)} & 0 \\ 0 & 0 & |C|^{\frac{1}{2}} \end{bmatrix} \cdot \begin{bmatrix} 0 & I & 0 \\ I & 0 & 0 \\ 0 & 0 & S_C \end{bmatrix} .$$

$$\cdot \begin{bmatrix} B_1^* & 0 & 0 \\ 0 & P^{\ker C}_{\mathcal{R}(B_1^*)} & 0 \\ S_C\vartheta^* & 0 & |C|^{\frac{1}{2}} \end{bmatrix}$$

Since the operator defined by the 3 x 3 block-matrix in (3.7) is surjective, using [5, Corollary 1.3], it follows the existence of the unitary operator W as in (3.7). From here we obtain that, in this case,

$$(3.10)\qquad \varkappa^-(H) = \varkappa^-(C) + \mathbf{rank}(B_1).$$

Let us assume now that $a_1 = a - \vartheta S_C \vartheta^* \neq 0$. Denoting $\beta_1 = |a_1|^{-\frac{1}{2}} B_1$, it is easy to see that the following factorization holds

$$(3.11)\qquad H = \begin{bmatrix} |a_1|^{\frac{1}{2}} & 0 & \vartheta S_C \\ \beta_1^* S_a & |\beta_1| & 0 \\ 0 & 0 & |C|^{\frac{1}{2}} \end{bmatrix} \cdot \begin{bmatrix} S_{a_1} & 0 & 0 \\ 0 & S_{-\beta_1^* a_1^{-1}\beta_1} & 0 \\ 0 & 0 & S_C \end{bmatrix} .$$

$$\cdot \begin{bmatrix} |a_1|^{\frac{1}{2}} & S_a\beta_1 & 0 \\ 0 & |\beta_1| & 0 \\ S_C\vartheta^* & 0 & |C|^{\frac{1}{2}} \end{bmatrix}$$

From here, reasoning exactly as above, we get the existence of the unitary operator W as in (3.8) and, as consequence, the following identity holds

(3.12) $\varkappa^-(H) = \varkappa^-(C) + \varkappa^-(a - \vartheta S_C \vartheta^*) + \varkappa^+(a - \vartheta S_C \vartheta^*) \cdot \mathbf{rank}(B_1)$.

Finally, it remains to notice that the general formula (3.6) follows from its particular cases (3.10) and (3.12).■

Coming back to the hermitian matrix H in (3.1), the calculation of $\varkappa^-(H)$ can be performed by iterating the scheme obtained in Lemma 3.2, times $\mathbf{dim}(\mathcal{H}_1)$. Of course, there are particular cases (e.g. if $B = |A|^{\frac{1}{2}} \Gamma |C|^{\frac{1}{2}}$) when $\varkappa^-(H)$ can be simply calculated, but in general this is complex.

In this paper we will need the description of selfadjoint 2x2 block-matrices having finite negative signature. The next example shows that, even for the case of finite negative signature, if at least one of the spaces $\mathcal{H}_1$ or $\mathcal{H}_2$ is infinite dimensional, the situation is far more complicated, in general, than that described in Lemma 3.2 (compare also with Lemma 3.1).

3.3. EXAMPLE. Let $\mathcal{H}$ be an infinite dimensional Hilbert space and $C \in \mathcal{L}(\mathcal{H})$, $C \geq 0$ be such that $\mathcal{R}(C)$ is dense in $\mathcal{H}$ but it is not closed (e.g. if $\mathcal{H}$ is separable and $\{e_n\}_{n \geq 1}$ is an orthonormal basis of $\mathcal{H}$, defining $Ce_n = \lambda_n e_n$, where $\lambda_n > 0$ and $\lambda_n \to 0$ $(n \to \infty)$, the operator C has the required properties). Let x be a vector in $\mathcal{H}$ such that $x \notin \mathcal{R}(C^{\frac{1}{2}})$ and consider the selfadjoint operator $H \in \mathcal{L}(\mathbf{C} \oplus \mathcal{H})$ defined by

$$H = \begin{bmatrix} -1 & x \\ (\cdot, x) & C \end{bmatrix}.$$

It is easy to see that $\varkappa^-(H) = 1$ (e.g. performing a Schur reduction), but the analog of the operator ϑ (see (3.4)) does not exist at all in this case.

§ 4. Row extensions

In this Section we investigate the Problem $(*)_r$ (see the Introduction) in case the space $\mathcal{K}'_1$ is an anti-Hilbert space (i.e. a Krein space such that its inner product is negative definite). The results are formally close to that obtained in [6, Section 3] for the case of Hilbert space $\mathcal{K}'_1$. For this reason we will leave more details to be worked by the reader.

Let $\mathcal{K}_1$ and $\mathcal{K}_2$ be Krein spaces onto which we fix the f.s. J_1 and respectively J_2 and consider an operator $T \in \mathcal{L}(\mathcal{K}_1, \mathcal{K}_2)$ and a Hilbert space $\mathcal{H}$. Denote $\mathcal{K}'_1 = [-\mathcal{H}]$, i.e. -I is (the unique) f.s. of $\mathcal{K}'_1$, and then consider the Krein space $\tilde{\mathcal{K}}_1 = \mathcal{K}_1[+]\mathcal{K}'_1$ onto which we fix the f.s. $\tilde{J}_1 = J_1 \oplus -I$. With respect to J_1, J_2 and $\tilde{J}_1$ will be considered defect operators, defect spaces, link operators, etc. (see Section 2).

In the following we will use also this notation: for a Hilbert space $\mathcal{G}$ we denote by $[\mathcal{G} \oplus \mathcal{G}]$ the Krein space whose indefinite inner product is determined by the symmetry

$$J_o = \begin{bmatrix} 0 & I \\ I & 0 \end{bmatrix}. \tag{4.1}$$

4.1. LEMMA. Let $T_r \in \mathcal{L}(\tilde{\mathcal{H}}_1, \mathcal{H}_2)$ be a row extension of T, i.e. $T_r|\mathcal{H}_1 = T$, and assume that

$$\mathcal{R}(P^{\mathcal{H}_2}_{\mathcal{D}_{T^*}} T_r|\mathcal{H}'_r) \subseteq \mathcal{R}(D_{T^*}). \tag{4.2}$$

Then there exist uniquely determined operators $\Lambda \in \mathcal{L}(\mathcal{H}'_1, \ker D_{T^*})$, $\Gamma \in \mathcal{L}(\ker \Lambda, \mathcal{D}_{T^*})$, and $\Delta \in \mathcal{L}(\mathcal{R}(\Lambda^*), \mathcal{D}_{T^*})$ such that

$$T_r|\mathcal{H}'_1 = \begin{bmatrix} D_{T^*}\Gamma & D_{T^*}\Delta \\ 0 & \Lambda \end{bmatrix} \tag{4.3}$$

and the following identities hold

$$\varkappa^-[I - T_r T_r^{\#}] = \varkappa^-[I - \Gamma\Gamma^{\#}], \tag{4.4}$$

$$\varkappa^-[I - T_r^{\#} T_r] = \varkappa^-[I - T^{\#}T] + \varkappa^-[I - \Gamma^{\#}\Gamma] + \mathbf{rank}(\Lambda). \tag{4.5}$$

Moreover, if $\varkappa^-[I - T_r T_r^{\#}]$ is finite then there exists a unique unitary operator $U_*(T_r)$ (between Pontryagin spaces) such that

$$U_*(T_r) : \mathcal{D}_{T_r^*} \longrightarrow \mathcal{D}_{\Gamma^*}[+]\mathcal{R}(\Lambda^*),$$

(4.6)

$$U_*(T_r)D_{T_r^*} = \begin{bmatrix} D_{\Gamma^*}D_{T^*} & 0 \\ \Delta^* D_{T^*} & \Lambda^* \end{bmatrix} = D_{*,r},$$

and, correspondingly, if $\varkappa^-[I - T_r^{\#} T_r]$ is finite then Λ has finite rank and there exists a unique unitary operator $U(T_r)$ (between Pontryagin spaces) such that

$$U(T_r) : \mathcal{D}_{T_r} \to \mathcal{D}_T[+][\mathcal{R}(\Lambda) + \mathcal{R}(\Lambda)][+]\mathcal{D}_\Gamma$$

$$(4.7)\qquad U(T_r)D_{T_r} = \begin{bmatrix} D_T & 0 & -J_T L_{T^*}\Delta & -J_T L_{T^*}\Gamma \\ 0 & -P^{\mathcal{K}_2}_{\mathcal{R}(\Lambda)} J_2 T & -\tfrac{1}{2}\Lambda^{*-1}E & -\Lambda^{*-1}\Delta^* J_{T^*}\Gamma \\ 0 & 0 & \Lambda & 0 \\ 0 & 0 & 0 & D_\Gamma \end{bmatrix} = D_r$$

<u>where we denoted</u>

$$(4.8)\qquad E = I + \Lambda^* P^{\mathcal{K}_2}_{\mathcal{R}(\Lambda)} J_2 \Lambda + \Delta^* J_{T^*} \Delta.$$

Proof. Denote $\Lambda = P^{\mathcal{K}_2}_{\ker D_{T^*}} T_r|\mathcal{K}'_1$. Then, from the inclusion (4.2) it follows that the identities

$$(4.9)\qquad D_{T^*}\Gamma = P^{\mathcal{K}_2}_{\mathcal{D}_{T^*}} T_r|\ker\Lambda,\quad D_{T^*}\Delta = P^{\mathcal{K}_2}_{\mathcal{D}_{T^*}} T_r|\overline{\mathcal{R}(\Lambda^*)},$$

uniquely determines operators $\Gamma \in \mathcal{L}(\ker\Lambda, \mathcal{D}_{T^*})$ and $\Delta \in \mathcal{L}(\overline{\mathcal{R}(\Lambda^*)}, \mathcal{D}_{T^*})$, hence the representation (4.3) holds.

Using the representation (4.3) it follows the factorization

$$(4.10)\qquad J_2 - T_r\tilde{J}_1T_r^* = D^*_{*,r}\begin{bmatrix} J_{\Gamma^*} & 0 \\ 0 & I|\overline{\mathcal{R}(\Lambda^*)} \end{bmatrix} D_{*,r},$$

where $D_{*,r} : \mathcal{K}_2 \to \mathcal{D}_{\Gamma^*}[+]\overline{\mathcal{R}(\Lambda^*)}$ is defined at (4.6). Observing that the operator $D_{*,r}$ has dense range, the correctness of the identity (4.4) follows by means of an argument of Pontryagin Lemma type (e.g. as in the proof of [5, Corollary 1.3]). If we assume, in addition, that $\varkappa^-[I - T_rT_r^{\#}]$ is finite, then application of [5, Corollary 1.3] to the factorization (4.10) shows that the operator $U_*(T_r)$ as in (4.6) is correctly defined and unitary.

Let us assume now that the operator Λ (hence also Λ^*) has closed range. Then, using the identities (2.16) - (2.18) one verifies the correctness of the factorization

$$(4.11)\qquad \tilde{J}_1 - T_r^*J_2T_r = D_r^*\begin{bmatrix} J_T & 0 & 0 \\ 0 & J_o & 0 \\ 0 & 0 & J_\Gamma \end{bmatrix} D_r,$$

where J_o is the symmetry defined in (4.1) with respect to the Krein space $[\mathcal{R}(\Lambda)\oplus\mathcal{R}(\Lambda)]$ and the operator $D_r : \tilde{\mathcal{K}}_1 \to \mathcal{D}_T[+][\mathcal{R}(\Lambda)\oplus\mathcal{R}(\Lambda)][+]\mathcal{D}_\Gamma$ is defined at (4.7). Since D_r has dense range we conclude as above that the identity (4.5) holds.

If Λ has not closed range then **rank**(Λ) is necessarily infinite. The examination of the block-matrix representation of $\tilde{J}_1 - T_r^* J_2 T_r$ shows that the following inequality holds

$$\varkappa^-[I - T_r^{\#} T_r] \geq \mathbf{rank}(\Lambda), \tag{4.12}$$

hence $\varkappa^-[I - T_r^{\#} T_r]$ is infinite and this proves that (4.5) holds also in this case.

If $\varkappa^-[I - T_r^{\#} T_r]$ is finite then from (4.5) it follows that Λ has finite rank, in particular we have the factorization (4.11). Using [5, Corollary 1.3] and the surjectivity of D_r, this implies the existence of the unitary operator $U(T_r)$ as in (4.7). ■

4.2. COROLLARY. _Let T_r be a row extension of T such that the inclusion in (4.2) holds and $\varkappa^-[I - T_r^{\#} T_r]$ is finite. Then, with respect to the representation (4.3), we have_

$$\mathbf{ker}\, D_{T_r} = \{ D_T^{-1} J_T L_{T^*} \Gamma\beta \oplus -J_1 T^* \Lambda^{*-1} \Delta^* J_{T^*} \Gamma \beta \oplus \alpha \oplus \beta \mid \tag{4.13}$$

$$\mid \alpha \in P^{\mathcal{K}_1}_{\mathbf{ker}\, D_T} T J_1 \mathbf{ker}\, \Lambda^*,\ \beta \in \mathbf{ker}\, D_\Gamma,\ L_{T^*} \Gamma\beta \in \mathcal{R}(D_T) \}.$$

Proof. From Lemma 4.1 it follows the existence of the unitary operator $U(T_r)$ as in (4.7) and this implies $\mathbf{ker}\, D_{T_r} = \mathbf{ker}\, D_r$. Now (4.13) is a direct consequence of the 4×4 block-matrix of D_r (see (4.7)). ■

4.3. REMARK. Reasoning as in the proof of [2, Proposition 3.1] it follows that $I - T^{\#}T$ has closed range if and only if $I - TT^{\#}$ has closed range. Moreover, if these hold then D_{T^*} (equivalently, D_T) has closed range and in this case (4.2) hold for any row extension T_r of T. In particular, this is true if T is of finite rank, or, more particularly, if either $\mathcal{K}_1$ or $\mathcal{K}_2$ are finite dimensional.

In this paper we are mainly interested to describe those row extensions T_r of T such that $\varkappa^-[I - T_r^{\#} T_r]$ and $\varkappa^-[I - T_r T_r^{\#}]$ are finite. The next example shows that if the hypothesis from Remark 4.3 are not imposed, a description similar to that obtained in Lemma 4.1 is not possible.

4.4. EXAMPLE. Let $\mathcal{G}$ be an infinite dimensional Hilbert space and $T \in \mathcal{L}(\mathcal{G})$ be such that $T^*T \leq I$, $\mathcal{R}(I - T^*T)$ is dense in $\mathcal{G}$ but not closed (e.g. if $\{g_n\}_{n \geq 1}$ is an orthonormal basis of $\mathcal{G}$, $\{\lambda_n\}_{n \geq 1} \subset \mathbf{C}$ such that $\lambda_n < 1$ and $|\lambda_n| \to 1$ $(n \to \infty)$, defining $Tg_n = \lambda_n g_n$, T has the desired properties). Let $x \in \mathcal{G}$ be such that $x \notin \mathcal{R}(D_{T^*})$ and take the row extension

$$T_r = [T \quad x] : \mathcal{G}[+][-\mathbf{C}] \to \mathcal{G}.$$

Then $\varkappa^-[I - T_r T_r^{\#}] = 0$, hence (via (2.5)) we have $\varkappa^-[I - T_r^{\#} T_r] = 1$. However, the analog of the operator Γ (see (4.3)) does not exist at all in this case.

4.5. COROLLARY. Let $T_r \in \mathcal{L}(\tilde{\mathcal{K}}_1, \mathcal{K}_2)$ be a row extension of T such that (4.2) holds. Then

$$(4.14) \qquad \varkappa^-[I - T_r^{\#} T_r] + \varkappa^-[I - TT^{\#}] = \varkappa^-[I - T_r T_r^{\#}] + \varkappa^-[I - T^{\#}T] + \varkappa^-[\mathcal{K}'_1].$$

Proof. From Lemma 4.1 we have the identities (4.4) and (4.5). According to (2.5) we also have the identity

$$(4.15) \qquad \varkappa^-[I - \Gamma\Gamma^{\#}] + \mathbf{dim\,ker}\ \Lambda = \varkappa^-[I - \Gamma^{\#}\Gamma] + \varkappa^-[I - TT^{\#}]$$

and taking into account that

$$(4.16) \qquad \varkappa^-[\mathcal{K}'_1](= \mathbf{dim}\ \mathcal{K}'_1) = \mathbf{dim\,ker}\ \Lambda + \mathbf{rank}(\Lambda),$$

all these imply (4.14).■

4.6. REMARK. If $\mathcal{K}_1$ and $\mathcal{K}_2$ are Potryagin spaces then the identity (4.14) is true without the assumption that (4.2) holds. In this case, (4.14) follows simply from (2.5) written for T and T_r, add and reduce $\varkappa^-[\mathcal{K}_1]$ and $\varkappa^-[\mathcal{K}_2]$.

As a conclusion of the results obtained in this section we can solve the Problem $(*)_r$ in a certain case.

4.7. THEOREM. Let $\varkappa_1, \varkappa_2$ be nonnegative integer numbers, $\mathcal{K}'_1$ be a finite dimansional anti-Hilbert space and assume that for $T \in \mathcal{L}(\mathcal{K}_1, \mathcal{K}_2)$, the operator $I - T^{\#}T$ has closed range. Then, the Problem $(*)_r$ has solutions if and only if the following hold

$$(4.17) \qquad \varkappa_1 + \varkappa^-[I - TT^{\#}] = \varkappa_2 + \varkappa^-[I - T^{\#}T] + \varkappa^-[\mathcal{K}'_1],$$

$$(4.18) \qquad \mathbf{max}\{0, \varkappa^-[I - TT^{\#}] - \varkappa^-[\mathcal{K}'_1]\} \leq \varkappa_2 \leq \varkappa^-[I - TT^{\#}].$$

Moreover, assuming that (4.17) and (4.18) are fulfilled then the formula (4.3) establishes a bijective correspondence between the set of solutions T_r of Problem $(*)_r$ and the set of triplets $\{\Lambda, \Gamma, \Delta\}$ such that $\Lambda \in \mathcal{L}(\mathcal{K}'_1, \mathbf{ker}\ D_{T^*})$ and $\Gamma \in \mathcal{L}(\mathbf{ker}\ \Lambda, \mathcal{D}_{T^*})$ satisfy

$$\varkappa^-[I - \Gamma\Gamma^{\#}] = \varkappa_2$$

and $\Delta \in \mathcal{L}(\mathcal{R}(\Lambda^*), \mathcal{D}_{T^*})$ is arbitrary.

Proof. Let T_r be a row extension of T such that $\varkappa^-[I - T_r^{\#}T_r] = \varkappa_1$ and $\varkappa^-[I - T_rT_r^{\#}] = \varkappa_2$. Then (4.17) must hold due to the Corollary 4.5. The inequalities from (4.18) are obtained using [6, Lemma 2.5].

Conversely, assume that $\varkappa_1$ and $\varkappa_2$ satisfy (4.17) and (4.18). Then it is easy to see that (4.8) implies

$$0 \leq \varkappa^-[I - TT^{\#}] - \varkappa_2 \leq \varkappa^-[\mathcal{K}'_1],$$

hence there exists $\Gamma \in \mathcal{L}(\mathcal{K}'_1, \mathcal{D}_{T^*})$ such that

$$\varkappa^-[I - \Gamma\Gamma^{\#}] = \varkappa_2. \tag{4.19}$$

Take the row extension T_r of T

$$T_r = [T \quad D_{T^*}\Gamma]. \tag{4.20}$$

From (4.4) and (4.19) we obtain $\varkappa^-[I - T_rT_r^{\#}] = \varkappa_2$ and then, using this and (4.17), we obtain $\varkappa^-[I - T_r^{\#}T_r] = \varkappa_1$, hence T_r defined in (4.20) is a solution of Problem $(*)_r$.

The second part of the statement is a direct consequence of Lemma 4.1.∎

4.8. REMARK. As a consequence of (4.17) and (4.18), in order that the Problem $(*)_r$ has solutions, one must have the following inequalities

$$\begin{aligned} &\varkappa^-[I - T^{\#}T] + \mathbf{max}\{0, \varkappa^-[\mathcal{K}'_1] - \varkappa^-[I - TT^{\#}]\} \leq \\ &\leq \varkappa_1 \leq \varkappa^-[I - T^{\#}T] + \varkappa^-[\mathcal{K}'_1]. \end{aligned} \tag{4.21}$$

Conversely, (4.17) and (4.21) imply (4.18).

§ 5. The lifting problem revisited

In this Section we will solve the Problem $(*)$ (see the Introdction) in case $\mathcal{K}'_1$ and $\mathcal{K}'_2$ are anti-Hilbert spaces. The approach is to consider a solution $\widetilde{T}$ of it, successively as a column extension of a row extension of T, and to use the results obtained in the previous sections. The limitations of this approach, which were pointed out in Examples 3.3 and 4.4, lead us to solve completely only the finite dimensional case. In this situation, we will produce an iterative scheme to be applied for a finite number of steps.

Let $\mathcal{K}_1$ and $\mathcal{K}_2$ be finite dimesional Krein spaces, $\mathcal{K}'_1$ and $\mathcal{K}'_2$ finite dimensional anti-Hilbert spaces and denote $\widetilde{\mathcal{K}}_1 = \mathcal{K}_1[+]\mathcal{K}'_1$ and $\widetilde{\mathcal{K}}_2 = \mathcal{K}_2[+]\mathcal{K}'_2$. Consider also $T \in \mathcal{L}(\mathcal{K}_1, \mathcal{K}_2)$ and $\varkappa_1, \varkappa_2$ two nonnegative integers.

5.1. LEMMA. With the hypothesis made above, the Problem (*) has solutions if and only if the following conditions hold:

$$\varkappa_1 + \varkappa^-[I - TT^\#] + \varkappa^-[\mathcal{K}'_2] = \varkappa_2 + \varkappa^-[I - T^\# T] + \varkappa^-[\mathcal{K}'_1], \tag{5.1}$$

$$\max\{0, \varkappa^-[I - T^\# T] - \varkappa^-[\mathcal{K}'_2] + \max\{0, \varkappa^-[\mathcal{K}'_1] - \varkappa^-[I - TT^\#]\}\} \leq \tag{5.2}$$
$$\leq \varkappa_1 \leq \varkappa^-[I - T^\# T] + \varkappa^-[\mathcal{K}'_1].$$

Proof. This is a consequence of the first part of Theorem 4.7 and Remark 4.8.■

5.2. REMARK. In Lemma 5.1 the condition (5.2) can be replaced by its dual one

$$\max\{0, \varkappa^-[I - TT^\#] - \varkappa^-[\mathcal{K}'_1] + \max\{0, \varkappa^-[\mathcal{K}'_2] - \varkappa^-[I - T^\# T]\}\} \leq \tag{5.3}$$
$$\leq \varkappa_2 \leq \varkappa^-[I - TT^\#] + \varkappa^-[\mathcal{K}'_2].$$

In order to describe the set of all solutions of the Problem (*) (for this particular case), in the following we will assume that the nonnegative integer numbers $\varkappa_1$ and $\varkappa_2$ satisfy the conditions (5.1) and (5.2) (or, equivalently, (5.1) and (5.3)). Then, according to Lemma 5.1, there exist solutions of the Problem (*) and let $\tilde{T} \in \mathcal{L}(\tilde{\mathcal{K}}_1, \tilde{\mathcal{K}}_2)$ be one of them, i.e.

$$P_{\mathcal{K}_2}^{\tilde{\mathcal{K}}_2} \tilde{T} | \mathcal{K}_1 = T, \quad \varkappa^-[I - \tilde{T}^\# \tilde{T}] = \varkappa_1, \quad \varkappa^-[I - \tilde{T}\tilde{T}^\#] = \varkappa_2. \tag{5.4}$$

Let us consider the operator $T_r \in \mathcal{L}(\mathcal{K}_1, \tilde{\mathcal{K}}_2)$ defined by

$$T_r = P_{\mathcal{K}_2}^{\tilde{\mathcal{K}}_2} T. \tag{5.5}$$

Then T_r is a row extension of T and according to Theorem 4.7 and Remark 4.8 we have

$$\varkappa^-[I - T_r^\# T_r] + \varkappa^-[I - TT^\#] = \varkappa^-[I - T_r T_r^\#] + \varkappa^-[I - T^\# T] + \varkappa^-[\mathcal{K}'_1], \tag{5.6}$$

$$\max\{0, \varkappa^-[\mathcal{K}'_1] - \varkappa^-[I - TT^\#]\} + \varkappa^-[I - T^\# T] \leq \tag{5.7}$$
$$\leq \varkappa^-[I - T_r^\# T_r] \leq \varkappa^-[I - T^\# T] + \varkappa^-[\mathcal{K}'_1],$$

$$\max\{0, \varkappa^-[I - TT^\#] - \varkappa^-[\mathcal{K}'_1]\} \leq \varkappa^-[I - T_r T_r^\#] \leq \varkappa^-[I - TT^\#], \tag{5.8}$$

and

$$P_{\mathcal{K}_2}^{\tilde{\mathcal{K}}_2} \tilde{T} | \mathcal{K}'_1 = T_r \mathcal{K}'_1 = \begin{bmatrix} D_{T^*}\Gamma & D_{T^*}\Delta \\ 0 & \Lambda \end{bmatrix}, \tag{5.9}$$

where $\Lambda \in \mathcal{L}(\mathcal{K}'_1, \ker D_{T^*})$, $\Gamma \in \mathcal{L}(\ker \Lambda, \mathcal{D}_{T^*})$ and $\Delta \in \mathcal{L}(\mathcal{R}(\Lambda^*), \mathcal{D}_{T^*})$ are such that

$$\varkappa^-[I - T_r T_r^\#] = \varkappa^-[I - \Gamma\Gamma^\#], \tag{5.10}$$

or, equivalently,

$$\varkappa^-[I - T_r^\# T_r] = \varkappa^-[I - T^\# T] + \varkappa^-[I - \Gamma^\#\Gamma] + \mathbf{rank}(\Lambda). \tag{5.11}$$

From (5.5) it follows that $\tilde{T}$ is a column extension of T_r hence, using the dual variant of Theorem 4.7, we have

$$P^{\tilde{\mathcal{K}}_2}_{\mathcal{K}_2}\tilde{T} = \begin{bmatrix} \Gamma' D_{T_r} & 0 \\ \Delta' D_{T_r} & \Lambda' \end{bmatrix}, \tag{5.12}$$

where $\Lambda' \in \mathcal{L}(\ker D_{T_r}, \mathcal{K}'_2)$, $\Gamma' \in \mathcal{L}(\mathcal{D}_{T_r}, \ker \Lambda'^*)$, and $\Delta' \in \mathcal{L}(\mathcal{D}_{T_r}, \mathcal{R}(\Lambda'))$ satisfy

$$\varkappa_1 = \varkappa^-[I - \Gamma'^\#\Gamma'], \tag{5.13}$$

or, equivalently,

$$\varkappa_2 = \varkappa^-[I - T_r T_r^\#] + \varkappa^-[I - \Gamma'\Gamma'^\#] + \mathbf{rank}(\Lambda'). \tag{5.14}$$

In order to describe Λ' it is more convenient to extend it on $\tilde{\mathcal{K}}_1$ by $\Lambda' \mathcal{D}_{T_r} = 0$. Taking into account the calculation of $\ker D_{T_r}$ from Corollary 4.2, it follows that, with respect to the decomposition

$$\tilde{\mathcal{K}}_1 = \mathcal{D}_T \oplus \ker D_T \oplus \mathcal{R}(\Lambda^*) \oplus \ker \Lambda, \tag{5.15}$$

Λ' must have the row block-matrix

$$\Lambda' = [L\Gamma^* L_{T^*} J_T D_T^{-1} \qquad K \qquad -L\Gamma^* J_{T^*}\Delta\Lambda^{-1} T J_1 \quad L] \tag{5.16}$$

where $K \in \mathcal{L}(\ker D_T, \mathcal{K}'_2)$ and $L \in \mathcal{L}(\ker \Lambda, \mathcal{K}'_2)$ satisfy

$$\mathcal{R}(K^*) \subseteq P^{\mathcal{K}_1}_{\ker D_T} T J_1 \ker \Lambda^*, \qquad \mathcal{R}(L^*) \subseteq \ker D \quad . \tag{5.17}$$

From (5.16) we get

$$\ker \Lambda'^* = \ker L^* \cap \ker K^*, \quad \mathcal{R}(\Lambda') = \mathcal{R}(L) \vee \mathcal{R}(K). \tag{5.18}$$

Further on, in order to obtain a description of $\Gamma' D_{T_r}$ and $\Delta' D_{T_r}$ we have to take into account the unitary operator $U(T_r)$ defined in (4.7). Define Γ'' and Δ'' by

$$(5.19)\qquad \begin{aligned} &\Gamma'' : \mathcal{D}_T[+][\mathcal{R}(\Lambda)\oplus\mathcal{R}(\Lambda)][+]\,\mathcal{D}_\Gamma \to \mathbf{ker}\, L^* \cap \mathbf{ker}\, K^* \\ &\Gamma' = \Gamma'' U(T_r), \end{aligned}$$

and

$$(5.20)\qquad \begin{aligned} &\Delta'' : \mathcal{D}_T[+][\mathcal{R}(\Lambda)\oplus\mathcal{R}(\Lambda)][+]\,\mathcal{D}_\Gamma \to \mathcal{R}(L)\vee\mathcal{R}(K), \\ &\Delta' = \Delta'' U(T_r). \end{aligned}$$

Since $U(T_r)$ is unitary, the signatures of defect of Γ'' are equal to those of Γ'. With respect to the decomposition of the domain of Γ'' and

$$(5.21)\qquad \mathbf{ker}\, L^* \cap \mathbf{ker}\, K^* = \mathbf{ker}\, H^* \oplus \mathcal{R}(H)$$

we consider the representation

$$(5.22)\qquad \Gamma'' = \begin{bmatrix} \Gamma''_{11} & 0 & \Gamma''_{13} & \Gamma''_{14} \\ \Gamma''_{21} & H & \Gamma''_{23} & \Gamma''_{24} \end{bmatrix}$$

and then define

$$(5.23)\qquad \begin{aligned} &G : \mathcal{D}_T \oplus \mathcal{D}_\Gamma \to \mathbf{ker}\, H^*, \\ &G = [\Gamma''_{11} \quad \Gamma''_{14}], \end{aligned}$$

and

$$(5.24)\qquad \begin{aligned} &R : \mathcal{D}_T \oplus \mathcal{D}_\Gamma \to \mathcal{R}(H), \\ &R = [\Gamma''_{21} \quad \Gamma''_{24}]. \end{aligned}$$

Now recall that the f.s. on $\mathcal{K}'_2$ (and on any subspace of it) is $-I$ and the f.s. on $\mathcal{D}_T[+][\mathcal{R}(\Lambda)\oplus\mathcal{R}(\Lambda)][+]\,\mathcal{D}_\Gamma$ is

$$(5.25)\qquad J = J_T \oplus \begin{bmatrix} 0 & I \\ I & 0 \end{bmatrix} \oplus J_\Gamma .$$

Representing $-I - \Gamma'' J \Gamma''^*$ with respect to (5.21) by

$$(5.26)\qquad -I - \Gamma'' J \Gamma''^* = \begin{bmatrix} A & B \\ B^* & C \end{bmatrix}$$

we have

$$(5.27)\qquad A = -I - G(J_T \oplus J_\Gamma)G^*,$$

$$(5.28)\qquad B = -\Gamma''_{13}H^* - G(J_T \oplus J_\Gamma)R^*,$$

(5.29) $$C = -I - R(J_T \oplus J)R^* - H\Gamma''^*_{23} - \Gamma''_{23}H^*.$$

From (5.10), (5.14), (5.19), (5.26), and (5.27) it follows

(5.30) $$\varkappa_2 = \varkappa^-[I - \Gamma\Gamma^*] + \mathbf{rank}([K \quad L]) + $$
$$+ \varkappa^-\left(\begin{bmatrix} -I - G(J_T \oplus J_\Gamma)G^* & B \\ B^* & C \end{bmatrix}\right),$$

where B and C are given by (5.28) and (5.29). With these operators we solve the equations (5.28) and (5.29) with respect to Γ''_{13} and Γ''_{23} (for (5.29) we use [15]) and obtain

(5.31) $$\Gamma''_{13} = [-B - G(J_T \oplus J_\Gamma)R^*H^{*-1} \qquad X],$$

and

(5.32) $$\Gamma''_{23} = [-\tfrac{1}{2}(I + R(J_T \oplus J_\Gamma)R^* + C + iS)H^{*-1} \qquad Y],$$

where these row matrices have to be considered with respect to the decomposition

(5.33) $$\mathcal{R}(\Lambda) = \mathcal{R}(H^*) \oplus \mathbf{ker}\, H,$$

$S \in \mathcal{L}(\mathcal{R}(H))$, $S = S^*$ and $X \in \mathcal{L}(\mathbf{ker}\, H, \mathbf{ker}\, H^*)$, $Y \in \mathcal{L}(\mathbf{ker}\, H, \mathcal{R}(H))$ are arbitrary.

Finally, representing Δ'' (see (5.20)) with respect to the decomposition of its domain by

$$\Delta'' = [D_1 \qquad D_2 \qquad D_3 \qquad D_4],$$

and using (4.7) and the formulae obtained up-to-now, we have

(5.34) $$P^{\tilde{\mathcal{K}}_2}_{\mathcal{K}'_2}\tilde{T}\,\mathcal{K}_1 = \begin{bmatrix} G|\mathcal{D}_T & 0 \\ R|\mathcal{D}_T & H \\ D_1 & D_2 \end{bmatrix} \cdot \begin{bmatrix} D_T & 0 \\ 0 & -P^{\mathcal{K}_2}_{\mathcal{R}(\Lambda)}J_2T \end{bmatrix} + [L\Gamma^*L_{T^*}J_TD_T^{-1} \qquad K],$$

and

(5.35) $$P^{\tilde{\mathcal{K}}_2}_{\mathcal{K}'_2}\tilde{T}\,\mathcal{K}'_1 = \begin{bmatrix} G|\mathcal{D}_T & 0 & \Gamma''_{13} & G|\mathcal{D}_\Gamma \\ R|\mathcal{D}_T & H & \Gamma''_{23} & R|\mathcal{D}_\Gamma \\ D_1 & D_2 & D_3 & D_4 \end{bmatrix}.$$

$$\cdot \begin{bmatrix} D_T & 0 & -J_T L_{T^*} \Delta & -J_T L_{T^*} \Gamma \\ 0 & -P_{\mathcal{R}(\Lambda)} J_2 T & -\frac{1}{2} \Lambda^{*-1}(I + \Lambda^* P_{\mathcal{R}(\Lambda)}^{\mathcal{K}_2} J_2 \Lambda + \Delta^* J_{T^*} \Delta & - \Lambda^{*-1} \Delta^* J_{T^*} \Gamma \\ 0 & 0 & & 0 \\ 0 & 0 & 0 & D_\Gamma \end{bmatrix} +$$

$$+ \quad [\, -L \Gamma^* J_{T^*} \Delta \Lambda^{-1} T J_1 \qquad L],$$

where Γ''_{13} and Γ''_{23} are given by (5.31) and (5.32).

We have proved the following

5.3. THEOREM. Assume that $\tilde{\mathcal{K}}_1$ and $\tilde{\mathcal{K}}_2$ are finite dimensional, $\mathcal{K}'_1$ and $\mathcal{K}'_2$ are anti-Hilbert spaces and the nonnegative integers $\varkappa_1$ and $\varkappa_2$ are satisfying (5.1) and (5.2). Then the formulae (5.9), (5.34), and (5.35) establish a bijective correspondence between the set of solutions $\tilde{T}$ of the Problem (*) and the set of parameters Γ, Δ, Λ, L, K, G, R, etc. satisfying the condition (5.30).

Let us investigate now the problem of simplicity of the parameters:

- some of the parameters as Δ, R, X, Y, D_k are arbitrary, except domain and range; these are the simplest ones.

- the parameters K, L are controlled by their rank and for these we can use the recent results obtained in [11], [19] (see also [9])

- the parameters B and C satisfy the condition

$$\varkappa^-\left(\begin{bmatrix} -I - G(J_T \oplus J_\Gamma)G^* & B \\ B^* & C \end{bmatrix}\right) = \varkappa_2 - \varkappa^-[I - \Gamma\Gamma^*] +$$

$$+ \,\mathbf{rank}([K \quad L]),$$

hence their explicit form will be obtained by iterating Lemma 3.2.

- the parameters Γ are controlled by their negative signatures of defect. For these we can iterate Theorem 5.3 on spaces of lower dimensions.

REFERENCES

1. **ARSENE, GR. ; CEAUŞESCU, Z. ; CONSTANTINESCU, T :** Schur analysis of some completion problems, Linear Algebra Appl. **109**(1988), 1-35.

2. **ARSENE, GR. ; CONSTANTINESCU, T ; GHEONDEA, A.:** Lifting of operators and prescribed numbers of negative squares, Michigan Math. J. **34**(1987), 201-216.

3. **AZIZOV, T.YA. ; IOKHVIDOV, I.S. :** Foundations of the theory of linear operators in spaces with indefinite metric (Russian), Nauka, Moskow 1986.

4. **BOGNAR, J :** Indefinite inner product spaces, Springer Verlag, Berlin 1974.

5. **CONSTANTINESCU, T. ; GHEONDEA, A. :** Minimal signature in lifting of operators. I, J. Operator Theory, **22**(1989) 345-367.

6. **CONSTANTINESCU,T. ; GHEONDEA,A. :** Minimal signature in lifting of operators. II, J. Functional Analysis (to appear).

7. **DAVIS, CH. :** J-unitary dilation of a general operator, Acta Sci. Math., **33**(1970), 75-86.

8. **DAVIS, CH. :** A factorization of an arbitrary m x n contractive operator matrix, in Toeplitz Centennial, Operator Theory : Advances and Applications vol. 4, Birkhäuser Verlag, Basel 1982.

9. **DAVIS, CH. :** Completing a matrix so as to minimize the rank, in Topics in operator theory and interpolation, Operator Theory : Advances and Applications vol. 29, p. 87-95, Birkhauser Verlag, Basel-Boston-Berlin 1988.

10. **DRITSCHEL, M.A. :** A lifting theorem for bicontractions in Krein spaces, J. Functional Analysis (to appear).

11. **GOHBERG,I. ; KAASHOEK, M.A. :** Minimal representations of semi-separable kernels and systems with separable boundary conditions, J. Math. Anal. Appl. **124**(1987), 436-458.

12. **IOKHVIDOV, I.S. ; KREIN, M.G. :** Spectral theory of operators in spaces with an indefinite metric I. (Russian), Trudy Moskov. Mat. Obshch., **5**(1956), 367-432.

13. **IOKHVIDOV, I.S. ; KREIN, M.G. ; LANGER, H. :** Introduction to the spectral theory of operators in spaces with an indefinite metric, Akademie - Verlag, Berlin 1982.

14. **KREIN, M.G. ; SHMULYAN, YU.L. :** On fractional-linear transformations with operator coefficients (Russian), Mat. Issled., **2**(1967), 64-96.

15. **LANCASTER, P. ; ROZSA, P. :** On the matrix equation $AX + X^*A^* = C$, SIAM J. Alg. Disc. Meth., **4**(1983), 432-436.

16. **SHMULYAN, YU.L. :** An operator Hellinger integral (Russian), Mat. Sb., **91**(1959), 381-430.

17 **SHMULYAN, YU.L. ; YANOVSKAYA, R.N. :** On matrices whose entries are contractions, Izv. Vysch. Ucheb. Zaved. Matematica **7**(1981), 201-212.

18. **SZ.-NAGY, B. ; FOIAŞ, C. :** Harmonic analysis of operators on Hilbert space, North Holland, Amsterdam 1970.

19. **WOERDERMANN, H.J. :** The lower order of lower triangular operators and minimal rank extensions, Integral Equations Operator Theory **10**(1987), 859-879.

Department of Mathematics, INCREST
Bd. Pacii 220, 79622 Bucharest, Romania

Operator Theory:
Advances and Applications, Vol. 48

LOCAL TOPOLOGY AND A SPECTRAL THEOREM

THOMAS JECH[1]

1. Introduction.

The concepts of continuity and convergence pervade the study of functional analysis, and are based on the precise formulation of the (vague) concept of *closeness*. Traditionally, the concept of closeness has been defined in terms of topology. It is well known, however, that not every version of convergence can be expressed in the topological language.

In this note we present an approach somewhat more general than topological spaces, and illustrate this approach on three examples: the space of continuous functions, the space of measurable functions and the commutative von Neumann algebra. For the latter example we state a version of the Spectral Theorem.

The theory presented here was influenced by Takeuti's Boolean valued analysis [**2**] and is loosely related to our Boolean linear spaces [**1**]. I wish to thank Professor Louis de Branges for his invitation to submit my contribution to the volume honoring Ernst Hellinger.

2. An example.

Let E be the space of all (equivalence classes of) real-valued, Lebesgue measurable functions on $\mathbf{R}$, with the equivalence relation "$f(x) = g(x)$ a.e."

Let us consider the convergence

$$\lim_n f_n(x) = f(x) \text{ almost everywhere.}$$

[1]Supported by NSF Grant DMS-8614447. I wish to thank the All Souls College in Oxford for its hospitality while I was a Visiting Fellow during the Michaelmas term 1988 when this work was done.

It is known that this notion of convergence is not determined by any topology on E, because there exists a sequence $\{f_n\}$ with the property that every subsequence has a subsequence converging to 0 yet the limit $\lim_n f_n$ does not exist.

We shall show how the concept of almost everywhere convergence can be defined from a partially ordered system of topologies, in fact from a system of seminorms. This will motivate a more general concept of *local topology*. Let $\mathcal{P}$ be the collection of all measurable subsets of $\mathbf{R}$, of positive measure. For each $P \in \mathcal{P}$ and every $f \in E$, let

$$\|f\|_P = \text{ess.}\sup_{x \in P} |f(x)|.$$

For each $P \in \mathcal{P}$, $\|\ \|_P$ is a seminorm (with values including ∞), and defines a topology of uniform convergence on E, namely $\lim_n f_n(x) = f(x)$ uniformly a.e. on P. The system $\{\|\ \|_P : P \in \mathcal{P}\}$ determines convergence a.e. as follows:

PROPOSITION. *A sequence $\{f_n\}$ converges to f a.e. if and only if*

$$\forall P\, \forall \varepsilon > 0\, \exists Q \subseteq P\, \exists m\, \forall n \geq m \|f_n - f\|_Q < \varepsilon.$$

PROOF: Assume first that the condition fails, and let P be a set of positive measure and $\varepsilon > 0$ be such that for all $Q \subseteq P$ of positive measure, $\|f_n - f\|_Q \geq \varepsilon$ for infinitely many n. It follows that for almost all $x \in P, |f_n(x) - f(x)| \geq \varepsilon/2$ for infinitely many n, and therefore $\lim_n f_n(x) = f(x)$ fails almost everywhere on P.

Conversely, assume that the condition holds, and let $k \geq 1$ be an interger. For every $P \in \mathcal{P}$ there is a $Q \subseteq P$ such that for eventually all $n, \|f_n - f\|_Q < \frac{1}{k}$. It follows that for almost all $x, |f_n(x) - f(x)| < \frac{1}{k}$ holds for eventually all n; let D_k be the set of all such x. Consequently, $\lim_n f_n(x) = f(x)$ holds for every $x \in \bigcap_k D_k$. □

In this example, we employ a class of topologies that forms a partially ordered set with respect to fineness. The general framework introduced in the next section involves a partially ordered system of "approximations" to a topology, which we call a *local topology.*

3. Local topology.

We consider a "space" E (a set of "points") and define a "local topology" on E. The "localization" is represented by a partially ordered set $(P, \leq)$. The idea is to define, for a point a, a set of points X and each $p \in P$, the concept "a is p-close to X", in a similar way as "a is close to X" is defined in terms of a topology on E, with the intention that if a is p-close to X and if $q \leq p$, then a is q-close to X. We do this by introducing a p-base of a local topology, for each $p \in P$. But first it is necessary to consider "local equality", namely relations *a is p-equal to b*, a system of equivalence relations on E.

DEFINITION: For each $p \in P$, $=_p$ is an equivalence relation on E, and

(E1) if $a =_p b$ and $q \leq p$ then $a =_q b$

(E2) if $a \neq b$ then for some $p \in P$, $a \neq_p b$

(E3) if $a \neq_p b$ then there is some $q \leq p$ such that $a \neq_r b$ for all $r \leq q$

EXAMPLE. $E = \mathbf{C}[0,1]$, *the space of all continuous functions on* $[0,1]$.

Let P be the set of all nonempty open intervals $p \subset [0,1]$ and let

$$p \leq q \quad \text{if} \quad p \subseteq q.$$

Define

$$f =_p g \quad \text{if} \quad f(x) = g(x) \text{ for all } x \in p.$$

Conditions (E1)–(E3) are satisfied: For example, for (E3) assume that $f(x) \neq g(x)$ for some $x \in p$. Then there is a $q \subseteq p$ such that $f(x) \neq g(x)$ for all $x \in q$, and so $f \neq_r g$ for all $r \subseteq q$.

EXAMPLE.

Let P be the set of all measurable subsets of $\mathbf{R}$, of positive measure, and again let

$$p \leq q \quad \text{if} \quad p \subseteq q.$$

Define

$$f =_p g \quad \text{if} \quad f(x) = g(x) \text{ a.e. on } p.$$

Again, the equivalence relations $=_p$ satisfy the conditions (E1)–(E3).

We shall now introduce *local topology* on E. For a set $X \subseteq E$ and $p \in P$, let

$$X^p = \{z : z =_p x \text{ for some } x \in X\}.$$

DEFINITION: For each $p \in P$, the *p-base for a local topology on* E is a set $\mathcal{B}_p$ of nonempty subsets of E, such that the following conditions are satisfied, for all p and q in P:

(LT1) If $U \in \mathcal{B}_p$, then $U^p = U$.

(LT2) If $U \in \mathcal{B}_p$ and if $q \leq p$, then $U^q \in \mathcal{B}_q$.

(LT3) If $U \in \mathcal{B}_p, V \in \mathcal{B}_p$, and if $a \in U \cap V$, then there exists a $q \leq p$ and some $W \in \mathcal{B}_q$ such that $a \in W$ and $W \subseteq U^q \cap V^q$.

(LT4) If $a \neq_p b$ then there exists a $q \leq p$ and some U and V in $\mathcal{B}_q$ such that $a \in U$ and $b \in V$ and for every $r \leq q$ and for all $x \in U$ and all $y \in V$, $x \neq_r y$.

EXAMPLE.

Back to our example $E = \mathbf{C}[0,1]$: For every $f \in E, p \in P$ and $\varepsilon > 0$, let $U_p(f,\varepsilon)$ be as follows:

$$U_p(f,\varepsilon) = \{g : \sup_{x\in p} |f(x) - g(x)| < \varepsilon\}$$

and let

$$\mathcal{B}_p = \{U_p(f,\varepsilon) : f \in E, \varepsilon > 0\}.$$

We claim that $\{\mathcal{B}_p : p \in P\}$ is a local topology.

(LT1) is clearly satisfied as $(U_p(f,\varepsilon))^p = U_p(f,\varepsilon)$.

(LT2): We shall show that if $q \subseteq p$ then $(U_p(f,\varepsilon))^q = U_q(f,\varepsilon)$. If $g \in U_p(f,\varepsilon)$ and $h = g$ on q then $h \in U_q(f,\varepsilon)$. Conversely, if $h \in U_q(f,\varepsilon)$, then there exists a continuous function g such that $g = h$ on q and $\sup_{x\in q} |f(x) - g(x)| < \varepsilon$.

(LT3): Let $g \in U_p(f_1,\varepsilon_1) \cap U_p(f_2,\varepsilon_2)$. If we let, for $i = 1,2$, $\delta_i = \varepsilon_i - \sup_p |g(x) - f_i(x)|$, and $\delta = \min\{\delta_1,\delta_2\}$, then $U_p(g,\delta) \subseteq U_p(f_1,\varepsilon_1) \cap U_p(f_2,\varepsilon_2)$.

(LT4): Let p, f and g be such that $f(x) \neq g(x)$ for some $x \in p$. There exists $q \subseteq p$ and an $\varepsilon > 0$ such that $|f(x) - g(x)| \geq \varepsilon$ on q. If we let $U = U_p(f,\varepsilon/2)$ and $V = U_p(g,\varepsilon/2)$, then U and V satisfy the requirement, because for all $h \in U$ and $k \in V$, $h(x) \neq k(x)$ for all $x \in q$. □

REMARKS:

1. When P is trivial, i.e. has a single point, then local topology is just a topology on E.

2. In the example $E = \mathbf{C}[0,1]$ above, each $\mathcal{B}_p$ is a topology base. In general, this is not necessarily the case. The systems $\mathcal{B}_p$, together with the partial ordering of P, approximate a topology; the "limit" of the system is a local topology.

4. Continuity and convergence.

Let $\{\mathcal{B}_p : p \in P\}$ be a local topology on E. For each $a \in E$ we let

$$\mathcal{B}_p(a) = \{U \in \mathcal{B}_p : a \in U\}$$

and call the elements of $\mathcal{B}_p(a)$ *p-neighborhoods* of a.

DEFINITION: Let F be a function from E into E and let $a \in E$. We say that F is *continuous* at a if for every p and every $U \in \mathcal{B}_p(F(a))$ there exists a $q \leq p$ and a $V \in \mathcal{B}_q(a)$ such that $F(V) \subseteq U^q$.

REMARKS:

1. If F is continuous and if $a_1 =_p a_2$ then $F(a_1) =_p F(a_2)$.

PROOF: If $F(a_1) \neq_p F(a_2)$ then by (LT4) there are $q \leq p$ and $U \in \mathcal{B}_q(a_1)$ such that $a_2 \notin U^r$ for any $r \leq q$. By continuity at a_1, let $r \leq q$ and $V \in \mathcal{B}_r(a_1)$ be such that $F(V) \subseteq U^r$. It follows that $a_2 \neq_r a_1$ and so $a_2 \neq_p a_1$. □

2. We can similarly define continuous functions from E_1 into E_2, from $\mathbf{R}$ into E, and continuous functions of more variables.

3. An alternative definition of continuity uses the concept of "a is p-close to X":

Let $X \subseteq E$, $a \in E$ and $p \in P$. We say that a is *p-close to X* if for every $q \subseteq p$ and every $U \in \mathcal{B}_q(a)$ there is an $r \leq q$ such that $U^r \cap X \neq \emptyset$.

Then we can define:

F is *continuous* at $a \in E$ if for all p and all X whenever a is p-close to X then $F(a)$ is p-close to $F(X)$.

We now turn to convergence:

DEFINITION: Let $\{a_n\}_{n=1}^{\infty}$ be a sequence in E, and let $a \in E$. We say that a_n *converges* to a,

$$\lim_{n\to\infty} a_n = a$$

if for every $p \in P$ and every p-neighborhood U of a there exists some $q \leq p$ such that eventually all a_n belong to U^q.

REMARK: A limit, if it exists, is unique.

PROOF: Assume that a_n converges both to a and to b. If $a \neq b$ then $a \neq_p b$ for some p, and by (LT4) there exists a $q \leq p$, $U \in \mathcal{B}_q(a)$ and $V \in \mathcal{B}_q(b)$ such that U^r and V^r are disjoint for all $r \leq q$. An appeal to the definition yields a contradiction. □

We shall illustrate the concept of convergence on our example:

EXAMPLE. *A sequence $\{f_n\}$ converges to f in $\mathbf{C}[0,1]$ if and only if the set $\{x \in [0,1] : \lim f_n(x) = f(x)\}$ is comeager.*

PROOF: Assume that $\lim f_n = f$. For every open interval p and every $k \geq 1$ there exists $q \subseteq p$ such that $f_n \in (U_p(f, 1/k))^q$ for eventually all n. We have $(U_p(f,1/k))^q = U_q(f, 1/k)$, and it follows that the set

$$\{x : \text{ for eventually all } n \ |f_n(x) - f(x)| < 1/k\}$$

contains an open dense set, and so $\{x : \lim f_n(x) = f(x)\}$ is comeager.

Conversely, assume that $\lim f_n \neq f$. Then there is an open set p and an $\varepsilon > 0$ such that for all $q \subseteq p$ and for every n, there exists some $m \geq n$ such that $f_m \notin U_q(f, 2\varepsilon)$. Hence for every n, the set D_n of all $x \in p$ for which there exists some $m \geq n$ such that $|f_m(x) - f(x)| > \varepsilon$ is dense in p (and open). So the set $A = \bigcap_n D_n$ is nonmeager and for all $x \in A, \lim f_n(x) \neq f(x)$. □

5. Locally normed linear spaces.

Many linear spaces such as the space of measurable functions described in Section 2 have a local topology given by a partially ordered system of seminorms. We call such space *locally normed.*

DEFINITION: Let E be a (real) vector space, and let $(P, \leq)$ be a partially ordered set. A system of seminorms $\{\| \ \|_p : p \in P\}$ makes E a *locally normed linear space* if it satisfies the following:

(LN1) $0 \leq \|a\|_p \leq \infty$
$\|a+b\|_p \leq \|a\|_p + \|b\|_p$
$\|\lambda a\|_p = |\lambda| \cdot \|a\|_p$

(LN2) If $p \leq q$ then $\|a\|_p \leq \|a\|_q$.

(LN3) If $a \neq 0$ then for some p, $\|a\|_p > 0$

(LN4) If $\|a\|_p > 0$ then there exists some $q \leq p$ and $\varepsilon > 0$ such that $\|a\|_r \geq \varepsilon$ for all $r \leq q$.

(LN5) If $p \leq q$ and $\|a\|_p < \varepsilon$ then there exists some a' such that $\|a'\|_q < \varepsilon$ and $\|a' - a\|_p = 0$.

We note that the space of measurable functions from Section 2 is locally normed, by the seminorms $\|f\|_p = \operatorname{ess\,sup}_p |f(x)|$. Similarly, the space $\mathbf{C}[0,1]$ from Section 3 is locally normed, by $\|f\|_p = \sup_p |f(x)|$ (where p is an open set). In the next section we show that every Archimedean Riesz space is locally normed.

Local norms define local topology bases as follows:

For $a, b \,\varepsilon\, E$ and $p \in P$, let

$$a =_p b \quad \text{if} \quad \|a-b\|_p = 0.$$

Then $=_p$ is an equivalence relation on E and satisfies (E1), (E2) and (E3) (the proof of (E3) uses (LN4)).

Let

$$U_p(a, \varepsilon) = \{b : \|a-b\|_p < \varepsilon\}$$

and let

$$\mathcal{B}_p = \{U_p(a,\varepsilon) : a \in E, \varepsilon > 0\}.$$

We shall verify that the $\mathcal{B}_p$ satisfy (LT1)–(LT4).

(LT1): If $b \in U_p(a, \varepsilon)$ and $\|c = b\|_p = 0$ then $c \in U_p(a,\varepsilon)$, and so $(U_p(a,\varepsilon))^p = U_p(a,\varepsilon)$.

(LT2): We claim that if $p \leq q$ then $(U_q(a,\varepsilon))^p = U_p(a,\varepsilon)$. First, if $b \in (U_q(a,\varepsilon))^p$ then for some c, $\|b-c\|_p = 0$ and $\|c-a\|_q < \varepsilon$. Therefore $\|b-a\|_p \leq \|b-c\|_p + \|c-a\|_p < \varepsilon$ by (LN1) and (LN2).

Conversely, let $b \in U_p(a,\varepsilon)$, i.e. $\|b-a\|_p < \varepsilon$. By (LN5) there exists some c such that $\|c-a\|_q < \varepsilon$ and $\|c-b\|_p = 0$, and hence $b \in (U_q(a,\varepsilon))^p$.

(LT3): Let $a \in U \cap V$ where $U = U_p(b,\varepsilon_1)$ and $V = U_p(c,\varepsilon_2)$. If we let $\delta > 0$ be such that $\|a-b\|_p + \delta \leq \varepsilon_1$, and $\|a-c\|_p + \delta \leq \varepsilon_2$, then $U_p(a,\delta) \subseteq U \cap V$.

(LT4): Let a, b be such that $\|a-b\|_p > 0$. By (LN4) get a $q \leq p$ and $\varepsilon > 0$ such that $\|a-b\|_r \geq 2\varepsilon$ for all $r \leq q$, and let $U = U_q(a,\varepsilon)$ and $V = U_q(b,\varepsilon)$. Then for all $x \in U, y \in V$ and all $r \leq q$ we have $\|x-y\|_r > 0$. □

It is easy to see that for locally normed linear spaces, the definitions of continuity and convergence from Section 4 take the following form:

PROPOSITION. *A function $F : E \to E$ is continuous at a iff*

$$\forall p \in P \;\; \forall \varepsilon > 0 \;\; \exists q \leq p \; \exists \delta > 0 \; \forall x (\|x-a\|_q < \delta \Rightarrow \|F(x) - F(a)\|_q < \varepsilon).$$

A sequence $\{a_n\}$ converges to a iff

$$\forall p \in P \;\; \forall \varepsilon > 0 \;\; \exists q \leq p \; (\textit{for eventually all } n, \|a_n - a\|_q < \varepsilon).$$

6. Archimedean Riesz spaces.

We shall show that every Archimedean Riesz space (vector lattice) is locally normed. To simplify the argument, we assume that the space has a unit, that is an element 1 with the property that $1 \wedge a > 0$ for all $a > 0$. (A modification of the argument works in the general case.)

Thus let E be an Archimedean Riesz space with unit 1. For P take the positive cone of E, the set $\{p \in E : p > 0\}$, partially ordered by the lattice order. For $a \in E$ and $p \in P$ let

$$\|a\|_p = \inf\{\lambda > 0 : p \perp (|a| - \lambda \cdot 1)^+\}$$

(using the convention that $\inf \emptyset = \infty$). It is rather clear that $\| \;\; \|_p$ are seminorms satisfying (LN1) and (LN2). We shall verify (LN3)–(LN5).

(LN3): Here we use that E is Archimedean. If $a > 0$ we claim that $\|a\|_a > 0$. Since E is Archimedean, there exists some $\lambda > 0$ such that $a \not\leq \lambda \cdot 1$. Hence $a \wedge (a - \lambda \cdot 1)^+ > 0$, and so $\lambda \leq \|a\|_a$.

(LN4): Let $\|a\|_p > 0$, and assume, without loss of generality, that $a > 0$. Let $\varepsilon = \frac{1}{2}\|a\|_p$, and let $q = p \wedge (a - \varepsilon \cdot 1)^+$. Since $\varepsilon < \|a\|_p$, we have $q \neq 0$. Now if $r \leq q$ $(r \in P)$, then $r \wedge (a - \varepsilon \cdot 1)^+ = r > 0$, and so $\varepsilon \leq \|a\|_r$.

(LN5): Let $\|a\|_p < \varepsilon$. First consider the case when $a > 0$. Let $\lambda = \|a\|_p$, and let $a' = a \wedge \lambda \cdot 1$. It is clear that $\|a'\|_q \leq \lambda < \varepsilon$ for all q, and since $p \wedge (a - a') = 0$, we have $\|a - a'\|_p = 0$.

In the general case, let again $\lambda = \|a\|_p$; then $a' = \lambda \cdot 1 \wedge a \vee -\lambda \cdot 1$ will do. □

7. Spectral theorem.

We shall now apply local topology to commutative operator algebras. Let $\mathfrak{A}$ be either a commutative von Neumann algebra, or the algebra of all (possibly unbounded) normal operators associated with a commutative von Neumann algebra. Let P be the set of all nonzero projection operators in $\mathfrak{A}$. For a, b in $\mathfrak{A}$ and $p \in P$ we let

$$a =_p b \quad \text{if} \quad pa = pb.$$

For $p \in P$, $a \in \mathfrak{A}$ and $\varepsilon > 0$ we let

$$b \in U_p(a, \varepsilon) \quad \text{if} \quad \forall q \leq p \, |qa - qb| < \varepsilon q.$$

It is not difficult to verify that the local neighborhood systems $\{U_p(a, \varepsilon) : \varepsilon > 0\}$ form a local topology on $\mathfrak{A}$.

THEOREM. *For every continuous function $F\colon \mathbf{R} \to \mathbf{R}$ there exists a unique continuous $\hat{F}\colon \mathfrak{A} \to \mathfrak{A}$ such that $\hat{F}(\lambda I) = F(\lambda)I$ for all $\lambda \in \mathbf{R}$.*

The existence of an $\hat{F}$ extending F is of course well known. Our claim is that $\hat{F}$ is continuous, and that such a continuous extension is unique.

Both the existence and the uniqueness of a continuous extension is proved, in a more general setting, in [**1**, Theorem 10.27]. As the general result uses a somewhat different definition of continuity, I shall sketch the proof that for the local topology on $\mathfrak{A}$, the present definition of continuity is equivalent to the definition used in [**1**], namely the definition using nets.

LEMMA. *A function $F\colon \mathfrak{A} \to \mathfrak{A}$ is continuous if and only if for every net $\{a_n\}_n$, if a_n converges to a then $F(a_n)$ converges to $F(a)$.*

PROOF: Assume that F is continuous, and let $\{a_n\}_n$ be a net converging to a. Let $U \in \mathcal{B}_p(F(a))$. First, by continuity, there is some $r \leq p$ and a $V \in \mathcal{B}_r(a)$ such that $F(V) \subseteq U^r$. Since a_n converges to a, there is a $q \leq r$ such that $a_n \in V^q$ for eventually all n. Since $x =_q y$ implies $F(x) =_q F(y)$, we have $F(V^q) \subseteq (U^r)^q = U^q$, proving that $F(a_n)$ converges to $F(a)$.

Conversely, assume that F is not continuous at a. There exists a p and some $U \in \mathcal{B}_p$ such that

$$\forall q \leq p \; \forall V \in \mathcal{B}_q(a) \; \exists a \in V \text{ such that } F(a) \notin U^q.$$

We claim that there is a directed set D and for each $d \in D$ some $a_d \in \mathfrak{A}$ such that $\{a_d\}_{d \in D}$ converges to a while $F(a_d)$ does not converge to $F(a)$. The set D consists of functions $d = \{(q, V_q) : q \in A_d\}$ where A_d is a maximal antichain in P, and for each q, $V_q \in \mathcal{B}_q(a)$. To get a bigger d in D, refine the partition and take $V_r \subseteq V_q^r$ for $r \le q$. To get a_d, let a_q be, for each $q \in A_d$, an element of V_q such that $F(a_q) \notin U^q$, and let a_d be the element of $\mathfrak{A}$ defined by

$$qa_d = qa_q \quad \text{for all } q \in A_d.$$

That the net $\{a_d\}_{d\in D}$ converges to a while $\{F(a_d)\}_d$ does not converge to $F(a)$ is a routine verification. □

We conclude with some remarks about continuity of functions $F\colon \mathfrak{A} \to \mathfrak{A}$. First, it is not difficult to see that F is continuous at $a \in \mathfrak{A}$ if and only if

$$\forall\varepsilon > 0\ \forall p\ \exists q \subseteq p\ \exists\delta > 0\ \forall x\,(\text{if } |qx - qa| \le \delta I \text{ then } |qF(x) - qF(a)| \le \varepsilon I).$$

Somewhat less immediate is the following equivalence:

DEFINITION: F is *uniform* if $\forall P\,\forall x\,\forall y(px = py$ implies $pF(x) = pF(y))$.
F is *norm-continuous* at a if $\forall\varepsilon\,\exists\delta\,\forall x(|x - a| \le \delta$ implies $|F(x) - F(a))| \le \varepsilon)$.

By Remark 1 in Section 3 every continuous function is uniform.

THEOREM. *F is continuous if and only if there is a maximal antichain A in P and a collection $\{F_p\colon p \in A\}$ of uniform norm-continuous functions such that $F = \sum_{p\in A} F_p$.*

We omit the proof and refer the reader to [**1**, Section 7] for similar arguments (P is weakly distributive).

As a final comment we make the following observation. Let p and $q = I - p$ be nonzero projections. There exists a continuous function $F\colon \mathfrak{A} \to \mathfrak{A}$ such that $F(p) = q$ and $F(q) = p$, namely the function $F(x) = I - x$. On the other hand, let p_1, p_2 and p_3 be projections such that $p_i \cdot p_j = \delta_{ij} I$ and $p_1 + p_2 + p_3 = I$. We claim that there is no uniform function $F\colon \mathfrak{A} \to \mathfrak{A}$ such that $F(p_1) = p_2$, $F(p_2) = p_3$ and $F(p_3) = p_1$.

If such an F existed, we would have (using $p_i =_{p_j} \delta_{ij} I$)

$$p_2 = F(p_1) = p_1 F(p_1) + p_2 F(p_1) + p_3 F(p_1) = p_1 F(I) + p_2(F(0)) + p_3 F(0)$$

and so

$$F(0) =_{p_3} 0,$$

while

$$p_3 = F(p_2) = p_1 F(0) + p_2 F(I) + p_3 F(0)$$

and so

$$F(0) =_{p_3} I,$$

getting a contradiction.

References

1. T. Jech, *Boolean-linear spaces*, Advances in Mathematics **81** (1990), 117–197.
2. G. Takeuti, "Two applications of logic to mathematics," Princeton University Press, Princeton, NJ, 1978.

Department of Mathematics, The Pennsylvania State University, 215 McAllister Building, University Park, PA 16802, U.S.A.

Operator Theory:
Advances and Applications, Vol. 48

Isometries from the Laguerre Shift

THOMAS L. KRIETE III*

1. Introduction

The Laguerre shift is the operator L defined on $L^2(0,\infty)$ by

$$(Lf)(x) = f(x) - \int_0^x e^{\frac{t-x}{2}} f(t)\, dt.$$

It was discovered by von Neumann **[vN]** that L is actually a simple unilateral shift which acts by shifting the orthonormal basis $\{\Phi_n\}_{n=0}^{\infty}$ for $L^2(0,\infty)$ consisting of the Laguerre functions. Here, $\Phi_n(x) = e^{-x/2}L_n(x)$, where L_n is the nth Laguerre polynomial of order zero. For a modern point of view, see **[R-R, pp. 17-18]**. The operator L is thus unitarily equivalent to the canonical shift operator S on the Hardy space H^2 of the unit circle, given by $(Sf)(e^{i\theta}) = e^{i\theta}f(e^{i\theta})$. The corresponding unitary equivalence $e^{in\theta} \to \Phi_n(x)$, $n \geq 0$, can be realized (after accounting for a change of variable from the real line to the unit circle) as the classical Fourier transform.

In this note we consider operators of the form LU where U is a unitary multiplication operator on $L^2(0,\infty)$, $(Uf)(x) = u(x)f(x)$, with u a measurable unimodular function on $(0,\infty)$. Clearly, any such operator LU is an isometry and thus has a Wold decomposition **[H, R-R]**. The shift part of this decomposition is clearly of multiplicity one again, so the only question is: What is the unitary part of LU? Recall that U has a unique decomposition $U = U_{ac} \oplus U_{sing}$, where U_{ac} and U_{sing} are, respectively, absolutely continuous and singular unitary operators with respect to Lebesgue measure $d\theta$ on the unit circle. We prove the following:

---

* This research was supported in part by the National Science Foundation.

Theorem. *LU is unitarily equivalent to $S \oplus U_{ac}$.*

The proof which I will give is somewhat indirect; it involves computing the Hellinger-Hahn spectral invariants of the unitary part of LU and observing that they match up with those of U_{ac}.

It is interesting to compare the above result with a theorem of Joseph A. Ball. Ball considered the compression T of LU to $L^2(0,b)$ for finite $b > 0$. Since T is a contraction, it can be decomposed uniquely as $T = T_0 \oplus V$, where V is unitary and T_0 is completely non-unitary. Let $h\,d\theta$ (where h is in $L^1(d\theta)$) and r be, respectively, a scalar spectral measure and the multiplicity function of the absolutely continuous part of $U|L^2(0,b)$. Further let Y denote the set of points $e^{i\theta}$ on the unit circle for which $h(e^{i\theta}) > 0$ and $r(e^{i\theta}) \geq 2$, and write χ_Y for the characteristic function of Y. In a special case of a very general development, Ball proved the following [**B, p. 66**]:

Theorem. *(J. Ball) The unitary operator V is absolutely continuous, having $\chi_Y\,d\theta$ as a scalar spectral measure and with multiplicity function n satisfying $n(e^{i\theta}) = r(e^{i\theta}) - 1 \quad d\theta$-a.e. on Y.*

An analogous result describing the self-adjoint part of $A + iJ$, where A is a self-adjoint multiplication operator and J is the Volterra integration operator, both acting on $L^2(0,b)$, can be found in [**K2**].

One can ask why the unitary parts in Ball's result and the present theorem have multiplicity differing by one. An 'explanation' which I find satisfying is as follows. In Ball's case, the operator T_0 has the same spectrum as $U|L^2(0,b)$, and indeed could be thought of as *obtaining* its spectrum from $U|L^2(0,b)$, in the process absorbing the entire singular part of that operator. Absolutely continuous spectra being less digestible, T_0 absorbs only the largest possible multiplicity one piece (thus filling out its spectrum) and leaving the rest for V. By contrast, the completely non-unitary part of the uncompressed operator LU is a shift, and is thus already 'filled up,' spectrally speaking. The process of forming LU discards the singular spectrum (somewhat mysteriously) and the shift part absorbs nothing, thus leaving a complete copy of U_{ac} for the unitary part.

The de Branges-Rovnyak model and the Sz.-Nagy-Foias model form the essential machinery of Ball's theorem and its analogue in [**K2**], respectively. The proof which I offer here is elementary in that neither functional model appears; only the most basic properties of H^2 will be needed. The method is an adaptation of some ideas from [**K2**].

2. Multiplicity Theory

This section briefly reviews spectral multiplicity theory, both in general and how one computes it for a multiplication operator. We restrict attention to unitaries acting on separable Hilbert spaces.

Let us write D for the open unit disk $\{z : |z| < 1\}$ in the complex plane and ∂D for the unit circle. Consider a direct integral space

$$\mathcal{D} = \int_{\partial D} \oplus H_\lambda \, d\nu(\lambda),$$

where ν is a finite positive Borel measure on ∂D and $\{H_\lambda : \lambda \in \partial D\}$ is a Borel measurable field of separable Hilbert spaces. Then $\mathcal{D}$ is a separable Hilbert space with a canonical unitary operator M acting on it, defined by $(Mf)(\lambda) = \lambda f(\lambda)$ for f in $\mathcal{D}$. Without loss of generality, it can be assumed that dim $H_\lambda \geq 1$ $\nu-a.e.$ In this case ν is a *scalar spectral measure* for M, and the function n, defined by $n(\lambda) =$ dim H_λ, is the *multiplicity function.* Multiplicity theory tells us that, up to unitary equivalence, every separably acting unitary operator has the form of such an M. Furthermore, given two such operators M_1 and M_2, associated in the same way with measures ν_1 and ν_2 and having multiplicity functions n_1 and n_2, respectively, it transpires that M_1 is unitarily equivalent to M_2 exactly when ν_1 and ν_2 have the same null sets and $n_1 = n_2$, ν_1 (or ν_2)$-a.e.$ Thus, given a unitary operator W, it makes sense to refer to the multiplicity function and the class of scalar spectral measures for W; these objects determine W up to unitary equivalence. Following standard usage, we speak of "the" multiplicity function for W, understanding that it is only determined $\nu-a.e.$ This formulation of multiplicity theory can be found in [**D**].

Consider now a unitary operator W acting on a separable Hilbert space H and having spectral measure E :

$$W = \int_{\partial D} \lambda \, dE(\lambda).$$

It is easily seen that a finite positive Borel measure ν is a scalar spectral measure for W if and only if it has the same null sets as E. It is also possible to describe the multiplicity function in terms of E, as follows. For each pair x, y of vectors in H, consider the complex Borel measure $\nu_{x,y}$ on ∂D given by $\nu_{x,y}(C) = \langle E(C)x, y\rangle$. Fix a scalar spectral measure ν for W; then $\nu_{x,y}$ is absolutely continuous with respect to ν, and we may choose a representative $d\nu_{x,y}/d\nu$ of the Radon-Nikodym derivative. Further, let $\mathcal{S}$ be any countable spanning subset of H. The following algorithm for finding $n(\lambda)$ appears in [**K3**].

General Principle. *For ν-almost every λ, $n(\lambda)$ is equal to the supremum of those non-negative integers k for which there exist $x_1, x_2, \cdots, x_k$ in $\mathcal{S}$ such that the matrix*

$$\left[\frac{d\nu_{x_i,x_j}}{d\nu}(\lambda)\right]_{i,j=1}^{k}$$

is invertible.

It is easy to see that this matrix is *self-adjoint* for ν-almost every λ.

Recall that W is said to be *absolutely continuous* (respectively, *singular*) if its scalar spectral measures are absolutely continuous (respectively, singular) with respect to Lebesgue measure $d\theta$ on ∂D. Moreover, there is always a unique decomposition $H = H_{ac} \oplus H_{sing}$ of H into reducing subspaces for W with $W_{ac} = W|H_{ac}$ being absolutely continuous and $W_{sing} = W|H_{sing}$ being singular.

Let us turn to our operator U on $L^2(0,\infty)$. The spectral invariants of multiplication operators were worked out in [**A-K**] (see also [**A, K3**]), where the underlying measure space is taken to be finite. However, it is easy to convert U into an operator of this type. Consider the finite measure $m = e^{-x}dx$ on $(0,\infty)$. The unitary operator $\tilde{U} : f(x) \to u(x)f(x)$ on $L^2(m)$ is clearly unitarily equivalent to U. The spectrum of U is the essential range of the multiplier u; a scalar spectral measure ν for $\tilde{U}$, and thus for U, is given by $\nu(C) = m(u^{-1}(C))$, for C a Borel set in ∂D.

The multiplicity function n of $\tilde{U}$ and U can be described as follows. For λ in ∂D, let $I(\delta,\lambda)$ be the closed arc in ∂D of length δ and center λ. For λ in the essential range of u and S a Borel set in $(0,\infty)$, we let

$$D(S,\lambda) = \liminf_{\delta\to 0} \frac{m(S \cap u^{-1}(I(\delta,\lambda)))}{m(u^{-1}(I(\delta,\lambda)))} .$$

For such λ, the *essential preimage* of λ under u, denoted $u_e^{-1}(\lambda)$, is the set of those points x in $(0,\infty)$ such that $D(V,\lambda) > 0$ whenever V is an open set in $(0,\infty)$ containing x. For points λ outside the essential range of u, we take $u_e^{-1}(\lambda)$ to be the empty set. Let us write $\#u_e^{-1}(\lambda)$ for the number of points in $u_e^{-1}(\lambda)$ when finite, and put $\#u_e^{-1}(\lambda) = \infty$ otherwise. The main result of [**A-K**], applied to $\tilde{U}$, tells us that

$$n(\lambda) = \#u_e^{-1}(\lambda) \qquad \nu\text{-}a.e.$$

3. Proof of the Theorem

Let us write M and N for the shift subspace and unitary subspace, respectively, of LU. Thus, $L^2(0,\infty) = M \oplus N$, M and N reduce LU, $B = LU|M$ is unitarily equivalent to the H^2 shift S, and $W = LU|M$ is unitary. Our object is to show that the multiplicity theory of W matches that of U_{ac}.

We will need a particular unitary equivalence $\mathcal{F}$ between S and B. We think of H^2 both as a space of analytic functions in D and as a subspace of $L^2(d\theta/2\pi)$. The operator $\mathcal{F}$ just mentioned will map H^2 into $L^2(0,\infty)$, and is constructed from a family φ_x, $x \geq 0$, of bounded analytic functions on D, defined by

$$\varphi_x(z) = \exp\left\{-\frac{1}{2}\int_0^x \frac{u(t)+z}{u(t)-z}\,dt\right\}.$$

For z in D, let h_z be the function

$$h_z(x) = \frac{\overline{\varphi_x(z)}}{1-\bar{z}u(x)}, \qquad x > 0.$$

Note that $h_0(x) = \varphi_x(0) = e^{-x/2}$. Also, for z in D, we have $|\varphi_x(z)|^2 \leq e^{-ax}$, where $a = (1-|z|)/(1+|z|)$, and so each h_z lies in $L^2(0,\infty)$. It is easy to check that

$$\frac{d}{dx}\overline{\varphi_x(w)}\varphi_x(z) = (\bar{w}z - 1)h_w(x)\overline{h_z(x)}\ ,$$

and thus for $0 \leq t < \infty$ and z, w in D,

$$\int_t^\infty h_w(x)\overline{h_z(x)}\,dx = \frac{\overline{\varphi_t(w)}\varphi_t(z)}{1-\bar{w}z}\ . \tag{3.1}$$

On setting $t = 0$, we have

$$\int_0^\infty h_w(x)\overline{h_z(x)}\,dx = \frac{1}{1-\bar{w}z}\ . \tag{3.2}$$

The right side here is exactly $k_w(z)$, where k_w is the Szegö kernel function at w for H^2. It is well known that $\langle f, k_z\rangle = f(z)$ for any f in H^2, and in particular, $\langle k_w, k_z\rangle = k_w(z)$. Thus, (3.2) states precisely that for any z, w in D, $\langle h_w, h_z\rangle = \langle k_w, k_z\rangle$, the first inner product being taken in $L^2(0,\infty)$ and the second in H^2. Since the kernels k_w span H^2, there is a unique isometry $\mathcal{F} : H^2 \to L^2(0,\infty)$ with $\mathcal{F}\, k_w = h_w$, w in D. This isometry is a "generalized Fourier transform" of the general type studied by several authors **[Ah-C, K1, L, B]** in the 1970's.

Lemma 3.1. *The range of $\mathcal{F}$ is exactly M, and, moreover, $\mathcal{F}\, S = B\mathcal{F}$.*

Proof. Let us compute the action of $(LU)^* = U^*L^*$ on h_w. Note that L^* has the form

$$(L^*f)(x) = f(x) - \int_x^\infty e^{\frac{x-t}{2}} f(t)\, dt.$$

Now $h_0(t) = e^{-t/2}$, and so by Eqn. (3.1),

$$\int_x^\infty e^{\frac{x-t}{2}} h_w(t)\, dt = e^{x/2} \int_x^\infty h_w(t)\overline{h_0(t)}\, dt = e^{x/2}\overline{\varphi_x(w)\varphi_x(0)} = \overline{\varphi_x(w)}.$$

It follows that $(L^*h_w)(x) = h_w(x) - \overline{\varphi_x(w)} = \bar{w}u(x)h_w(x)$, and so $(LU)^*h_w = \bar{w}h_w$. Since $|w| < 1$, h_w must lie in the shift subspace M. The eigenvalues of $B^* = (LU)^*|M$ are of course simple, the eigenvectors necessarily corresponding to the kernel functions k_w in H^2, which are the eigenvectors of S^*. Therefore M is the span of $\{h_w : w \in D\}$, that is, $M = \mathcal{F}\, H^2$. Finally, for any w in D, $\mathcal{F}\, S^*k_w = \mathcal{F}(\bar{w}k_w) = \bar{w}h_w = B^*h_w = B^*\mathcal{F}\, k_w$, and so $\mathcal{F}\, S^* = B^*\mathcal{F}$, from which the last conclusion follows. ■

Note that $\mathcal{F}^* : L^2(0,\infty) \to H^2$ is computable: For f in $L^2(0,\infty)$,

$$(3.3)\quad (\mathcal{F}^*f)(z) = \langle \mathcal{F}^*f, k_z\rangle = \langle f, \mathcal{F}\, k_z\rangle = \langle f, h_z\rangle = \int_0^\infty f(x)\frac{\varphi_x(z)}{1 - z\overline{u(x)}}\, dx, \qquad |z| < 1.$$

Now for $t > 0$, let χ_t denote the characteristic function of (t,∞) and consider the unit vector w_t in $L^2(0,\infty)$ defined by

$$w_t(x) = e^{t/2}\chi_t(x)e^{-x/2}.$$

We decompose w_t orthogonally as $w_t = a_t + y_t$ with a_t in M and y_t in N. Denote the set of positive rational numbers by $\mathbf{Q}$; then $\{w_t : t \in \mathbf{Q}\}$ is clearly a countable spanning set in $L^2(0,\infty)$, whence the set

$$\mathcal{S} = \{y_t : t \in \mathbf{Q}\}$$

is countable and spans the unitary subspace N. We will use this set in our general principle from §2 to compute the multiplicity theory of W.

We need two computational lemmas.

Lemma 3.2. $\mathcal{F}^* a_t = \varphi_t, \qquad t > 0.$

Proof. Since $\mathcal{F}$ is isometric and a_t is the projection of w_t onto the range M of $\mathcal{F}$, it is enough to show that $\mathcal{F}^* w_t = \varphi_t$. But then we see from Eqns. (3.3) and (3.1) that

$$(\mathcal{F}^* w_t)(z) = \int_0^\infty w_t(x)\overline{h_z(x)}\,dx = e^{t/2}\int_t^\infty h_0(x)\overline{h_z(x)}\,dx = \varphi_t(z)$$

for $|z| < 1$. ■

The second lemma is a variant of an identity of Ball [**B, p. 47**]. The proof here, a straightforward calculation, is left for the interested reader (hint: first differentiate $e^{x/2}\varphi_x(z)$).

Lemma 3.3. *For* $t > 0$ *and* $|z| < 1$, $(I - \bar{z}LU)\chi_t h_z = \overline{\varphi_t(z)} w_t$.

Let us turn to the spectral theory of W. We write E for the spectral measure of W,

$$W = \int_{\partial D} \lambda \, dE(\lambda),$$

and define the measures $\nu_{s,t}$ by $\nu_{s,t}(C) = \langle E(C)y_s, y_t\rangle$, $s, t > 0$. For $|z| < 1$ and $t > 0$, we have an orthogonal decomposition

$$(I - \bar{z}LU)^{-1} w_t = (I - \bar{z}B)^{-1} a_t + (I - \bar{z}W)^{-1} y_t, \tag{3.4}$$

the first summand on the right lying in M and the second in N. We write P_z for the Poisson kernel at z :

$$P_z(e^{i\theta}) = \frac{1 - |z|^2}{|e^{i\theta} - z|^2} \; .$$

Then by the spectral theorem, Eqn. (3.4), Lemma 3.3 and Eqn. (3.1), in that order, we find

$$\begin{aligned}\int_{\partial D} P_z \, d\nu_{t,t} &= (1 - |z|^2)\|(I - \bar{z}W)^{-1} y_t\|^2 \leq (1 - |z|^2)\|(I - \bar{z}LU)^{-1} w_t\|^2 \\ &= \frac{1 - |z|^2}{|\varphi_t(z)|^2}\int_t^\infty |h_z(x)|^2\,dx = 1\end{aligned}$$

for $|z| < 1$ and $t > 0$. It follows that all of the measures $\nu_{t,t}$ are absolutely continuous with respect to $d\theta$. Since $\{y_t : t \in \mathbf{Q}\}$ spans N, *the operator* W *is absolutely continuous.*

The main remaining task is to find the multiplicity function n for W. The reader will see that the idea is quite simple, but is necessarily accompanied by some bookkeeping with sets of measure zero. Let ν be a scalar spectral measure for W. By the chain rule we have

$$\frac{d\nu_{s,t}}{d\theta} = \frac{d\nu_{s,t}}{d\nu}\frac{d\nu}{d\theta} \qquad d\theta\text{−}a.e.$$

It is then clear from the General Principle of §2 that for $d\nu$-almost every λ in ∂D, $n(\lambda)$ can be taken to be the supremum of those non-negative integers k for which there exist $t_1, t_2, ..., t_k$ in $\mathbf{Q}$ such that the matrix

$$\left[\frac{d\nu_{t_i,t_j}}{d\theta}(\lambda)\right]_{i,j=1}^{k} \tag{3.5}$$

is invertible; clearly we may assume that $t_1 < t_2 < \cdots < t_k$.

Our first job is then to compute $d\nu_{s,t}/d\theta$. From the spectral theorem and the orthogonal decomposition (3.4), we have

$$\begin{aligned} \int_{\partial D} P_z\, d\nu_{s,t} &= (1-|z|^2)\langle (I-\bar{z}W)^{-1}y_s, (I-\bar{z}W)^{-1}y_t\rangle \\ &= (1-|z|^2)\langle (I-\bar{z}LU)^{-1}w_s, (I-\bar{z}LU)^{-1}w_t\rangle \\ &\quad - (1-|z|^2)\langle (I-\bar{z}B)^{-1}a_s, (I-\bar{z}B)^{-1}a_t\rangle. \end{aligned} \tag{3.6}$$

If we assume that $s \geq t$, Lemma 3.3 and Eqn. (3.1) imply that the first term on the right is exactly

$$\frac{1-|z|^2}{\overline{\varphi_s(z)}\varphi_t(z)} \int_s^\infty |h_z(x)|^2\, dx = \frac{\varphi_s(z)}{\varphi_t(z)}\ .$$

To calculate the second term, recall that $\mathcal{F}^*$ is isometric on its range M and use the intertwining relation $\mathcal{F}^*B = S\mathcal{F}^*$ plus Lemma 3.2. The second term then equals

$$\begin{aligned} &(1-|z|^2)\langle \mathcal{F}^*(I-\bar{z}B)^{-1}a_s, \mathcal{F}^*(I-\bar{z}B)^{-1}a_t\rangle \\ &\qquad = (1-|z|^2)\langle (I-\bar{z}S)^{-1}\varphi_s, (I-\bar{z}S)^{-1}\varphi_t\rangle = \frac{1}{2\pi}\int_{\partial D} P_z\varphi_s\overline{\varphi_t}\, d\theta. \end{aligned}$$

On substituting this into Eqn. (3.6) and letting z tend to λ in ∂D non-tangentially, we arrive by Fatou's Theorem at the fundamental relation (which holds whenever $s \geq t$)

$$2\pi\frac{d\nu_{s,t}}{d\theta}(\lambda) = \frac{\varphi_s(\lambda)}{\varphi_t(\lambda)} - \varphi_s(\lambda)\overline{\varphi_t(\lambda)} = \varphi_s(\lambda)\overline{\varphi_t(\lambda)}Q(t,\lambda) \tag{3.7}$$

$d\theta-a.e.$, where

$$Q(t,\lambda)=\frac{1-|\varphi_t(\lambda)|^2}{|\varphi_t(\lambda)|^2}\ .$$

Putting it another way, we have shown that if $s \geq t$,

$$(3.8)\qquad \langle E(C)y_s,y_t\rangle=\frac{1}{2\pi}\int_C \varphi_s(e^{i\theta})\overline{\varphi_t(e^{i\theta})}Q(t,e^{i\theta})\,d\theta$$

for every Borel set C in ∂D.

The functions φ_t are in H^2, so by keeping λ out of an appropriate Lebesgue null set G, we can guarantee that $\varphi_t(\lambda)$ is non-zero for all t in $\mathbf{Q}$. We may also assume that (3.7) holds whenever s and t are in $\mathbf{Q}$, $s \geq t$, and λ is outside of G. Since $\{y_t : t \in \mathbf{Q}\}$ spans N, it is clear from the formula (3.8) that the set

$$\Lambda=\bigcup_{t\in\mathbf{Q}}\{\lambda\in\partial D : |\varphi_t(\lambda)|<1\}$$

is a carrier for any scalar spectral measure for W, and moreover, $E(C)=0$ exactly when $C\cap\Lambda$ has Lebesgue measure zero.

Let us consider how this relates to U_{ac}. Recall from §2 that a scalar spectral measure ρ for U is given by $\rho(C)=m(u^{-1}(C))$, where $m=e^{-x}dx$. Let $\rho=gd\theta+\alpha$ be the Lebesgue decomposition of ρ, so that g is in $L^1(d\theta)$ and α is singular with respect to $d\theta$. Clearly, $gd\theta$ is a scalar spectral measure for U_{ac}. We put $\Sigma=\{\lambda \text{ in } \partial D : 0<g(\lambda)<\infty\}$. This set carries the spectral measure F of U_{ac} and $F(C)=0$ exactly when $C\cap\Sigma$ has Lebesgue measure zero.

Lemma 3.4. *$\Lambda\backslash\Sigma$ and $\Sigma\backslash\Lambda$ have Lebesgue measure zero.*

Proof. For every positive number t, we define measures μ_t on ∂D by $\mu_t(C)=|(0,t)\cap u^{-1}(C)|$, where $|K|$ denotes the Lebesgue measure of the set K. By a change of variables we see that

$$|\varphi_t(z)|^2=\exp(-\int_{\partial D}P_z\,d\mu_t).$$

Consider the Lebesgue decomposition $\mu_t=b_td\theta+\gamma_t$, where b_t is in $L^1(d\theta)$ and γ_t is singular with respect to $d\theta$. By Fatou's theorem, we have $|\varphi_t|^2=\exp(-2\pi b_t)$ $d\theta-a.e.$ on ∂D, so, except for a Lebesgue null set, Λ can be characterized as the set of λ for which $b_t(\lambda)>0$ for some t in $\mathbf{Q}$.

Now we can clearly find a set J in ∂D of Lebesgue measure zero and which carries α and every γ_t with t in $\mathbf{Q}$. For $C\subset\partial D\backslash J$, we have

$$\int_C b_t\,d\theta=|(0,t)\cap u^{-1}(C)| \qquad \text{and} \qquad \int_C g\,d\theta=\int_{u^{-1}(C)}e^{-x}\,dx.$$

The lemma easily follows. ■

Remark. From what we have done so far, it is clear that the following are equivalent:

(i) LU is a shift.

(ii) $\mathcal{F}$ maps H^2 onto $L^2(0,\infty)$.

(iii) Each φ_t is an inner function.

(iv) The measure ρ is singular with respect to $d\theta$.

To prove our theorem, it suffices to argue that $n(\lambda) = \# u_e^{-1}(\lambda) \quad d\theta - a.e.$ on Λ. To this end, we return to the positive rationals $t_1 < t_2 < \cdots < t_k$ and the matrix (3.5). Let G be the Lebesgue null set defined previously; without loss of generality we may assume that the matrix (3.5) is self-adjoint for λ in $\Lambda \backslash G$. For such λ, invertibility of this matrix is clearly equivalent to invertibility of the self-adjoint matrix $A = [a_{ij}]_{i,j=1}^k$ with $a_{ij} = Q(t_i, \lambda)$, $i \le j$. From the definition of φ_t, we see that for $d\theta$-almost every λ in ∂D, $|\varphi_t(\lambda)|$ is non-increasing as t ranges through $\mathbf{Q}$; enlarge G if necessary to include all λ which are exceptions. Thus, for λ outside of G we have

$$0 \le Q(t_1, \lambda) \le Q(t_2, \lambda) \le \cdots \le Q(t_k, \lambda). \tag{3.9}$$

Now A is an "L-shaped" matrix and the analysis in [**Br-H-S, p. 131**] yields

$$\det\ A = Q(t_1, \lambda) \prod_{i=2}^{k} (Q(t_i, \lambda) - Q(t_{i-1}, \lambda)).$$

Thus, A is invertible exactly when each inequality in (3.9) is strict, that is, when $|\varphi_{t_1}(\lambda)| < 1$ and $|\varphi_{t_i}(\lambda)| < |\varphi_{t_{i-1}}(\lambda)|$ for $i = 2, 3, \ldots, k$.

Now recall the quantity $D(S, \lambda)$ which defines the essential preimage.

Lemma 3.5. *There is a Lebesgue null set $K \subset \Lambda$ such that for all λ in $\Lambda \backslash K$ and all positive rationals $s < t$, the following is true: $|\varphi_t(\lambda)| < |\varphi_s(\lambda)|$ if and only if $D((s,t), \lambda) > 0$, and furthermore, $|\varphi_t(\lambda)| < 1$ if and only if $D((0,t), \lambda) > 0$.*

Proof. The Lebesgue differentiation theorem allows us to evaluate $g(\lambda)\, d\theta$-almost everywhere as a limit of symmetric difference quotients. Lemma 3.4 implies that $g(\lambda) > 0 \quad d\theta - a.e.$ on Λ, and so

$$D((s,t), \lambda) = \frac{1}{g(\lambda)} \liminf_{\delta \to 0} \frac{m((s,t) \cap u^{-1}(I(\delta, \lambda)))}{2\delta} \qquad d\theta - a.e. \text{ on } \Lambda.$$

For any Borel set C, we have $|(s,t) \cap u^{-1}(C)| = \mu_t(C) - \mu_s(C)$, which implies, in conjunction with the definition of m, that

$$e^{-t}(\mu_t(C) - \mu_s(C)) \leq m((s,t) \cap u^{-1}(C)) \leq \mu_t(C) - \mu_s(C).$$

Moreover, $b_t(\lambda)$ (from the proof of Lemma 3.4) is $d\theta - a.e.$ equal to the limit, as $\delta \to 0$, of $\mu_t(I(\delta,\lambda))/2\delta$. Therefore $e^{-t}(b_t(\lambda) - b_s(\lambda)) \leq g(\lambda)D((s,t),\lambda) \leq b_t(\lambda) - b_s(\lambda)$ $\quad d\theta - a.e.$ on Λ. The first assertion of the lemma now follows since $|\varphi_t|^2 = \exp(-2\pi b_t)$. For the second assertion, set $s = 0$. ■

Now we enlarge the Lebesgue null set G to include the sets $\Lambda \backslash \Sigma$ and K from the last two lemmas. Lemma 3.5 implies that when λ lies in $\Lambda \backslash G$, the matrix A is invertible exactly when $D((0,t_1),\lambda) > 0$ and $D((t_{i-1},t_i),\lambda) > 0$ for $i = 2,3,...,k$.

To finish the proof, we can proceed exactly as in the two concluding paragraphs of **[K3]**. For completeness the argument is included here. Fix λ in $\Lambda \backslash G$ and suppose the essential preimage $u_e^{-1}(\lambda)$ contains at least k points. We can choose our positive rationals $t_1 < t_2 < \cdots < t_k$ so that each of the intervals $(0,t_1)$ and (t_{i-1},t_i), $i = 2,...,k$, contains at least one point of $u_e^{-1}(\lambda)$. Then $D((0,t_1),\lambda)$ and $D((t_{i-1},t_i),\lambda))$, $i = 2,...,k$, are all positive, and so the corresponding matrix A is invertible. It follows that $n(\lambda) \geq k$, whence $n(\lambda) \geq \#u_e^{-1}(\lambda)$.

For the reverse inequality, we first invoke Lemma 7.4 of **[K2]** or Lemma 4 of **[K3]**, which tells us that the set of points λ in Λ for which $u_e^{-1}(\lambda)$ contains a rational number must have Lebesgue measure zero. Without loss of generality, assume this set is contained in G. Suppose now that λ is in $\Lambda \backslash G$ and $\#u_e^{-1}(\lambda) = p < \infty$. If $k > p$ and $t_1 < t_2 < \cdots < t_k$ are in $\mathbf{Q}$, then at least one of the intervals $(0,t_1]$, $[t_1,t_2],...,[t_{k-1},t_k]$, call it I, does not intersect $u_e^{-1}(\lambda)$. An elementary compactness argument **[K3, Lemma 3]** then shows that $D(I,\lambda) = 0$, and so the corresponding matrix has det $A = 0$. By our general principle, we must have $n(\lambda) \leq p$ as desired.

We have shown that $n(\lambda) = \#u_e^{-1}(\lambda)$ $\quad d\theta - a.e.$ on Λ, and the theorem is proved.

References

[A] M. B. Abrahamse, *Multiplication operators* in *Hilbert Space Operators*, Springer Lecture Notes, Vol. 693, Springer-Verlag, New York, 1977, pp. 17-36.

[A-K] M. B. Abrahamse and T. L. Kriete, *The spectral multiplicity of a multiplication operator*, Indiana U. Math. Jour. **22** (1973), pp. 845-857.

[Ah-C] P. R. Ahern and D. N. Clark, *On functions orthogonal to invariant subspaces*, Acta Math., **124** (1970), pp. 191-204.

[B] J. A. Ball, *Factorization and model theory for contraction operators with unitary part*, Memoirs Amer. Math. Soc., Vol. 13, no. 198 (1978).

[Br-H-S] A. Brown, P. R. Halmos, A. L. Shields, *Cesàro operators*, Acta Sci. Math. (Szeged) **26** (1965), pp. 125-137.

[D] J. Dixmier, *Les algèbres d'opérateurs dans l'espace Hilbertien (Algébras de von Neumann)*, Cahiers scientifiques, fasc. 25, Gauthier-Villars, Paris, 1957.

[H] P. R. Halmos, *Shifts on Hilbert spaces*, J. reine angew. Math. **208** (1961), pp. 102-112.

[K1] T. L. Kriete, *Fourier transforms and chains of inner functions*, Duke Math. Jour., **40** (1973), pp. 131-143.

[K2] ———, *Canonical models and the self-adjoint parts of dissipative operators*, Jour. Func. Anal. **23** (1976), pp. 39-94.

[K3] ———, *An elementary approach to the multiplicity theory of multiplication operators*, Rocky Mtn. Jour. Math., **16** (1986), pp. 23-32.

[L] A. Lubin, *Isometries of *-invariant subspaces*, Trans. Amer. Math. Soc. **190** (1974), pp. 405-415.

[vN] J. von Neumann, *Zur Theorie der unbeschränkten Matrizen*, J. f. Math. **161** (1929), pp. 208-236; Collected Works, Vol. II, no. 3.

[R-R] M. Rosenblum and J. Rovnyak, *Hardy Classes and Operator Theory*, Oxford Univ. Press, New York, 1985.

Department of Mathematics
Mathematics–Astronomy Building
University of Virginia
Charlottesville, Virginia 22903–3199
U. S. A.

Operator Theory:
Advances and Applications, Vol. 48

A second order differential operator depending nonlinearly on the eigenvalue parameter

H. Langer, R. Mennicken and M. Möller

ABSTRACT. In this note we study the spectral properties of the differential equation (0.2-3) below containing the floating singularity $(u-\lambda)^{-1}$. With this equation a λ-linear eigenvalue problem is associated. If the functions u, q are real and satisfy some sign conditions, the corresponding operator is definitizable in a Krein space.

0. INTRODUCTION

We consider the differential equation

$$y'' + (\lambda + \frac{q}{u-\lambda})y = f \tag{0.1}$$

and the corresponding homogeneous equation

$$y'' + (\lambda + \frac{q}{u-\lambda})y = 0 \tag{0.2}$$

on the interval [0,1] with boundary conditions

$$y(0) = y(1) = 0 \tag{0.3}$$

in the Hilbert space $L^2(0,1)$. Here q and u are essentially bounded functions on $[0,1]$, which have to satisfy some additional conditions, and $\lambda \in \mathbb{C}$ is the eigenvalue parameter.

Equations of this type (with a so-called "floating singularity" $(u-\lambda)^{-1}$) were studied e.g. in [Ba]; by a simple transformation of the eigenvalue parameter also the problems considered in [S] can be given the form (0.2-3) with the additional property that the functions q, u are in fact constants. Differential equations with floating singularities arise also in connection with expansions in special functions (see, e.g., [MMS]).

The methods used here to study the eigenvalue problem (0.2-3) have much in common with those from the theory of operator polynomials of second degree in λ (see, e.g., [KL], [R]). Namely, with the original problem a λ-linear spectral problem of an operator acting in a larger space is associated. If, in particular the functions q, u are real (and satisfy some additional conditions), this larger space can be chosen to be a Krein space and the corresponding operator is selfadjoint and definitizable in this Krein space. Thus the theory of definitizable operators (see [L]) yields the existence of a spectral function, and the eigenfunctions or "generalized" eigenfunctions of the original problem (0.2-3) can be classified into those of positive, negative or neutral type. It follows that under some conditions on q and u, these eigenfunctions of positive type (or of negative type) span a subspace of $L^2(0,1)$ of at most finite codimension, see Section 2 below. In Section 1 the linearization of the problem (0.1-3) is done and its spectral properties are studied under more general assumptions on q and u.

The method used in this note also applies if the boundary conditions (0.3) are replaced by any other selfadjoint boundary conditions of the operator $\frac{d^2}{dx^2}$ on $[0,1]$, and also if $\frac{d^2}{dx^2}$ is replaced by a more general differential operator.

We would like to point out that the definitizable operators mentioned above can have critical points which are embedded in the continuous spectrum (which is necessarily of negative type) and have a finite index of positivity. The appearance of these singularities is excluded by the assumptions of [Ba]. These questions will be considered in a subsequent note.

1. Linearization of $T(\lambda)$

1.1. Let $q, u \in L^\infty(= L^\infty(0,1))$. In the space $L^2(= L^2(0,1))$ we consider the differential equations (0.1-2) with the boundary conditions (0.3). If the expression on the left hand side of the equations (0.1-2) is denoted by

$$T(\lambda)y := y'' + (\lambda + \frac{q}{u-\lambda})y,$$

then these equations become $T(\lambda)y = f$ and $T(\lambda)y = 0$, respectively. The expression $T(\lambda)$ makes sense for all values of the spectral parameter λ for which $q(u-\lambda)^{-1} \in L^\infty$, in particular for all λ such that $\operatorname{ess\,inf}|u-\lambda| > 0$. Then the operator $T(\lambda)$ is defined on the set $\mathcal{D}$ of all $y \in L^2$ which have an absolutely continuous derivative y' with $(y')' \in L^2$, and which satisfy the boundary conditions (0.3). Let D be the following operator:

$$Dy := -y'' \qquad (y \in \mathcal{D}).$$

It is selfadjoint in L^2 with spectrum

$$\sigma(D) = \{k^2\pi^2 : k = 1, 2, \ldots\}. \tag{1.1}$$

If in L^2 we introduce the operator of multiplication by the function u, also denoted by u, then ([W], p. 51)

$$\sigma(u) = \{\lambda \in \mathbb{C} : \text{ meas } \{\xi : |u(\xi) - \lambda| \leq \epsilon\} > 0 \text{ for all } \epsilon > 0\}$$

or, equivalently,

$$\rho(u) = \{\lambda \in \mathbb{C} : \text{ ess inf } |u - \lambda| > 0\}.$$

By $\stackrel{w}{\to}$ and $\stackrel{s}{\to}$ we denote the weak and strong convergence in L^2, respectively. The spectrum of u can also be characterized as follows.

LEMMA 1.1. *The point $\lambda_0 \in C$ belongs to $\sigma(u)$ if and only if there exists a sequence $(f_n) \subset L^2$ such that*

$$\|f_n\| = 1, f_n \stackrel{w}{\to} 0 \text{ and } (u - \lambda_0)f_n \stackrel{s}{\to} 0.$$

PROOF. The sufficiency is clear. In order to prove the necessity, denote

$$\Gamma_n := \{\xi : |u(\xi) - \lambda_0| \leq \frac{1}{n}\}.$$

Then $\text{meas}\,\Gamma_n > 0$ and the sequence (Γ_n) is nonincreasing. If $\lim_{n\to\infty}(\text{meas}\Gamma_n) = 0$ then λ_0 is not an eigenvalue of u and we choose

$$f_n(\xi) := \begin{cases} \dfrac{1}{(\text{meas}\Gamma_n)^{\frac{1}{2}}} & \xi \in \Gamma_n, \\ 0 & \text{otherwise,} \end{cases} \qquad n = 1, 2, \dots .$$

It follows that $\|f_n\| = 1$ and $\int_0^1 |u - \lambda_0|^2 |f_n|^2 d\xi \leq \frac{1}{n^2} \to 0$. Also, $f_n \stackrel{w}{\to} 0$ as otherwise a subsequence would weakly converge to some $f_0 \neq 0$ which would be an eigenvector to λ_0, but this is impossible. If $\lim_{n\to\infty}$ $(\text{meas}\,\Gamma_n) > 0$, then λ_0 is an eigenvalue of infinite multiplicity of u and we can choose an orthonormal sequence (f_n) of eigenvectors of u to the eigenvalue λ_0.

On $\rho(u)$ the equations (0.1), (0.2) can be transformed into a problem which contains λ linearly as follows. Setting

$$y_1 := y, \quad y_2 := -\frac{q}{u - \lambda} y,$$

the equation (0.2) is equivalent to

$$-y_1'' + y_2 = \lambda y_1, \quad q y_1 + u y_2 = \lambda y_2,$$

and with $\mathbf{y} = \begin{bmatrix} y_1 \\ y_2 \end{bmatrix}$ the problem (0.2), (0.3) becomes

$$\begin{bmatrix} D & 1 \\ q & u \end{bmatrix} \mathbf{y} - \lambda \mathbf{y} = 0. \tag{1.2}$$

Similarly (0.1), (0.3) are equivalent to

$$\begin{bmatrix} D & 1 \\ q & u \end{bmatrix} \mathbf{y} - \lambda \mathbf{y} = \begin{bmatrix} -f \\ 0 \end{bmatrix}. \tag{1.3}$$

In the following, by $\mathbf{L}$ we denote the operator in $L^2 \oplus L^2$, defined on $\mathfrak{D}(\mathbf{L}) = \mathcal{D} \oplus L^2$ by the matrix

$$\mathbf{L} = \begin{bmatrix} D & 1 \\ q & u \end{bmatrix}.$$

We mention, that this linearization $\mathbf{L}$ represents the operator function T only if $q(\xi) \neq 0$ in "sufficiently many" points ξ. Indeed, assume, e.g. that $q = 0$. Then T does not depend on u at all. On the other hand $\mathbf{L}$ is of the form

$$\mathbf{L} = \begin{bmatrix} D & 1 \\ 0 & u \end{bmatrix}$$

hence u is an essential component of $\mathbf{L}$.

Recall (see, e.g. [MM]) that $\lambda_0 \in \rho(u)$ is called an eigenvalue of the operator function T if there exists an element $y_0 \in \mathcal{D}, y_0 \neq 0$, such that $T(\lambda_0)y_0 = 0$. If, in this case, there are elements $y_1, y_2, \ldots, y_k$ in $\mathcal{D}$ such that

$$T(\lambda_0)y_0 = 0, \quad T(\lambda_0)y_1 + \frac{1}{1!}\dot{T}(\lambda_0)y_0 = 0, \ldots,$$
$$T(\lambda_0)y_k + \frac{1}{1!}\dot{T}(\lambda_0)y_{k-1} + \ldots + \frac{1}{k!}\overset{(k)}{T}(\lambda_0)y_0 = 0,$$

then $y_0, y_1, \ldots, y_k$ are said to form a Jordan chain corresponding to the eigenvalue λ_0. Here $\dot{T}(\lambda)$ etc. denotes the formal derivative of $T(\lambda)$ with respect to λ, that is

$$\dot{T}(\lambda) = I + \frac{q}{(u-\lambda)^2}, \quad \ddot{T} = \frac{2q}{(u-\lambda)^3} \qquad \text{etc.}$$

The set of all eigenvalues of T is denoted by $\sigma_p(T)$. If A is a linear operator in some Hilbert space $H, \sigma_p(A)$ denotes also the set of eigenvalues of A; the essential spectrum $\sigma_{ess}(A)$ is the set of all points of the spectrum $\sigma(A)$ which are not isolated normal eigenvalues of finite algebraic multiplicity (here an eigenvalue $\lambda_0 \in \sigma_p(A)$ is called normal if $H = H_0 \dot{+} \mathfrak{L}_{\lambda_0}(A)$, where $\mathfrak{L}_{\lambda_0}(A)$ is the algebraic eigenspace of A at λ_0 and $A - \lambda_0$ maps H_0 bijectively onto itself).

PROPOSITION 1.2. (1) *If $\lambda_0 \in \sigma_p(T) \cap \rho(u)$ and $y_0, y_1, \ldots, y_k$ is a Jordan chain of T at λ_0, then $\lambda_0 \in \sigma_p(\mathbf{L})$ and the vector functions*

$$\mathbf{y}_0 = \begin{bmatrix} y_0 \\ -\frac{q}{u-\lambda_0} y_0 \end{bmatrix}, \mathbf{y}_1 = \begin{bmatrix} y_1 \\ -\frac{q}{u-\lambda_0} y_1 - \frac{q}{(u-\lambda_0)^2} y_0 \end{bmatrix}, \ldots,$$
$$\mathbf{y}_k = \begin{bmatrix} y_k \\ -\frac{q}{u-\lambda_0} y_k - \frac{q}{(u-\lambda_0)^2} y_{k-1} - \ldots - \frac{q}{(u-\lambda_0)^{k+1}} y_0 \end{bmatrix} \tag{1.4}$$

form a Jordan chain of $\mathbf{L}$ *at* λ_0. *Conversely, if* $\lambda_0 \in \sigma_p(\mathbf{L}) \cap \rho(u)$ *with a corresponding Jordan chain* $\mathbf{y}_0, \mathbf{y}_1, \ldots, \mathbf{y}_k$, *then this Jordan chain is of the form* (1.4) *and the first components of these vector functions form a Jordan chain of* T *at* λ_0.

(2) *For* $\lambda \in \rho(u) \cap \rho(D)$ *we have* $\lambda \in \sigma_p(\mathbf{L})$ *if and only if* $1 \in \sigma_p((u-\lambda)^{-1}q(D-\lambda)^{-1})$.

(3) $\sigma(u) \subset \sigma_{ess}(\mathbf{L})$.

(4) *If* $\rho(u)$ *is connected, then* $\sigma_{ess}(\mathbf{L}) \subset \sigma(u)$.

PROOF. (1) The fact that $\sigma_p(T) \cap \rho(u) = \sigma_p(\mathbf{L}) \cap \rho(u)$ and the relation between corresponding eigenfunctions follow from the above transformation from (0.2),(0.3) into (1.2); the relations between the Jordan chains are easy to check.

(2) This assertion is proved by a straightforward calculation.

(3) First let λ be an isolated eigenvalue in $\sigma(u)$. Then the set

$$\Gamma_\lambda := \{\xi \in (0,1) : u(\xi) = \lambda\}$$

has a positive measure. By Egorov's theorem there is a compact subset M of Γ_λ with positive measure such that $q|_M$ is continuous. The elements in the range of $\mathbf{L} - \lambda$ are of the form

$$\begin{bmatrix} (D-\lambda)f_1 + f_2 \\ qf_1 + (u-\lambda)f_2 \end{bmatrix} \quad \text{with } f_1 \in \mathcal{D}, f_2 \in L^2.$$

Their second component, restricted to M, is $qf_1|_M$. The set $\{qf_1|_M : f_1 \in \mathcal{D}\}$ is a linear manifold of $L^2(M)$ consisting of continuous functions and thus has infinite codimension. This shows that the range of $\mathbf{L} - \lambda$ has infinite codimension which proves that $\lambda \in \sigma_{\text{ess}}(\mathbf{L})$.

An accumulation point of $\sigma(u)$ is also an accumulation point of $\sigma(u) \cap \rho(D)$ because $\sigma(D)$ is a discrete subset of $\mathbb{C}$. Thus we have proved that an accumulation point of $\sigma(u)$ belongs to $\sigma_{\text{ess}}(\mathbf{L})$ if we show that $\sigma(u) \cap \rho(D) \subset \sigma(\mathbf{L})$. To this end, choose $\lambda \in \sigma(u) \cap \rho(D)$, suppose that $\lambda \in \rho(\mathbf{L})$ and define

$$S(\lambda) := u - \lambda - q(D-\lambda)^{-1}.$$

Since

$$(\mathbf{L} - \lambda)\begin{bmatrix} y_1 \\ y_2 \end{bmatrix} = \begin{bmatrix} 0 \\ g \end{bmatrix}$$

implies that $S(\lambda)y_2 = g$, the assumption $\lambda \in \rho(\mathbf{L})$ yields that $S(\lambda)$ has a bounded inverse. On the other hand, with the sequence (f_n) from Lemma 1.1, the compactness of $q(D-\lambda)^{-1}$ implies that

$$S(\lambda)f_n \xrightarrow{s} 0 \quad (n \to \infty).$$

Hence $S(\lambda)$ is not boundedly invertible, a contradiction.

(4) The function $\lambda \mapsto (u-\lambda)^{-1}q(D-\lambda)^{-1}$ is finitely meromorphic on $\rho(u)$, its values being compact operators. For $\lambda_0 \in \rho(u) \cap \rho(D)$ with sufficiently large imaginary part we have

$$\|(u-\lambda_0)^{-1}q(D-\lambda_0)^{-1}\| \leq \frac{\text{ess sup}\,|q|}{\text{dist}(\sigma(u),\lambda_0)\cdot\text{dist}(\sigma(D),\lambda_0)} < 1\,. \tag{1.5}$$

Thus $(I-(u-\lambda_0)^{-1}q(D-\lambda_0)^{-1})^{-1}$ exists as a bounded everywhere defined operator, whence the operator function

$$R(\lambda) := \left(I - (u-\lambda)^{-1}q(D-\lambda)^{-1}\right)^{-1} \quad (\lambda \in \rho(u)) \tag{1.6}$$

is finitely meromorphic on $\rho(u)$, see [GS], Lemma 2.1. If the domain of holomorphy of R is denoted by $\mathfrak{D}_R$, it is easy to check that for $\lambda \in \mathfrak{D}_R \cap \rho(D)$ we have

$$(\mathbf{L}-\lambda)^{-1} = \begin{bmatrix} (D-\lambda)^{-1}R(\lambda) & -(D-\lambda)^{-1}R(\lambda)(u-\lambda)^{-1} \\ -R(\lambda)(u-\lambda)^{-1}q(D-\lambda)^{-1} & R(\lambda)(u-\lambda)^{-1} \end{bmatrix}.$$

It follows that the operator function $(\mathbf{L}-\lambda)^{-1}$ is finitely meromorphic on $\rho(u)$ which implies $\rho(u) \cap \sigma_{\text{ess}}(\mathbf{L}) = \emptyset$.

REMARK 1. If $\rho(u)$ is not connected, the statements (3) and (4) read as follows: Let Ω_0 be the union of all (bounded) components Ω of $\rho(u)$ for which $1 \in \sigma_p((u-\lambda)^{-1}q(D-\lambda)^{-1})$ for all $\lambda \in \Omega \cap \rho(D)$. Then

$$\sigma_{\text{ess}}(\mathbf{L}) = \Omega_0 \cup \sigma(u).$$

REMARK 2. If we observe (1.1), Proposition 1.2, (2) and the inequality (1.5), it follows that $\lambda \in \rho(\mathbf{L})$ if

$$\text{ess sup}\,|q| < \text{dist}(\lambda, \sigma(u)) \cdot \text{dist}(\lambda, (\pi^2, \infty));$$

in particular, $\sigma(\mathbf{L})$ is always bounded from the left.

If $\lambda \in \sigma(u)$ then $\lambda \in \sigma(\mathbf{L})$ by Proposition 1.2, (3). More information is given by the

LEMMA 1.3. *Let λ be an isolated eigenvalue of u and assume that $q(\xi) \neq 0$ a.e. on $\Gamma_\lambda = \{\xi \in (0,1) : u(\xi) = \lambda\}$. Then $\lambda \notin \sigma_p(\mathbf{L})$.*

PROOF. The set Γ_λ has positive measure because λ is an isolated eigenvalue of u. Let V be the set of those $\xi \in (0,1)$ for which there is an $\epsilon_\xi > 0$ such that

$$\text{meas}\,(\Gamma_\lambda \cap (\xi - \epsilon_\xi, \xi + \epsilon_\xi)) = 0.$$

From Lindelöf's covering theorem we infer that meas $(\Gamma_\lambda \cap V) = 0$. Thus, by changing the values of u on the set $\Gamma_\lambda \cap V$, we may assume that $\Gamma_\lambda \cap V = \emptyset$, i.e.

$$\text{meas}\,(\Gamma_\lambda \cap (\xi - \epsilon, \xi + \epsilon)) > 0$$

for all $\xi \in \Gamma_\lambda$ and $\epsilon > 0$.

Now consider an element $\mathbf{y} = \begin{bmatrix} y_1 \\ y_2 \end{bmatrix} \in \mathfrak{D}(\mathbf{L})$ such that

$$(D-\lambda)y_1 + y_2 = 0, \quad qy_1 + (u-\lambda)y_2 = 0. \tag{1.7}$$

The second equation implies $y_1(\xi) = 0$ a.e. on Γ_λ. Thus, for every $\xi_0 \in \Gamma_\lambda$ and every $\epsilon > 0$ there are infinitely many points ξ in $\Gamma_\lambda \cap (\xi_0 - \epsilon, \xi_0 + \epsilon)$ for which $y_1(\xi) = 0$. Hence $y_1(\xi_0) = 0$ and $y_1'(\xi_0) = 0$ as y_1 is differentiable. By the continuity of y_1 and y_1' we obtain that $y_1(\xi) = 0$ and $y_1'(\xi) = 0$ for all $\xi \in \overline{\Gamma}_\lambda$. Let I_1 be a component of $(0,1)\backslash\overline{\Gamma}_\lambda$. Since $\Gamma_\lambda \neq \emptyset$, at least one boundary point ξ_1 of I_1 belongs to $\overline{\Gamma}_\lambda$. Hence $y_1(\xi_1) = y_1'(\xi_1) = 0$, and on the interval I_1 the function y_1 satisfies the differential equation

$$y_1'' + (\lambda + \frac{q}{u - \lambda})y_1 = 0.$$

As the coefficients of this equation are bounded in I_1, the uniqueness theorem yields $y_1 = 0$ and, by (1.7), $y_2 = 0$ on I_1, whence the lemma is proved.

1.2. Let again $q, u \in L^\infty$. For simplicity we assume that $\rho(u)$ is connected. If $\lambda \in \rho(u)$ then there exists an $N(\lambda) < \infty$ such that $|u - \lambda|^{-1} \leq N(\lambda)$ a.e. on $[0,1]$. Therefore, if $\lambda \in \rho(u)$ we can consider the integral equations

$$\varphi(x;\lambda) = 1 - \int_0^x \Big(\lambda + \frac{q(\xi)}{u(\xi) - \lambda}\Big)\varphi(\xi;\lambda)(x - \xi)d\xi,$$

$$\psi(x;\lambda) = x - \int_0^x \Big(\lambda + \frac{q(\xi)}{u(\xi) - \lambda}\Big)\psi(\xi;\lambda)(x - \xi)d\xi.$$

They define functions $\varphi(\cdot;\lambda), \psi(\cdot;\lambda)$ with absolutely continuous derivatives:

$$\varphi'(x;\lambda) = -\int_0^x \Big(\lambda + \frac{q(\xi)}{u(\xi) - \lambda}\Big)\varphi(\xi;\lambda)d\xi,$$

$$\psi'(x;\lambda) = 1 - \int_0^x \Big(\lambda + \frac{q(\xi)}{u(\xi) - \lambda}\Big)\psi(\xi;\lambda)d\xi,$$

and $\varphi(\cdot;\lambda), \psi(\cdot;\lambda)$ satisfy the equation (0.2) and the initial conditions

$$\begin{aligned} \varphi(0;\lambda) &= 1, \quad \varphi'(0;\lambda) = 0, \\ \psi(0;\lambda) &= 0, \quad \psi'(0;\lambda) = 1. \end{aligned} \tag{1.8}$$

Moreover, if $x \in [0,1]$ is fixed, the functions $\varphi(x;\cdot), \psi(x;\cdot), \varphi'(x;\cdot), \psi'(x;\cdot)$ are holomorphic on $\rho(u)$, and for the Wronskian we have $W(\varphi,\psi) = 1$. As in the case of Sturm-Liouville equations the Titchmarsh-Weyl coefficient m can be introduced (see, e.g. [T]):

$$m(\lambda) := -\frac{\varphi(1;\lambda)}{\psi(1;\lambda)} \qquad (\lambda \in \rho(u)).$$

Then m is meromorphic on $\rho(u)$, and the function $\chi(\cdot;\lambda) = \varphi(\cdot;\lambda) + m(\lambda)\psi(\cdot;\lambda)$ is a solution of (0.2) with the property $\chi(1;\lambda) = 0$, that means it satisfies the boundary condition at $x = 1$.

The following proposition is proved in the same way as for Sturm-Liouville equations.

PROPOSITION 1.4. *The eigenvalues of the operator function T in $\rho(u)$ are the poles of the Titchmarsh-Weyl function m in $\rho(u)$. If $\lambda \in \rho(u)\backslash\sigma_p(T)$ the inverse $T(\lambda)^{-1}$ is given by the relation*

$$(T(\lambda)^{-1}f)(x) = -\int_0^1 G(x,\xi;\lambda)f(\xi)d\xi \qquad (x \in [0,1]),$$

where

$$G(x,\xi;\lambda) = \begin{cases} \psi(x;\lambda)\chi(\xi;\lambda) & x \leq \xi, \\ \psi(\xi;\lambda)\chi(x;\lambda) & x \geq \xi. \end{cases}$$

The above considerations imply that each eigenvalue λ_0 of T in $\rho(u)$ is geometrically simple; the length of the corresponding maximal Jordan chain is the order of the pole of m at λ_0. Further, the identity

$$\int_0^1 (\chi''(\xi;\lambda) + \frac{q(\xi)}{u(\xi)-\lambda}\chi(\xi;\lambda))\overline{\chi}(\xi;\lambda)d\xi =$$

$$= \chi'(.;\lambda)\overline{\chi}(.;\lambda)|_0^1 - \int_0^1 |\chi'(\xi;\lambda)|^2 d\xi + \int_0^1 \frac{q(\xi)}{u(\xi)-\lambda}|\chi(\xi;\lambda)|^2 d\xi,$$

the fact that $\chi(1;\lambda) = 0$ and the initial conditions (1.8) imply that

$$-\lambda \int_0^1 |\chi(\xi;\lambda)|^2 d\xi = -m(\lambda) - \int_0^1 |\chi'(\xi;\lambda)|^2 d\xi + \int_0^1 \frac{q(\xi)}{u(\xi)-\lambda}|\chi(\xi;\lambda)|^2 d\xi,$$

whence

$$m(\lambda) = \int_0^1 (\lambda + \frac{q(\xi)}{u(\xi)-\lambda})|\chi(\xi;\lambda)|^2 d\xi - \int_0^1 |\chi'(\xi;\lambda)|^2 d\xi. \tag{1.9}$$

2. The case of real functions u and q

2.1. In the rest of this note, besides $q, u \in L^\infty$, we suppose additionally that q and u are real and that

$$\text{ess inf } |q| > 0 \tag{2.1}$$

The reality of q and u implies that $T(\lambda)^* = T(\overline{\lambda})$. The condition (2.1) yields the boundedness of the symmetric operator

$$\mathbf{G}: \ = \begin{bmatrix} 1 & 0 \\ 0 & q^{-1} \end{bmatrix} \qquad \text{in } L^2 \oplus L^2.$$

Moreover, as $q \in L^\infty$, the operator $\mathbf{G}$ is also boundedly invertible.

We introduce in $L^2 \oplus L^2$ the inner product

$$[\mathbf{f}, \mathbf{g}] := (\mathbf{G}\mathbf{f}, \mathbf{g}) \quad (\mathbf{f}, \mathbf{g} \in L^2 \oplus L^2).$$

The space $L^2 \oplus L^2$ equipped with this inner product is denoted by $\mathbf{H}$. It is a Krein space or, if $q > 0$, even a Hilbert space, and the operator $\mathbf{L}$ is selfadjoint in this Krein or Hilbert space as

$$\mathbf{G}\mathbf{L} = \begin{bmatrix} D & 1 \\ 1 & q^{-1}u \end{bmatrix}$$

has this property in $L^2 \oplus L^2$. We mention that $\mathbf{G}\mathbf{L}$ is nonnegative in $L^2 \oplus L^2$ or, equivalently, $\mathbf{L}$ is nonnegative in the Krein or Hilbert space $\mathbf{H}$, if

$$\text{ess } \inf q^{-1}u \geq 1/\pi^2.$$

More can be said if the function q is positive or negative. For the rest of this section 2.1 we suppose that

$$\text{ess } \inf q > 0.$$

Then $\mathbf{L}$ is a selfadjoint operator in the Hilbert space $\mathbf{H}$. This case has many features of the usual spectral theory of selfadjoint differential operators in Hilbert space. We mention only the following two statements:

(i) The Titchmarsh-Weyl coefficient m is holomorphic in $\mathbb{C}^+ \cup \mathbb{C}^-$ and

$$\operatorname{Im} m(\lambda)/ \operatorname{Im} \lambda \geq 0 \quad (\lambda \in \mathbb{C}^+ \cup \mathbb{C}^-).$$

Indeed, the relations (1.9) and $\chi(\xi; \overline{\lambda}) = \overline{\chi(\xi; \lambda)}$ imply

$$\frac{m(\lambda) - \overline{m(\lambda)}}{\lambda - \overline{\lambda}} = \int_0^1 (1 + \frac{q(\xi)}{|u(\xi) - \lambda|^2}) |\chi(\xi; \lambda)|^2 d\xi,$$

for $\lambda \notin \sigma(u) \cup \sigma(T)$.

(ii) Each isolated eigenvalue of u is an accumulation point of isolated eigenvalues of $\mathbf{L}$.

Indeed, let λ be an isolated eigenvalue of u. Since $\sigma(u) = \sigma_{\text{ess}}(\mathbf{L})$ by Proposition 1.2, (3) and (4), the point λ is also isolated in $\sigma_{\text{ess}}(\mathbf{L})$. According to Lemma 1.3 we have $\lambda \notin \sigma_p(\mathbf{L})$, and the claim follows from well-known properties of the spectrum of selfadjoint operators in Hilbert space.

2.2. In this section we suppose that ess sup $q < 0$. Then $\mathbf{H}$ is a Krein space.

LEMMA 2.1. *Let $q, u \in L^\infty$ be real and suppose that* ess sup $q < 0$. *If β is chosen such that $\beta >$* ess sup u, *then the Hermitian sesquilinear form*

$$\{\mathbf{f}, \mathbf{g}\} := [(\mathbf{L} - \beta)\mathbf{f}, \mathbf{g}] \qquad (\mathbf{f}, \mathbf{g} \in \mathfrak{D}(\mathbf{L}))$$

has a finite number, say κ, of negative squares on $\mathfrak{D}(\mathbf{L})$.

Recall that this means that each linear manifold $\mathfrak{L}$ of $\mathfrak{D}(\mathbf{L})$ such that $\{\mathbf{f},\mathbf{f}\} < 0$ for all $\mathbf{f} \in \mathfrak{L}, \mathbf{f} \neq 0$, is of dimension $\leq \kappa$ and for at least one such linear manifold the dimension is κ.

PROOF. First we note that

$$\mathbf{G}(\mathbf{L}-\beta) = \begin{bmatrix} D-\beta & 1 \\ 1 & (u-\beta)/q \end{bmatrix}.$$

Further, we have $\alpha :=$ ess $\sup q/(u-\beta) > 0$, and the operator

$$\begin{bmatrix} \alpha & 1 \\ 1 & (u-\beta)/q \end{bmatrix}$$

is nonnegative in $L^2 \oplus L^2$ as

$$\alpha \frac{u-\beta}{q} \geq \alpha \text{ ess } \inf \frac{u-\beta}{q} = 1.$$

Denote by P the orthogonal projection in L^2 onto the linear span of the eigenvectors of D corresponding to the eigenvalues $< \beta + \alpha$, and define

$$D_0 := (I-P)D + (\alpha+\beta)P.$$

Then $D_0 \geq \alpha + \beta$ and $D - D_0$ is of finite rank:

$$\tilde{\kappa} := \dim \mathfrak{R}(D-D_0) = \#(\sigma(D) \cap (0, \alpha+\beta)). \tag{2.2}$$

Now we have $D_0 - \beta \geq \alpha$ and

$$\begin{aligned} \mathbf{G}(\mathbf{L}-\beta) &= \begin{bmatrix} D_0-\beta & 1 \\ 1 & (u-\beta)/q \end{bmatrix} + \begin{bmatrix} D-D_0 & 0 \\ 0 & 0 \end{bmatrix} \\ &\geq \begin{bmatrix} \alpha & 1 \\ 1 & (u-\beta)/q \end{bmatrix} + \begin{bmatrix} D-D_0 & 0 \\ 0 & 0 \end{bmatrix} \end{aligned} .$$

As the first term on the right-hand side is nonnegative, it follows that $\mathbf{G}(\mathbf{L}-\beta)$ has at most $\tilde{\kappa}$ negative eigenvalues and the lemma is proved.

REMARK 1. In the proof of Lemma 2.1, the number $\tilde{\kappa}$ depends on the choice of β. If, with the above notation, we put

$$c := \beta - \text{ ess } \sup u \; (> 0),$$

then α above can be replaced by

$$\tilde{\alpha} := \frac{\text{ess } \sup |q|}{c} \geq \alpha.$$

We have

$$\tilde{\alpha} + \beta = c + \text{ ess sup}\, u + \frac{\text{ess sup}\,|q|}{c},$$

and this value becomes minimal if $c = \sqrt{\text{ess sup}\,|q|}$, that is, $\alpha + \beta$ in the definition of P and in (2.2) can be replaced by

$$2\sqrt{\text{ess sup}\,|q|} + \text{ ess sup}\, u.$$

REMARK 2. If β can be chosen such that

$$(\pi^2 - \beta) \text{ ess inf} \frac{u - \beta}{q} \geq 1,$$

then $\mathbf{L} - \beta$ is nonnegative in the Krein space $\mathbf{H}$. If, in particular, π^2 ess $\inf u/q \geq 1$, then $\mathbf{L}$ is nonnegative in $\mathbf{H}$.

The Lemma 2.1, Proposition 1.2 and [L], p. 11f, imply the following result; there and in the sequel κ is the same as in Lemma 2.1.

PROPOSITION 2.2. *Under the assumptions of Lemma 2.1, the operator* $\mathbf{L}$ *is definitizable with a definitizing polynomial* p *of the form* $p(\lambda) = (\lambda - \beta)r(\lambda)\overline{r}(\lambda)$, r *being a polynomial with deg* $r \leq \kappa$.

The Proposition 2.2 has immediate consequences about the spectrum of $\mathbf{L}$ in the Krein space $\mathbf{H}$. In order to formulate these we denote the spectral function of $\mathbf{L}$ by $\mathbf{E}$ (see, e.g., [L]), the algebraic eigenspace corresponding to the eigenvalue λ of $\mathbf{L}$ by $\mathfrak{L}_\lambda$, the signature of the inner product $[\cdot,\cdot]$ on $\mathfrak{L}_\lambda$ by $(\kappa_-(\lambda), \kappa_0(\lambda), \kappa_+(\lambda))$ and $\tilde{\kappa}_\pm(\lambda) := \kappa_\pm(\lambda) + \kappa_0(\lambda)$. Recall that $\lambda \in \sigma_p(\mathbf{L})$ is said to be of positive (negative, neutral resp.) type if the corresponding eigenvectors $\mathbf{y}$ are positive (negative, neutral resp.), that is $[\mathbf{y},\mathbf{y}] > 0$ $(< 0, = 0$ resp.). More generally, a subset $\triangle$ of the spectrum of $\mathbf{L}$ is said to be of positive (negative resp.) type if $\mathbf{E}(\triangle)$ is defined and its range is a positive (negative resp.) subspace.

PROPOSITION 2.3. *Under the assumptions of Lemma 2.1, the nonreal spectrum of* $\mathbf{L}$ *consists of at most* κ *pairs of normal eigenvalues* $\lambda, \overline{\lambda}$ *with finite-dimensional algebraic eigenspaces. If* β *is chosen such that it is not an eigenvalue of* $\mathbf{L}$ *(which is possible as the spectrum of* $\mathbf{L}$ *is discrete outside of* $\sigma(u)$*), then it holds*

$$\sum_{\lambda\in\sigma_p(\mathbf{L})\cap(-\infty,\beta)} \tilde{\kappa}_+(\lambda) + \sum_{\lambda\in\sigma(\mathbf{L})\cap(\beta,\infty)} \tilde{\kappa}_-(\lambda) + \sum_{\lambda\in\sigma(\mathbf{L})\cap\mathbb{C}^+} \dim \mathfrak{L}_\lambda = \kappa \quad ; \tag{2.3}$$

in particular, with the exception of finitely many points (which appear in the corresponding sum in (2.3)) *the eigenvalues of* $\mathbf{L}$ *in* (β,∞) *are of positive type and the spectrum of* $\mathbf{L}$ *in* $(-\infty,\beta)$ *with the exception of the points which appear in the corresponding sum in* (2.3) *is of negative type.*

Observe that the assumption $\beta >$ ess sup u implies according to Proposition 1.2, (4), that the spectrum of $\mathbf{L}$ in (β,∞) is discrete. If u is a simple function (that is there exists a finite decomposition of $[0,1]$ into measurable sets such that u is constant a.e. on each of these sets) also $\sigma(L)\cap(-\infty,\beta)$ is discrete with possible exception of the points of $\sigma(u)$. These exceptional points do not belong to $\sigma_p(\mathbf{L})$ by Lemma 1.3, hence they are spectral points of negative type and not critical points of $\mathbf{L}$. Thus, using the properties of the spectral function of a definitizable operator, it follows (for the definition of a Riesz basis see [GK]):

PROPOSITION 2.4. *If, additionally to the assumptions of Lemma* 2.1, *u is a simple function, then the eigenvectors and associated vectors of* $\mathbf{L}$ *form a Riesz basis of* $\mathbf{H}$.

2.3. Now we formulate some consequences of these statements for the pencil T. First we observe that for the eigenvectors of $\mathbf{L}$ the indefinite inner product $[.,.]$ can be expressed in terms of T as follows. If $\lambda_0\in\sigma_p(T)\cap\rho(u)\cap\mathbb{R}$ and y_0 is a corresponding eigenvector, then with the eigenvector

$$\mathbf{y}_0=\begin{bmatrix} y_0 \\ -\frac{q}{u-\lambda_0}y_0\end{bmatrix}$$

of $\mathbf{L}$ we have

$$[\mathbf{y}_0,\mathbf{y}_0]=(\dot{T}(\lambda_0)y_0,y_0) \tag{2.4}$$

as both expressions equal

$$\int_0^1 |y_0(\xi)|^2(1+\frac{q(\xi)}{|u(\xi)-\lambda|^2})d\xi.$$

The eigenvalues of T are geometrically simple, hence they can be classified according to the sign of the expressions in (2.4): The eigenvalue λ_0 of T is said to be of positive (negative, neutral resp.) type if $(\dot{T}(\lambda_0)y_0,y_0)>0\quad(<0,=0$ resp.). From (2.4) it follows that this holds if and only if λ_0 is of positive (negative, neutral resp.) type for $\mathbf{L}$.

PROPOSITION 2.5. *Under the assumptions of Lemma* 2.1 *the eigenvectors of* T *corresponding to eigenvalues of positive type in* (β,∞) *form a Riesz basis of a subspace of* L^2 *of codimension*

$$\sum_{\lambda\in\sigma_p(\mathbf{L})\cap(-\infty,\beta)}\tilde{\kappa}_+(\lambda)\ +\sum_{\lambda\in\sigma(\mathbf{L})\cap\mathbb{C}^+}\dim\mathfrak{L}_\lambda\ +\sum_{\substack{\lambda\in\sigma_p(\mathbf{L})\cap(\beta,\infty),\\ \kappa_-(\lambda)>0}}\kappa_+(\lambda). \tag{2.5}$$

If, additionally, u is a simple function, then the eigenvectors of T *corresponding to eigenvalues of negative type in* $(-\infty,\beta)$ *form a Riesz basis of a subspace of* L^2 *of codimension not greater than*

$$\sum_{\lambda\in\sigma_p(\mathbf{L})\cap(\beta,\infty)}\kappa_-(\lambda)\ +\sum_{\lambda\in\sigma(\mathbf{L})\cap\mathbb{C}^+}\dim\mathfrak{L}_\lambda\ +\sum_{\substack{\lambda\in\sigma_p(\mathbf{L})\cap(-\infty,\beta),\\ \kappa_+(\lambda)>0}}\kappa_-(\lambda). \tag{2.6}$$

PROOF. We consider the canonical decomposition

$$\mathbf{H}=\mathbf{K}_+[\dot{+}]\mathbf{K}_-$$

of the Krein space $\mathbf{H}$, where $\mathbf{K}_+$ is the first component of the representation $\mathbf{H} = L^2[+]L^2$, equipped with the usual Hilbert inner product of L^2, and $\mathbf{K}_-$ is the second component, equipped with the inner product $(q^{-1}.,.)$. Let e.g. $\mathfrak{L}$ be the closed linear span of the positive eigenvectors of $\mathbf{L}$ corresponding to eigenvalues in (β, ∞). Then $\mathfrak{L}$ is a uniformly positive subspace of the Krein space $\mathbf{H}$ (observe that, according to Remark 2 after Proposition 1.2 $\sigma(\mathbf{L})$ is bounded from below, hence ∞ is not a critical point of $\mathbf{L}$), and the defect of $\mathfrak{L}$ from a maximal positive subspace is given by (2.5). It is well-known (see, e.g., [Br]) that $\mathfrak{L}$ admits a representation

$$\mathfrak{L} = \left\{ \begin{bmatrix} x_+ \\ Kx_+ \end{bmatrix} : \mathbf{x}_+ \in \mathfrak{L}_+ \subset \mathbf{K}_+ \right\},$$

where $\mathfrak{L}_+$ is a subspace of $\mathbf{K}_+$ of codimension given by (2.5) and K is a strict contraction from $\mathbf{K}_+$ into $\mathbf{K}_-$ equipped with the Hilbert inner product $-(q^{-1}.,.)$. It follows that for $\mathbf{x} \in \mathfrak{L}, \mathbf{x} = \begin{bmatrix} x_+ \\ Kx_+ \end{bmatrix}$ we have

$$(1 - \|K\|^2)\|x_+\|^2 \leq [x, x] \leq 2\|x_+\|^2.$$

Therefore, if on $\mathfrak{L}_+$ an inner product is defined by the relation

$$(x_+, y_+) := \left[\begin{bmatrix} x_+ \\ Kx_+ \end{bmatrix}, \begin{bmatrix} y_+ \\ Ky_+ \end{bmatrix} \right],$$

the norm of this inner product is equivalent to the L^2-norm on $\mathfrak{L}_+$. Moreover, the eigenvectors of T corresponding to different eigenvalues are orthogonal with respect to this inner product, and the first statement of Proposition 2.5 follows. The second statement is proved analogously.

COROLLARY 2.6. *Suppose that* ess sup $q < 0$, ess sup $u < 0$ *and* $q^{-1}u > 1/\pi^2$ *(or, more generally, that there exists a number* β *such that*

$$(\pi^2 - \beta) \text{ ess inf} \frac{u - \beta}{q} > 1).$$

Then the spectrum of T *is real and the eigenvectors of* T *of positive type form a Riesz basis of* L^2. *If, additionally,* u *is a simple function also the eigenvectors of* T *of negative type form a Riesz basis of* L^2.

REFERENCES

[Ba] Je. P. Bogomolova, Some questions of the spectral analysis of a nonselfadjoint differential operator with a "floating singularity" in the coefficient. Diff. Uravnenija 21, 11 (1985), 1843–1849 [Russian].

[Br] J. Bognár, Indefinite inner product spaces. Springer-Verlag, Berlin-Heidelberg-New York, 1974.

[GK] I. C. Gohberg, M. G. Krein, Introduction to the theory of linear nonselfadjoint operators. Translations of Mathematical Monographs, vol. 18, AMS, Providence, RI, 1969.

[GS] I. C. Gohberg, Je. I. Sigal, An operator generalization of the logarithmic residue theorem and the theorem of Rouché. Mat. Sb. 84, 126 (1971), 607–629 [Russian]; Engl. transl.: Math. USSR Sb. 13 (1971), 603–625.

[KL] M. G. Krein, H. Langer, On some mathematical principles in the linear theory of damped oscillations of continua. Integral Equations Operator Theory 1 (1978), 364–399 and 539–566.

[L] H. Langer, Spectral functions of definitizable operators in Krein spaces. Lecture Notes in Mathematics 948 (1982), 1–46.

[MM] R. Mennicken, M. Möller, Equivalence of boundary eigenvalue operator functions and their characteristic matrix functions. Math. Ann. 279 (1987), 267–275.

[MMS] R. Mennicken, M. Möller, A. Sattler, Carlitz series as eigenfunction expansions. In preparation.

[R] L. Rodman, Operator polynomials. Birkhäuser-Verlag, Basel-New York, 1989.

[S] K. A. Shukurov, On a spectral problem on a finite interval. Izvestija AN Turkmenskoj SSR, serija fiziko-tekhnicheskikh, khimicheskikh i geologicheskikh Nauk 1 (1988), 73–75 [Russian].

[T] E. C. Titchmarsh, Eigenfunction expansions associated with second order differential equations, I. Oxford University Press, London, 1962.

[W] J. Weidmann, Linear operators in Hilbert spaces. Springer-Verlag, New York-Heidelberg-Berlin, 1980.

H. Langer, R. Mennicken and M. Möller
NWF I - Mathematik, Universität Regensburg,
Universitätsstraße 31, D-8400 Regensburg, FRG

Operator Theory:
Advances and Applications, Vol. 48

EXPOSED POINTS IN H^1, II

Donald Sarason

The spaces $\mathcal{H}(b)$ of L. de Branges are used to study the exposed points of the unit ball of H^1.

1. INTRODUCTION

In Part I [**13**] a conjecture about the structure of the exposed points in the unit ball of the Hardy space H^1 was formulated and a small portion of it was established. The present paper contains further progress toward proving the conjecture. As in [**13**], a key role is played by de Branges's spaces $\mathcal{H}(b)$.

The next section contains background material, the statements of the main results (consisting of two theorems), and the plan for the remainder of the paper. As will be explained more fully below, Daniel Hitt deserves credit for the key idea underlying the proof of Theorem 1.

NOTATIONS. The Hardy spaces in this paper are the classical ones in the unit disk, D, and the Lebesgue spaces are those of normalized Lebesgue measure on ∂D. As usual, functions in the Hardy spaces will be identified with their boundary functions.

If φ is a function in L^2, then T_φ, the Toeplitz operator with symbol φ, is the operator on H^2 that sends the function g to the Fourier projection of φg, that is, to the holomorphic function in D defined by

$$(T_\varphi g)(z) = \frac{1}{2\pi}\int_{\partial D} \frac{\varphi(e^{i\theta})g(e^{i\theta})}{1-e^{-i\theta}z}\,d\theta.$$

If φ is in L^∞ one can write $T_\varphi g = P(\varphi g)$, where P is the orthogonal projection in L^2 with range H^2.

The norm and inner product in L^2 will be denoted by $\|\cdot\|_2$ and $\langle\cdot,\cdot\rangle$, respectively. The kernel function in H^2 for the point w of D will be denoted by k_w ($k_w(z) = (1-\bar{w}z)^{-1}$).

The definition of exposed point, although not needed explicitly below, is worth recording: A point in the unit ball of a Banach space is called an exposed point of the unit ball if there is a real linear functional on the space that equals 1 at the point but has value less than 1 elsewhere in the unit ball. (The definition was misstated in [**13**].)

2. DE BRANGES'S SPACES AND EXPOSED POINTS IN H^1

For b a nonconstant function in the unit ball of H^∞, the space $\mathcal{H}(b)$, as a vector space, is the range of the operator $(1 - T_bT_{\bar{b}})^{1/2}$; it is given the Hilbert space structure that makes the preceding operator a coisometry. The norm and inner product in $\mathcal{H}(b)$ will be denoted by $\|\cdot\|_b$ and $\langle\cdot,\cdot\rangle_b$, respectively. Thus, for example, if g is in H^2 and is orthogonal to the kernel of $(1-T_bT_{\bar{b}})^{1/2}$, then $\|(1-T_bT_{\bar{b}})^{1/2}g\|_b = \|g\|_2$. The kernel function in $\mathcal{H}(b)$ for the point w of D is the function $k_w^b = (1 - \overline{b(w)}b)k_w$. This and other basic properties of the spaces $\mathcal{H}(b)$ can be found in [**6**], where the spaces were first studied, and in [**11**].

A representation of $\mathcal{H}(b)$ in terms of Cauchy integrals is needed. If b is as above, then the function $\frac{1+b}{1-b}$ has a positive real part in D and so can be written as a Hergoltz integral:

$$\frac{1+b(z)}{1-b(z)} = \int_{\partial D} \frac{e^{i\theta}+z}{e^{i\theta}-z}\, d\mu_b(e^{i\theta}) + ic,$$

where c is a real constant and μ_b is a finite positive Borel measure on ∂D, uniquely determined by b. For q in $L^2(\mu_b)$ let the function V_bq in D be defined by

$$(V_bq)(z) = (1-b(z)) \int_{\partial D} \frac{q(e^{i\theta})}{1-e^{-i\theta}z}\, d\mu_b(e^{i\theta}).$$

The operator V_b vanishes on the orthogonal complement of $H^2(\mu_b)$ (the closure of the polynomials in $L^2(\mu_b)$).

PROPOSITION 1. *V_b is an isometry of $H^2(\mu_b)$ onto $\mathcal{H}(b)$.*

The proof, which can be found in [**13**], shows that $V_bk_w = (1 - \overline{b(w)})^{-1}k_w^b$.

As noted in [**13**], the earliest version of Proposition 1 seems to be due to D. N. Clark [**8**]; it pertains to the case where b is an inner function, in which case $\mathcal{H}(b)$ is just the orthogonal complement of the invariant subspace bH^2. Since the writing of [**13**] I have learned, thanks to T. L. Kriete and J. A. Ball, that a vector-valued version having Proposition 1 as a very special case appears in Ball's dissertation [**3**]. The dissertation was not published in its entirety, but the relevant part appears in a joint paper of Ball and A. Lubin [4]. A vector-valued result encompassing Proposition 1 appears also in a paper of D. Alpay and H. Dym [1]. The announcement [7] of Clark mentions a near relative of Proposition 1 pertaining to Sz.–Nagy–Foias model spaces.

For the remainder of this paper we fix a nonconstant outer function f in H^2 such that $\|f\|_2 = 1$ and $f(0) > 0$. By a theorem of K. de Leeuw and W. Rudin [9], f^2 is, to within a unimodular multiplicative constant, the most general nonconstant extreme point in the unit ball of H^1. Our aim is to understand what it means, in terms of the structure of f, for f^2 to be an exposed point of the unit ball of H^1 (hereafter referred to simply as an exposed point).

Let the function b in the unit ball of H^∞ be defined by

$$\frac{1+b(z)}{1-b(z)} = \frac{1}{2\pi}\int_{\partial D} \frac{e^{i\theta}+z}{e^{i\theta}-z}|f(e^{i\theta})|^2\, d\theta,$$

so that $d\mu_b = |f|^2 d\theta/2\pi$. The outer function $a = (1-b)f$ is then also in the unit ball of H^∞, and $|a|^2 + |b|^2 = 1$ almost everywhere on ∂D. (Further details can be found in [13].) Associated with the function a is the space $\mathcal{M}(a)$, which consists, as a vector space, of the range of the operator T_a. (Its Hilbert space structure, which will not be of importance here, is the one that makes T_a an isometry of H^2 onto $\mathcal{M}(a)$.) The following result from [13] is the link between de Branges's spaces and exposed points in H^1.

PROPOSITION 2. *The space $\mathcal{M}(a)$ is contained in $\mathcal{H}(b)$. The function f^2 is an exposed point if and only if $\mathcal{M}(a)$ is dense in $\mathcal{H}(b)$.*

For λ in ∂D, let $f_\lambda = a/(1-\lambda b)$. These functions are outer functions, they are in the unit ball of H^2, and they have norm 1 for almost all λ [13]. The conjecture advanced in [13] is that the following three conditions are equivalent:

(i) f^2 is an exposed point.

(ii) There is no nonconstant inner function u such that $f/(1-u)$ is in H^2.

(iii) $\|f_\lambda\|_2 = 1$ for all λ in ∂D.

The implication (i) $\Rightarrow$ (ii) lies near the surface and has been long known. It is discussed in [13], which also contains a proof of the implication (i) $\Rightarrow$ (iii). The first main result of this paper is a proof of the implication (ii) $\Rightarrow$ (iii).

THEOREM 1. *If there is a* λ *in* ∂D *such that* $\|f_\lambda\|_2 < 1$, *then there is a nonconstant inner function* u *such that* $f/(1-u)$ *is in* H^2.

As mentioned in Section 1, the proof is based on an idea of D. Hitt. Here is Hitt's observation: *If* u *and* v *are nonconstant inner functions such that* $\mu_u \leq \mu_v$, *then the quotient* $\frac{1-v}{1-u}$ *is in* H^2. The result is a simple consequence of Proposition 1 (which Hitt rediscovered for the case of an inner function, being unaware of Clark's paper). Namely, from Proposition 1 and the remark following it one sees that if $\mu_u \leq \mu_v$ then $\frac{1-v}{1-u}$ equals

$$(1-\overline{u(0)})(1-\overline{u(0)}u)^{-1}V_v(d\mu_u/d\mu_v).$$

Thus, the reasoning shows more generally that $\frac{1-v}{1-u}$ will be in H^2 whenever μ_u is absolutely continuous with respect to μ_v and $d\mu_u/d\mu_v$ is in $L^2(\mu_v)$. (The converse of the last statement is true also; the explanation is postponed to the end of the section.)

To use Hitt's idea to prove Theorem 1 one needs to know that the operator $T_{f_\lambda}T_{\bar{f}}$ is bounded when $\lambda \neq 1$; this is established in Section 3. The proof of Theorem 1 is given in Section 4, and Section 5 contains a discussion of certain questions raised by the boundedness result of Section 3.

The boundedness of the operators $T_{f_\lambda}T_{\bar{f}}$ is closely related to the following result from **[13]**.

PROPOSITION 3. *For* λ *in* ∂D, *the operator* $T_{1-\lambda b}T_{\bar{f}_\lambda}$ *is an isometry of* H^2 *into* $\mathcal{H}(b)$; *its range is all of* $\mathcal{H}(b)$ *if and only if* $\|f_\lambda\|_2 = 1$.

This result also lies behind Proposition 2. One proves it by composing the map $V_{\lambda b}$, going from $H^2(\mu_{\lambda b})$ to $\mathcal{H}(b)$, with the map of multiplication by $1/f_\lambda$, going from H^2 to $H^2(\mu_{\lambda b})$. The proposition implies, in particular, that $T_{1-\lambda b}T_{\bar{f}_\lambda}$ is a contraction of H^2 into itself.

Once Theorem 1 has been established, all that will be needed to establish the conjecture is a proof that (iii) implies (i). The second main result of this paper is a special case of that implication. In case f^2 is not an exposed point, we let $\mathcal{K}(b)$ denote the orthogonal complement of $\mathcal{M}(a)$ in $\mathcal{H}(b)$. The orthogonal projection in $\mathcal{H}(b)$ with range $\mathcal{K}(b)$ will be denoted by P_0. The functions 1 and b belong to $\mathcal{H}(b)$ **[11]** and, as will be shown later, their projections onto $\mathcal{K}(b)$ are nonzero.

THEOREM 2. *If f^2 is not exposed, and if $P_0 1$ and $P_0 b$ are linearly dependent, then there is a unique λ in ∂D such that $\|f_\lambda\|_2 < 1$.*

The proof is given in Section 6, which also contains examples fulfilling the hypotheses. Section 7, the final section, makes a few remarks on the still unproved portions of the conjecture.

To conclude this section, a proof will be briefly sketched of the converse of the extended version of Hitt's observation. The converse states: *If u and v are nonconstant inner functions such that $\frac{1-v}{1-u}$ is in H^2, then μ_u is absolutely continuous with respect to μ_v and $d\mu_u/d\mu_v$ is in $L^2(\mu_v)$.* To prove this one notes that, if $\frac{1-v}{1-u}$ is in H^2, then $\frac{1-v}{1-u}(1-\overline{u(0)}u)$ is in $\mathcal{H}(v)$, for its product with $\bar{v}$ equals $\frac{\bar{v}-1}{\bar{u}-1}(\bar{u}-\overline{u(0)})$, which is orthogonal to H^2. Then, by Proposition 1, the function $\frac{1-v}{1-u}(1-\overline{u(0)}u)$ equals $V_v q$ for a function q in $L^2(\mu_v)$. After straightforward manipulations one arrives at the equality

$$\frac{1+u(z)}{1-u(z)} = (1-\overline{u(0)})^{-1}\int_{\partial D}\left(\frac{e^{i\theta}+z}{e^{i\theta}-z}\right)q(e^{i\theta})d\mu_v(e^{i\theta}) + i\,\mathrm{Im}\left(\frac{1+\overline{u(0)}}{1-\overline{u(0)}}\right),$$

showing that $d\mu_u = (1-\overline{u(0)})^{-1}q\,d\mu_v$.

3. THE BOUNDEDNESS OF CERTAIN OPERATORS

For λ in ∂D let W_λ denote the operator $T_{1-\lambda b}T_{\bar{f}_\lambda}$, regarded as an operator from H^2 to $\mathcal{H}(b)$. Its adjoint, W_λ^*, thus goes from $\mathcal{H}(b)$ to H^2.

PROPOSITION 4. *If λ and η are in ∂D then*

$$W_\lambda^* W_\eta = (1-\bar{\lambda}\eta)T_{f_\lambda}T_{\bar{f}_\eta} + \bar{\lambda}\eta.$$

In particular, if $\lambda \neq \eta$ then $T_{f_\lambda}T_{\bar{f}_\eta}$ is bounded, with norm at most $2/|\lambda-\eta|$.

It is interesting to note, in connection with Proposition 4, that the function $(1-\bar{\lambda}\eta)f_\lambda\bar{f}_\eta + \bar{\lambda}\eta$ is unimodular on ∂D; in fact, it equals $(1-\overline{\lambda b})(1-\eta b)/(1-\lambda b)(1-\overline{\eta b})$. To someone having experience with Toeplitz operators it might seem surprising, then, that the operator $(1-\bar{\lambda}\eta)T_{f_\lambda}T_{\bar{f}_\eta} + \bar{\lambda}\eta$ can contrive to be unitary, as it does when f_λ and f_η both have unit norm in H^2—at least, I was surprised. The proposition seems to reflect a remarkable cooperative behavior amongst the functions f_λ.

It is worth noting, also, that the functions f_λ are uniformly bounded in supremum norm only in the (in many respects trivial) case where $\|b\|_\infty < 1$. However, it is possible for each f_λ to be bounded even though $\|b\|_\infty = 1$; that occurs, for example, when $b(z) = -z(z+1)/2$ (in which case $a(z) = (1-z)/2$).

To prove Proposition 4 one evaluates the inner product $\langle W_\lambda^* W_\eta k_w, k_z\rangle$ for two points w and z in the disk. We have

$$\begin{aligned} \langle W_\lambda^* W_\eta k_w, k_z\rangle &= \langle W_\eta k_w, W_\lambda k_z\rangle_b \\ &= \langle (1-\eta b)T_{\bar{f}_\eta} k_w, (1-\lambda b)T_{\bar{f}_\lambda} k_z\rangle_b \qquad (1) \\ &= \overline{f_\eta(w)} f_\lambda(z)\langle (1-\eta b)k_w, (1-\lambda b)k_z\rangle_b . \end{aligned}$$

Use will be made of the equality

$$\langle k_w, k_z\rangle_b = (1-\bar{w}z)^{-1}\left(1 + \frac{\overline{b(w)}b(z)}{\overline{a(w)}a(z)}\right)$$

from [**11**]. (More precisely, the equality can be established by a slight modificaiton of the proof of Proposition 1 in [**11**], which pertains to the case $z = w$. The equality is also easily deduced from Proposition 8 below.) If $b(w) \neq 0$ then the function $(1-\eta b)k_w$ is a linear combination of k_w^b and k_w, and similarly for $(1-\lambda b)k_z$ if $b(z) \neq 0$. Inserting these linear combinations into the right side of (1), and using the value above for $\langle k_w, k_z\rangle_b$ together with the reproducing property of the functions k_w^b and k_z^b, one obtains after a somewhat lengthy but straightforward calculation, whose inclusion here seems unnecessary, the equality

$$\langle W_\lambda^* W_\eta k_w, k_z\rangle = (1-\bar{w}z)^{-1}[(1-\bar{\lambda}\eta)f_\lambda(z)\overline{f_\eta(w)} + \bar{\lambda}\eta].$$

The right side here equals

$$\langle [(1-\bar{\lambda}\eta)T_{f_\lambda}T_{\bar{f}_\eta} + \bar{\lambda}\eta]k_w, k_z\rangle,$$

so it follows that the two operators $W_\lambda^* W_\eta$ and $(1-\bar{\lambda}\eta)T_{f_\lambda}T_{\bar{f}_\eta} + \bar{\lambda}\eta$ agree on all the functions k_w for which $b(w) \neq 0$. These functions span H^2, and a standard limit argument completes the proof.

4. PROOF OF THEOREM 1

Assume $\|f_\lambda\|_2 < 1$. Because the function $(1-|b|^2)/|1-\lambda b|^2$ $(= \operatorname{Re}\frac{1+\lambda b}{1-\lambda b})$ in D is the Poisson integral of the measure $\mu_{\lambda b}$, the absolutely continuous component of that measure is $|f_\lambda|^2 d\theta/2\pi$. Since $\mu_{\lambda b}$ is a probability measure, it thus has a nonzero singular component, say ν. The function u such that $\frac{1+u}{1-u}$ is the Herglotz integral of ν is then an inner function. Let h be the image of $d\nu/d\mu_{\lambda b}$ under $V_{\lambda b}$. Then h also equals $(1-\lambda b)V_u 1/(1-u)$, and $V_u 1 = (1-\overline{u(0)})^{-1}(1-\overline{u(0)}u)$. Moreover h, being in $\mathcal{H}(b)$, is by Proposition 3 equal to $T_{1-b}T_{\bar{f}}g$ for some g in H^2. We conclude that

$$\frac{(1-\lambda b)(1-\overline{u(0)}u)}{(1-\overline{u(0)})(1-u)} = (1-b)T_{\bar{f}}g,$$

and so

$$\frac{f}{1-u} = \frac{a}{(1-u)(1-b)} = \frac{(1-\overline{u(0)})(1-\overline{u(0)}u)^{-1}a}{1-\lambda b}T_{\bar{f}}g$$
$$= (1-\overline{u(0)})(1-\overline{u(0)}u)^{-1}T_{f_\lambda}T_{\bar{f}}g.$$

By Proposition 4 the function on the right side is in H^2, and the proof is complete.

5. FURTHER DISCUSSION OF PROPOSITION 4

Two questions are suggested immediately by Proposition 4.

QUESTION 1. If g and h are in H^∞, does the inequality $\|T_g T_{\bar{h}}\| \leq \|gh\|_\infty$ necessarily hold?

QUESTION 2. If g and h are in H^2 and gh is bounded, is $T_g T_{\bar{h}}$ necessarily bounded?

Both answers are negative, suggesting that the boundedness of $T_{f_\lambda}T_{\bar{f}_\eta}$ for $\lambda \neq \eta$ depends, in case at least one of the two factors is unbounded, on some sort of subtle cooperation between them.

One can obtain a counterexample for Question 1 by taking g to be any nonconstant invertible function in H^∞ and taking h to be $1/g$. The operator $T_g T_{1/\bar{g}}$ is the inverse of $T_{\bar{g}/g}$; if its norm were bounded by 1 the latter operator, since its norm is 1, would be unitary. However, the only unitary Toeplitz operators are scalar multiples of the identity.

One can obtain a counterexample for Question 2 by taking g to be an outer function in H^2 such that $1/g$ is in H^2 but such that $T_{\bar{g}/g}$ is not invertible, and taking h to be $1/g$. Then $T_g T_{1/\bar{g}}$ is not bounded, for if it were it would be the inverse of $T_{\bar{g}/g}$.

To guarantee the noninvertibility of $T_{\bar{g}/g}$ one arranges to have $|g|^2$ violate Muckenhoupt's condition (A_2). (A discussion of these matters can be found in [**10**].) It is possible to do that with a bounded g such that $1/g$ is in H^p for all finite p.

A more substantial question than the two above is that of finding a structural characterization of those pairs of functions g, h in H^2 for which $T_g T_{\bar{h}}$ is bounded. The boundedness of gh is a necessary condition, as one sees from the equality

$$T_{\bar{g}/g} T_g T_{\bar{h}} T_{h/\bar{h}} = T_{\bar{g}h}.$$

The question is therefore equivalent to that of characterizing the boundedness of the semicommutator $T_g T_{\bar{h}} - T_{g\bar{h}}$, assuming the boundedness of $T_{g\bar{h}}$. It most likely is related to the problem of characterizing the compactness of the semicommutator $T_\varphi T_\psi - T_{\varphi\psi}$, which, at least for bounded φ and ψ, is studied in [**2**] and [**14**]. Progress on the boundedness problem should give further insight into the compactness problem.

Another question that might be interesting is whether, when g and h are functions in H^2 such that $T_g T_{\bar{h}}$ is bounded, the latter operator belongs to the Toeplitz algebra (the norm-closed algebra generated by the bounded Toeplitz operators). It does, of course, if g and h are bounded, but the answer is not apparent if one of them is unbounded.

J. Barría and P. R. Halmos [**5**] have introduced the notion of an asymptotic Toeplitz operator. The operator T on H^2 is a Toeplitz operator if and only if it satisfies $S^*TS = T$, where S is the unilateral shift operator (the operator on H^2 of multiplication by z). Barría and Halmos call T an asymptotic Toeplitz operator if $\lim_{n\to\infty} S^{*n}TS^n$ exists strongly; the limit is then a Toeplitz operator whose symbol is called the symbol of T. They show that every operator in the Toeplitz algebra is an asymptotic Toeplitz operator. It is thus natural to ask whether $T_g T_{\bar{h}}$ is an asymptotic Toeplitz operator when g and h are functions in H^2, at least one of which is unbounded, such that $T_g T_{\bar{h}}$ is bounded. I do not know the answer. However, it is fairly easy to see that, under the stated conditions, $T_g T_{\bar{h}}$ is a weak asymptotic Toeplitz operator, in the sense that $\lim_{n\to\infty} S^{*n} T_g T_{\bar{h}} S^n$ exists weakly, and in fact equals $T_{g\bar{h}}$. Here is a sketch of the proof. Let $e_0, e_1, \ldots$ be the usual basis vectors for H^2 ($e_j(z) = z^j$). Because of the assumption that $T_g T_{\bar{h}}$ is bounded, it is enough to show that $\lim_{n\to\infty}\langle S^{*n} T_g T_{\bar{h}} S^n e_j, e_k\rangle = \langle T_{g\bar{h}} e_j, e_k\rangle$ for each j and k. The equality

$$S^*T_gT_{\bar{h}}S - T_gT_{\bar{h}} = S^*g \otimes S^*h$$

is easily verified, from which one obtains

$$S^{*n}T_gT_{\bar{h}}S^n = T_gT_{\bar{h}} + \sum_{m=1}^{n} S^{*m}g \otimes S^{*m}h.$$

Thus, letting $A_n = S^{*n}T_gT_{\bar{h}}S^n - T_{g\bar{h}}$, we have

$$A_n = T_gT_{\bar{h}} - T_{g\bar{h}} + \sum_{m=1}^{n} S^{*m}g \otimes S^{*n}h.$$

From this it is easy to deduce the equality

$$\begin{aligned}\langle A_ne_j, e_k\rangle &= \sum_{m=n+1}^{\infty} \hat{g}(j+m)\overline{\hat{h}(k+m)} \\ &= \langle S^{*(j+n+1)}g, S^{*(k+n+1)}h\rangle.\end{aligned}$$

For fixed j and k, the right side clearly tends to 0 as $n \to \infty$.

6. PROOF OF THEOREM 2

We assume f^2 is not an exposed point. It is known that $\mathcal{H}(b)$ is invariant under S, the unilateral shift operator, and that S acts as a bounded operator in $\mathcal{H}(b)$ [**11**]. The restriction $S \mid \mathcal{H}(b)$ will be denoted by Y. Since $\mathcal{M}(a)$ is obviously Y-invariant, its closure is a Y-invariant subspace of $\mathcal{H}(b)$; the assumption that f^2 is not an exposed point guarantees that it is a proper subspace.

If λ is in ∂D then, by Proposition 3, the inequality $\|f_\lambda\|_2 < 1$ holds if and only if the range of $T_{1-\lambda b}T_{\bar{f}_\lambda}$ (which is closed by Proposition 3) is a proper subspace of $\mathcal{H}(b)$. That range is the Y-invariant subspace generated by $1-\lambda b$, for if one applies $T_{1-\lambda b}T_{\bar{f}_\lambda}$ to the function z^n, one gets the product of $1-\lambda b$ with a polynomial of degree n, and those functions span the range of $T_{1-\lambda b}T_{\bar{f}_\lambda}$ (in the $\mathcal{H}(b)$ norm). We note for future reference that the range of $T_{1-\lambda b}T_{\bar{f}_\lambda}$ always contains $\mathcal{M}(a)$, as one sees from the equality $T_{1-\lambda b}T_{\bar{f}_\lambda}T_{f_\lambda/\bar{f}_\lambda} = T_a$ [**13**].

The constant function 1 is not in the closure of $\mathcal{M}(a)$ because the Y-invariant subspace generated by it is all of $\mathcal{H}(b)$ [**12**]. We shall see shortly that b also is not in the closure of $\mathcal{M}(a)$. The hypotheses of Theorem 2 thus guarantee the existence of a unique nonzero scalar λ such that $1-\lambda b$ is in the closure of $\mathcal{M}(a)$ and hence generates a proper Y-invariant subspace of $\mathcal{H}(b)$. It will be shown that necessarily $|\lambda|=1$, and the desired inequality $\|f_\lambda\|_2<1$ will follow by the remarks in the preceding paragraph.

The following result from [**12**] will be used.

PROPOSITION 5. *The function h in H^2 belongs to $\mathcal{H}(b)$ if and only if there is a function h^+ in H^2 such that $T_{\bar b}h = T_{\bar a}h^+$. The function h^+ is uniquely determined by h. If h_1 and h_2 are in $\mathcal{H}(b)$, then*

$$\langle h_1,h_2\rangle_b = \langle h_1,h_2\rangle + \langle h_1^+,h_2^+\rangle.$$

This leads to a criterion for membership in $\mathcal{K}(b)$.

PROPOSITION 6. *The function h in $\mathcal{H}(b)$ belongs to $\mathcal{K}(b)$ if and only if $T_{\bar a}h + T_{b\bar a/a}h^+ = 0$.*

To see this let g be any function in H^2. One easily verifies that $(ag)^+ = T_{\bar b a/\bar a}g$, so by Proposition 5,

$$\begin{aligned}\langle h,ag\rangle_b &= \langle h,ag\rangle + \langle h^+, T_{\bar b a/\bar a}g\rangle\\ &= \langle T_{\bar a}h + T_{b\bar a/a}h^+, g\rangle,\end{aligned}$$

from which the desired conclusion is immediate.

It follows easily from the next proposition that b is not in the closure of $\mathcal{M}(a)$.

PROPOSITION 7. *The function h in $\mathcal{H}(b)$ is orthogonal to the Y-invariant subspace generated by b if and only if $h^+=0$.*

To establish this we shall use the formula

$$Y^*h = S^*h + \langle h,b\rangle_b S^*b,$$

whose derivation can be found (in a slightly disguised form) in [**11**]. Suppose first that h is orthogonal to $Y^n b$ for all nonnegative integers n. From the formula above for Y^* we obtain

$$\begin{aligned}Y^{*(n+1)}h &= S^*Y^{*n}h + \langle Y^{*n}h,b\rangle_b S^*b\\ &= S^*Y^{*n}h + \langle h,Y^n b\rangle_b S^*b = S^*Y^{*n}h,\end{aligned}$$

so by induction we find that $Y^{*n}h = S^{*n}h$ for all n. That together with the easily verified equalities $b^+ = \overline{a(0)}^{-1} - a$ and $(S^{*n}h)^+ = S^{*n}h^+$ gives

$$\begin{aligned} 0 = \langle S^{*n}h, b\rangle_b &= \langle S^{*n}h, b\rangle + \langle S^{*n}h^+, \overline{a(0)}^{-1} - a\rangle \\ &= \langle T_{\bar b}h, z^n\rangle + \frac{1}{a(0)}\langle h^+, z^n\rangle - \langle T_{\bar a}h^+, z^n\rangle \\ &= \frac{1}{a(0)}\langle h^+, z^n\rangle, \end{aligned}$$

implying that $h^+ = 0$, as desired. That proves one half of Proposition 7, and one can deduce the other half (which is not needed for Theorem 2) by a straightforward reversal of the preceding argument.

It is now easily seen that no nonzero function in $\mathcal{K}(b)$ is orthogonal to the Y-invariant subspace generated by b, so, as promised earlier, b cannot possibly be in the closure of $\mathcal{M}(a)$. Namely, if h is in $\mathcal{K}(b)$ and h is orthogonal to the Y-invariant subspace generated by b, then $h^+ = 0$ by Proposition 7, so $T_{\bar a}h = -T_{b\bar a/a}h^+ = 0$ by Proposition 6, which implies $h = 0$ since $T_{\bar a}$ has a trivial kernel (a being an outer function).

One additional proposition is needed for the proof of Theorem 2.

PROPOSITION 8. *For h in $\mathcal{H}(b)$ and w in D,*

$$\langle h, k_w\rangle_b = h(w) + \frac{b(w)}{a(w)}h^+(w)$$

$$\langle h, bk_w\rangle_b = \frac{h^+(w)}{a(w)}.$$

The first formula follows from Proposition 5 together with the easily verified equality $(k_w)^+ = \overline{b(w)}k_w/\overline{a(w)}$. The second formula follows from the first one together with the equality $k_w^b = (1 - \overline{b(w)}b)k_w$. (Notice that the first formula yields easily the value of $\langle k_w, k_z\rangle_b$ needed in the proof of Proposition 4.)

Now for the proof of Theorem 2: Let λ be the unique complex number, which exists in virtue of the hypotheses of the theorem and is nonzero, such that the function $1 - \lambda b$ is in the closure of $\mathcal{M}(a)$. As mentioned earlier, the main step is to show that $|\lambda| = 1$.

The spectrum of the operator Y is known to be the closed unit disk [**12**]. Therefore, for w in D, the function $(1-\lambda b)k_w$ can be rewritten as $(1-\bar{w}Y)^{-1}(1-\lambda b)$. The closure of $\mathcal{M}(a)$ is Y-invariant and therefore also invariant under $(1-\bar{w}Y)^{-1}$ for w in D (since the Neumann series for $(1-\bar{w}Y)^{-1}$ converges to it in operator norm). Consequently, each function $(1-\lambda b)k_w$ lies in the closure of $\mathcal{M}(a)$ and so is orthogonal to $\mathcal{K}(b)$. This implies in conjunction with Proposition 8 that $ah=(\bar{\lambda}-b)h^+$ for all h in $\mathcal{K}(b)$.

Suppose that $|\lambda|>1$, and let h be any function in $\mathcal{K}(b)$. The function $g=(\bar{\lambda}-b)^{-1}h$ is then in H^2, and $h^+=ag$. The equality $T_{\bar{a}}h+T_{b\bar{a}/a}h^+=0$ from Proposition 6 can thus be rewritten as $T_{\bar{a}}h=-T_{\bar{a}}bg$, which implies that $h=-bg$ since the kernel of $T_{\bar{a}}$ is trivial. But also $h=(\bar{\lambda}-b)g$, so it follows that $h=0$, in contradiction to the nontriviality of $\mathcal{K}(b)$. We conclude that $|\lambda|\le 1$.

Suppose now that $|\lambda|<1$, and again let h be any function in $\mathcal{K}(b)$. As noted above, $h^+=ah/(\bar{\lambda}-b)$, and therefore

$$-\bar{b}h+\bar{a}h^+=-\bar{b}h+\frac{|a|^2h}{\bar{\lambda}-b}=\frac{|a|^2h+|b|^2h-\bar{\lambda}\bar{b}h}{\bar{\lambda}-b}=\frac{(1-\overline{\lambda b})h}{\bar{\lambda}-b}.$$

Thus

$$\frac{h}{\bar{\lambda}-b}=\frac{-\bar{b}h+\bar{a}h^+}{1-\overline{\lambda b}}.$$

The function $-\bar{b}h+\bar{a}h^+$ is in $(H^2)^\perp$ because of the equality $T_{\bar{b}}h=T_{\bar{a}}h^+$, so, because $|\lambda|<1$, we can conclude that the function $h/(\bar{\lambda}-b)$ is in $(H^2)^\perp$. On the other hand, because $(\bar{\lambda}-b)h^+=ah$ and a is outer, the inner factor of $\bar{\lambda}-b$ must divide the inner factor of h. Since $h/(\bar{\lambda}-b)$ is in L^2 we can conclude that it is in H^2. Since $h/(\bar{\lambda}-b)$ is in both H^2 and $(H^2)^\perp$ it is 0, so $h=0$, and again we have a contradiction to the nontriviality of $\mathcal{K}(b)$. We can conclude that $|\lambda|=1$.

As noted at the beginning of the section, it now follows that $\|f_\lambda\|_2<1$, so Theorem 2 is proved except for the uniqueness of λ. To see that λ is unique, let η be any point on the unit circle different from λ. Then the Y-invariant subspace generated by $1-\lambda b$ and $1-\eta b$ together is all of $\mathcal{H}(b)$ since it contains the constant function 1. The Y-invariant subspaces generated by $1-\lambda b$ and $1-\eta b$ separately contain the closure of $\mathcal{M}(a)$ (as remarked earlier), and the one generated by $1-\lambda b$ is contained in the closure of $\mathcal{M}(a)$. It follows that the Y-invariant subspace generated by $1-\eta b$ must be all of $\mathcal{H}(b)$, and so $\|f_\eta\|_2=1$. The proof of Theorem 2 is now complete.

To produce a simple example in which the hypotheses of Theorem 2 are satisfied, let u be any inner function vanishing at the origin, and let $b = -u(1+u)/2$. That gives $a = (1-u)/2$ and $f = (1-u)/(2+u+u^2)$. That $\|f\|_2 = 1$ will be verified shortly; at least one knows at the outset that $\|f_\lambda\|_2 = 1$ for almost all λ [**13**]. Since $1+b = (2-u-u^2)/2 = (1-u)(2+u)/2$, the function $1+b$ lies in $\mathcal{M}(a)$, so the hypotheses of Theorem 2 do hold, or at least we shall be able to so assert once we know that $\|f\|_2 = 1$. But the last equality in fact follows from the uniqueness part of the theorem (just apply the theorem with f replaced by any f_λ of unit norm, of which we know there is an abundant supply).

7. CONCLUDING REMARKS

What remains in order to establish the conjecture? (I am certain it is true.) After the results above it remains to show that if f^2 is not an exposed point and if the functions $P_0 1$ and $P_0 b$ are linearly independent, then there is a λ in ∂D such that $\|f_\lambda\|_2 < 1$. In an attempt to get a handle on that problem, I have tried to analyze the operator $P_0 Y \mid \mathcal{K}(b)$. This operator is a rank-one perturbation of a unitary operator if $P_0 1$ and $P_0 b$ are linearly independent. (It is unitary in the contrary case, the case addressed by Theorem 2.) The analysis can be carried quite far, but at this point the most it has led to seems to be a reduction of the general case to the case where $f/\bar{f}$ is an inner function. While the latter case has some simplifying features, it is not clear how one can take advantage of them.

One source of difficulty is that, when f^2 is not an exposed point and $P_0 1$ and $P_0 b$ are linearly independent, there is no obvious candidate for a λ such that $\|f_\lambda\|_2 < 1$. I feel, though, that the missing implication (iii) $\Rightarrow$ (i) should hold in a strong form, namely, I would conjecture that the closure of $\mathcal{M}(a)$ is the intersection of the Y-invariant subspaces generated by the functions $1 - \lambda b$ individually ($\lambda \in \partial D$). This should be obtainable from a sufficiently penetrating analysis of the operator $P_0 Y \mid \mathcal{K}(b)$.

A proof of the implication (iii) $\Rightarrow$ (ii) would probably be worth attempting. Although that by itself would not settle what remains of the conjecture, it undoubtedly would provide important insight.

REFERENCES

1. D. Alpay and H. Dym, Hilbert spaces of analytic functions, inverse scattering, and operator models, I. Integral Equations and Operator Theory 7 (1984), 589–641.

2. S. Axler, S.–Y. A. Chang and D. Sarason, Products of Toeplitz operators. Integral Equations and Operator Theory 1 (1978), 285–309.

3. J. A. Ball, "Unitary Perturbations of Contractions." Doctoral Dissertation, University of Virginia, 1973.

4. J. A. Ball and A. Lubin, On a class of contractive perturbations of restricted shifts. Pacific J. Math. 63 (1976), 309–323.

5. J. Barría and P. R. Halmos, Asymptotic Toeplitz operators. Trans. Amer. Math. Soc. 273 (1982), 621–630.

6. L. de Branges and J. Rovnyak, "Square Summable Power Series." Holt, Rinehart and Winston, New York, 1966.

7. D. N. Clark, Extending Fourier transforms into Sz.–Nagy–Foias spaces. Bull. Amer. Math. Soc. 78 (1972), 65–67.

8. D. N. Clark, One dimensional perturbations of restricted shifts. J. Analyse Math. 25 (1972), 169–191.

9. K. de Leeuw and W. Rudin, Extreme points and extremum problems in H_1. Pacific J. Math. 8 (1958), 467–485.

10. D. Sarason, "Function Theory on the Unit Circle." Virginia Polytechnic Institute and State University, Blacksburg, Virginia, 1979.

11. D. Sarason, Shift-invariant spaces from the Brangesian point of view. The Bierberbach Conjecture – Proceedings of the Symposium on the Occasion of the Proof, Mathematical Surveys, Amer. Math. Soc., Providence, 1986, pp. 153–166.

12. D. Sarason, Doubly shift-invariant spaces in H^2. J. Operator Theory 16 (1986), 75–97.

13. D. Sarason, Exposed points in H^1, I. Operator Theory: Advances and Applications, Vol. 41 (1989), Birkhäuser Verlag, Basel, pp. 485–496.

14. A. L. Volberg, Two remarks concerning the theorem of S. Axler, S.-Y. A. Chang and D. Sarason. J. Operator Theory 7 (1982), 209–218.

Department of Mathematics
University of California
Berkeley, CA 94720

Operator Theory:
Advances and Applications, Vol. 48

FUNCTIONAL COMPLETENESS OF DEFINITE SUBSPACES IN KREIN SPACE

Yu. L. Shmul'yan*

If L is maximal definite subspace in Krein space H, then any linear continuous (with respect to intrinsic topology) functional on L can be represented in the form $[.,y]$ for some $y \in H$ (but not necessarily $y \in L$).

Let H be a Krein space [1,2] with Hilbert scalar product $(.,.)$ and indefinite scalar product $[.,.]$, $H_1 \oplus H_2$ - its canonical decomposition, P_1 and P_2 - orthoproectors to H_1 and H_2 respectively. A linear subset of H (not necessarily closed) will be called lineal.

Let L be a positive lineal in H. This lineal is pre-Hilbert space with respect to scalar product $[.,.]$. Corresponding topology of L will be called an intrinsic one or a τ-topology.

For any $y \in H$ the functional $f_y(x)=[x,y]$ $(x \in L)$ is linear, but not necessarily τ-continuous. Let $\tilde{L}$ be the set of all $y \in H$, for which f_y is τ-continuous on L.

In the present paper we give a characterization of the lineal $\tilde{L}$ in the case, when L is a maximal positive subspace.

1. By Cauchy-Bunyakowski inequality $L \subset \tilde{L}$. On the other hand, if $y \in L^{[\perp]}$, then f_y is a zero-functional on L. Hence $\tilde{L} \supset L[+]L^{[\perp]}$, but in general equality is not valid.

In the following part L is a maximal positive subspace in H. Let K be its angular operator. Then K is a contraction $H_1 \rightarrow H_2$. We define defect operators

$$D=(I_1-K^*K)^{1/2}, \quad D_*=(I_2-KK^*)^{1/2}.$$

It is well known [4], that

$$KD=D_*K. \tag{1}$$

*With deep regrets the editors inform the readers that Professor Shmul'yan died in Odessa on 8th March, 1990.

The subspace L is a set of all vectors $u=g+Kg$ ($g \in H_1$). If $u'=g'+Kg' \in L$, then

$$[u,u']=(g,g')-(Kg,Kg')=(Dg,Dg').$$

Consequently, the correspondence $u=g+Kg \longleftrightarrow Dg=DP_1u$ is an isometry between L and ImD ($\subset H_1$). In particular, $[u,u]^{1/2}=\|Dg\|$.

We shall use frequently next

LEMMA[3]. *Let T be a bounded operator in H. An element $z \in H$ belongs to ImT iff*

$$\sup_{g \in H} \frac{|(g,z)|}{\|T^*g\|} < \infty .$$

The next theorem gives a characterizatio. of lineal $\tilde{L}$.

THEOREM 1. *Let $y \in H$, $y_k=P_ky$ $(k=1,2)$. e following conditions are equivalent:*

1) $y_1-K^*y_2 \in ImD$; 2) $y_2-Ky_1 \in ImD_*$; 3) $y \in \tilde{}$.

PROOF. 1) $\Longrightarrow$ 2) : Let 1) be true. We hawe $y_1-K^*y_2=Du$ for some $u \in H_1$. Then by (1) $y_2-Ky_1=D_*^2$ -

$-K(y_1-K^*y_2)=D_*^2y_2-KDu=D_*(D_*y_2-Ku) \in ImD_*$, i.e. 2) is satisfied.

In the same way we can prove 2) $\Longrightarrow$ 1).

1) $\Longleftrightarrow$ 3). Let $y \in H$, $y_k=P_ky$ $(k=1,2)$, $x=x_1+Kx_1 \in L$.

We hawe $f_y(x)=[x,y]=(x_1,y_1)-(Kx_1,y_2)=(x_1,y_1-K^*y_2)$.

Hence

$$\frac{|f_y(x)|}{[x,x]^{1/2}} = \frac{|(x_1,y_1-K^*y_2)|}{\|Dx_1\|} .$$

By the lemma the boundness of f_y on L is equivalent to the condition $y_1-K^*y_2 \in ImD$.

2. The space L with norm $[x,x]^{1/2}$ in general is not complete and for this reason not each linear τ-continuous functional on L has the form f_y for some $y \in L$. We shall proof, that any linear τ-continuous functional F on L has a form f_y for some $y \in \tilde{L}$. This property expresses the functional completeness of L.

THEOREM 2. *Let F be a linear τ-continuous functional on L. Then there exists such* $y \in \tilde{L}$, *for which* $F=f_y$.

PROOF. We define a linear functional Φ on ImD by the next way: for $u=Dg (g \in H_1)$ we put $\Phi(u)=F(g+Kg)$. This functional is well defined and continuous, because

$|\Phi(u)| \leq \|F\| [g+Kg, g+Kg]^{1/2} = \|F\| \cdot \|Dg\| = \|F\| \cdot \|u\|$.

Hence, there exists some vector $y_1 \in H_1$ for which

$\Phi(u)=(u,y_1)$ $(\forall u \in \mathrm{ImD})$, $F(g+Kg)=(u,y_1)=(Dg,y_1)=[g+Kg,Dy_1]$,

i.e. $F=f_y$, where $y=Dy_1$ $(\in \tilde{L})$.

3. In the present part another characterisation of $\tilde{L}$ is given.

Let S be an operator in H with the matrix representation

$$\begin{bmatrix} I_1 & K^* \\ K & I_2 \end{bmatrix}$$

with respect to canonical decomposition $H=H_1 \oplus H_2$. This operator is nonnegative, because

$$(Sg,g)=\|g_1+K^*g_2\|^2+\|D_*g_2\|^2 \quad (g \in H) \tag{2}$$

THEOREM 3. $\tilde{L}=\mathrm{ImS}^{1/2}$

PROOF. Let $u \in \tilde{L}$, $u_k=P_k u$ $(k=1,2)$. By theorem 1 $u_2-Ku_1=D_*h$ for some $h \in H_2$. For any $g \in H$, $g_k=P_k g$ $(k=1,2)$ we put

$$v=P_1Sg=g_1+K^*g_2 \tag{3}$$

We have

$(u,g)=(u_1,g_1)+(u_2,g_2)=(u_1,v-K^*g_2)+(u_2,g_2)=(u_1,v)+(u_2-Ku_1,g_2)=$

$=(u_1,v)+(D_*h,g_2)=(u_1,v)+(h,D_*g_2)$.

Then by (2) and (3)

$|(u,g)|^2 \leq (\|u_1\|^2+\|h\|^2)(\|v\|^2+\|D_*g_2\|^2)=(\|u_1\|^2+\|h\|^2)(Sg,g)$.

Hence

$$\frac{|(u,g)|}{\|S^{1/2}g\|} \leq (\|u_1\|^2+\|h\|^2)^{1/2} \quad (g \in H_1)$$

and by the lemma $u \in \mathrm{ImS}^{1/2}$.

Conversely, let $u \in \mathrm{Im}S^{1/2}$, i.e.

$$\sup_g \frac{|(u,g)|^2}{(Sg,g)} < \infty.$$

In particular, we consider $g=-K^*g_2+g_2$, where g_2 runs H_2. By (2) $(Sg,g)=\|D_*g_2\|^2$. Further, $(u,g)=(u_1,-K^*g_2)+(u_2,g_2)=(u_2-Ku_1,g_2)$. Hence

$$\sup_{g_2} \frac{|(u_2-Ku_1,g_2)|}{\|D_*g_2\|} < \infty.$$

By the lemma $u_2-Ku_1 \in \mathrm{Im}D_*$, and by theorem 1 $u \in \tilde{L}$.

REFERENCES

1. Azizov, T.Ya., Iohvidov, I.S.: Foundations of theory of linear operators in the spaces with an indefinite metric, Moscow, "Nauka", 1986. (Russian)

2. Bogn`ar, J.: Indefinite inner product spaces, Springer-Verlag, Berlin-Heidelberg-New Jork, 1974.

3. Smul'yan, Yu.L.: Two-sided division in a ring of operators, Matem. zametki, 1(1967), 605-610. (Russian)

4. Sz-Nagy, B., Foias, C.: Harmonic analysis of operators in Hilbert space, Academiai Kiado, North Holland, Budapest - Amsterdam - London, 1970.

Departement of Mathematics,
Odessa Institute of Marine Engineers,
34, Mechnikov str.,
Odessa, 270029,
USSR

Operator Theory:
Advances and Applications, Vol. 48

Expansions in Generalized Eigenfunctions of Schrödinger Operators with Singular Potentials

Günter Stolz

We proof the existence of an expansion in generalized eigenfunctions for Schrödinger operators with singular electromagnetic potentials, satisfying conditions not going beyond those guaranteeing essential selfadjointness on $C_0^\infty(\mathbb{R}^\nu)$. The generalized eigenfunctions and their weak derivatives up to the second order are locally square integrable.

Under suitable conditions on the asymptotics of the electromagnetic potentials the generalized eigenfunctions lie in weighted Sobolev spaces. In particular for the case of polynomially increasing potentials a polynomially weighted Sobolev space can be chosen. In this situation we in fact show polynomial boundedness by means of regularity theory for solutions of partial differential equations.

1. Introduction

The beginning of the developement of the mathematical theory of expansions in generalized eigenfunctions has to be dated before the quantum mechanics area. Hellinger in 1909 used non square summable eigenvectors in the discription of the spectrum of bounded quadratic forms [He] and Weyl in 1910 gave the first eigenfunction expansions for ordinary differential operators [We].

But it was not until the quantum mechanical hypothesis of completeness of eigenfunctions that the question of having expansions in generalized eigenfunctions of self-adjoint differential operators gained its full importance.

Since the 1950s much work was done to settle with this question in a mathematically rigorous way. A survey of the literature is given in [S2], chapter C.5. There also the notions IP-expansion (for Ikebe and Povzner) and BGK-expansion (Berezanskii, Browder, Garding, Gelfand, Kac) are used to distinguish between the first type of expansions, which is an important tool in scattering theory, and the second type, which gives less information but exists for a much wider class of operators.

Results on the existence of BGK-expansions for Schrödinger operators have been proven in [S2] and [KS]. The main goal of this paper is to generalize these results by proving the existence of an expansion in generalized eigenfunctions for "almost every" Schrödinger operator, which can be defined as an essentially selfadjoint operator on $C_0^\infty(\mathbb{R}^\nu)$.

We base our investigations on an abstract expansion theorem of BGK-type, which is given with a short proof in [PSW]. Theorem 2.1 in Section 2 is just this theorem in a form adapted to our applications, Theorem 2.2 is a "localized" version. Henceforth the term "generalized eigenfunction" is understood in the sense of these theorems.

We will apply Theorems 2.1 and 2.2 to the case of Schrödinger operators

$$H = (-i\nabla - b)^2 + V \quad \text{in} \quad L_2(\mathbb{R}^\nu). \tag{1.1}$$

To describe V, we choose conditions of Stummel-type (e.g. [W1], Ch. 10.3). Therefore we define for measurable functions $q : \mathbb{R}^\nu \to \mathbb{C}$ and $\rho \in \mathbb{R}$

$$M_{q,\rho}(x) := \begin{cases} \left(\int\limits_{|x-y|\le 1} |q(y)|^2 \, |x-y|^{\rho-\nu} \, dy \right)^{1/2} & \text{for } \rho < \nu, \\ \left(\int\limits_{|x-y|\le 1} |q(y)|^2 \, dy \right)^{1/2} & \text{for } \rho \ge \nu. \end{cases}$$

Stummel spaces and local Stummel spaces are given by

$$M_\rho(\mathbb{R}^\nu) \;:=\; \{ q : \mathbb{R}^\nu \to \mathbb{C} : \; M_{q,\rho}(\cdot) \text{ is bounded} \},$$

$$M_{\rho,loc}(\mathbb{R}^\nu) := \{ q : \mathbb{R}^\nu \to \mathbb{C} : \; M_{q,\rho}(\cdot) \text{ is locally bounded} \}.$$

The Schrödinger operator (1.1) is essentially selfadjoint on $C_0^\infty(\mathbb{R}^\nu)$ under the conditions

(A) $\quad b \in C^1(\mathbb{R}^\nu)^\nu, \quad V = V_1 + V_2,$

$\quad V_1 \in M_\rho(\mathbb{R}^\nu), \quad V_2 \in M_{\rho,loc}(\mathbb{R}^\nu)$ for some $\rho < 4$,

$\quad V_2 \ge -C k_\mu$ for some $\mu \in [0, 2]$ and $C \ge 0$

([W1], Th. 10.23), where $k_\mu(x) := (1 + |x|^2)^{\mu/2}$. Subsequently H denotes the selfadjoint closure with

$$D(H) = \left\{ f \in L_2(\mathbb{R}^\nu) \cap W_{2,2,loc}(\mathbb{R}^\nu) : (-i\nabla - b)^2 f + Vf \in L_2(\mathbb{R}^\nu) \right\},$$

where $W_{2,2,loc}(\mathbb{R}^\nu) = \{g: \varphi g \in W_{2,2}(\mathbb{R}^\nu)$ for every $\varphi \in C_0^\infty(\mathbb{R}^\nu)\}$, $W_{2,2}(\mathbb{R}^\nu)$ being the Sobolev space of second order.

The following two theorems are our main results and will be proven in Section 3.

Theorem 1.1: *Let* H *satisfy* (A). *Then there exists an expansion in generalized eigenfunctions of* H, *which lie in* $W_{2,2,loc}(\mathbb{R}^\nu)$.

If we want to get results on asymptotic properties of the generalized eigenfunctions, we have to make corresponding assumptions on the asymptotics of b and V. Here we will investigate polynomial boundedness:

(B1) $|b| \le C_1 k_{s_1}$, $|\mathrm{div}\, b| \le C_2 k_{s_2}$, $|V_2| \le C_3 k_{s_3}$

for $C_i \ge 0$, $s_i \ge 0$, $i = 1, 2, 3$.

Theorem 1.2: *Let* H *satisfy* (A) *and* (B). *Then there exists an* $s > 0$ *and an expansion in generalized eigenfunctions of* H, *which lie in* $\{g : k_{-s} g \in W_{2,2}(\mathbb{R}^\nu)\}$.

Theorem 1.1 seems to be the first result on the existence of an eigenfunction expansion for Schrödinger operators under conditions not going beyond those guaranteeing essential selfadjointness on $C_0^\infty(\mathbb{R}^\nu)$. It generalizes the expansion for ordinary differential operators in [W2], Theorems 8.4 and 8.7 to Schrödinger operators in any dimension. Furthermore we get generalized eigenfunctions in $W_{2,2,loc}(\mathbb{R}^\nu)$, whereas in [W2] the assumptions about the coefficients of the differential operator are so weak that only the local L_1-property of the second derivative of the generalized eigenfunctions is guaranteed.

Theorem 1.2 extends the results of [S2] and [KS] in the following respects:

(i) Magnetic fields are included,

(ii) V(x) may behave like $-|x|^2$ as $|x| \to \infty$ (in [S2], Theorem B.12.2 only $-|x|$ is allowed, whereas in [KS] relative boundedness of the quadratic form corresponding to V_- with respect to $-\Delta$ is needed,

(iii) the theory directly gives generalized eigenfunctions in a polynomially weighted Sobolev space (in [S2], Theorem C.5.4 and [KS], Theorem 1.2 one only gets a polynomially weighted L_2-space, and one has to apply regularity theory of weak solutions to get further information).

Section 4 deals with showing that the generalized eigenfunctions are in fact locally or polynomially bounded, not only in the L_2-mean. For technical reasons we need some slightly stronger conditions on the singularities of V:

(A') $b \in C_1(\mathbb{R}^\nu)^\nu$, $V = V_1 + V_2$,

$V_1 \in L_{p,loc,unif}(\mathbb{R}^\nu)$, $V_2 \in L_{p,loc}(\mathbb{R}^\nu)$ for $p \geq 2$, $p > \frac{\nu}{2}$,

$V_2 \geq -C k_\mu$ for $\mu \in [0, 2]$ and $C \geq 0$.

Here $L_{p,loc,unif}(\mathbb{R}^\nu)$ are the measurable functions f for which $\int_{|x-y|\leq 1} |f(y)|^p dy \leq C$ with C independent of x. That (A') is a special case of (A) follows by using Hölder's inequality.

With techniques of regularity theory we prove

Theorem 1.3: *Let* H *satisfy* (A').

a) The generalized eigenfunctions of H *are locally bounded.*

b) If furthermore (B) *is satisfied, then the generalized eigenfunctions are polynomially bounded, i.e. for some* $s > 0$ *they lie in* $\{g : k_{-s} g \in L_\infty(\mathbb{R}^\nu)\}$.

It seems to us that polynomial boundedness should be the best possible result on the asymptotics of generalized eigenfunctions under our conditions. For a proof of the boundedness of generalized eigenfunctions, usually taken for granted in physics, the class of potentials should be restricted to more physical situations.

We have not written down the theorems and proofs under the most general conditions on b, which can be handled by our methods. All results remain true if b is only assumed to be weakly differentiable and locally resp. polynomially bounded. In any case $\operatorname{div} b$ has to fulfill the same assumptions as V.

All the results given here are included in the thesis [St], where also a detailed discussion of mathematical and physical motivation for the study of generalized eigenfunctions is presented. In Section 5 we comment on some other results on eigenfunction expansions, which are given in [St].

2. The expansion theorem

For every selfadjoint operator H in a Hilbert space $\mathcal{H}$ there exists a *spectral measure* μ, i.e. a Borel measure μ on $\mathbb{R}$ such that $\mu(\Delta) = 0$ if and only if $E(\Delta) = 0$, where $E(\cdot)$ is the spectral resolution of H. With such a spectral measure a *spectral representation* of the following form can be given for H:

There exists an $N \in \mathbb{N} \cup \{\infty\}$ and a unitary operator

$$U = (U_j) : \mathcal{H} \to \bigoplus_{j=1}^{N} L_2(M_j, d\mu)$$

such that $UHU^{-1} = M_{id}$. The $M_j \subset \mathbb{R}$ are μ-measurable sets with $M_{j+1} \subset M_j$, $j = 1, 2, \dots$. $M_j \backslash M_{j+1}$ are exactly the sets of *spectral multiplicity* j of H. We call U a *μ-spectral representation* for H (see [PSW]).

Now we can give the expansion theorem of [PSW] in a form adapted to the treatment of selfadjoint differential operators (l.i.m. denotes limits in L_2-sense, $L_{2,g}(\mathbb{R}^\nu) := \{ f : gf \in L_2(\mathbb{R}^\nu) \}$).

Theorem 2.1: *Let* H *be a selfadjoint operator in* $L_2(\mathbb{R}^\nu)$, μ *a spectral measure for* H *and* U *a μ-spectral representation. Suppose that* $C_0^\infty(\mathbb{R}^\nu) \subset D(H)$ *and that* Hf *is of compact support for every* $f \in C_0^\infty(\mathbb{R}^\nu)$. *For a bounded, continuous function* $\gamma: \mathbb{R} \to \mathbb{C}$ *with* $0 < |\gamma|$ *on* $\sigma(H)$ *and a bounded, continuous and positive function* g *on* $\mathbb{R}^\nu$ *let* $\gamma(H)g$ *be a Hilbert-Schmidt operator.*

Then for $j \in \{1, \dots, N\}$ *there exist μ-measurable functions* $\varphi_j : M_j \to L_{2,g}(\mathbb{R}^\nu)$ *such that for every* $f \in L_2(\mathbb{R}^\nu)$ *and μ-almost every* $\lambda \in M_j$

a) $$(U_j f)(\lambda) = \underset{R\to\infty}{\text{l.i.m.}} \int_{|x|\le R} \overline{\varphi_j(\lambda, x)}\, f(x)\, dx,$$

b) $\varphi_j(\lambda, \cdot)$ *is a weak solution of* $H\varphi = \lambda\varphi$, *i.e.*

$$\int \overline{\varphi_j(\lambda, x)}\, (H\psi)(x)\, dx = \lambda \int \overline{\varphi_j(\lambda, x)}\, \psi(x)\, dx \quad \textit{for every} \quad \psi \in C_0^\infty(\mathbb{R}^\nu),$$

c) $$f(x) = \underset{n\to N, E\to\infty}{\text{l.i.m.}} \sum_{j=1}^{n} \int_{\{|\lambda|\le E\} \cap M_j} (U_j f)(\lambda)\, \varphi_j(\lambda, x)\, d\mu(\lambda).$$

If furthermore A *is an injective operator in* $L_2(\mathbb{R}^\nu)$ *with bounded inverse and* $\gamma(H)gA$ *is the restriction of a Hilbert-Schmidt operator to* D(A), *then the* $\varphi_j(\lambda)$ *lie in the range of* $\frac{1}{g}(A^{-1})^*$.

Proof: Without restriction we assume $g \le 1$. Our assumptions are a special case of the situation of Theorems 1 and 2 in [PSW]. The *T-triple* from there is chosen by taking for T^{-1} the multiplication with g, hence $\mathcal{H}_-(T) = L_{2,g}(\mathbb{R}^\nu)$ and $\mathcal{H}_+(T) = L_{2,1/g}(\mathbb{R}^\nu)$. As a generalized scalar product ([PSW], Lemma 1) we get the ordinary L_2-scalar product, the factors weighted with g and 1/g respectively. With these specifications our statements are identical to those of Theorems 1 and 2 in [PSW]. □

Theorem 2.1 does not cover situations, where $\gamma(H)g$ is Hilbert-Schmidt only for every $g \in C_0^\infty(\mathbb{R}^\nu)$. Next we show that in this case generalized eigenfunctions in $L_{2,loc}(\mathbb{R}^\nu)$ exist. The method of proof is that of Theorem 1 in [PSW] modified by the fact, that the generalized eigenfunctions are first constructed locally; by a uniqueness argument we get the global existence. This method was first used in the proof of Theorem 8.4 in [W2] for ordinary differential operators.

Theorem 2.2: *Let* H *be selfadjoint in* $L_2(\mathbb{R}^\nu)$, $C_0^\infty(\mathbb{R}^\nu) \subset D(H)$, Hf *be of compact support for every* $f \in C_0^\infty(\mathbb{R}^\nu)$, μ *a spectral measure for* H *and* U *a* μ*-spectral representation. For a bounded, continuous function* $\gamma : \mathbb{R} \to \mathbb{C}$ *with* $0 < |\gamma|$ *on* $\sigma(H)$ *and every* $g \in C_0^\infty(\mathbb{R}^\nu)$ *let* $\gamma(H)g$ *be a Hilbert-Schmidt operator. Then there exist* μ*-measurable functions* $\varphi_j : M_j \to L_{2,loc}(\mathbb{R}^\nu)$, *which satisfy a), b) and c) of Theorem 2.1.*

If A *is an injective operator in* $L_2(\mathbb{R}^\nu)$ *with bounded inverse and* $\gamma(H)gA$ *the restriction of a Hilbert-Schmidt operator to* D(A) *for every* $g \in C_0^\infty(\mathbb{R}^\nu)$, *then the generalized eigenfunctions lie locally in the range of* $(A^{-1})^*$, *i.e. for every* n, j *and* μ*-almost every* λ *there exists* $h_{n,j,\lambda} \in R((A^{-1})^*)$ *such that*

$$\varphi_j(\lambda, x) = h_{n,j,\lambda}(x) \quad \textit{for} \quad |x| \le n.$$

Proof: For each $n \in \mathbb{N}$ choose $\psi_n \in C_0^\infty(\mathbb{R}^\nu)$ such that $0 \le \psi_n \le 1$, $\psi_n(x) = 1$ for $|x| \le n$ and $\psi_n(x) = 0$ for $|x| \ge n+1$. Then $\gamma U_j \psi_n = U_j \gamma(H) \psi_n$ is Hilbert-Schmidt as an operator from $L_2(\mathbb{R}^\nu)$ to $L_2(M_j, d\mu)$, i.e. there exist μ-measurable functions $u_j^n(\cdot) : M_j \to L_2(\mathbb{R}^\nu)$ with $\|u_j^n(\cdot)\| \in L_2(M_j, d\mu)$ and

$$\gamma(\lambda)(U_j \psi_n f)(\lambda) = \langle u_j^n(\lambda), f \rangle \quad \text{for} \quad f \in L_2(\mathbb{R}^\nu) \text{ and } \mu\text{-a.e. } \lambda \in M_j.$$

Let $w_j^n(\lambda) := \gamma(\lambda)^{-1} u_j^n(\lambda)$; then $w_j^n(\cdot) : M_j \to L_2(\mathbb{R}^\nu)$ and

$$(U_j \psi_n f)(\lambda) = \langle w_j^n(\lambda), f \rangle \quad \text{for} \quad f \in L_2(\mathbb{R}^\nu),\ \mu\text{-a.e. } \lambda \in M_j.$$

If $n_1 \ge n$ and $\operatorname{supp} f \subset \{x \in \mathbb{R}^\nu : |x| \le n\}$, then we have

$$0 = (U_j \psi_{n_1} f)(\lambda) - (U_j \psi_n f)(\lambda) = \langle w_j^{n_1}(\lambda) - w_j^n(\lambda), f \rangle,$$

and therefore $w_j^n(\lambda, x) = w_j^{n_1}(\lambda, x)$ almost everywhere in $M_j \times \{|x| \le n\}$ and we can define $\varphi_j(\cdot) : M_j \to L_{2,loc}(\mathbb{R}^\nu)$ by

$$\varphi_j(\lambda, x) = w_j^n(\lambda, x) \quad \text{a.e. in } M_j \times \{|x| \le n\}.$$

Now part a) of the Theorem follows for $f \in L_2(\mathbb{R}^\nu)$ with compact support. For general f the limit has to be taken.

In particular for $\psi \in C_0^\infty(\mathbb{R}^\nu)$ it follows that

$$\int_{\mathbb{R}^\nu} \overline{\varphi_j(\lambda, x)}\, (H\psi)(x)\, dx = (U_j H \psi)(\lambda) = \lambda (U_j \psi)(\lambda) = \lambda \int_{\mathbb{R}^\nu} \overline{\varphi_j(\lambda, x)}\, \psi(x)\, dx,$$

and therefore b) is proven.

Let $v \in L_2(\mathbb{R}^\nu)$ have compact support and $g = (g_j) \in \oplus_j L_2(M_j, d\mu)$ with $g_j = 0$ for $j > j_0$ and $g_j(\lambda) = 0$ for $|\lambda| > R$, $j = 1, \dots, j_0$. Then

$$\langle U^{-1}g, v\rangle = \langle g, Uv\rangle = \sum_{j=1}^{j_0} \int_{M_j} \overline{g_j(\lambda)}\,(U_j v)(\lambda)\,d\mu(\lambda)$$

$$= \sum_{j=1}^{j_0} \int_{M_j} \overline{g_j(\lambda)} \int_{\mathbb{R}^\nu} \overline{\varphi_j(\lambda, x)}\, v(x)\,dx\,d\mu(\lambda)$$

$$= \int_{\mathbb{R}^\nu} \overline{\sum_{j=1}^{j_0} \int_{M_j} g_j(\lambda)\,\varphi_j(\lambda, x)\,d\mu(\lambda)}\, v(x)\,dx,$$

hence

$$(U^{-1}g)(x) = \sum_{j=1}^{j_0} \int_{M_j} g_j(\lambda)\,\varphi_j(\lambda, x)\,d\mu(\lambda) \quad \text{for a.e. } x \in \mathbb{R}^\nu,$$

Taking the limit for an arbitrary g and then inserting $g = Uf$ gives c).

Under the stronger assumptions in the last part of the Theorem we proceed as above to get functions $\tilde{u}_j^n(\cdot): M_j \longrightarrow L_2(\mathbb{R}^\nu)$ with $\|\tilde{u}_j^n(\cdot)\| \in L_2(M_j, d\mu)$ and

$$\gamma(\lambda)(U_j\psi_n A g)(\lambda) = \langle \tilde{u}_j^n(\lambda), g\rangle \quad \text{for } g \in D(A).$$

Let $\tilde{w}_j^n(\lambda) = \gamma(\lambda)^{-1}\tilde{u}_j^n(\lambda)$ and $w_j^n(\lambda)$ as above. Then for every $f \in L_2(\mathbb{R}^\nu)$

$$\langle w_j^n(\lambda), f\rangle = (U_j\psi_n A)(A^{-1}f)(\lambda) = \langle \tilde{w}_j^n(\lambda), A^{-1}f\rangle = \langle (A^{-1})^* \tilde{w}_j^n(\lambda), f\rangle$$

and for $|x| \le n$ and μ-a.e. λ

$$\varphi_j(\lambda, x) = w_j^n(\lambda, x) = ((A^{-1})^* \tilde{w}_j^n(\lambda))(x),$$

hence we can choose $h_{n,j,\lambda} = (A^{-1})^* \tilde{w}_j^n(\lambda)$. □

3. Eigenfunction expansions for Schrödinger operators

In the proofs of Theorems 1.1 and 1.2 we will need the boundedness of $-\Delta\varphi(H - z)^{-1}$ for $\varphi \in C_0^\infty(\mathbb{R}^\nu)$ resp. of $-\Delta k_{-s}(H - z)^{-1}$ for some $s > 0$. Both statements follow from the next theorem. $B(L_2(\mathbb{R}^\nu))$ denotes the bounded and everywhere defined linear operators on $L_2(\mathbb{R}^\nu)$.

Theorem 3.1: *Let* H *satisfy* (A). *For* $g \in C^2(\mathbb{R}^\nu)$ *let*

$$(1 + k_{\mu/2}|b| + |b|^2 + |\operatorname{div} b| + |V_2|)\,g, \quad (|b| + k_{\mu/2})\,|\nabla g| \ \textit{and}\ \Delta g$$

be bounded functions on $\mathbb{R}^\nu$. *Then* $-\Delta g(H-z)^{-1} \in B(L_2(\mathbb{R}^\nu))$ *for every* $z \in \rho(H)$, *in particular* $gf \in W_{2,2}(\mathbb{R}^\nu)$ *for every* $f \in D(H)$.

Proof: We have to show

$$\|-\Delta g f\| \le C(\|Hf\| + \|f\|) \tag{3.1}$$

for some $C > 0$ and every $f \in C_0^\infty(\mathbb{R}^\nu)$, the latter being a core of H. For such f we compute

$$-\Delta gf = gHf - (\Delta g)f + 2\sum_{j=1}^{\nu}\big(b_j g + (D_j g)\big)D_j f - (b^2 + i\,\mathrm{div}\, b + V)gf, \tag{3.2}$$

where D_j means $\frac{1}{i}\frac{\partial}{\partial x_j}$. From the assumptions on g it follows that

$$\|gHf - (\Delta g)f - (b^2 + i\,\mathrm{div}\, b + V_2)gf\| \le C_1(\|Hf\| + \|f\|).$$

V_1 is bounded with respect to $-\Delta$ with relative bound 0 ([W1], Theorem 10.17), in particular there is a $C_2 > 0$ such that

$$\|V_1 gf\| \le \frac{1}{2}\|-\Delta gf\| + C_2\|gf\|.$$

Inserting the last two estimates in (3.2) we get

$$\|-\Delta gf\| \le C_3(\|Hf\| + \|f\|) + 4\Big\|\sum_{j=1}^{\nu}\big(b_j g + (D_j g)\big)D_j f\Big\|. \tag{3.3}$$

To treat the last term we estimate

$$\begin{aligned}\|\big(b_j g + (D_j g)\big)D_j f\| &\le \|\big(b_j g + (D_j g)\big)(D_j - b_j)f\| + \|\big(b_j g + (D_j g)\big)b_j f\| \\ &\le C_4\|k_{-\mu/2}(D_j - b_j)f\| + C_5\|f\|.\end{aligned} \tag{3.4}$$

Now we use Auxiliary Theorem 10.26 in [W1], which says

$$\sum_{j=1}^{\nu}\|\chi_{\{|x|\le 1\}}(D_j - b_j)f\|^2 \le 2\|Hf\|^2 + C_6\|f\|^2$$

and

$$\sum_{j=1}^{\nu}\|\chi_{\{n\le|x|\le n+1\}}(D_j - b_j)f\|^2 \le 2\|\chi_{\{n-1\le|x|\le n+2\}}Hf\|^2 + C_7 k_\mu(n)\|\chi_{\{n-1\le|x|\le n+2\}}f\|^2$$

for every n. The second estimate is in fact a slight modification of the result in [W1], which is shown with the same proof. We get

$$\begin{aligned}&\Big(\sum_{j=1}^{\nu}\|k_{-\mu/2}(D_j - b_j)f\|\Big)^2 \le C(\nu)\sum_{j=1}^{\nu}\|k_{-\mu/2}(D_j - b_j)f\|^2 \\ &\le C(\nu)\sum_{j=1}^{\nu}\Big\{\|\chi_{\{|x|\le 1\}}(D_j - b_j)f\|^2 + \sum_{n=1}^{\infty}k_{-\mu}(n)\|\chi_{\{n\le|x|\le n+1\}}(D_j - b_j)f\|^2\Big\} \\ &\le C(\nu)\Big\{2\|Hf\|^2 + C_6\|f\|^2 + \sum_{n=1}^{\infty}\big(2k_{-\mu}(n)\|\chi_{\{n-1\le|x|\le n+2\}}Hf\|^2 + C_7\|\chi_{n-1\le|x|\le n+2\}}f\|^2\big)\Big\}\end{aligned}$$

$$\leq C(\nu)\left\{2\|Hf\|^2 + C_6\|f\|^2 + 6\|Hf\|^2 + 3C_7\|f\|^2\right\}$$
$$\leq C_8\left(\|Hf\| + \|f\|\right)^2.$$

Inserting this in (3.4) we get (3.1) from (3.3). □

Now we give the main step in the proof of Theorem 1.2 by finding the Hilbert-Schmidt operators, which we need to apply Theorem 2.1. The proof is based on some properties of the spaces $B_p(\mathcal{H})$, i.e. the compact operators in a Hilbert space $\mathcal{H}$, which have singular values forming an ℓ_p-sequence. We freely use the following facts (see for example [Sl]):

(i) $B_2(\mathcal{H})$ is the space of Hilbert-Schmidt operators.

(ii) If $S \in B(\mathcal{H})$ and $T \in B_p(\mathcal{H})$, $1 \leq p < \infty$, then $ST \in B_p(\mathcal{H})$ and $TS \in B_p(\mathcal{H})$.

(iii) Let $1 \leq p, q, r < \infty$ with $\frac{1}{p} + \frac{1}{q} = \frac{1}{r}$, $S \in B_p(\mathcal{H})$ and $T \in B_q(\mathcal{H})$, then $ST \in B_r(\mathcal{H})$.

(iv) Let $r \geq 2$, $f, g \in L_r(\mathbb{R}^\nu)$, $\mathcal{F}$ the Fourier transform on $L_2(\mathbb{R}^\nu)$, then $\overline{f\mathcal{F}^{-1}g\mathcal{F}} \in B_r(L_2(\mathbb{R}^\nu))$.

Theorem 3.2: *Let* H *satisfy* (A) *and* (B), $z \in \rho(H)$ *and* N_n *for every* $n \in \mathbb{N}$ *be defined by* $N_1 := \max\{s_1 + \frac{\mu}{2}, 2s_1, s_2, s_3\}$, $N_{n+1} := N_1 + N_n + \max\{1, s_1\}$. *Then*

a) $k_{-s}(H - z)^{-n}$ *is a Hilbert-Schmidt operator for* $n > \max\{1, \frac{\nu}{2} - 1\}$ *and* $s \geq N_n + 2$,

b) $(-\Delta + 1)k_{-s}(H - z)^{-n}$ *is a Hilbert-Schmidt operator for* $n > \max\{2, \frac{\nu}{2} + 1\}$ *and* $s \geq N_n$.

Remark: At least for small ν the choice of n and s is not optimal, which does not effect our purposes.

Proof: $p_j := \mathcal{F}^{-1}x_j\mathcal{F}$ denotes the the j-th component of the momentum operator, $|p| := \left(\Sigma_{j=1}^{\nu} p_j^2\right)^{1/2}$, $p_{j,0}$ means the restriction of p_j to $C_0^\infty(\mathbb{R}^\nu)$. We abbreviate $B(L_2(\mathbb{R}^\nu))$ and $B_p(L_2(\mathbb{R}^\nu))$ by B and B_p. We leave out the (easier) proof for $\nu = 1$ (see [St]) and suppose $\nu \geq 2$.

a) By induction on $n \in \mathbb{N}$ we proof the following three-part statement, whose first part includes the desired Hilbert-Schmidt property:

$$A_n \begin{cases} (1) & k_{-s}(H-z)^{-n} \in B_q \quad \text{for} \quad q > \max\left\{\frac{\nu}{n+1}, \left(\frac{n-1}{\nu}+\frac{1}{2}\right)^{-1}\right\}, \quad q \ge 1 \quad \text{and} \quad s \ge N_n+2, \\ (2) & k_{-s}(H-z)^{-n} \in B_q \quad \text{for} \quad q > \max\left\{\frac{\nu}{n}, \left(\frac{n-1}{\nu}+\frac{1}{2}\right)^{-1}\right\}, \quad q \ge 1 \quad \text{and} \quad s \ge N_n+1, \\ (3) & \overline{k_{-s}p_{j,0}(H-z)^{-1}}(H-z)^{-n+1} \in B_q \\ & \qquad \text{for} \quad q > \max\left\{\frac{\nu}{n}, \left(\frac{n-1}{\nu}+\frac{1}{2}\right)^{-1}\right\}, \quad q \ge 1 \quad \text{and} \quad s \ge N_n+1, \end{cases}$$

First step: Proof of A_1.

The choice of N_1 is made in such a way that $g = k_{-s}$ for $s \ge N_1$ can be used in Theorem 3.1 to get

$$(-\Delta+1)k_{-s}(H-z)^{-1} \in B \quad \text{for} \quad s \ge N_1.$$

Further

$$k_{-2}(-\Delta+1)^{-1} \in B_q \quad \text{for} \quad q > \tfrac{\nu}{2}, \quad q \ge 2$$

by property (iv) above. We get $A_1(1)$ by taking the product of these terms.

From $|p| \le -\Delta+1$ it follows that

$$(|p|+1)k_{-s}(H-z)^{-1} \in B \quad \text{for} \quad s \ge N_1,$$

which together with

$$k_{-1}(|p|+1)^{-1} \in B_q \quad \text{for} \quad q > \nu, \quad q \ge 2,$$

implies $A_1(2)$.

Let $s \ge N_1+1$ and $j \in \{1, \dots, \nu\}$. On $(H-z)C_0^\infty(\mathbb{R}^\nu)$ we get

$$\begin{aligned} p_j k_{-s}(H-z)^{-1} &= p_j k_{-1} k_{-s+1}(H-z)^{-1} \\ &= (D_j k_{-1})k_{-s+1}(H-z)^{-1} + k_{-1}p_j k_{-s+1}(H-z)^{-1}. \end{aligned} \tag{3.5}$$

Actually (3.5) holds on all of $L_2(\mathbb{R}^\nu)$, because $(H-z)C_0^\infty(\mathbb{R}^\nu)$ is dense and both sides are bounded.

By $|D_j k_{-1}| \le Ck_{-2}$ and $A_1(1)$ it follows

$$(D_j k_{-1})k_{-s+1}(H-z)^{-1} \in B_q \quad \text{for} \quad q > \max\left\{\tfrac{\nu}{2}, 2\right\}.$$

Using Theorem 3.1 we have

$$k_{-1}p_j k_{-s+1}(H-z)^{-1} = \left(k_{-1}p_j(p^2+1)^{-1}\right)(p^2+1)k_{-s+1}(H-z)^{-1} \in B_q \quad \text{for} \quad q > \nu, \ q \ge 2.$$

Inserting in (3.5) this implies

$$p_j k_{-s}(H-z)^{-1} \in B_q \quad \text{for} \quad q > \max\{\nu, 2\}.$$

Next we find by first restricting to $(H-z)C_0^\infty(\mathbb{R}^\nu)$ and then passing to the closure

$$\overline{k_{-s}p_{j,0}(H-z)^{-1}} = p_j k_{-s}(H-z)^{-1} + (D_j k_{-s})(H-z)^{-1}.$$

$A_1(3)$ follows by applying $A_1(1)$ to the second term on the right hand side.

Second step: Computation of the commutator $[k_{-s}, (H-z)^{-1}]$.

For every $s > 0$ we have on $C_0^\infty(\mathbb{R}^\nu)$

$$(H-z)k_{-s} = k_{-s}(H-z) + 2\sum_{j=1}^{\nu}(D_j k_{-s})p_{j,0} - (\Delta k_{-s}) - 2\sum_{j=1}^{\nu} b_j(D_j k_{-s}).$$

Multiplying from both sides with $(H-z)^{-1}$ and taking the closure we get

$$[k_{-s}, (H-z)^{-1}] = \overline{(H-z)^{-1}\Big(2\sum_{j=1}^{\nu}(D_j k_{-s})p_{j,0} - (\Delta k_{-s}) - 2\sum_{j=1}^{\nu} b_j(D_j k_{-s})\Big)(H-z)^{-1}}, \tag{3.6}$$

Third step: Proof of A_n from $A_1, \ldots, A_{n-1}$.

To show $A_n(1)$ let $s \geq N_n + 2$ and r_1, r_2 be chosen such that $r_1 + r_2 = s$, $r_2 \geq N_1 + 2$, $r_1 \geq N_{n-1} + \max\{1, s_1\}$. The choice of r_1 guarantees

$$2\sum_{j=1}^{\nu}\overline{(D_j k_{-r_1})p_{j,0}(H-z)^{-1}} - \Big((\Delta k_{-r_1}) + 2\sum_{j=1}^{\nu} b_j(D_j k_{-r_1})\Big)(H-z)^{-1} \in B, \tag{3.7}$$

where the first term can be treated with $A_1(3)$ and the second term is a product of two bounded operators. Using (3.6) we get

$$\begin{aligned}
k_{-s}(H-z)^{-n} &= k_{-r_2}\big(k_{-r_1}(H-z)^{-1}\big)(H-z)^{-n+1} \\
&= k_{-r_2}\Bigg\{(H-z)^{-1}k_{-r_1} + \\
&\quad \overline{\Big((H-z)^{-1}\Big(2\sum_{j=1}^{\nu}(D_j k_{-r_1})p_{j,0} - (\Delta k_{-r_1}) - 2\sum_{j=1}^{\nu} b_j(D_j k_{-r_1})\Big)(H-z)^{-1}\Big)}\Bigg\}(H-z)^{-n+1} \\
&= k_{-r_2}(H-z)^{-1}\Bigg\{k_{-r_1}(H-z)^{-n+1} + 2\sum_{j=1}^{\nu}\overline{(D_j k_{-r_1})p_{j,0}(H-z)^{-1}}\,(H-z)^{-n+1} \\
&\qquad - \Big((\Delta k_{-r_1}) + 2\sum_{j=1}^{\nu} b_j(D_j k_{-r_1})\Big)(H-z)^{-n}\Bigg\} \\
&=: k_{-r_2}(H-z)^{-1}\big[T_1 + T_2 + T_3\big],
\end{aligned} \tag{3.8}$$

the right hand side being defined everywhere by (3.7). T_1 and T_3 lie in B_q for $q > \max\left\{\frac{\nu}{n-1}, \left(\frac{n-2}{\nu} + \frac{1}{2}\right)^{-1}\right\}$ by $A_{n-1}(2)$. From $A_{n-1}(3)$ we get the same property for T_2, hence also for $T_1 + T_2 + T_3$. $k_{-r_2}(H-z)^{-1}$ is in B_q for $q > \max\{\frac{\nu}{2}, 2\}$ by $A_1(1)$. Combining this in (3.8) it follows that $k_{-s}(H-z)^{-n} \in B_q$ for

$$\frac{1}{q} < \frac{1}{\max\left\{\frac{\nu}{n-1}, \left(\frac{n-2}{\nu}+\frac{1}{2}\right)^{-1}\right\}} + \frac{1}{\max\left\{\frac{\nu}{2}, 2\right\}}$$

$$= \min\left\{\frac{n-1}{\nu}+\frac{2}{\nu},\ \frac{n-1}{\nu}+\frac{1}{2},\ \frac{n-2}{\nu}+\frac{1}{2}+\frac{2}{\nu},\ \frac{n-2}{\nu}+\frac{1}{2}+\frac{1}{2}\right\} = \min\left\{\frac{n+1}{\nu},\ \frac{n-1}{\nu}+\frac{1}{2}\right\}. \tag{3.9}$$

Here we used $\nu \geq 2$ in order to estimate $\frac{n-2}{\nu}+1 \geq \frac{n-1}{\nu}+\frac{1}{2}$. $A_n(1)$ is proved.

In the proof of $A_n(2)$ we choose $r_2 \geq N_1 + 1$, so we get $k_{-r_2}(H-z)^{-1} \in B_q$ for $q > \max\{\nu, 2\}$ from $A_1(2)$. The rest of the proof is identical with that of $A_n(1)$.

The proof of $A_n(3)$ is similar to that of $A_1(3)$: If $s = r_1 + r_2$, $r_1 \geq N_{n-1} + \max\{1, s_1\}$, $r_2 \geq N_1 + 1$ we have

$$\begin{aligned} p_j k_{-s}(H-z)^{-n} &= \overline{p_{j,0} k_{-s}(H-z)^{-1}}\, (H-z)^{-n+1} \\ &= \overline{\left((D_j k_{-r_2}) k_{-r_1}(H-z)^{-1} + k_{-r_2} p_{j,0} k_{-r_1}(H-z)^{-1}\right)}\, (H-z)^{-n+1} \\ &= (D_j k_{-r_2}) k_{-r_1}(H-z)^{-n} + k_{-r_2} p_j k_{-r_1}(H-z)^{-n} \end{aligned}$$

and as in (3.8)

$$k_{-r_2} p_j k_{-r_1}(H-z)^{-n} = \overline{k_{-r_2} p_{j,0} k_{-r_1}(H-z)^{-1}}\, (H-z)^{-n+1} =$$

$$\overline{k_{-r_2} p_{j,0}(H-z)^{-1}\left[k_{-r_1} + 2\sum_{j=1}^{\nu} \overline{(D_j k_{-r_1}) p_{j,0}(H-z)^{-1}} - \left((\Delta k_{-r_1}) + 2\sum_{j=1}^{\nu} b_j (D_j k_{-r_1})\right)(H-z)^{-1}\right]}$$
$$\times\ (H-z)^{-n+1}$$

$$= \overline{k_{-r_2} p_{j,0}(H-z)^{-1}}\ (T_1 + T_2 + T_3).$$

Altogether

$$\begin{aligned} \overline{k_{-s} p_{j,0}(H-z)^{-1}}\ (H-z)^{-n+1} &= \overline{\left(p_{j,0} k_{-s}(H-z)^{-1} - (D_j k_{-s})(H-z)^{-1}\right)}\ (H-z)^{-n+1} \\ &= p_j k_{-s}(H-z)^{-n} - (D_j k_{-s})(H-z)^{-n} \\ &= (D_j k_{-r_2}) k_{-r_1}(H-z)^{-n} + \overline{k_{-r_2} p_{j,0}(H-z)^{-1}}\ (T_1 + T_2 + T_3) - (D_j k_{-s})(H-z)^{-n}. \end{aligned}$$

The first and third term on the right hand side have the desired B_q-property by $A_n(1)$. The second term is the product of $\overline{k_{-r_2} p_{j,0}(H-z)^{-1}}$, which by $A_1(3)$ belongs to B_q for $q > \max\{\nu, 2\}$, and $T_1 + T_2 + T_3$, which was treated above. A computation as in (3.9) completes the proof of $A_n(3)$ and therefore of part a) of the Theorem.

b) We take $r_1 \geq N_{n-1} + \max\{1, s_1\}$, $r_2 \geq N_1$ with $s = r_1 + r_2$. As above

$$(-\Delta+1)k_{-s}(H-z)^{-n} = (-\Delta+1)k_{-r_2}(k_{-r_1}(H-z)^{-1})(H-z)^{-n+1}$$

$$= (-\Delta+1)k_{-r_2}(H-z)^{-1}(T_1+T_2+T_3).$$

$(-\Delta+1)k_{-r_2}(H-z)^{-1}$ is bounded by Theorem 3.1 and

$$T_1+T_2+T_3 \in B_q \quad \text{for} \quad q > \max\left\{\frac{\nu}{n-1}, \left(\frac{n-2}{\nu}+\frac{1}{2}\right)^{-1}\right\}$$

as proved in part a). For $n > \max\{2, \frac{\nu}{2}+1\}$ we can choose $q = 2$. □

To complete the **proof of Theorem 1.2**, we take $A = -\Delta+1$, $g = k_{-s}$ and $\gamma(H) = (H-z)^{-n}$ for some $z \in \rho(H)$ in Theorem 2.1. If s and n are given by Theorem 3.2 b) we get that $(-\Delta+1)k_{-s}(H-\overline{z})^{-n}$ and hence

$$\overline{(H-z)^{-n}k_{-s}(-\Delta+1)} = \left((-\Delta+1)k_{-s}(H-\overline{z})^{-n}\right)^*$$

are Hilbert-Schmidt operators. From Theorem 2.1 we get an expansion in generalized eigenfunctions $\varphi_j(\lambda)$, which are in the range of $k_{-s}(-\Delta+1)^{-1}$, i.e. in $k_s W_{2,2}(\mathbb{R}^\nu)$. □

Remark: If our only aim would be to prove that the generalized eigenfunctions are in $k_s L_2(\mathbb{R}^\nu)$ for some small s, we would only need Theorem 3.2 a) in connection with the first part of Theorem 2.1.

The proof of Theorem 1.1 works in a similar fashion. Instead of Theorem 3.2 we have

Theorem 3.3: *Let* H *satisfy* (A) *and* $z \in \rho(H)$. *Then*

a) $\varphi(H-z)^{-n}$ *is a Hilbert-Schmidt operator for* $n > \max\{1, \frac{\nu}{2}-1\}$ *and every* $\varphi \in C_0^\infty(\mathbb{R}^\nu)$,

b) $(-\Delta+1)\varphi(H-z)^{-n}$ *is a Hilbert-Schmidt operator for* $n > \max\{2, \frac{\nu}{2}+1\}$ *and every* $\varphi \in C_0^\infty(\mathbb{R}^\nu)$.

Proof: We find a) as a consequence of the first part of the following statement:

$$B_n \begin{cases} (1) \quad \varphi(H-z)^{-n} \in B_q \quad \text{for} \quad q > \max\left\{\frac{\nu}{n+1}, \left(\frac{n-1}{\nu}+\frac{1}{2}\right)\right\}, \quad q \geq 1 \text{ and } \varphi \in C_0^\infty(\mathbb{R}^\nu), \\ (2) \quad \varphi(H-z)^{-n} \in B_q \quad \text{for} \quad q > \max\left\{\frac{\nu}{n}, \left(\frac{n-1}{\nu}+\frac{1}{2}\right)^{-1}\right\}, \quad q \geq 1 \text{ and } \varphi \in C_0^\infty(\mathbb{R}^\nu), \\ (3) \quad \overline{\varphi\, p_{j,0}(H-z)^{-1}}(H-z)^{-n+1} \in B_q \\ \qquad \text{for} \quad q > \max\left\{\frac{\nu}{n}, \left(\frac{n-1}{\nu}+\frac{1}{2}\right)^{-1}\right\}, \quad q \geq 1 \text{ and } \varphi \in C_0^\infty(\mathbb{R}^\nu). \end{cases}$$

This can be shown in almost the same way as in the proof of Theorem 3.2. We only remark the following changes:

First we get boundedness of $(-\Delta+1)\varphi(H-z)^{-1}$ for every $\varphi \in C_0^\infty(\mathbb{R}^\nu)$ from Theorem 3.1. To see this define

$$\tilde{V}_1 := V_1 + \chi_{\operatorname{supp}\varphi} V_2 \quad \text{and} \quad \tilde{V}_2 := (1-\chi_{\operatorname{supp}\varphi})V_2,$$

χ_M being the characteristic function of M. Using $\tilde{V}_1$ and $\tilde{V}_2$ in place of V_1 and V_2 we see that (A) as well as all other assumptions of Theorem 3.1 are satisfied.

Second we replace factorizations of the type $k_{-s} = k_{-r_1} k_{-r_2}$ by $\varphi = \psi\varphi$, where for a given $\varphi \in C_0^\infty(\mathbb{R}^\nu)$ we choose another $\psi \in C_0^\infty(\mathbb{R}^\nu)$ such that $\psi(x) = 1$ for $x \in \operatorname{supp}\varphi$.

The same remarks apply to the proof of b). □

Using Theorem 3.3 instead of Theorem 3.2 and Theorem 2.2 instead of Theorem 2.1 we can now **prove Theorem 1.1** in the same way as Theorem 1.2.

4. Polynomial boundedness of generalized eigenfunctions

The main step in the proof of local or polynomial boundedness of generalized eigenfunctions will be the following Theorem 4.1 on local L_p-properties of weak solutions of the Schrödinger equation. Theorem 1.3 will then be proven by an iterative application of Theorem 4.1.

Let $u:\mathbb{R}^\nu \to \mathbb{C}$ be a weak solution of

$$\left(\tfrac{1}{i}\nabla - b\right)^2 u + Vu = Eu,$$

i.e. for every $\varphi \in C_0^\infty(\mathbb{R}^\nu)$

$$\int_{\mathbb{R}^\nu} \left\{\left(\tfrac{1}{i}\nabla - b(y)\right)^2 \varphi(y) + \left(V(y)-E\right)\varphi(y)\right\} u(y)\,dy =$$

$$\int_{\mathbb{R}^\nu} \left\{-\Delta\varphi(y) + 2ib(y)\cdot\nabla\varphi(y) + \left(i\operatorname{div} b(y) + b(y)^2 + V(y) - E\right)\varphi(y)\right\} u(y)\,dy = 0.$$

We will suppose

(A″) $V \in L_{p,loc}(\mathbb{R}^\nu)$ with $p > \frac{\nu}{2}$ and $p \geq 2$, $b \in C^1(\mathbb{R}^\nu)^\nu$.

Let ω_ν and V_ν be the surface area and the volume of the unit ball in $\mathbb{R}^\nu$, $K_\delta := \{x: |x| < \delta\}$, χ_δ the characteristic function of K_δ and $\|f\|_{\delta,p} := \|\chi_\delta f\|_p$. Further let $\eta := \min\left\{\frac{1}{\nu}, \frac{2}{\nu} - \frac{1}{p}\right\}$, $\tilde{V} := -i\operatorname{div} b + b^2 + V - E$,

$$K(\xi) := \frac{1}{(2-\nu)\omega_\nu} |\xi|^{2-\nu} \quad \text{for } \xi \in \mathbb{R}^\nu,\ \nu \geq 3$$

the fundamental solution of the Laplacian and

$$f(x) := \int_{K_{2\delta}} K(x-y)\big(\tilde{V}(y)\,u(y) - 2i\,b(y)\cdot(\nabla u)(y)\big)\,dy$$

the Newtonian potential of $\chi_{2\delta}(\tilde{V}u - 2ib\cdot\nabla u)$.

Theorem 4.1: *Let* $\nu \geq 3$. *If* $u \in L_{\alpha_1,loc}(\mathbb{R}^\nu)$ *for* $2 \leq \alpha_1 \leq \infty$ *and* u *is weakly differentiable with* $\nabla u \in L_{\beta_1,loc}(\mathbb{R}^\nu)^\nu$ *for* $\frac{1}{\beta_1} = \frac{1}{\alpha_1} + \frac{1}{\nu}$, *then*

a) $u \in L_{\alpha_2,loc}(\mathbb{R}^\nu)$ *and*

$$\|u\|_{\delta,\alpha_2} \leq C(\nu,\alpha_1,\alpha_2,p)\,\delta^{\nu(\frac{1}{\alpha_2}-\frac{1}{\alpha_1})}\Big[\big(1+\delta^{2-\frac{\nu}{p}}\|\tilde{V}\|_{2\delta,p}\big)\|u\|_{2\delta,\alpha_1} + \delta\,\|b\|_{2\delta,\infty}\|\nabla u\|_{2\delta,\beta_1}\Big]$$

for every α_2 *with* $\frac{1}{\alpha_2} > \frac{1}{\alpha_1} - \eta$ *and every* $\delta > 0$ $(\alpha_2 \leq \infty$ *if* $\frac{1}{\alpha_1} < \eta)$,

b) $\nabla u \in L_{\beta_2,loc}(\mathbb{R}^\nu)$ *and*

$$\|\nabla u\|_{\delta,\beta_2} \leq C(\nu,\beta_1,\beta_2,p)\,\delta^{\nu(\frac{1}{\beta_2}-\frac{1}{\beta_1})}\Big[\big(1+\delta^{2-\frac{\nu}{p}}\|\tilde{V}\|_{2\delta,p}\big)\|u\|_{2\delta,\alpha_1} + \delta\,\|b\|_{2\delta,\infty}\|\nabla u\|_{2\delta,\beta_1}\Big]$$

for every β_2 *with* $\frac{1}{\beta_2} > \frac{1}{\beta_1} - \eta$ *and every* $\delta > 0$.

Proof: a) For $\alpha_2 < \alpha_1$ we have by Hölder's inequality

$$\|u\|_{\delta,\alpha_2} \leq \big(V_\nu\delta^\nu\big)^{\frac{1}{\alpha_2}-\frac{1}{\alpha_1}}\|u\|_{\delta,\alpha_1};$$

therefore we may assume $\alpha_2 \geq \alpha_1$. By the definition of f and using Green's representation for functions in $C_0^\infty(K_{2\delta})$ we find that $u - f$ is weakly harmonic on $K_{2\delta}$. So by Weyl's lemma $u - f$ is harmonic on $K_{2\delta}$. From the mean value property we get

$$\begin{aligned}\|u-f\|_{\delta,\alpha_2} &= \Big\{\int_{K_\delta}\Big|\frac{1}{V_\nu\delta^\nu}\int_{|y-x|<\delta}(u-f)(y)\,dy\Big|^{\alpha_2}dx\Big\}^{1/\alpha_2}\\ &\leq \Big\{\int_{K_\delta}\Big(\frac{1}{V_\nu\delta^\nu}\int_{K_{2\delta}}|(u-f)(y)|\,dy\Big)^{\alpha_2}dx\Big\}^{1/\alpha_2} = (V_\nu\delta^\nu)^{\frac{1}{\alpha_2}-1}\|u-f\|_{2\delta,1}.\end{aligned} \tag{4.1}$$

On the other hand we will show

$$\|f\|_{2\delta,\alpha_2} \leq C(\nu,\alpha_1,\alpha_2,p)\,\delta^{\nu(\frac{1}{\alpha_2}-\frac{1}{\alpha_1})}\big(\delta^{2-\frac{\nu}{p}}\|\tilde{V}\|_{2\delta,p}\|u\|_{2\delta,\alpha_1} + \delta\,\|b\|_{2\delta,\infty}\|\nabla u\|_{2\delta,\beta_1}\big). \tag{4.2}$$

To prove (4.2) we write $f = f_1 - f_2$, where

$$f_1(x) := \int_{K_{2\delta}} K(x-y)\,\tilde{V}(y)\,u(y)\,dy, \qquad f_2(x) := \int_{K_{2\delta}} K(x-y)\,2i\,b(y)\cdot(\nabla u)(y)\,dy.$$

If q_1 is defined by $\frac{1}{\alpha_2} + 1 = \frac{1}{q_1} + \frac{1}{p} + \frac{1}{\alpha_1}$ we get from the inequalities of Young and Hölder

$$\|f_1\|_{2\delta,\alpha_2} \le \|K\|_{4\delta,q_1} \|\tilde{V}\|_{2\delta,p} \|u\|_{2\delta,\alpha_1} = \frac{\omega_\nu^{\frac{1}{q_1}-1}}{\nu-2} (4\delta)^{\frac{\nu}{q_1}-\nu+2} \|\tilde{V}\|_{2\delta,p} \|u\|_{2\delta,\alpha_1},$$

and with $\frac{1}{\alpha_2} + 1 = \frac{1}{q_2} + \frac{1}{\beta_1}$

$$\|f_2\|_{2\delta,\alpha_2} \le 2\|K\|_{4\delta,q_2} \|b\|_{2\delta,\infty} \|\nabla u\|_{2\delta,\beta_1} = \frac{2\omega_\nu^{\frac{1}{q_2}-1}}{\nu-2} (4\delta)^{\frac{\nu}{q_2}-\nu+2} \|b\|_{2\delta,\infty} \|\nabla u\|_{2\delta,\beta_1}.$$

These estimates yield (4.2).

Now we use (4.1) and again Hölder's inequality to find

$$\begin{aligned}
\|u\|_{\delta,\alpha_2} &\le \|u - f\|_{\delta,\alpha_2} + \|f\|_{\delta,\alpha_2} \le (V_\nu \delta^\nu)^{\frac{1}{\alpha_2}-1} \|u - f\|_{2\delta,1} + \|f\|_{2\delta,\alpha_2} \\
&\le (V_\nu \delta^\nu)^{\frac{1}{\alpha_2}-1} \left((V_\nu (2\delta)^\nu)^{1-\frac{1}{\alpha_1}} \|u\|_{2\delta,\alpha_1} + (V_\nu (2\delta)^\nu)^{1-\frac{1}{\alpha_2}} \|f\|_{2\delta,\alpha_2} \right) + \|f\|_{2\delta,\alpha_2} \\
&\le C(\nu,\alpha_1,\alpha_2) \left(\delta^{\nu(\frac{1}{\alpha_2}-\frac{1}{\alpha_1})} \|u\|_{2\delta,\alpha_1} + \|f\|_{2\delta,\alpha_2} \right).
\end{aligned}$$

Inserting (4.2) completes the proof of a).

b) By an argument similar to the one given at the beginning of the proof of a), it is sufficient to look at β_2 with $\frac{1}{\beta_1} - \max\left\{\frac{1}{\nu} - \frac{1}{p}, 0\right\} \ge \frac{1}{\beta_2}$.

We observe that f has a weak derivative, which is found by differentiating under the integral sign:

$$(\nabla f)(x) = \int_{K_{2\delta}} (\nabla K)(x-y) \left(\tilde{V}(y) u(y) - 2i\, b(y) \cdot (\nabla u)(y) \right) dy \quad \text{for a.e. } x \in K_{2\delta}.$$

To see this, we have to show that

$$\int_{K_{2\delta}} (\nabla\varphi)(x) f(x)\, dx = \int_{K_{2\delta}} \int_{K_{2\delta}} \varphi(x) (\nabla K)(x-y) \left(\tilde{V}(y) u(y) - 2i\, b(y) \cdot (\nabla u)(y) \right) dy\, dx$$

for every $\varphi \in C_0^\infty(K_{2\delta})$. Using $(\nabla K)(x) = \frac{x}{\omega_\nu |x|^\nu} \in L_{1,loc}(\mathbb{R}^\nu)^\nu$ we prove this by first restricting the integration to those x and y with $|x-y| > \varepsilon$ and then passing with ε to 0. Similarly we get ∇f_1 and ∇f_2.

If now q_1 and q_2 are defined by $\frac{1}{\beta_2} + 1 = \frac{1}{q_1} + \frac{1}{p} + \frac{1}{\alpha_1}$ and $\frac{1}{\beta_2} + 1 = \frac{1}{q_2} + \frac{1}{\beta_1}$ we get

$$\|\nabla f_1\|_{2\delta,\beta_2} \le \|\nabla K\|_{4\delta,q_1} \|\tilde{V}\|_{2\delta,p} \|u\|_{2\delta,\alpha_1} = \omega_\nu^{\frac{1}{q_1}-1} (4\delta)^{\frac{\nu}{q_1}-\nu+1} \|\tilde{V}\|_{2\delta,p} \|u\|_{2\delta,\alpha_1},$$

$$\|\nabla f_2\|_{2\delta,\beta_2} \le 2\|\nabla K\|_{4\delta,q_2} \|b\|_{2\delta,\infty} \|\nabla u\|_{2\delta,\beta_1} = 2\omega_\nu^{\frac{1}{q_2}-1} (4\delta)^{\frac{\nu}{q_2}-\nu+1} \|b\|_{2\delta,\infty} \|\nabla u\|_{2\delta,\beta_1},$$

and therefore

$$\|\nabla f\|_{2\delta,\beta_2} \le C(\nu,\beta_1,\beta_2,p)\,\delta^{\nu\left(\frac{1}{\beta_2}-\frac{1}{\beta_1}\right)}\left(\delta^{2-\frac{\nu}{p}}\|\tilde{V}\|_{2\delta,p}\|u\|_{2\delta,\alpha_1}+\delta\|b\|_{2\delta,\infty}\|\nabla u\|_{2\delta,\beta_1}\right). \tag{4.3}$$

The components of $\nabla(u-f)$ are harmonic. So using a gradient estimate and the mean value property of harmonic functions we find

$$\|\nabla(u-f)\|_{\delta,\beta_2} \le (V_\nu\delta^\nu)^{\frac{1}{\beta_2}}\|\nabla(u-f)\|_{\delta,\infty} \le (V_\nu\delta^\nu)^{\frac{1}{\beta_2}}\frac{2\nu}{\delta}\|u-f\|_{3\delta/2,\infty}$$

$$\le (V_\nu\delta^\nu)^{\frac{1}{\beta_2}}\frac{2\nu}{\delta}\frac{1}{V_\nu(\delta/2)^\nu}\|u-f\|_{2\delta,1} = C(\nu,\beta_2)\,\delta^{\nu\left(\frac{1}{\beta_2}-1\right)-1}\|u-f\|_{2\delta,1}.$$

It follows that

$$\|\nabla u\|_{\delta,\beta_2} \le \|\nabla(u-f)\|_{\delta,\beta_2}+\|\nabla f\|_{\delta,\beta_2}$$

$$\le C(\nu,\beta_2)\,\delta^{\nu\left(\frac{1}{\beta_2}-1\right)-1}\|u-f\|_{2\delta,1}+\|\nabla f\|_{2\delta,\beta_2}.$$

We finish the proof by estimating $\|u-f\|_{2\delta,1}$ as in a) and $\|\nabla f\|_{2\delta,\beta_2}$ according to (4.3). □

The restriction to $\nu \ge 3$ in the above theorem was made only in order to have a unique form of the fundamental solution K. With minor changes the Theorem extends to dimensions 1 and 2.

Proof of Theorem 1.3: a) If u is a generalized eigenfunction we have by Theorem 1.1 $u \in L_{2,loc}(\mathbb{R}^\nu)$ and $\nabla u \in L_{2,loc}(\mathbb{R}^\nu)^\nu \subset L_{1/(1/2+1/\nu),loc}(\mathbb{R}^\nu)^\nu$, so Theorem 4.1 can be applied with $\alpha_1 = 2$ and $\beta_1 = \left(\frac{1}{2}+\frac{1}{\nu}\right)^{-1}$. We get $u \in L_{\alpha_2,loc}(\mathbb{R}^\nu)$ for $\frac{1}{\alpha_2} > \frac{1}{2}-\eta$ and $\nabla u \in L_{\beta_2,loc}(\mathbb{R}^\nu)^\nu$ for $\frac{1}{\beta_2} > \frac{1}{2}+\frac{1}{\nu}-\eta$. After $N > \frac{1}{2\eta}$ iterations of this step we arrive at $u \in L_{\infty,loc}(\mathbb{R}^\nu)$.

b) We follow the proof of a). Theorem 1.2 gives polynomial boundedness of $\|u\|_{\delta,2}$ and $\|\nabla u\|_{\delta,2}$ with respect to δ. This also holds for $\|\nabla u\|_{\delta,\beta_1}$ by Hölder's inequality. By conditions (A') and (B) all other terms which appear during the iteration are polynomially bounded, so we get the same for $\|u\|_{\delta,\infty}$.

5. Remarks on other results

In the following we list some other results on eigenfunction expansions, which can be proved with the methods of Sections 2 to 4. More details are given in [St].

(i) Without assuming any asymptotic bounds of the electromagnetic field except the condition $V_2 \ge -Ck_2$ used to guarantee selfadjointness, we get generalized eigenfunctions which are locally bounded (Theorems 1.1 and 1.3 a). Polynomially bounded fields give rise to polynomially bounded generalized eigenfunctions (Theorems 1.2 and 1.3 b). This correlation extends to other situations, for example linear exponential boundedness, i.e. boundedness by a function of the form $C e^{a|\cdot|}$, carries over from the fields to the generalized eigenfunctions.

(ii) Having established local boundedness in Section 4, it is not difficult to show other regularity properties of generalized eigenfunctions or more generally of weak solutions of the Schrödinger equation by our methods:

— Let u be a weak solution of $(\frac{1}{i}\nabla - b)^2 u + Vu = Eu$ and suppose $u \in W_{2,1,loc}(\mathbb{R}^\nu)$. Under condition (A″) of Section 4 we find *Hölder continuity* for u with exponent $\min\{1, 2 - \frac{\nu}{p}\}$ by again using that $u - f$ is harmonic, such that only f has to be dealt with. The inhomogeneous equation $(\frac{1}{i}\nabla - b)^2 u + Vu = g$ can be treated in the same way. Results of this type can also be found in [Si] and [Wi].

— Using the gradient estimate for harmonic functions, we can prove *gradient estimates* for weak solutions of $(-\Delta + V)u = Eu$ if V satisfies (A″): Let Ω and Ω' be subsets of $\mathbb{R}^\nu$, $\Omega \subset \Omega'$, Ω bounded and open, Ω' compact and $\frac{1}{r} > \frac{1}{p} - \frac{1}{\nu}$. Then we have

$$\|\nabla u\|_{\Omega',r} \le C(\nu, r, p, \Omega, \Omega', V - E)\|u\|_{\Omega,\infty},$$

where $\|f\|_{M,p} := \|\chi_M f\|_p$. The dependence of C on Ω, Ω' and $V - E$ can be given explicitly, for example in the case $p > \nu$ we get

$$\|\nabla u\|_{\Omega',\infty} \le \frac{C(\nu, p)}{d}\left(1 + \varnothing(\Omega)^{2 - \frac{\nu}{p}}\|V - E\|_{\Omega,p}\right)\|u\|_{\Omega,\infty},$$

where $d := \mathrm{dist}(\partial\Omega, \Omega')$ and $\varnothing(\Omega) := \max\{|x - y| : x, y \in \Omega\}$.

— From Theorem 4.1 a) we easily get a *mean value inequality* for weak solutions of $(-\Delta + V)u = Eu$:

$$|u(0)|^2 \le C(\nu, p)\delta^{-\nu}(1 + \delta^{2 - \frac{\nu}{p}}\|V - E\|_{\delta,p})^{2N} \int_{K_\delta} |u(y)|^2\, dy,$$

where $N = \min\{n \in \mathbb{N} : n > \frac{1}{2\eta}\}$, $\eta = \min\left\{\frac{1}{2}, \frac{2}{\nu} - \frac{1}{p}\right\}$.

(iii) Let M(H) be the set of values $\lambda \in \mathbb{R}$ such that there exists a polynomially bounded weak solution u of $Hu = \lambda u$. Using the existence of an expansion in polynomially bounded generalized eigenfunctions for H, it easily follows that $\sigma(H) \subset \overline{M(H)}$. This as well as a proof of the opposite inclusion $\overline{M(H)} \subset \sigma(H)$ can be found in [S2]. To sum up one could say that the spectrum is the closure of the set of "generalized eigenvalues".

(iv) Our methods of Sections 2 and 3 can be reformulated to yield results on eigenfunction expansions for Dirac operators with singular potentials. Let T be the *free*

Dirac operator in $L_2(\mathbb{R}^3)^4$ (see e.g. [W1], Ch. 10.6) and V the operator of multiplication by a measurable Hermitean 4×4-matrix-valued function $v = v_1 + v_2 + v_3$, where

$$|v_1(x)| \le \frac{C}{|x|} \quad \text{for some} \quad C < \frac{1}{2},$$

$$|v_2(\cdot)| \in M_\rho(\mathbb{R}^3) \quad \text{for some} \quad \rho < 2,$$

$$|v_3(\cdot)| \le C' k_s \quad \text{for some} \quad s > 0.$$

Then there exists an expansion in generalized eigenfunctions of $S := T + V$, which lie in $k_r W_{2,1}(\mathbb{R}^3)^4$ for $r \ge 3s + 2$. Here the notion of an expansion in generalized eigenfunctions has to be adapted to operators in $L_2(\mathbb{R}^3)^4$. We also get that $\sigma(S) = \overline{M(S)}$, where $M(S)$ is the set of those $E \in \mathbb{R}$ for which a weak solution $u \ne 0$ of $Su = Eu$ exists with $u \in W_{2,1,loc}(\mathbb{R}^3)^4 \cap k_r L_2(\mathbb{R}^3)^4$ for some $r > 0$.

(v) Another Hilbert space in which an expansion theorem can be formulated is $\ell_2(\mathbb{Z}^\nu)$. Given a selfadjoint operator H in $\ell_2(\mathbb{Z}^\nu)$ in its matrix representation $H(\cdot,\cdot)$ we define $M(H)$ to be the set of those $E \in \mathbb{R}$ for which a nontrivial solution of

$$\sum_{\beta \in \mathbb{Z}^\nu} H(\alpha, \beta) u(\beta) = E u(\alpha) \quad \text{for every} \quad \alpha \in \mathbb{Z}^\nu \tag{5.1}$$

with $|u| \le p$ for some polynomial $p: \mathbb{Z}^\nu \to \mathbb{R}$ exists. To have finite sums in (5.1) we assume $\ell_{2,0}(\mathbb{Z}^\nu) := \{f \in \ell_2(\mathbb{Z}^\nu): f(\alpha) \ne 0 \text{ for finitely many } \alpha\} \subset D(H)$ and $H\ell_{2,0}(\mathbb{Z}^\nu) \subset \ell_{2,0}(\mathbb{Z}^\nu)$. The fact that multiplication by the discrete version of k_{-s} itself is a Hilbert-Schmidt operator for $s > \frac{\nu}{2}$ leads to the existence of an expansion in generalized eigenfunctions in $M(H)$ for every such operator H. We further have $\sigma(H) \subset \overline{M(H)}$. $\sigma(H) = \overline{M(H)}$ can be shown for *difference operators*

$$H := \sum_{|\gamma| \le M} c_\gamma T_\gamma,$$

where T_γ is translation by $\gamma \in \mathbb{Z}^\nu$: $(T_\gamma u)(\beta) = u(\beta - \gamma)$. $c_\gamma: \mathbb{Z}^\nu \to \mathbb{C}$ are functions such that c_γ is bounded for every $\gamma \ne 0$ and $c_{-\gamma} = T_{-\gamma}\overline{c}_\gamma$ for every γ, the last assumption guaranteeing selfadjointness.

Acknowledgement

The author wishes to thank Professor J. Weidmann for his permanent interest and many helpful discussions on the subject of this work.

References

[He] Hellinger, E.: Neue Begründung der Theorie der quadratischen Formen von unendlichvielen Veränderlichen. Journal für die reine und angewandte Mathematik 136 (1909) 210 - 271

[KS] Kovalenko, V. F.; Semenov, Yu. A.: Some Problems on Expansion in Generalized Eigenfunctions of the Schrödingeroperator with strongly singular Potentials. Russ. Math. Surveys 33 : 4 (1978) 119 - 157

[PSW] Poerschke, Th.; Stolz, G.; Weidmann, J.: Expansions in Generalized Eigenfunctions of Selfadjoint Operators. Math. Z. 202 (1989) 397 - 408

[Si] Simader, C. G.: An elementary proof of Harnack's inequality for Schrödinger operators and related topics. Preprint, University of Bayreuth 1989, to appear in Math. Z.

[S1] Simon, B.: Trace Ideals and their applications. Cambridge University Press, London 1979

[S2] Simon, B.: Schrödinger Semigroups. Bull. Am. Math. Soc. 7 (1982) 447 - 526

[St] Stolz, G.: Entwicklung nach verallgemeinerten Eigenfunktionen von Schrödinger-operatoren. Thesis, University of Frankfurt 1989

[W1] Weidmann, J.: Linear Operators in Hilbert Spaces. Graduate Texts in Mathematics, Vol. 68, Springer-Verlag, New York, Heidelberg, Berlin 1980

[W2] Weidmann, J.: Spectral Theory of Ordinary Differential Operators. Springer, Lecture Notes in Mathematics 1258, Berlin, Heidelberg, New York 1987

[We] Weyl, H.: Über gewöhnliche Differentialgleichungen mit Singularitäten und die zugehörigen Entwicklungen willkürlicher Funktionen. Math. Annalen 68 (1910) 220 - 269

[Wi] Witte, J.: Über die Regularität der Spektralschar eines singulären elliptischen Differentialoperators. Math. Z. 107 (1968) 116 - 126

Johann Wolfgang Goethe - Universität
Fachbereich Mathematik
Postfach 11 19 32
Robert - Mayer - Str. 6 - 10
D-6000 Frankfurt 1

Operator Theory:
Advances and Applications, Vol. 48

DUALITY AND HANKEL OPERATORS

Sun Shunhua*

Dedicated to the Memory of **Professor Ernst Hellinger**

A duality approach is exhibited in this paper to characterize the Hankel operators that are bounded, compact, or nuclear. This answers, in a sense, a question asked by S. Axler in [1] and [2]. As a by-product, a simplified proof of Peller's criteria for nuclearity of Hankel operators on Hardy space is given. Moreover, a sharp sufficient condition for products of Hankel operators on the Bergman space to be nuclear is also obtained.

INTRODUCTION. All the known results about Hankel operators suggest that there should be a unified duality approach that would provide an intrinsic description of which Hankel operators are bounded, compact, or nuclear. This is why S. Axler asked the following question [2]: **Is there a theorem that would characterize the boundedness and compactness of Hankel operators on a large class of function spaces, using only duality?**

In Section 1 of this paper a general duality approach for a Hilbert space operator to be in some operator ideal class is introduced. Based upon this general duality machinery, an intrinsic proof is given in Section 2 to characterize the Hankel operators on the Hardy and Bergman space that are bounded or compact. In Section 3, the same approach provides a simple proof of Peller's criteria [5] for Hankel operators on the Hardy space to be nuclear. Finally, a sharp sufficient condition for the product of Hankel operators on the Bergman space to be nuclear is established in Section 4, using the duality theorem given in Section 1.

Although almost all that we will prove in Sections 2—4 are known results, the reader will find that the approach given in this paper is unified and that it works in other contexts, in particular for Hankel operators on the unit ball in $\mathcal{C}^N$.

The preliminary idea of this paper was clarified during the author's visits to SUNY at Stony Brook, Purdue University, University of Iowa, Virginia Polytech Institute, Texas A & M University, and Cleveland State University during the Fall of 1988, hosted by Professors R. Douglas, L. DeBranges, P. Muhly, R. Olin,

* This work supported by NNSFC.

G. Chen, D. Larson and P. Ghatage et al, respectively. The author would like to express his thanks to all of the professors mentioned above. The author is also grateful to Professor Yu Dahai and Mr. Dechao Zheng for many helpful conversations.

SECTION 1. DUALITY AND MAPPING OF FUNCTION SPACES INTO $\mathcal{L}(\mathcal{H})$ OR VICE VERSA

DEFINITION 1.1. Let S_∞ be the compact operator ideal of $\mathcal{L}(\mathcal{H})$, where $\mathcal{H}$ is a separable Hilbert space, and let $\mathcal{A}$ be a normed function space and $L : \mathcal{A} \to S_\infty$ a bounded linear map, i.e., $\|L(g)\|_{S_\infty} \leqq M\|g\|_{\mathcal{A}}$ $\forall g \in \mathcal{A}$ with M constant independent of $g \in \mathcal{A}$.

DEFINITION 1.2. By dual pairing for $(\mathcal{A}, \mathcal{A}^*)$ we now define dual operator L^* and second dual operator L^{**} as follows

$$\begin{cases} L^* & : S_\infty^* = S_1 \to \mathcal{A}^* \text{ by } \langle L(g), A\rangle_{(S_\infty, S_1)} = \langle g, L^*(A)\rangle_{(\mathcal{A},\mathcal{A}^*)} \ \ \forall g \in \mathcal{A},\ A \in S_1, \\ L^{**} & : \mathcal{A}^{\#*} \to \mathcal{L}(\mathcal{H}) = (S_1)^* \text{ by } \langle g, L^*(A)\rangle = \langle L^{**}(g), A\rangle \ \ \forall g \in \mathcal{A}^{\#*},\ A \in S_1 \end{cases} \tag{$*$}$$

where $\mathcal{A}^{\#} \equiv \|\cdot\|_{\mathcal{A}^*} - Cl\{L^*(S_1)\}$, the closure of $L^*(S_1)$ w.r.t. the norm of $\mathcal{A}^*$, and S_1 is the trace class operators.

REMARK 1.1. With canonically embedding $\mathcal{A} \to \mathcal{A}^{\#*}$, it is evident that $L^{**} \mid \mathcal{A} = L$. And hence L^{**} can be viewed as an extension of L to $\mathcal{A}^{\#*}$.

LEMMA A. Assume ${}_{\#}\mathcal{A}^{\#} = \|\cdot\|_{\mathcal{A}^{\#*}} - Cl\{\mathcal{A}\}$, the closure of $\mathcal{A}$ w.r.t. the norm of the dual of $\mathcal{A}^{\#}$, then $L({}_{\#}\mathcal{A}^{\#}) \subset S_\infty$ and $L(\mathcal{A}^{\#*}) \subset \mathcal{L}(\mathcal{H})$.

PROOF. Actually, for $g \in \mathcal{A}$, we have

$$\begin{aligned} \|L(g)\|_{S_\infty} &= \sup_{A \in (S_1)_1} |\langle L(g), A\rangle| \\ &= \sup_{A \in (S_1)_1} |\langle g, L^*(A)\rangle_{(\mathcal{A},\mathcal{A}^*)}| \\ &\leqq \|L^*\| \sup_{\phi \in (\mathcal{A}^{\#})_1} |\langle g, \phi\rangle_{(\mathcal{A},\mathcal{A}^*)}| = \|L^*\|\ \|g\|_{{}_{\#}\mathcal{A}^{\#}}. \end{aligned} \tag{1.1}$$

Since $\mathcal{A}$ is dense in ${}_{\#}\mathcal{A}^{\#}$ w.r.t. $\|\cdot\|_{{}_{\#}\mathcal{A}^{\#}}$ and $L(\mathcal{A}) \subset S_\infty$, it is clear that $L({}_{\#}\mathcal{A}^{\#}) \subset S_\infty$.

As to the assertion $L(\mathcal{A}^{\#*}) \subset \mathcal{L}(\mathcal{H})$, this is nothing but the duality construction given in $(*)$.

REMARK 1.2. If ${}_{\#}\mathcal{A}^{\#}$ is defined only as the predual of $\mathcal{A}^{\#}$, then $\mathcal{A}$ would not be necessarily dense in ${}_{\#}\mathcal{A}^{\#}$. And if it were so defined, then Lemma A would be not true in general.

DEFINITION 1.3. Let L be a linear bounded map of S_∞ ($\subset \mathcal{L}(\mathcal{H})$) into $\mathcal{A}$, a normed space. By a similar construction, we can define:

$$L^* : \mathcal{A}^* \to S_1 \text{ by } \langle L(A), g\rangle = \langle A, L^*(g)\rangle \;\; \forall A \in S_\infty,\; g \in \mathcal{A}^*. \tag{**}$$

The following is routine

LEMMA B. If L is a linear bounded mapping of S_∞ into $\mathcal{A}$, as given in Definition 1.3, then $L^*(\mathcal{A}) \subset S_1$.

SECTION 2. BOUNDEDNESS AND COMPACTNESS OF HANKEL OPERATORS

A unified approach is now at hand to characterize which Hankel operators on some natural Hilbert spaces will be bounded, compact, or nuclear. However, only a few examples will be provided in this section to explain how the general duality approach works. As to the nuclearity, it will be given in the next two sections.

As usual, $A(D)$ stands for the disk algebra over the unit disk D in $\mathcal{C}$, $H^2(D)$ the Hardy space on the unit disk and $L^2_a(D)$ the Bergman space on the unit disk. For $\phi \in L^\infty(\partial D)$, define the Hankel operator, H_ϕ, on Hardy space from $H^2(D)$ into $H^{2\perp}(D)$ by $H_\phi h = (I-P)\{\phi h\}$ $\forall h \in H^2(D)$, where P is the orthogonal projection of $L^2(\partial D)$ onto $H^2(D)$. Similarly, define the Hankel operator on Bergman space, still denoted by H_ϕ, by $H_\phi h = (I-P)\{\phi h\}$ $\forall h \in L^2_a(D)$, where $\phi \in L^\infty(D)$, P is the orthogonal projection of $L^2(D)$ onto $L^2_a(D)$.

PROPOSITION 2.1. Let $\mathcal{A} = \overline{A(D)}$, the complex conjugate of $A(D)$, and define L mapping of $\mathcal{A}$ into S_∞ (of $H^2(D)$ into $H^{2\perp}(D)$ or $L^2_a(D)$ into $L^{2\perp}_a(D)$ respectively) by:

$$L(g) = H_g \quad \text{for Hardy or Bergman space} \quad \forall g \in \overline{A(D)},$$

then (cf. Def. 1.1 for the definition of $\mathcal{A}^\#$)

$$\mathcal{A}^\# = \begin{cases} H^1(d) & \text{if } \mathcal{H} = H^2(D), \\ L^1_a(d) & \text{if } \mathcal{H} = L^2_a(D). \end{cases} \tag{2.1}$$

PROOF. Embed $\mathcal{A} \to C(\partial D)$ in the case $\mathcal{H} = H^2(D)$. The Hahn–Banach theorem and the Riesz representation theorem show that $\mathcal{A}^\# \subset$ {Regular Borel measures M on ∂D}. Moreover, each $A \in S_1$

$(H^{2\perp}(D) \to H^2(D))$ has expression $A = \sum_{i=1}^{\infty} \lambda_i e_i \otimes d_i$, where $\lambda_j \geqq 0$, $\sum_{j=1}^{\infty} \lambda_j < \infty$ and $\{e_j\}_1^\infty$ and $\{d_j\}_1^\infty$ are orthonormal sequences in $H^2(D)$ and $H^{2\perp}(D)$ respectively, so

$$\begin{aligned}\langle L^*(A), g\rangle = \langle A, L(g)\rangle = Tr\{AL(g)\} &= \sum_{j=1}^{\infty} \langle d_j, L(g)e_j\rangle \\ = \sum_{j=1}^{\infty} \lambda_j \int_{\partial D} d_j \overline{g e_j} \frac{d\theta}{2\pi} &= \int_{\partial D} \left(\sum_{j=1}^{\infty} \lambda_j d_j \bar{e}_j \right) \bar{g} \frac{d\theta}{2\pi}.\end{aligned} \tag{2.2}$$

Hence $L^*(A) \in \overline{H^1(d)}$. Obviously $L^*(S_1)$ is dense in $\overline{H^1(D)}$ and the norm $\|\cdot\|_{C^*(\partial D)}$ is equal to $\|\cdot\|_{H^1}$. This means $\mathcal{A}^\# = H^1(D)$ with the obvious identification.

As to the Bergman space, the proof is a little bit different, because the function $\sum_{j=1}^{\infty} \lambda_j e_j \bar{d}_j$, appearing in (2.2) is not automatically in $L_a^1(D)$ in general and the orthogonal projection $P : L^2(D) \to L_a^2(D)$ is not bounded from $L^1(D)$ into $L_a^1(D)$. However, it is a fact that

$$\begin{aligned}\|L^*(A)\|_{L_a^1(D)} = \left\| P\left\{ \sum_{j=1}^{\infty} \lambda_j e_j \bar{d}_j \right\} \right\|_{L^1} &\leqq C\|A\|_{S_1} \\ = C\sum_{j=1}^{\infty} \lambda_j \qquad &\forall A \in S_1(L_a^2 \to L_a^2).\end{aligned} \tag{2.3}$$

Actually

$$\begin{aligned}\left\| P\left\{ \sum_{j=1}^{\infty} \lambda_j e_j \bar{d}_j \right\} \right\|_{L^1} &= \sup_{\psi \in (L^\infty(D))_1} \left| \int_D \psi(z) \int_D \frac{\Sigma \lambda_j e_j(w) \overline{d_j(w)}}{(1 - z\overline{w})^2} dA(w) dA(z) \right| \\ &= \sup_{\psi \in (L^\infty(D))_1} \left| \int_D \sum_{j=1}^{\infty} \lambda_j e_j(w) \overline{d_j(w)} \int_D \frac{\psi(z)}{(1 - z\overline{w})^2} dA(z) dA(w) \right| \\ &= \sup_{\psi \in (L^\infty(D))_1} \left| \sum_{j=1}^{\infty} \lambda_j \langle H_{\overline{P\psi}} e_j, d_j\rangle \right| \\ &= \sup_{\psi \in (L^\infty(D))_1} \left| Tr\left\{ H_{\overline{P\psi}} \cdot \sum_{j=1}^{\infty} \lambda_j e_j \otimes d_j \right\} \right| \\ &\leqq \sup_{\psi \in (L^\infty(D))_1} \left\| H_{\overline{P\psi}} \right\| \left\| \sum_{j=1}^{\infty} \lambda_j e_j \otimes d_j \right\|_{S_1} \\ &\quad \text{(by Theorems 2.7 \& 5.3 of [2])} \\ &\leqq M \left\| \sum_{j=1}^{\infty} \lambda_j e_j \otimes d_j \right\|_{S_1} \\ &= M \sum_{j=1}^{\infty} \lambda_j,\end{aligned} \tag{2.4}$$

where M is a constant.

The proof of the proposition is thus completed.

REMARK 2.1. It is not too hard to prove the following stronger result: $L^*(S_1)$ is closed, i.e., $L^*(S_1)$ is exactly equal to H^1 or L^1_a respectively. This means that the norm of Hankel operator H_ϕ with ϕ co-analytic is equivalent to the norm of ϕ being viewed as an element of $(L^*(S_1))^*$. Of course, the latter is well known.

PROPOSITION 2.2. Let $\mathcal{A}$ and L be as that given in Proposition 2.1, then

$$\mathcal{A}^{\#*} = \begin{cases} BMOA & \text{if } \mathcal{H} = H^2(D), \\ B & \mathcal{H} = L^2_a(D), \end{cases} \tag{2.5}$$

and

$$_{\#}\mathcal{A}^{\#} = \begin{cases} VMOA & \text{if } \mathcal{H} = H^2(D), \\ B_0 & \mathcal{H} = L^2_a(D), \end{cases} \tag{2.6}$$

where B (B_0) is the (little) Block space. (cf. [1], [2], and [4] for the definitions of the spaces B, B_0, $BMOA$ and $VMOA$).

PROOF. It is a fact that $H^1(D))^* = BMOA$ is just the Fefferman's duality theorem [4], and a fact that $(VMOA)^* = H^1(D)$ follows from [4] p. 276, Exercise 14a. Also, it is well known by [2] that $(L^1_a(D))^* = B$ and that $(B_0)^* = L^1_a(D)$.

THEOREM 2.3. [1],[2],[6]. If $\phi \in \mathcal{H}$ $(= H^2(D)$ or $L^2_a(D)$ respectively) then $H_{\overline{\phi}}$ is bounded if and only if $\phi \in \mathcal{A}^{\#*}$ $(= BMOA$ or B respectively), and $H_{\overline{\phi}}$ is compact if and only if $\phi \in {}_{\#}\mathcal{A}^{\#}$ $(= VMOA$ or B_0 respectively).

PROOF. From Lemma A, only the "only if" part needs to be proved. To save space, we give the proof here only for the Bergman space case. Actually, what we give below is just the repeat of Axler's argument of the corresponding part.

If $H_{\overline{\phi}} \in \mathcal{L}(L^2_a(D), L^{2\perp}_a(D))$ with $\phi \in L^2_a(D)$, then

$$\begin{aligned} \|H_{\overline{\phi}}\|^2 &\geqq \sup_{\lambda\in D} \left\| H_{\overline{\phi}}\left\{\frac{K_\lambda}{\|K_\lambda\|}\right\}\right\|^2 \\ &= \sup_{\lambda\in D}\left\{(1-|\lambda|^2)^2(\langle T^*_{\overline{\phi}}K_\lambda, T^*_{\overline{\phi}}K_\lambda\rangle_{L^2_a(D)} - \langle T_{\overline{\phi}}K_\lambda, T_{\overline{\phi}}K_\lambda\rangle_{L^2_a(D)})\right\} \\ &= \sup_{\lambda\in D}(1-|\lambda|^2)^2\|(\phi-\phi(\lambda))K_\lambda\|^2 \\ &\cong \|\phi\|_B \end{aligned} \tag{2.7}$$

where $K_\lambda(z) = \frac{1}{(1-\overline{\lambda}z)^2}$. The last equality of (2.7) is given in [1], p. 327.

This concludes the proof.

SECTION 3. THE NUCLEARITY OF HANKEL OPERATORS

We now consider the nuclearity of Hankel operators. To show how our approach works, we restrict ourselves only to the Hardy space case. To this end, we need to define a normed function space $\mathcal{A}$ and a linear bounded mapping of $S_\infty(H^2 \to H^{2\perp})$ into $\mathcal{A}$. First, let

$$\mathcal{A} = \{\overline{\phi} : \phi \text{ analytic on } D \text{ with } \|\overline{\phi}\|_{\mathcal{A}} = \sup_{z\in D}\{(1-|z|^2)|\phi(z)|)\} < \infty \text{ and } \lim_{|z|\to 1}(1-|z|^2)|\phi(z)| = 0\}. \quad (3.1)$$

i.e.,

$$\mathcal{A} = \{\overline{\phi} : \phi = \psi'(z) \text{ for some } \psi \in B_0 \text{ and } \|\phi\|_{\mathcal{A}} \equiv \|\psi\|_{B_0}\}. \quad (3.2)$$

Second, for any finite rank operator in $\mathcal{L}(H^{2\perp}, H^2)$, say $A = \sum_{j=1}^{N} \lambda_j e_j \otimes d_j$ with $\{e_j\}_1^N \subset H^2$ and $\{d_j\}_1^N \subset H^{2\perp}$, define L by

$$L(A) = \sum \lambda_j \overline{V(\overline{d}_j)(\overline{\alpha})} e_j(\alpha) \in \mathcal{A} \quad (3.3)$$

where

$$V\left(\sum_{j=0}^{\infty} a_j z^j\right) = \sum_{j=1}^{\infty} a_j z^{j-1}.$$

LEMMA 3.1. The operator L defined by (3.3) is bounded, and it can be extended to a bounded mapping of $S_\infty(H^{2\perp} \to H^2)$ onto $\mathcal{A}$.

PROOF. First, for a finite rank operator $A = \sum_{j=1}^{N} \lambda_j e_j \otimes d_j \in \mathcal{L}(H^{2\perp}, H^2)$, we may assume that $\{e_j\}$ and $\{\overline{d}_j\}$ are all polynomials, so $L(A) = \sum \lambda_j \overline{V(\overline{d}_j)(\overline{\alpha})} e_j(\alpha) \equiv P(\alpha)$ is a polynomial in $\alpha \in D$. Second, for the rank one operator $\overline{zK_{\overline{\alpha}}(z)} \otimes K_\alpha(z) (\in S_1(H^2 \to H^{2\perp}))$ where $K_\alpha(z) = \frac{1}{1-\overline{\alpha}z}$, compute the following dual pairing

$$\begin{aligned} Tr\{A \cdot \overline{zK_{\overline{\alpha}}} \otimes K_\alpha\} &= \langle A, \overline{zK_{\overline{\alpha}}} \otimes K_\alpha \rangle_{(S_\infty, S_1)} \\ &= \sum \lambda_j \langle e_j, K_\alpha \rangle_{H^2} \cdot \langle \overline{zK_{\overline{\alpha}}}, d_j \rangle_{H^{2\perp}} \\ &= \sum \lambda_j e_j(\alpha) \overline{\langle \overline{d}_j, zK_{\overline{\alpha}} \rangle_{H^2}} \\ &= \sum \lambda_j e_j(\alpha) \overline{V(\overline{d}_j)(\overline{\alpha})} \\ &= P(\alpha) = L(A). \end{aligned} \quad (3.4)$$

Noticing that $\|\overline{zK_{\overline{\alpha}}} \otimes K_\alpha\|_{S_1} = \|K_\alpha\|_{H^2} \cdot \|zK_{\overline{\alpha}}\|_{H^2} = \frac{1}{1-|\alpha|^2}$, we have

$$\begin{aligned} \|L(A)\|_{\mathcal{A}} &= \sup_{\alpha \in D}\{(1-|\alpha|^2)|Tr(A \cdot \overline{zK_{\overline{\alpha}}} \otimes K_\alpha)|\} \\ &= \sup_{\alpha\in D}\left|Tr\left(A \cdot \frac{zK_{\overline{\alpha}} \otimes K_\alpha}{\|zK_{\overline{\alpha}} \otimes K_\alpha\|_{S_1}}\right)\right| \leqq \sup_{\|Q\|_{S_1} \leqq 1} |Tr(A \cdot Q)| \\ &= \|A\|. \end{aligned} \tag{3.5}$$

In other words, L is bounded from the set of finite rank operators into $\mathcal{A}$ So L can be extended to a bounded mapping of S_∞ into $\mathcal{A}$. By using a modified version of (2.7), it can be shown that $\mathcal{L}$ is actually onto $\mathcal{A}$. We omit the details here.

PROPOSITION 3.2. $\mathcal{A}^* = A_1^1(D)$, the Besov space B_p^s with $p = s = 1$ and consisting of analytic function.

PROOF. The following formula can be easily verified if any one of the terms appeared in chain of equalities below makes sense (cf. [2], p. 14):

$$\begin{aligned} \frac{1}{2\pi}\int_{\partial D} \phi\overline{g} d\theta &= \int_D \phi\overline{(zg)'} dA \\ &= \int_D (V\phi)(z)\overline{(zg)''}(1-|z|^2 dA(z) \end{aligned} \tag{3.6}$$

where ϕ and g are analytic on D with $\phi(0) = 0$ or $g(0) = 0$ assumed. So if $(zg)'' \in L_a^1(D)$ which is equivalent to $g \in A_1^1(D)$ (cf. [5], p. 473 by taking $n = 2$, $s = p = 1$), then

$$\left|\frac{1}{2\pi}\int_0^{2\pi} \phi\overline{g} d\theta\right| \leqq \|V\phi\|_{\mathcal{A}} \cdot \|(zg)''\|_{L_a^1}, \quad \forall \phi \in \mathcal{A}, \tag{3.7}$$

i.e., $A_1^1(D) \subset \mathcal{A}^*$. By noticing that $(B_0)^* = L_a^1$, $(L_a^1)^* = B$ and $\|\phi\|_{\mathcal{A}} = \|\psi\|_B$ $\forall\phi \in \mathcal{A}$, where $\psi(z) = \int_0^z \phi(s)ds$, it is easy to check that $\mathcal{A}^* \subseteq A_1^1(D)$. We are done.

THEOREM 3.1. (Peller's criteria). If $\overline{\phi}$ is analytic on D, then H_ϕ is nuclear if and only if $\overline{\phi} \in A_1^1(D)$.

PROOF. First we show that $L^*g = \Omega H_{zg}$ if $\overline{g} \in A_1^1(D) = \mathcal{A}^*$, where $\Omega : H^{2\perp} \to H^{2\perp}$ is a "conjugate unitary" operator defined by $\Omega\left(\sum_{n<0} a_n z^n\right) = \sum_{n<0} \overline{a}_n z^n$. Indeed

$$
\begin{aligned}
\langle L\left(\sum \lambda_j e_j \otimes d_j\right), \overline{g}\rangle_{(\mathcal{A},\mathcal{A}^*)} &= \langle \sum \lambda_j \overline{V(\overline{d}_j)(\overline{z})} e_j(z), \overline{g}(z)\rangle_{L^2(\partial D)} \\
&= \sum \lambda_j \langle e_j, V(\overline{d}_j)(\overline{z})\overline{g}\rangle_{L^2(\partial D)} \\
&= \sum \lambda_j \langle e_j, \overline{d}_j(\overline{z})\overline{zg}\rangle_{L^2(\partial D)} \\
&= \sum \lambda_j \langle H_{zg} e_j, \overline{d}_j(\overline{z})\rangle_{L^2(\partial D)} \\
&= \sum \lambda_j \langle H_{zg} e_j, \Omega d_j\rangle_{L^2(\partial D)} \\
&= Tr\{\Omega H_{zg}(\sum \lambda_j e_j \otimes d_j)\}
\end{aligned}
\tag{3.8}
$$

This is equivalent to $L^*(\overline{g}) = \Omega H_{zg}$ as asserted. Since Ω is "conjugate unitary" on $H^{1\perp}$, Lemma B implies that H_g is nuclear if $\overline{g} \in \mathcal{A}^* = A_1^1(D)$.

Since the above procedure is reversible and $\|\psi\|_B = \|\phi\|_{\mathcal{A}} = \|\psi\|_{(L_a^1)^*}$ where $\psi(z) = \int_0^z \phi(s)ds$, from Lemma 3.1 we easily see that if H_g is nuclear, then $\overline{g} \in \mathcal{A}^* = A_1^1(D)$.

The proof of the theorem is thus finished.

SECTION 4. THE NUCLEARITY OF THE PRODUCT OF HANKEL OPERATORS ON BERGMAN SPACE

In [8], Dechao Zheng proved that for $\overline{\phi}, \overline{\psi} \in B$ then $H_\phi^* H_\psi$ is compact if and only if

$$\lim_{|\lambda|\to 1} \{(1-|\lambda|^2 \min(|\psi'(\lambda)|, |\phi'(\lambda)|\} = 0. \tag{4.1}$$

Moreover, it is well known [3] that for $\overline{\phi} \in L_a^2(D)$, $H_\phi^* H_\phi \in S_1$ if and only if

$$\int_D |\phi'(z)|^2 dA(z) < \infty. \tag{4.2}$$

This leads to a natural conjecture that if $\overline{\phi}' \in L_a^p(D)$ and $\overline{\psi}' \in L_a^q(D)$ with $\frac{1}{p} + \frac{1}{q} = 1$, then $H_\phi^* H_\psi \in S_1$. It will be proved in this section that it is indeed true.

Our approach to this conjecture relies on Lemma B. To this end, we define first a function space $\mathcal{A}$ which is the completion of $A(D) \times \overline{A(D)}$ w.r.t. a norm $\|\cdot\|_{\mathcal{A}}$ given as follows:

$$\mathcal{A} = \|\cdot\|_{\mathcal{A}} - Cl\{f(z,w) \in A(D) \times \overline{A(D)} \text{ with norm } \|f\|_{\mathcal{A}} = \sup_{\alpha,\beta \in D} |(1-|\alpha|^2)(1-|\beta|^2) f(\alpha,\beta)|\} \tag{4.3}$$

and then define an operator $L : S_\infty(L_a^2(D)) \to \mathcal{A}$ by

$$L\left(\sum_{j=1}^{\infty} \lambda_j e_j \otimes d_j\right) = \sum_{j=1}^{\infty} \lambda_j e_j(z)\overline{d_j(w)} \tag{4.4}$$

where $\{e_j\}$ and $\{d_j\}$ are orthonormal sequences in $L_a^2(D)$. It is easy to check that L is bounded from S_∞ into $\mathcal{A}$. Actually, as shown in the proof of Lemma 3.1,

$$\begin{aligned} \|L(\Sigma\lambda_j e_j \otimes d_j)\|_{\mathcal{A}} &= \sup_{\alpha,\beta\in D} |(1-|\alpha|^2)(1-|\beta|^2)\Sigma\lambda_j e_j(\alpha)\overline{d}_j(\beta)| \\ &= \sup_{\alpha,\beta\in D} |\Sigma\lambda\langle e_j, K_\alpha/\|K_\alpha\|\rangle\langle K_\beta/\|K_\beta\|, d_j\rangle| \\ &= \sup_{\alpha,\beta\in D} \left|Tr\left\{(\Sigma\lambda_j e_j \otimes d_j)\cdot\left(\frac{K_\beta}{\|K_\beta\|}\otimes\frac{K_\alpha}{\|K_\alpha\|}\right)\right\}\right| \\ &\leqq \|\Sigma\lambda_j e_j \otimes d_j\|_{S_\infty}\left\|\frac{K_\beta}{\|K_\beta\|}\otimes\frac{K_\alpha}{\|K_\alpha\|}\right\|_{S_1} \\ &= \|\Sigma\lambda_j e_j \otimes d_j\|_{S_\infty} \end{aligned} \tag{4.5}$$

where $K_\alpha(z) = \frac{1}{(1-\overline{\alpha}z)^2}$ is the reproducing kernel of $L_a^2(D)$ and $\|K_\alpha\|_{L_a^2} = \frac{1}{1-|\alpha|^2}$. Thus Lemma B yields to $L^*(\mathcal{A}^*) \subset S_1$. In order to relate the product of Hankel operators to L^*, we note that for $\overline{g}$, $\overline{f}$ in Bloch space, we have [1]:

$$H_g^* H_f = T_{\overline{g}f} - T_{\overline{g}}T_f \tag{4.6}$$

where T_f is the Toeplitz operator on Bergman space with symbol f. Therefore

$$\begin{aligned} Tr\{(H_g^*H_f)\cdot(h\otimes\psi)\} &= \langle H_g^*H_f h, \psi\rangle_{L^2(D)} \\ &= \int_D f(z)\overline{g}(z)h(z)\overline{\psi}(z)dA(z) \\ &\quad - \iiint\limits_{D\times D\times D} \frac{f(s)h(s)\overline{g}(w)\overline{\psi}(w)}{(1-z\overline{s})^2(1-\overline{z}w)^2}dA(s)dA(w)dA(z) \\ &= \iint\limits_{D\times D}\left\{\int_D \frac{(f(z)-f(s))(\overline{g}(z)-\overline{g}(w))}{(1-z\overline{s})^2(1-\overline{z}w)^2}dA(z)\right\}h(s)\overline{\psi}(w)dA(s)dA(w) \\ &= \langle G, L(h\otimes\psi)\rangle_{(\mathcal{A},\mathcal{A}^*)} \\ &= \langle L^*(G), h\otimes\psi\rangle_{(S_\infty,S_1)} \quad \forall h,\psi\in L_a^2(D) \end{aligned} \tag{4.7}$$

where

$$G(s,w) = \int_D \frac{(f(z)-f(s))(\overline{g}(z)-\overline{g}(w))}{(1-z\overline{s})^2(1-\overline{z}w)^2}dA(z). \tag{4.8}$$

A slightly different formula was given in [8], p. 107, the reader may consult [8] for details of the computation appearing in (4.7).

We are now ready to give the main result of this section.

THEOREM 4.1. If $\overline{g}' \in L^p_a(D)$ and $\overline{f}' \in L^q_a(D)$ where $p, q > 1$ and $\frac{1}{p} + \frac{1}{q} = 1$, then $H^*_g H_f \in S_1$, i.e., $H^*_g H_f$ is nuclear.

PROOF. For $F(s,w) \in A(D) \times \overline{A(D)}$, applying formula (3.6) twice (cf. [2], p. 14 also), we have

$$\begin{aligned}\iint\limits_{D\times D} & G(s,w) s\overline{w} F(s,w) dA(s) dA(w) = \\ &= \iint\limits_{D\times D} \left(\frac{\partial^2}{\partial \overline{s} \partial w} G(s,w)\right) \left(1 - |s|^2\right) \left(1 - |w|^2\right) \cdot F(s,w) dA(s) dA(w)\end{aligned} \tag{4.9}$$

From (4.9) and the fact that the $\mathcal{A}$-norm of the function $(s\overline{w}F(s,w))$ is equivalent to $\|F\|_{\mathcal{A}}$ $\forall F \in A(D) \times \overline{A(D)}$, in order to show $G \in \mathcal{A}^*$ it suffices to prove $\frac{\partial^2 G(s,w)}{\partial \overline{s} \partial w} \in L^1(D \times D)$. Once we prove this, then Lemma B implies $H^*_g H_f \in S_1$ because the formula (4.7) tells us $L^*(G) = H^*_g H_f$.

The remainder of the proof consists of showing $\frac{\partial^2 G(s,w)}{\partial \overline{s} \partial w} \in L^1(D \times D)$. Direct computation gives

$$\begin{aligned}\frac{\partial^2 G(s,w)}{\partial \overline{s} \partial w} &= \int_D \frac{f'(s)\overline{g}'(w)}{(1 - z\overline{s})^2 (1 - \overline{z} w)^2} dA(z) \\ &\quad - 2\int_D \overline{z} \frac{f'(s)(\overline{g}(z) - \overline{g}(w))}{(1 - z\overline{s})^2 (1 - \overline{z} w)^3} dA(z) \\ &\quad - 2\int_D z \frac{(f(z) - f(s))\overline{g}'(w)}{(1 - z\overline{s})^3 (1 - \overline{z} w)^2} dA(z) \\ &\quad + 4\int_D \frac{(f(z) - f(s))(\overline{g}(z) - \overline{g}(w))}{(1 - z\overline{s})^3 (1 - \overline{z} w)^3} dA(z) \\ &= (I) - 2(II) - 2(III) + 4(IV).\end{aligned} \tag{4.10}$$

In order to estimate the terms (I)–(IV), we need the following lemmas:

LEMMA 4.1. Define $\hat{T}_s f(z) = \int_D \frac{(1-|w|^2)^s f(w)}{|1-z\overline{w}|^{2+s}} dA(w)$, then $\|\hat{T}_s f\|_{L^p(D)} < C_{s,p} \|f\|_{L^p(D)}$ $\forall f \in L^p(D)$, if $(1+s) > \frac{1}{p}$, $p \geqq 1$, where s is real and $C_{s,p}$ constant independent of the function $f \in L^p(D)$.

PROOF. This is indeed a special case of Theorem 7.1.5 of [7]. And the proof itself given there contains the lemma.

LEMMA 4.2. If $g' \in L^p_a(D)$, $p > 1$, then $\int_D \left|\frac{g(z)-g(w)}{(1-\overline{z}w)^3}\right| dA(w) \in L^p(D)$.

PROOF. Using the reproducing kernel of $L_a^2(D)$, we have

$$
\begin{aligned}
g(z)-g(w) &= \int_D \left(\frac{1}{(1-z\overline{\rho})^2}-\frac{1}{(1-w\overline{\rho})^2}\right) g(\rho)dA(\rho) \\
&= (z-w)\int_D \frac{(1-w\overline{\rho})+(1-z\overline{\rho})}{(1-z\overline{\rho})^2(1-w\overline{\rho})^2}\overline{\rho} g(\rho)dA(\rho) \\
\text{(by (3.6))} &= (z-w)\int_D \frac{(1-w\overline{\rho})+(1-z\overline{\rho})}{(1-z\overline{\rho})^2(1-w\overline{\rho})^2}(1-|\rho|^2)g'(\rho)dA(\rho) \\
&= (z-w)\int_D \left(\frac{1}{(1-w\overline{\rho})^2(1-z\overline{\rho})}+\frac{1}{(1-z\overline{\rho})^2(1-w\overline{\rho})}\right)(1-|\rho|^2)g'(\rho)dA(\rho) \\
&= (z-w)((I')+(II')).
\end{aligned}
\tag{4.11}
$$

Furthermore

$$
\begin{aligned}
|I'| &= \left|\int_D \frac{(1-|\rho|^2)g'(\rho)}{(1-w\overline{\rho})^2(1-z\overline{\rho})}dA(\rho)\right| \leqq \text{(Holder inequality)} \\
&\left(\int_D \frac{(1-|\rho|^2)^2}{|1-w\overline{\rho}|^4}|g'(\rho)|dA(\rho)\right)^{1/2}\cdot\left(\int_D \frac{|g'(\rho)|}{|1-\overline{z}\rho|^2}dA(\rho)\right)^{1/2} \\
&= \left(\hat{T}_2|g'|(w)\right)^{1/2}\cdot\left(\hat{T}_0|g'|(z)\right)^{1/2},
\end{aligned}
\tag{4.12}
$$

and similarly

$$
|II'| \leqq \left(\hat{T}_2|g'|(z)\right)^{1/2}\cdot\left(\hat{T}_0|g'|(w)\right)^{1/2}. \tag{4.13}
$$

And so

$$
\begin{aligned}
\int_D \left|\frac{g(z)-g(w)}{(1-\overline{z}w)^3}\right| dA(w) &\leqq \int_D \frac{1}{|1-\overline{z}w|^2}\frac{|z-w|}{|1-\overline{z}w|}(|I'|+|II'|)dA(w) \\
&\leqq \int_D \frac{1}{|1-zw|^2}(|I'|+|II'|)dA(w) \\
&\leqq \int_D \frac{1}{|1-zw|^2}\{\left(\hat{T}_2|g'|(w)\right)^{1/2}\cdot\left(\hat{T}_0|g'|(z)\right)^{1/2} \\
&\quad + \left(\hat{T}_2|g'|(z)\right)^{1/2}\cdot\left(\hat{T}_0|g'|(w)\right)^{1/2}\}dA(w) \\
&= \left(\hat{T}_0|g'|(z)\right)^{1/2}\cdot\hat{T}_0\left((\hat{T}_2|g'|)^{1/2}\right)(z) \\
&\quad + \left(\hat{T}_2|g'|(z)\right)^{1/2}\cdot\hat{T}_0\left((T_0|g'|)^{1/2}\right)(z).
\end{aligned}
\tag{4.14}
$$

It follows from Lemma 4.1 that the first term of the most right hand of (4.14) has the L_p-estimation as follows:

$$
\begin{aligned}
&\int_D \left(\hat{T}_0|g'|(z)\right)^{p/2}\cdot\left(\hat{T}_0(\hat{T}_2|g|')^{1/2}\right)^p(z)dA(z) \\
&\leqq \|\hat{T}_0|g'|\,\|_{L^p}^{p/2}\cdot\|\hat{T}_0\left((\hat{T}_2|g'|)^{1/2}\right)\|_{L^{2p}}^p \\
&\leqq C_{0,p}^{p/2}\|g'\|_{L^p}^{p/2}\cdot C_{0,2p}^p\|(\hat{T}_2|g'|)^{1/2}\|_{L^{2p}}^p \\
&= C_{0,p}^{p/2}\|g'\|_{L_p}^{p/2}\cdot C_{0,2p}^p\|\hat{T}_2|g'|\,\|_{L^p}^{p/2} \\
&\leqq C_{0,p}^{p/2}\cdot C_{0,2p}^p C_{2,p}^{p/2}\|g'\|_{L_p}^{p/2}\cdot\|g'\|_{L_p}^{p/2} \\
&= C_{0,p}^{p/2}\cdot C_{0,2p}^p C_{2,p}^{p/2}\|g'\|_{L_p}^p \equiv C_1\|g'\|_{L_p}^p.
\end{aligned}
\tag{4.15}
$$

Similarly

$$\| \left(\hat{T}_2|g'|\right)^{1/2} \cdot \hat{T}_0 \left((\hat{T}_0|g'|)^{1/2}\right) \|_{L_p} < C_2 \|g'\|_{L_p}, \tag{4.16}$$

where C_2 is also a constant.

Putting (4.14), (4.15), and (4.16) together, we obtain the lemma.

We now return to the proof of Theorem 4.1. As indicated at the very beginning of the proof, we only need to show $\frac{\partial^2 G(s,w)}{\partial \bar{s} \partial w} \in L^1(D \times D)$. To this end, it suffices to prove that each term of (I), (II), (III), and (IV) appearing in (4.10) is in $L^1(D \times D)$. We provide only the estimation of (IV) here since others can be done similarly. From Lemma 4.2, we have

$$\begin{aligned}
\iint_{D \times D} |(IV)| dA(s) dA(w) &< \int_D \{ \int_D \left| \frac{f(z) - f(s)}{(1 - \bar{z}s)^3} \right| dA(s) \\
&\quad \cdot \int_D \left| \frac{\bar{g}(z) - \bar{g}(w)}{(1 - \bar{z}w)^3} \right| dA(w) \} dA(z) \\
&\leqq \text{(Holder inequality)} \left\| \int_D \left| \frac{f(\cdot) - f(s)}{(1 - \cdot\bar{s})^3} \right| dA(s) \right\|_{L^q(D)} \\
&\quad \cdot \left\| \int_D \left| \frac{\bar{g}(\cdot) - \bar{g}(w)}{(1 - \cdot w)^3} \right| dA(w) \right\|_{L^p(D)} \\
&\leqq C \|f'\|_{L^q} \|g'\|_{L^p} \qquad \text{(by Lemma 4.2)}
\end{aligned} \tag{4.17}$$

where C is a constant.

The proof of Theorem 4.1 is thus completed.

REMARK 4.1. Theorem 4.1 is sharp. As is well known, if $p = q = 2$ then the condition given in Theorem 4.1 is not only sufficient but also necessary. This suggests a natural conjecture: For $\bar{g}, \bar{f} \in L^2_a(D)$, the product $H^*_g H_f$ of Hankel operators on Bergman space is nuclear if and only if $\bar{g}' \in L^p_a(D)$ and $\bar{f}' \in L^q_a(D)$ for some p and q with $\frac{1}{p} + \frac{1}{q} = 1$ and $p, q > 1$.

REMARK AND ACKNOWLEDGEMENT. As pointed out by the referee, the conjecture above in not true in general. Moreover, V. V. Peller also exhibited an argument via Besov space theory together with his criteria of Hankel operators being of Schartten p–class to show that the conjecture is not true in general. The author would like to thank the referee and V. V. Peller for bringing to his attention the comments just mentioned as well as the helpful discussions with V. V. Peller. Also the author would like to thank the Mathematics Department of Cleveland State University, where this revised version was done, for its hospitality.

REFERENCES

[1] Axler, S., The Bergman space, the Bloch space and commutators of multiplication operators, Duke Math. J. 53(1986), 315–332.

[2] Axler, S., Bergman space and their operators, Survey of some recent results in operator theory, Vol. 1, Eds. J. B. Conway and B. B. Morel.

[3] Berger, C. A. and Shaw, J. B., Interwing, analytic structure, and the trace norm formula, Lecture Notes in Math., Vol. 345, 1–6.

[4] Garnet, J.B., Bounded analytic functions, Academic Press, Inc., 1981.

[5] Peller, V. V., A description of Hankel operators of class $\mathcal{S}_p$ for $p > 0$, an investigation of the rate of rational approxation, and other applications, Math. USSR Sbornik 50(1985), 465–494.

[6] Power, S. C., Hankel operators on Hilbert space, Research Notes in Math., Ser 64, 1984.

[7] Rudin, W., Function theory in the unit ball of C^n, Springer–Verlag, 1980.

[8] Zheng, Dechao, Hankel operators and Toeplitz operators on the Bergman space, J. Functional Analysis, Vol. 83, No. 1(1989), 98–120.

Department of Mathematics
Sichuan University
Chengdu, 610064
People's Republic of China

Operator Theory:
Advances and Applications, Vol. 48

EIGENVALUE DISTRIBUTION OF NONSELFADJOINT TOEPLITZ MATRICES AND THE ASYMPTOTICS OF TOEPLITZ DETERMINANTS IN THE CASE OF NONVANISHING INDEX

Harold Widom[1)]

Dedicated to the memory of Ernst Hellinger.

A classical theorem of Szegö states that the eigenvalues of a finite selfadjoint Toeplitz matrix are in a particular sense distributed, asymptotically as the size of the matrix increases, as the values of its (real) generating function. In the nonselfadjoint case this is no longer always true, but the question is raised here of whether it is true if the generating function does not extend analytically to an annulus. A certain class of nonreal generating functions is considered and the question is investigated with the aid of the asymptotics of the associated Toeplitz determinants.

1. INTRODUCTION

The limiting distribution of the eigenvalues of selfadjoint Toeplitz matrices

$$T_n(f) = (\hat{f}(i-j)) \qquad i,j=0,\ldots,n\,,$$

with f a bounded real-valued function on the unit circle $\mathbb{T}$, is given by Szegö's classical theorem [S]: If $\lambda_{i,n}$ $(i=0,\ldots,n)$ are the eigenvalues of $T_n(f)$ repeated according to multiplicity and F is a continuous function defined on $\mathbf{R}$ then

---

[1)]Supported by NSF grants DMS-8700901 and 8822906.

$$\lim_{n\to\infty} \frac{1}{n+1} \sum_{i=0}^{n} F(\lambda_{i,n}) = \frac{1}{2\pi} \int_{-\pi}^{\pi} F(f(e^{i\theta}))\,d\theta .$$

Equivalently, if $\mu_{n,f}$ is the measure assigning each $\lambda_{i,n}$ measure $(n+1)^{-1}$ then $\mu_{n,f}$ converges weakly as $n \to \infty$ to the measure μ_f defined by

$$\mu_f(E) = \text{meas } f^{-1}(E)$$

where "meas" denotes normalized Lebesgue measure on $\mathbb{T}$. The support of μ_f is clearly $R(f)$, the essential range of f.

Very little is known in the nonselfadjoint case, which is very different. For in the simplest case $f(z) = z^k$ with $k \neq 0$ all $\lambda_{i,n}$ (eigenvalues are now repeated according to algebraic multiplicity) are zero so the limiting measure is a unit mass at $z = 0$, whereas μ_f is normalized Lebesgue measure on $\mathbb{T}$. The first significant result of any generality is due to Schmidt and Spitzer [S-S] who considered the case of a general Laurent polynomial

$$f(z) = \sum_{k=-N}^{N} \hat{f}(k)\, z^k$$

and determined the limiting set of the eigenvalues. This is the set of $\lambda \in \mathbb{C}$ such that there exist sequences $n_k \to \infty$ and i_k with

$$\lambda = \lim_{k\to\infty} \lambda_{i_k, n_k} .$$

The limiting set was a union of analytic arcs, in no obvious way related to R(f). Hirschman [H] completed this work by determining the limiting measure, whose support turned out to be precisely the limiting set. (This holds in the selfadjoint case for any continuous f.) The most general result along this line is due to Day [D], who considered the case of an arbitrary rational function with poles off $\mathbb{T}$. Again the limiting measure is generally unrelated to μ_f.

This should come as no surprise. For suppose f has no poles in the annulus $r_1 < |z| < r_2$, and define f_r by $f_r(z) = f(rz)$ for $r \in (r_1, r_2)$ and $z \in \mathbb{T}$. A simple computation shows that the matrices $T_n(f)$ and $T_n(f_r)$ are similar via the diagonal matrix $(r^i \delta_{i,j})$. Thus there is no more reason for the limiting measure to be μ_f than it is for it to be μ_{f_r} for any r. A similar argument applies to any f which extends analytically a little into either the interior or the exterior of $\mathbb{T}$.

But this is a very unusual circumstance and we conjecture that except in rare cases such as these the limiting measure exists and equals μ_f. If this does occur we say that the eigenvalues of $T_n(f)$ are *canonically distributed.* What "rare cases such as these" means is left deliberately vague. The most extreme conjecture would give it the meaning "except when f extends analytically to an annulus $r < |z| < 1$ or $1 < |z| < r$." We mean by this that there exists a function F which is bounded and analytic on the annulus such that f is almost everywhere equal to the nontangential limit of F on $\mathbb{T}$. A similar meaning will apply when $\mathbb{T}$ is replaced by any other smooth curve. Of course if f is continuous then the limit is uniform.

All approaches to the eigenvalue distribution problems in the nonselfadjoint case have been via the Toeplitz determinants

$$D_n(f) = \det T_n(f) .$$

The basic first-order results about these determinants say in various contexts that if f satisfies an index zero condition then with an appropriate branch of the logarithm we have

$$\lim_{n\to\infty} \frac{1}{n+1} \log D_n(f) = \log G(f) := \frac{1}{2\pi} \int \log f(e^{i\theta})\, d\theta . \tag{1.1}$$

This is true for example if f is continuous and nonzero and

$$i(f) := \frac{1}{2\pi} [\arg f(e^{i\theta})]_{-\pi}^{\pi}$$

is equal to 0. It follows from recent work of Böttcher and Silbermann [B-S2] that this is also true if f is bounded away from zero and smooth except for a single jump discontinuity, at $\theta = \theta_0$ say, and

$$\left| \frac{1}{2\pi} [\arg f(e^{i\theta})]_{\theta_0+}^{\theta_0+2\pi-} \right| < 1 . \tag{1.2}$$

It turns out that the limiting behavior of the measures $\mu_{n,f}$ associated with the eigenvalues is more or less equivalent to the limiting behavior of the absolute values of the determinants $D_n(f-\lambda)$ for almost all $\lambda \in \mathbb{C}$. One precise result is that if $f \in L_\infty$ and

$$\lim_{n\to\infty} \frac{1}{n+1} \log |D_n(f-\lambda)| = \log G(|f-\lambda|) \tag{1.3}$$

in measure for $\lambda \in \mathbb{C}$ then the eigenvalues of $T_n(f)$ are canonically distributed. (See Lemma 5.1.)

If f is continuous then it is only in rare cases (e.g., if R(f) has measure zero and either f is even or R(f) does not separate the plane) that we have $i(f-\lambda) = 0$ for almost every λ and that, using only results quoted above, we can establish canonical eigenvalue distribution. If f is smooth except for a single jump discontinuity then, using the Böttcher-Silbermann result, we deduce canonical eigenvalue distribution if (1.2), with f replaced by $f-\lambda$, holds for all $\lambda \notin R(f)$. In particular it holds for the functions $e^{ia\theta}$ $(0 < \theta < 2\pi)$ with $-1 < a < 1$ (but not, as we know, for $a = \pm 1$).

The main part of the present paper is devoted to the determination of the asymptotics of the determinants $D_n(f)$ if the nonzero function f, which may have arbitrary index, is continuous and piecewise C^∞ but not C^∞. The main result, Theorem 4.2, is an asymptotic formula of the form

$$D_n(f) = G(f_0)^{n+1} n^{-e} (c_n + o(1)) \tag{1.4}$$

where

$$f_0(z) = f(z) z^{-i(f)} \tag{1.5}$$

so that $i(f_0) = 0$, where e is an exponent depending only on the nature of the singularities of f, (points in no neighborhood of which f is C^∞), and $\{c_n\}$ is a certain bounded sequence given by a rather complicated but (in principle) perfectly explicit formula. (One could also extend (1.4) to a complete asymptotic expansion.)

Using this to establish canonical eigenvalue distribution requires knowing that for almost all $\lambda \notin R(f)$ the sequence $\{c_n\}$ appearing in formula (1.4) for the function $f-\lambda$ is bounded away from zero. (For then (1.3) follows easily.) Unfortunately we have not been able to prove that this is always the case although we have no doubt that it is. We prove it if there is only one singularity, or more than one if their "orders" satisfy a certain condition, or more than one if each bounded component of the complement of R(f) has enough "regular singular values." These terms will be defined below.

At the end of the paper we establish some results of a different kind. Assuming f describes a simple smooth curve we show that almost all the eigenvalues of $T_n(f)$ lie in a compact subset of the interior of R(f) if and only if f extends analytically to an annulus $r < |z| < 1$ or $1 < |z| < r$ (Theorem 6.1), and that if some limiting measure of the $\mu_{n,f}$ has support contained in R(f) then it must be μ_f (Theorem 6.2). For these we essentially use only the basic result (1.1) in the case of vanishing index.

In the course of the proof of Theorem 4.2 we have to determine the asymptotics of certain determinants, of fixed order, whose entries depend on the parameter n. They are determinants of sections of what we call "generalized moment matrices," matrices with i,j entry

$$\iint s^i t^j \, d\mu(s,t) \qquad\qquad i,j = 0,1,\cdots$$

where μ is a complex measure on $\mathbb{C} \times \mathbb{C}$ such that the integrals are absolutely convergent. We shall show (Lemma 2.1) that the determinant of the upper left hand $m \times m$ block of this matrix is equal to

$$\frac{1}{m!} \int \cdots \int \prod_{i>j} [(s_i - s_j)(t_i - t_j)] \, d\mu(s_0,t_0) \cdots d\mu(s_{m-1},t_{m-1}) . \tag{1.6}$$

where, in the product, the indices run over 0,...,m-1. (This is not difficult and is essentially a special case of [P-S], prob. 68. We give a different proof below.) This representation will enable us to deduce the desired asymptotics, which seems to us of independent interest. The simplest result of this type is that for any $p > 0$ we have as $n \to \infty$

$$\det((n+i+j)^{-p})_{i,j=0,\ldots,m-1}$$
$$\sim n^{-m^2-(p-1)m} \det\Big(\prod_{k=0}^{i+j-1} (p+k)\Big)_{i,j=0,\ldots,m-1} . \tag{1.7}$$

(Here as usual the tilde indicates that the ratio of the two sides tends to 1.)

In part 2 of the paper we derive formula (1.6), and a generalization of it, and show how (1.7) follows. Determinants of this kind arise because the evaluation of $D_n(f)$ can be reduced to that of $D_n(f_0)$ (where f_0 is given by (1.5)) and the determinant of a section of the matrix $T_n(f_0)^{-1}$. These sections turn out to be (to within an arbitrary degree of accuracy) sections of generalized moment matrices. Part 3 begins the study of the asymptotics of the determinants of these sections. The study is completed in part 4 by combining the results of the preceding two parts, and (1.4) is established. In part 5 we prove that (1.3) is sufficient for canonical eigenvalue distribution and show that this holds for our f under the conditions mentioned. Finally, in part 6, we prove the stated results for the case when f describes a simple smooth curve.

2. DETERMINANTS OF GENERALIZED MOMENT MATRICES

LEMMA 2.1 *Let* $d\mu(s,t)$ *be a measure on* $\mathbb{C} \times \mathbb{C}$ *such that for* $i,j = 0, \cdots, m-1$ *we have* $s^i t^j \in L_1(d\mu)$. *Then the determinant of the* $m \times m$ *matrix with* i,j *entry*.

$$\iint s^i t^j \, d\mu(s,t) \qquad\qquad i,j = 0, \cdots, m-1$$

is equal to the expression (1.6).

PROOF. If one column of a matrix is the sum of (column) vectors then the determinant of the matrix is the sum of the determinants of the matrices obtained by replacing that column by one of the vectors. This extends by bilinearity to the case when all the columns are given as sums and then, by a limiting argument, to the case when all the columns are given as integrals. If we write the j'th column of our matrix as

$$\iint s_j^i t_j^j \, d\mu(s_j, t_j)$$

we deduce that its determinant equals

$$\int \cdots \int \det(s_j^i t_j^j)_{i,j=0,\ldots,m-1} \, d\mu(s_0, t_0) \cdots d\mu(s_{m-1}, t_{m-1}) .$$

The determinant in the integral has t_j^j as a factor of the j'th column, and after factoring these out we are left with a Vandermonde determinant $\det(s_j^i)$. We deduce that the integrand equals

$$\prod_{j=0}^{m-1} t_j^j \prod_{i>j} (s_i - s_j) .$$

Now because of the symmetry of the measure, the last integral is unchanged if the pairs of variables (s_i, t_i) are replaced by $(s_{\sigma(i)}, t_{\sigma(i)})$ where σ is any permutation of 0, ... , m-1. It follows that the integral is unchanged if the integrand is replaced by

$$\frac{1}{m!} \sum_{\sigma} \Big[\prod_{j=0}^{m-1} t_{\sigma(j)}^j \prod_{i>j} (s_{\sigma(i)} - s_{\sigma(j)}) \Big] .$$

Since

$$\prod_{i>j} (s_{\sigma(i)} - s_{\sigma(j)}) = \operatorname{sgn}\sigma \prod_{i>j} (s_i - s_j)$$

the above is equal to

$$\frac{1}{m!} \prod_{i>j} (s_i - s_j) \sum_{\sigma} \operatorname{sgn} \sigma \prod_{j=0}^{m-1} t_{\sigma(j)}^{j}$$

The sum here equals the Vandermonde determinant

$$\det (t_i^j)_{i,j=0,\dots,m-1} = \prod_{i>j} (t_i - t_j)$$

and this establishes the lemma. □

The result can be generalized, as follows. First we note that the sum of generalized moment matrices is clearly also a generalized moment matrix. Second, if $\phi(s)$ and $\psi(t)$ are functions such that each $\phi(s)^i \psi(t)^j \in L_1(d\mu(s,t))$ then the matrix with i,j entry

$$\int\int \phi(s)^i \psi(t)^j \, d\mu(s,t)$$

is the generalized moment matrix associated with the measure ν on $\mathbb{C} \times \mathbb{C}$ determined by

$$\nu(E \times F) = \mu(\phi^{-1}(E) \times \psi^{-1}(F)).$$

Putting these things together leads to the following generalization of Lemma 2.1, whose proof we leave to the reader.

LEMMA 2.2. *Suppose we have a finite set of measures* $d\mu_k(s,t)$ *and functions* $\phi_k(s)$, $\psi_k(t)$ *such that each* $\phi_k(s)^i \psi_k(t)^j \in L_1(d\mu_k(s,t))$. *Then the determinant of the* $m \times m$ *matrix with* i,j *entry*

$$\sum_k \int\int \phi_k(s)^i \psi_k(t)^j \, d\mu_k(s,t) \qquad (i,j=0,\dots,m-1)$$

is equal to

$$\frac{1}{m!} \sum_{k_0,\dots,k_{m-1}} \int \cdots \int \prod_{i>j} [(\phi_{k_i}(s_i) - \psi_{k_j}(s_j))(\phi_{k_i}(t_i) - \psi_{k_j}(t_j))] \cdot d\mu_{k_0}(s_0,t_0) \cdots d\mu_{k_{m-1}}(s_{m-1},t_{m-1}) \tag{2.1}$$

where the sum is taken overall m-tuples $k_0, \dots, k_{m-1}$.

As an application we prove (1.7). Similar but more complicated computations will have to be made later. We use the identity

$$\int_0^\infty e^{-nt}\, t^{p-1}\, dt = \frac{\Gamma(p)}{n^p}$$

to write

$$(n+i+j)^{-p} = \Gamma(p)^{-1} \int_0^\infty e^{-(i+j)t}\, e^{-nt}\, t^{p-1}\, dt\,.$$

Thus we have the case of Lemma 2.2 with a single measure $d\mu(s,t)$ equal to the measure

$$\Gamma(p)^{-1}\, e^{-nt}\, t^{p-1}\, dt$$

on the diagonal of $\mathbf{R}^+ \times \mathbf{R}^+$, and with $\phi(t) = \psi(t) = e^{-t}$. We conclude that the determinant in question equals

$$\frac{\Gamma(p)^{-m}}{m!} \int \cdots \int \prod_{i>j} (e^{-t_i} - e^{-t_j})^2\, e^{-n\Sigma t_i}\, \Pi\, t_i^{p-1}\, dt_0 \cdots dt_{m-1}\,,$$

the region of integration being $\mathbf{R}^+ \times \cdots \times \mathbf{R}^+$. We make the variable changes $t_i \to t_i/n$ and find that this equals

$$n^{-pm}\, \frac{\Gamma(p)^{-m}}{m!} \int \cdots \int \prod_{i>j} (e^{-t_i/n} - e^{-t_j/n})^2\, e^{-\Sigma t_i}\, \Pi\, t_i^{p-1}\, dt_0 \cdots dt_{m-1}\,. \tag{2.2}$$

As $n \to \infty$ the first factor in the integrand is

$$\prod_{i>j} (n^{-1}(t_j - t_i) + O(n^{-2}))^2 = n^{-m^2+m} \left[\prod_{i>j} (t_i - t_j)^2 + O(n^{-1})\right]$$

and so (2.2) is equal to

$$n^{-m^2-(p-1)m}\, \frac{\Gamma(p)^{-m}}{m!} \int \cdots \int \prod_{i>j} (t_i - t_j)^2\, e^{-\Sigma t_i}\, \Pi\, t_i^{p-1}\, dt_0 \cdots dt_{m-1}$$

plus an error in which the power of n is reduced by 1. Since the integrand here is positive so is the integral, and we deduce that the original determinant is asymptotic to this expression. To evaluate it we work backwards, and observe that

$$\frac{\Gamma(p)^{-m}}{m!} \int \cdots \int \prod_{i>j} (t_i - t_j)^2\, e^{-\Sigma t_i}\, \Pi\, t_i^{p-1}\, dt_0 \cdots dt_{m-1}$$

is exactly what Lemma 2.2 tells us is the determinant of the $m \times m$ matrix whose i,j entry equals

$$\Gamma(p)^{-1}\int_0^\infty t^{i+j}e^{-t}t^{p-1}\,dt = \frac{\Gamma(p+i+j)}{\Gamma(p)} = \prod_{k=0}^{i+j-1}(p+k).$$

Thus (1.7) is established.

3. THE DETERMINANT OF A SECTION OF $T_n(f)^{-1}$

Suppose f is continuous and piecewise C^∞, but not C^∞. Then there are finitely many points $z_h = e^{i\theta_h}$ at each of which there is a smallest nonnegative integer α_h such that

$$f^{(\alpha_h)}(z_h+) \neq f^{(\alpha_h)}(z_h-).$$

(Here $f^{(\alpha)}$ means $\dfrac{d^\alpha}{d\theta^\alpha}f(e^{i\theta})$ and $z\pm$ means $e^{i\theta\pm}$.) Clearly $\alpha_h \geq 1$ since f is continuous; we call it the *order* of the singularity at z_h.

In this section we assume that f is in addition nonzero with index zero (it will eventually be the function f_0 in (1.5)) and begin the determination of the asymptotics of the determinant of a certain section of the matrix $T_n(f)^{-1}$. The entries of this inverse matrix will be expressed in terms of the Wiener-Hopf factors of f and we present some preliminary notation and lemmas.

The operators $g \to \tilde{g}, g \to g^{\pm}, g \to g_{\pm}$ on $L_2(\mathbb{T})$ are defined by

$$\tilde{g}(z) = -i\sum \operatorname{sgn} k\, \hat{g}(k)\, z^k \quad (\operatorname{sgn} 0 = 0)$$

$$g^+(z) = \sum_{k=0}^{\infty}\hat{g}(k)z^k,\quad g^-(z) = \sum_{k=-\infty}^{0}\hat{g}(k)\,z^k,\quad g_{\pm} = g - g^{\mp}.$$

The Wiener-Hopf factorization of f is

$$f = w^- w^+,\quad w^{\pm} = \exp\left(\tfrac{1}{2}\log f \pm \frac{i}{2}\widetilde{\log f}\right)$$

and we set

$$u = \frac{w^-}{w^+} = e^{-i\widetilde{\log f}},\quad v = \frac{w^+}{w^-} = e^{i\widetilde{\log f}}.$$

Finally we define the operators U and V on $L_2(\mathbb{T})$ by

$$Ug = (z^{-n}ug)_+,\quad Vg = (z^n vg)_-.$$

It is well-known that under our assumptions $T_n(f)$ is invertible for sufficiently large n. In [W1, § II] it is shown how the inversion is effected using the operators U and V, the norms of which tend to zero as $n \to \infty$, and we state the result as a lemma. We think of $T_n(f)$ as acting on the space $\mathbf{P}_n$ of polynomials in z of degree at most n.

LEMMA 3.1. *If* $p = T_n(f)^{-1} q$, *with* $p, q \in \mathbf{P}_n$, *then we can write*

$$f\,p = q + \phi\, w^- + z^n\, \psi\, w^+ \tag{3.1}$$

where $\phi \in L_2(\mathbb{T})_-$ *satisfies*

$$(I - VU)\,\phi = -(z^n v (z^{-n} q / w^+)^-)_- . \tag{3.2}$$

and $\psi \in L_2(\mathrm{T})_+$ *satisfies a similar equation.*

We shall not need the equation satisfied by ψ. This lemma can clearly be used to determine the entries of $T_n(f)^{-1}$; the i,j entry is $\hat{p}(i)$ where p corresponds to $q(z) = z^j$. Since it is a determinant of these entries we shall need, we go directly to that. Set

$$X = (x_{i,j}) = (T_n(f)^{-1})_{n-1,j} \qquad i,j = 0, \cdots, m-1 .$$

It is $\det X$ we are interested in. Write

$$Y = (y_{i,j}) \qquad i,j = 0, \ldots, m-1$$

where

$$y_{i,j} = (z^{-n} u (I - VU)^{-1} z^j)\hat{\ }(-i) . \tag{3.3}$$

LEMMA 3.2. $\det X = (-1)^m G(f)^{-m} \det Y$.

PROOF. We can write

$$(z^n v (z^{-n} q / w^+)^-)_- = (q / w^-)_- - VU (q / w^-) = (q / w^-)^+ + (I - VU)(q / w^-)$$

and so from (3.2)

$$\phi = -(I - VU)^{-1} (q/w^-)^+ - q/w^- .$$

Inserting this into (3.1) gives

$$z^{-n} w^- p = z^{-n} u (I - VU)^{-1} (q/w^-)^+ + \psi .$$

Let us write $p_j = T_n(f)^{-1} z^j$, so that $x_{i,j} = \hat{p}_j(n-i)$. The multiplicative identity for functions, $p_j = (w^-)^{-1}(w^- p_j)$, leads to a convolution identity for Fourier coefficients

$$\hat{p}_j(n-i) = \sum_{k=0}^{i} \widehat{(w^-)^{-1}}(k-i)\ \widehat{w^- p_j}(n-k) = \sum_{k=0}^{i} \widehat{(w^-)^{-1}}(k-i)\ \widehat{z^{-n} w^- p_j}(-k).$$

The reason the sums are truncated as they are is that $(w^-)^{-1} \in L_2(\mathbb{T})^-$ and (since $p_j \in \mathbf{P}_n$) $z^{-n} w^- p_j \in L_2(\mathbb{T})^-$. This shows that X is a product of two matrices, the left a lower triangular matrix all of whose diagonal entries equal $\widehat{(w^-)^{-1}}(0) = G(f)^{-1/2}$ and the right having i,j entry $\widehat{z^{-n} w^- p_j}(-k)$. Thus

$$\begin{aligned} \det X &= G(f)^{-\frac{m}{2}} \det(\widehat{z^{-n} w^- p_j}(-i)) \\ &= (-1)^m G(f)^{-\frac{m}{2}} \det([z^{-n} u (I - VU)^{-1} (z^j / w^-)^+]\hat{}(-i)). \end{aligned} \tag{3.4}$$

Finally we notice that

$$(z^j / w^-)^+ = \sum_{k=0}^{j} \widehat{(w^-)^{-1}}(k-j)\, z^j$$

and so the matrix in (3.4) can be written as a product of two matrices, the one on the left being Y and the one of the right being triangular with diagonal entries $G(f)^{-1/2}$, and we arrive at the assertion of the lemma. □

Since the norms of U and V tend to zero as $n \to \infty$ the Neumann series for $(I - VU)^{-1}$ inserted into (3.3) will give an asymptotic expansion for the matrix entries $y_{i,j}$. The operators U and V have matrix representations with entries

$$U_{i,j} = \begin{cases} \hat{u}(i-j+n) & \text{if } i > 0, \\ 0 & \text{if } i \le 0; \end{cases} \qquad V_{i,j} = \begin{cases} \hat{v}(i-j-n) & \text{if } i < 0, \\ 0 & \text{if } i \ge 0. \end{cases}$$

Also, multiplication by $z^{-n} u$ has matrix representation $(\hat{u}(i-j+n))$. Thus the Neumann series inserted into (3.3) gives

$$y_{i,j} = \hat{u}(-i-j+n) + \sum_{k,l > 0} \hat{u}(-i+k+n)\, \hat{v}(-k-l-n)\, \hat{u}(-j+l+n) + \cdots. \tag{3.5}$$

In the r'th term on the right side the summand has 2r - 1 factors.

To transform this into an asymptotic expansion in decreasing powers of n we must find such expansions for $\hat{u}(n)$ and $\hat{v}(-n)$. Recall that f has singularities at $z_h = e^{i\theta_h}$ with orders α_h $(h = 1, 2, \cdots)$.

LEMMA 3.3. *There are asymptotic expansions*

$$\hat{u}(n) \sim \sum_{p,q,h} u_{p,q,h} \frac{\log^q n}{n^{p+1}} z_h^n, \quad \hat{v}(-n) \sim \sum_{p,q,h} v_{p,q,h} \frac{\log^q n}{n^{p+1}} z_h^n \tag{3.6}$$

where the sum is taken over all h (*indexing the singularities*) *and integers* p,q *satisfying* $q \geq 0$, $p \geq \alpha_h(q+1)$. *The leading coefficients are given by*

$$u_{\alpha_h,0,h} = a_{\alpha_h} e^{-i\widetilde{\log f}(z_h)} \frac{f^{(\alpha_h)}(z_h+) - f^{(\alpha_h)}(z_h-)}{f(z_h)}$$

$$v_{\alpha_h,0,h} = (-1)^{\alpha-1} a_{\alpha_h} e^{i\widetilde{\log f}(z_h)} \frac{f^{(\alpha_h)}(z_h+) - f^{(\alpha_h)}(z_h-)}{f(z_h)}$$

where a_α *is a nonzero constant depending only on* α.

PROOF. We begin with the case where there is a single singularity which we take to be at $z=1$ $(\theta=0)$, with order α. We shall consistently abuse notation and write $f(\theta)$ wherever convenient instead of $f(e^{i\theta})$. We shall also write $\theta^{\pm} = \theta\,\chi_{\mathbf{R}^{\pm}}(\theta)$. In the neighborhood of $\theta = 0$ we have an expansion (an asymptotic expansion as $\theta \to 0\pm$ which persists under differentiation)

$$\log f(\theta) \sim \log f(0) + \sum_{k=1}^{\infty} b_k^{\pm} \theta^{\pm k} \tag{3.7}$$

where $b_k^+ = b_k^-$ for $k < \alpha$ and

$$b_\alpha^+ - b_\alpha^- = \frac{f^{(\alpha)}(0+) - f^{(\alpha)}(0-)}{\alpha!\, f(0)}.$$

The conjugate function $\tilde{g}$ of g is given by a principal value integral

$$\tilde{g}(\theta) = \frac{1}{2\pi} \int_{-\pi}^{\pi} g(\theta') \cot \tfrac{1}{2}(\theta - \theta')\, d\theta' \tag{3.8}$$

and it is easy to deduce that

$$\widetilde{\theta^{\pm}} = \pm \pi^{-1} \theta \log|\theta| \qquad (\text{mod}\, C^\infty). \tag{3.9}$$

(This means that the conjugate function of any function which is C^∞ outside a neighborhood of $\theta = 0$ and equal to $\theta^{\pm}$ inside it is equal to a C^∞ function outside a neighborhood of $\theta = 0$ and to $\pm\pi^{-1}\theta \log|\theta|$ inside it.) We see from (3.8) that the commutator of the conjugate function operator and multiplication by θ takes any function to a function which is C^∞ near $\theta = 0$. This show that (3.9) can be differentiated arbitrarily often, giving

$$\widetilde{\theta^{\pm k}} = \pm\pi^{-1}\theta^k \log|\theta| \quad (\text{mod } C^\infty).$$

Therefore from (3.7) we obtain

$$\widetilde{\log f}(\theta) = \pi^{-1}\sum_{k=\alpha}^{\infty}(b_k^+ - b_k^-)\,\theta^k \log|\theta| \qquad (\text{mod } C^\infty).$$

This in turn means that for any N the function on the left minus the sum of sufficiently many terms of the series belongs to C^N in a neighborhood of $\theta = 0$. Of course the left side belongs to C^∞ outside any neighborhood of 0.

Next we exponentiate and obtain an expansion for $u = e^{-i\widetilde{\log f}}$,

$$u(\theta) = \Big(1 + \sum_{\substack{k\ge\alpha \\ l\ge 1}} b_{k,l}\,\theta^{kl}\log^l|\theta|\Big)\,\phi(\theta),$$

where $\phi \in C^\infty$ in a neighborhood of 0 and

$$b_{\alpha,1} = -\frac{i}{\pi\alpha!}\,\frac{f^{(\alpha)}(0+) - f^{(\alpha)}(0-)}{f(0)}.$$

Notice that $\phi(0) = e^{-i\widetilde{\log f(0)}}$ so we can rewrite the above expansion as

$$u(\theta) = \sum_{\substack{p\ge\alpha l \\ l\ge 1}} c_{p,l}\,\theta^p \log^l|\theta| \qquad (\text{mod } C^\infty) \tag{3.10}$$

where

$$c_{\alpha,1} = \frac{-i}{\pi\alpha!}\,e^{-i\widetilde{\log f}(0)}\,\frac{f^{(\alpha)}(0+) - f^{(\alpha)}(0-)}{f(0)}. \tag{3.11}$$

So to determine the asymptotics of $\hat{u}(n)$ it remains only to determine the asymptotics of the Fourier coefficients of a function equal to $\theta^p\log^l|\theta|$ $(\text{mod } C^\infty)$. It is easily shown that mod $O(n^{-\infty})$ (i.e., with error $O(n^{-N})$ for any N) these coefficients are equal to the values at n of the Fourier transform of $\theta^p\log^l|\theta|$. (This transform is a distribution which equals a C^∞ function away from 0.) With a circumflex now denoting Fourier transform we see upon changing variables that

$$(\theta^p \log^l |\theta|)^\wedge(n) = n^{-p-1}\,[\,\theta^p(\log|\theta| - \log n)^l\,]^\wedge(1) \\ = n^{-p-1}\sum_{q=0}^{l} a_{p,l,q}\,\log^q n \tag{3.12}$$

where

$$a_{p,l,l} = (-1)^l\,\theta^p(1) = 0, \quad a_{p,l,l-1} = (-1)^{l-1}\,l\;\theta^p\log|\theta|\,(1) \neq 0\,.$$

These last assertions are easily verified.

Combining (3.10)-(3.12) we obtain the asserted asymptotic expansion for $\hat{u}(n)$ in this special case. Notice that $a_{p,l,q} \neq 0$ only when $q \leq l-1$ and so in (3.10) only when $p \geq \alpha(q+1)$. Notice also that the coefficient $u_{\alpha,0}$ in the expansion of $\hat{u}(n)$ equals $c_{\alpha,1}\,a_{\alpha,1,0}$ and so is of the stated form.

To extend this to the case where f has several singularities, at the points θ_h, we use the fact that in a neighborhood of θ_h the function $\log f$ is determined $(\bmod\, C^\infty)$ by the values of f in any neighborhood of θ_h. It follows from this (upon translating from $\theta = 0$ to $\theta = \theta_h$) that in a neighborhood of θ_h we have an expansion of the form (3.10) with θ replaced by $\theta - \theta_h$ and with $c_{\alpha,1}$ given by (3.11) with 0 replaced by θ_h. We then use the fact that translating a function by θ_h multiplies its sequence of Fourier coefficients by $\{e^{in\theta_h}\}$, we apply a partition of unity in the obvious way to separate out the singularities, and we obtain the stated expansion for $\hat{u}(n)$. The expansion for $\hat{v}(-n)$ is obtained from this by observing that $v(-\theta)$ is equal to the function $u(\theta)$ associated with $f(-\theta)$.

4. THE ASYMPTOTIC FORMULA

We determine now, with notation as in the last section, the asymptotics of det Y. It will then be but a short step to the derivation of (1.4). The result is a little complicated to state. First we introduce the quantities

$$F_h = e^{-i\,\widetilde{\log f}(z_h)}\;\frac{f^{(\alpha_h)}(z_h+) - f^{(\alpha_h)}(z_h-)}{f(z_h)} \tag{4.1}$$

which enter into the statement of Lemma 3.3. We also introduce an index set $\{1, 2, \ldots, H\}$ for the singularities z_h and define

$$e = \min\{\sum_{h=1}^{H}(k_h^2 + \alpha_h k_h) : k_h \in \mathbf{Z}^+,\ \sum k_h = m\}. \tag{4.2}$$

Here $\mathbb{Z}^+$ denotes the nonnegative integers and, recall, m is the order of the matrix Y.

LEMMA 4.1. *We have as* $n \to \infty$

$$\det Y = n^{-e} \sum c_{k_1,\dots,k_H} \prod_{h=1}^{H} (F_h^{k_h} z_h^{k_h n}) + O(n^{-e-1} \log n) \tag{4.3}$$

where each $c_{k_1,\dots,k_H}$ *is a nonzero constant depending only on the* k_h, α_h *and* z_h *and where the sum is taken over all* $\{k_1, \dots, k_H\}$ *such that the minimum in (4.2) is attained.*

PROOF in the case H = 1. We may assume the unique singularity to be at z = 1 and have order α and the assertion is that

$$\det Y = n^{-m^2-\alpha m}\, c\, (F^m + O(n^{-1} \log n)) \tag{4.4}$$

where c depends only on m and α. (We have written F for F_1.) Even this is rather complicated and we shall, as a first step, determine the asymptotics of

$$\det(\hat{u}(-i-j+n)) \qquad i,j = 0, \cdots, m-1 .$$

By Lemma 3.3

$$\hat{u}(-i-j+n) \sim \sum_{\substack{q \ge 0 \\ p \ge \alpha(q+1)}} u_{p,q} \frac{\log^q (n-i-j)}{(n-i-j)^{p+1}} \tag{4.5}$$

with $u_{\alpha,0} = a_\alpha F$. Clearly to determine the asymptotics of the determinant to within any given error $O(n^{-N})$ only finitely many terms of this series need be retained. When we do this, we obtain a generalized moment matrix. To see this we use the fact that

$$\int_0^\infty e^{-nt} t^p \log^q t \, dt = n^{-p-1} \int_0^\infty e^{-t} t^p (\log t - \log n)^q \, dt .$$

The integral on the right is a polynomial in log n of exact degree q. It follows that we can rewrite (4.5) as

$$\hat{u}(-i-j+n) \sim \sum_{\substack{q \ge 0 \\ p \ge \alpha(q+1)}} u_{p,q} \int_0^\infty e^{(i+j)t} e^{-nt} t^p \log^q t \, dt$$

where the $u_{p,q}$ are not exactly as before, but where still $u_{\alpha,0} = a_\alpha F$ with a different constant a_α.

We now apply Lemma 2.2 where the indices k there are replaced by the pairs p,q, the measure $d\mu_k(s,t)$ by the measure

$$u_{p,q}\, e^{-nt}\, t^p \log^q t\, dt$$

on the diagonal of $\mathbf{R}^+ \times \mathbf{R}^+$, and with $\phi(t) = \psi(t) = e^t$. The lemma gives

$$\det(\hat{u}(-i-j+n))$$

$$\sim \frac{1}{m!} \sum_{p_i,q_i} \prod_{i=0}^{m-1} u_{p_i,q_i} \int\cdots\int \prod_{i>j} (e^{t_i}-e^{t_j})^2\, e^{-n\sum t_i}\, \Pi\, (t_i^{p_i} \log^{q_i} t_i)\, dt_0 \cdots dt_{m-1},$$

where the sum is taken over all m-tuples of pairs (p_i,q_i) $(i=0,...,m-1)$ such that $p_i \geq \alpha(q_i+1)$ for each i. Making the variable changes $t_i \to t_i/n$ as we did earlier, in section 2, we find that the integral corresponding to the pairs p_i, q_i is asymptotically equal to $n^{-m^2-\sum p_i}$ times a polynomial of degree $\sum q_i$ in log n with highest coefficient

$$\int\cdots\int \prod_{i>j} (t_i-t_j)^2\, e^{-\sum t_i}\, \Pi\, t_i^{p_i}\, dt_0 \cdots dt_{m-1}.$$

The error is $O(n^{-m^2-\sum p_i - 1}(\log n)^{\sum q_i})$. Since each $p_i \geq \alpha(q_i+1)$ there is clearly a unique term of greatest order of magnitude, corresponding to $q_i = 0, p_i = \alpha$ for all i, and we conclude that

$$\det(\hat{u}(-i-j+n)) = \frac{1}{m!}\, u_{\alpha,0}^m\, n^{-m^2-\alpha m} \int\cdots\int \prod_{i>j} (t_i-t_j)^2\, e^{-\sum t_i}\, \Pi\, t_i^{\alpha}\, dt_0 \cdots dt_{m-1}$$
$$+\; O(n^{-m^2-\alpha m-1} \log n). \tag{4.6}$$

(This error is achieved when one $q_i = 1$ if $\alpha = 1$.)

In fact this will give the asymptotics of det Y as well. One might think this is trivial since to a first approximation the entries of Y are $\hat{u}(-i-j+n)$ and the first approximation of a determinant should be obtained by taking a first approximation of the entries. This is just not so. Consider, for example, that a first approximation to $\hat{u}(-i-j+n)$ is a constant times $n^{-\alpha-1}$ and this would yield an "approximating" determinant equal to 0. We must look more carefully at the entries of Y given by (3.5).

Suppose the expansions (3.6) are substituted into, say, the second term on the right side of (3.5). Then we obtain a multiple series which can be rearranged as

$$\sum u_{p_1,q_1}\, u_{p_2,q_2}\, u_{p_3,q_3} \sum_{k,l>0} \frac{\log^{q_1}(n-i+k)}{(n-i+k)^{p_1+1}} \; \frac{\log^{q_2}(n+k+l)}{(n+k+l)^{p_2+1}} \; \frac{\log^{q_3}(n-j+l)}{(n-j+l)^{p_3+1}}$$

where the outer sum is taken over all triples of pairs p_i, q_i $(i = 1,2,3)$ satisfying $p_i \geq \alpha(q_i+1)$. We apply the same trick as before and replace each factor in the inner sum by an integral, so that this sum itself becomes

$$\sum_{k,l>0} \iiint e^{-(n-i+k)s_1}\, e^{-(n+k+l)s_2}\, e^{-(n-j+l)s_3} \prod_{i=1}^{3} (s_i^{p_i} \log^{q_i} s_i)\, ds_1\, ds_2\, ds_3$$

$$= \iiint e^{is_1+js_3} \frac{e^{-n(s_1+s_2+s_3)}}{(e^{s_1+s_2}-1)(e^{s_2+s_3}-1)} \prod_{i=1}^{3} (s_i^{p_i} \log^{q_i} s_i)\, ds_1\, ds_2\, ds_3 \,.$$

As before the coefficients u_{p_i, q_i} are replaced by others, but this will not matter. Notice that we have a generalized moment matrix once again, this time with measure

$$d\mu(s_1, s_3) = \left[\int_0^\infty \frac{e^{-n(s_1+s_2+s_3)}}{(e^{s_1+s_2}-1)(e^{s_2+s_3}-1)} \prod_{i=1}^{3} (s_i^{p_i} \log^{q_i} s_i)\, ds_2 \right] ds_1\, ds_3 \qquad (4.7)$$

on $\mathbf{R}^+ \times \mathbf{R}^+$ and with functions $\phi(s_1) = e^{s_1}, \psi(s_3) = e^{s_3}$.

Similarly all terms obtained by substituting (3.6) into (3.5) give rise to (more complicated) generalized moment matrices. We can then apply Lemma 2.2 to obtain a complete expansion for det Y. We shall find that all terms beyond those arising from $\hat{u}(-i-j+n)$ alone make a smaller contribution than the "main" term given by (4.6).

Rather than consider the general term arising from application of Lemma 2.2, which would be extremely complicated to write down, we consider what are in a sense the next after those already considered. We assume the indices $k_1, \cdots, k_{m-1}$ in (2.1) correspond to terms from the expansion of $\hat{u}(-i-j+n)$, say

$$\int e^{(i+j)t}\, e^{-nt}\, t^{p_i'} \log^{q_i'} dt \qquad (i = 1,...,m-1)$$

whereas k_0 corresponds to (4.7), a term arising from the second term on the right side of (3.5). Apart from constant factors, what we obtain is

$$\int\cdots\int \prod_{i>j} (e^{t_i}-e^{t_j})^2 \prod [\,(e^{t_i}-e^{s_1})\,(e^{t_i}-e^{s_3})\,]\, e^{-n\sum t_i} \prod (t_i^{p_i'} \log^{q_i'} t_i)$$

$$\cdot \frac{e^{-n\sum s_i}}{(e^{s_1+s_2}-1)\,(e^{s_2+s_3}-1)} \prod (s_i^{p_i} \log^{q_i} s_i)\, dt_1 \cdots dt_{m-1}\, ds_1\, ds_2\, ds_3\,. \tag{4.8}$$

Here any index for the t-variables runs through $1,\ldots,m-1$ whereas any index for the s-variables runs through 1,2,3. As usual we make the substitutions $t_i \to t_i/n$, $s_i \to s_i/n$, and note now that

$$e^{(s_1+s_2)/n} - 1 \sim (s_1+s_2)/n\,, \quad e^{(s_2+s_3)/n} - 1 \sim (s_2+s_3)/n\,.$$

Counting carefully the number of factors we find that (4.8) is asymptotically equal to a positive constant times

$$(\log n)^{\sum q_i + \sum q_i'} \,/\, n^{m^2 + \sum p_i' + \sum p_i}\,.$$

Since each $p_i \ge \alpha(q_i+1)$, $p_i' \ge (q_i'+1)$ and there are $m-1$ p_i's and 3 q_i's this is $O(n^{-m^2-\alpha m-2\alpha})$.

All further terms that arise from application of Lemma 2.2 are no larger than this one, and so $\det Y$ is also given by the right side of (4.6). Recalling that $u_{\alpha,0} = a_\alpha F$ where a_α depends only on α we see that (4.4), and so the lemma in this special case, does hold.

PROOF in the general case. Once again we determine first the asymptotics of $\det(\hat{u}(-i-j+n))$, and then afterwards consider the contributions of further terms of (3.5). In this general case the asymptotics of $\hat{u}(n)$ given by (3.6) involve triples $p,\ q,\ h$ with $p \ge \alpha_h(q+1)$, $h = 1,\ldots,H$. In the application of Lemma 2.2 the index k is thought of as this triple, the measure $d\mu_k(s,t)$ is the measure

$$u_{p,q,h}\, z_h^n\, e^{-nt}\, t^p \log^q t\, dt$$

on the diagonal of $\mathbf{R}^+ \times \mathbf{R}^+$, and $\phi(t) = \psi(t) = z_h^{-1} e^t$. Thus (2.1) becomes

$$\frac{1}{m!} \sum_{p_i,q_i,h_i} \prod (u_{p_i,q_i,h_i}\, z_{h_i}^n) \int\cdots\int \prod_{i\neq j} (z_{h_i}^{-1} e^{t_i} - z_{h_j}^{-1} e^{t_j})\, e^{-n\sum t_i} \prod (t_i^{p_i} \log^{q_i} t)\, dt_0 \cdots dt_{m-1}. \tag{4.9}$$

Again we make the substitutions $t_i \to t_i/n$ but now note that as $n \to \infty$

$$z_{h_i}^{-1}\, e^{t_i/n} - z_{h_j}^{-1}\, e^{t_j/n} \sim n^{-1}\, z_{h_i}^{-1} (t_j - t_i) \quad \text{if } z_{h_i} = z_{h_j}$$

$$z_{h_i}^{-1} e^{t_i/n} - z_{h_j}^{-1} e^{t_j/n} \to z_{h_i}^{-1} - z_{h_j}^{-1} \quad \text{if } z_{h_i} \neq z_{h_j}.$$

Let us denote by E the set of ordered pairs (i, j) with $i \neq j$ such that $z_{h_i} = z_{h_j}$. Each of these pairs contributes a factor n^{-1} and we find that the integral in (4.9) is equal to $n^{-|E|-\Sigma p_i-m}$ times a polynomial of degree Σq_i in $\log n$ with highest coefficient

$$\int \cdots \int \prod_{(i,j)\in E} [z_{h_i}^{-1}(t_j - t_i)] \prod_{(i,j)\notin E} (z_{h_i}^{-1} - z_{h_j}^{-1})\, e^{-\Sigma t_i} \prod t_i^{p_i}\, dt_0 \cdots dt_{m-1} \tag{4.10}$$

plus an error $O(n^{-|E|-\Sigma p_i-m-1}(\log n)^{\Sigma q_i})$. Here $|E|$ denotes the cardinality of E. For any given choice of $h_0, \ldots, h_{m-1}$ the power of n^{-1} is smallest when each $p_i = \alpha_{h_i}$ and so also each $q_i = 0$. The coefficient of the integral in this case is

$$\frac{1}{m!} \prod (u_{\alpha_i,0,h_i}\, z_{h_i}^n). \tag{4.11}$$

The dominant terms in (4.9) are obtained when $h_0, \ldots, h_{m-1}$ are chosen so that $|E| + \Sigma\alpha_{h_i} + m$ is as small as possible, for this will be the power of n^{-1} which occurs when each $p_i = \alpha_{h_i}$. So we let e denote the minimum of $|E| + \Sigma\alpha_{h_i} + m$ over all possible choices of $h_0, \ldots, h_{m-1}$. Then (4.9) is asymptotically equal to n^{-e} times the sum, over all m-tuples $h_0, \ldots, h_{m-1}$ for which this minimum is achieved, of the coefficient (4.11) times the integral (4.10) with each $p_i = \alpha_{h_i}$. The error is $O(n^{-e-1}\log n)$, achieved when a single $q_i = 1$ if $\alpha_{h_i} = 1$.

We shall now show that this is precisely the form of the right side of (4.3). Suppose that we have chosen an m-tuple $h_0, \ldots, h_{m-1}$ and that among the h_i the number h $(1 \leq h \leq H)$ occurs precisely k_h times. Then $|E| = \sum_{h=1}^{H} k_h(k_h - 1)$ and $\Sigma\alpha_{h_i} = \Sigma\alpha_h k_h$. Thus, since $\Sigma k_h = m$, the definition of e given above agrees with that given in (4.2). Moreover the m-tuples $h_0, \ldots, h_{m-1}$ for which their minimum is achieved correspond to the H-tuples $k_1, \ldots, k_H$ for which their minimum is achieved, any H-tuple $\{k_h\}$ correspond to an m-tuple $\{h_i\}$ which is unique up to permutation σ such that $h_{\sigma(i)} = h_i$ for each σ. The integral (4.10) (with $p_i = \alpha_{h_i}$) is unchanged by such a permutation. Recalling that each $u_{\alpha_h,0,h}$ is equal to F_h times a nonzero constant depending only on α_h we see that we have shown $\det(\hat{u}(-i-j+n))$ to have precisely the form of the right side of (4.3).

The final step is as in the proof of the special case $H = 1$. We show that any summand in (2.1) in which a term of the right side of (3.5) beyond the first comes into play is of lower order of magnitude than n^{-e}. In the analogue of (4.8) here each e^{t_i} is replaced by $z_{h_i}^{-1}e^{t_i}$

$(i = 1, \ldots, m-1)$ and each e^{s_i} by, say, $z_{k_i}^{-1}e^{s_i}$ $(i = 1, 2, 3)$. We make the usual substitutions $t_i \to t_i/n$, $s_i \to s_i/n$ and find that the power of n^{-1} that arises is at least

$$\sum_{i=1}^{m-1} \alpha_{h_i} + \sum_{i=1}^{3} \alpha_{k_i} + m + 2 \qquad \text{(from } \Pi t_i^{p_i'},\ \Pi s_i^{p_i},\ dt_i \text{ and } ds_i)$$

$$+ \,|\{(i, j) : i \neq j,\ z_{h_i} = z_{h_j}\}| \qquad \text{(from the first product)}$$

$$+ \,|\{i : z_{h_i} = z_{k_1}\}| + |\{i : z_{h_i} = z_{k_3}\}| \qquad \text{(from the second product)}$$

$$-\,2 \qquad \text{(from the factors in the denominator).}$$

Suppose that $h_1, \ldots, h_{m-1}$ are fixed and that, to choose one of two possible cases,

$$\{i : z_{h_i} = z_{k_1}\}| + \alpha_{k_1} \leq |\{i : z_{h_i} = z_{k_3}\}| + \alpha_{k_3}.$$

Then the sum displayed above is not increased if we replace k_3 by k_1 (which is left unchanged). But

$$\sum_{i=1}^{m-1} \alpha_{h_i} + \alpha_{k_1} + |\{(i, j) : i \neq j,\ z_{h_i} = z_{h_j}\}| + 2\,|\{i : z_{h_i} = \alpha_{k_1}\}| + m$$

is precisely one of the quantities of which the minimum is e. It follows that the displayed sum is at least

$$e + \alpha_{k_1} + \alpha_{k_2} \geq e + 2.$$

The integral is therefore bounded by a constant time n^{-e-2} times a power of $\log n$. Thus this term, and similarly all other remaining terms, are $O(n^{-e-1})$. □

We can now derive the asymptotic formula (1.4), which we give a more precise form. Of course if $i(f) = 0$ we have Szegö's sharper asymptotic formula

$$D_n(f) \sim G(f)^{n+1}\, c(f) \tag{4.12}$$

where

$$c(f) = \exp\left[\sum_{k=1}^{\infty} k\, \widehat{\log f}(k)\, \widehat{\log f}(-k)\right]. \tag{4.13}$$

In our situation the Fourier coefficients of f are $O(n^{-2})$ and (4.12) can be strengthened to

$$D_n(f) = G(f)^{n+1}\, (c(f) + O(n^{-1})). \tag{4.14}$$

(This follows from either [B-S1, p. 136] or [W1, §III].) But we are mainly interested in the case where $i(f) \neq 0$. We define f_0 as in (1.5) and modify (4.1) by writing now

$$G_h = e^{i\,\mathrm{sgn}\, i(f)\,\log f_0(z_h)}\ \frac{f_0^{(\alpha_h)}(z_h+) - f_0^{(\alpha_h)}(z_h-)}{f_0(z_h)}. \tag{4.15}$$

Note that if $i(f) < 0$ then G_h is, up to a factor $z_h^{i(f)}$, the quantity F_h corresponding to f_0; and if $i(f) > 0$ it is equal to $(-1)^{\alpha_h - 1}$ times the G_h associated with $f(z^{-1})$ at the singularity z_h^{-1}.

THEOREM 4.2. *We have as* $n \to \infty$

$$D_n(f) = (-1)^{(n+1)m}\ G(f_0)^{n+1} \left\{ n^{-e}\ c(f_0)\ \sum c_{k_1,\ldots,k_H} \prod_{h=1}^{H} (G_h^{k_h} z_h^{k_h n}) + O(n^{-e-1}\log n) \right\}$$

where $c(f_0)$ *is given by* (4.13), *where* e *is given by* (4.2) *with* $m = |\, i(f)\,|$, *and where each* $c_{k_1,\ldots,k_H}$ *is a nonzero constant depending only on the* k_h, α_h *and* z_h. *The sum is taken over all* $\{k_1, \ldots, k_H\}$ *for which the minimum in* (4.2) *is achieved.*

PROOF. Let us assume $i(f) = -m < 0$. (In the other case we apply the result in this case to $f(z^{-1})$.) According to identity (2.5) on p. 140 of [B-S1] we have

$$D_n(f) = (-1)^{nm}\, D_{n+m}(f_0) \det X$$

where now

$$X = (x_{i,j}) = (T_{n+m}(f_0)^{-1})_{n+m-i,j} \qquad i, j = 0, \ldots, m-1.$$

The assertion of the theorem follows upon applying (4.12) and (4.13) to evaluate $D_{n+m}(f_0)$ and Lemmas 3.2 and 4.1 (with f replaced by f_0 and n by $n+m$) to evaluate $\det X$.

5. EIGENVALUE DISTRIBUTION

We begin with he assertion made in the introduction.

LEMMA 5.1. *Suppose* $f \in L_\infty$ *and that* (1.3) *holds in the sense of convergence in measure. Then the eigenvalues of* $T_n(f)$ *are canonically distributed.*

PROOF. The functions

$$\frac{1}{n+1} \log |D_n(f-\lambda)| = \frac{1}{n+1} \sum_{i=1}^{n} \log|\lambda_{i,n} - \lambda|$$

are uniformly integrable (with respect to two-dimensional Lebesgue measure) on bounded subsets of $\mathbb{C}$, i.e., for any compact set K and any $\varepsilon > 0$ there exists a δ such that if A is a subset of K of measure less than δ then the integral of the absolute value of any one of these functions over A is less than ε. This follows immediately from the local integrability of $\log|\lambda|$. It follows from this that if (1.3) holds in the sense of convergence in measure then for every continuous function ϕ with compact support we have

$$\lim_{n\to\infty} \int \frac{1}{n+1} \sum \log|\lambda_{i,n} - \lambda| \phi(\lambda)\, d\lambda = \int \log G(|f-\lambda|)\, \phi(\lambda)\, d\lambda,$$

where $d\lambda$ denotes two-dimensional Lebesgue measure. In particular the (distributions associated with) the functions

$$\frac{1}{n+1} \sum \log |\lambda_{i,n} - \lambda|$$

converge, in the sense of distributions, to $\log G(|f-\lambda|)$. Since the Laplacian Δ is continuous on distributions we deduce that

$$\frac{1}{n+1} \sum \Delta \log|\lambda_{i,n} - \lambda| \to \Delta \log G(|f-\lambda|)$$

in the sense of distributions. But since

$$\frac{1}{n+1} \sum \log |\lambda_{i,n} - \lambda| = \int \log |z-\lambda|\, d\mu_{n,f}(z),$$

$$\log G(|f-\lambda|) = \int \log |z-\lambda|\, d\mu_f(z),$$

and since for every measure ν with compact support

$$\Delta \int \log |z-\lambda|\, d\nu(z) = 2\pi\nu$$

(see, for example, [B, ch. IV, par. 2]) we deduce that $\mu_{n,f} \to \mu_f$ in the sense of distributions. It follows that any weak limit point of the sequence of measures $\{\mu_{n,f}\}$ must be μ_f and so in fact $\mu_{n,f} \to \mu_f$ weakly. □

It is now easy to establish canonical eigenvalue distributions in certain cases.

THEOREM 5.2. *Assume that for each nonzero* $m = |i(f-\lambda)|$ *with* $\lambda \notin R(f)$, *the minimum in* (4.2) *is achieved for a unique H-tuple* $k_1, \ldots, k_H$. *Then the eigenvalues of* $T_n(f)$ *are canonically distributed.*

PROOF. Take any $\lambda \notin R(f)$ and apply Theorem 4.2 to the function $f-\lambda$. The sum on the right side consists of a single term, and we find (since $|f_0| = |f-\lambda|$) that

$$|D_n(f-\lambda)| = G(|f-\lambda|)^{n+1}\{n^{-e}\,|c(f_0)|\;\,|c_{k_1,\ldots k_H}\Pi G_h^{k_h}| + o(n^e)\}.$$

Relation (1.3) follows since all the constants are nonzero. □

The hypothesis of the theorem is clearly satisfied if there is only one singularity, for the minimum is achieved by the unique "1-tuple" $\{m\}$. In the case $H=2$ and $m=1$, for example, there is a unique minimum if and only if $\alpha_1 \neq \alpha_2$ while for $m=2$ the condition is that α_1 and α_2 do not differ by 2. These statements are easily verified. The following lemma gives some information about the general case.

LEMMA 5.3. *Given* $\alpha_1, \ldots, \alpha_H$ *and* m, *the set of H-tuples* $\{k_1, \ldots, k_H\}$ *for which the the minimum in* (4.2) *is achieved has the following form. There exists an H-tuple* $\{\bar{k}_1, \ldots, \bar{k}_H\}$, *a subset* A *of* $\{1, \ldots, H\}$, *and an integer* $p \leq \min(m, |A|-1)$ *such that* $\{k_h\}$ *is minimizing if and only if each* k_h *equals either* $\bar{k}_h$ *or* $\bar{k}_h + 1$, *and*

$$\{h : k_h = \bar{k}_h + 1\}$$

is a subset of A *of cardinality* p.

REMARK. Uniqueness occurs when $p = 0$, in which case the set A is arbitrary. Otherwise A is, clearly, uniquely determined.

PROOF. If we write $\alpha_h = 2\beta_h$, complete the square, and replace k_h by $k_h' - \beta_h$ we see that the minimum is achieved in (4.2) exactly when

$$\min \sum k_h'^2$$

is achieved, where $\{k_h'\}$ runs through all H-tuples such that

$$\sum k_h' = m + \sum \beta_h, \qquad \text{each } k_h' - \beta_h \in \mathbb{Z}^+.$$

We can think of the minimizing problem in the following way. We start with the H-tuple $\{\beta_1, \ldots, \beta_H\}$ and we perform m steps, each of which consists of raising a component of the H-tuple by 1. (The same component can be raised at different steps.) We want to perform these steps in such a way that the sum of the squares of the components of the resulting H-tuple is as small as possible. If H were 2 and m were 1 it is clear, and easily proved, that if the components are equal then we can raise either of them by 1 whereas if they are unequal we must raise the smaller one. Using this fact we see that it is reasonable (and it can be shown—we leave this as an exercise) that all minimizing H-tuples are obtained as follows.

1. If $\min \beta_h$ occurs for a set of A of indices and $|A| \geq m$ then we raise by 1 any m of the β_h with $h \in A$. This gives all minimizing H-tuples. If $|A| = m$ then there is a unique one. Otherwise the p of the statement equals $m < |A|$.

2. If $\min \beta_h$ occurs for a set A_1 of indices and $|A_1| < m$ then we replace all β_h with $h \in A_1$ by $\beta_h + 1$, replace m by $m - |A_1|$, and go back to step 1.

The assertion of the lemma follows immediately from this description of the process. The $\bar{k}_h$ of the statement is equal to the smallest amount the h'th entry can get raised by the end ($\bar{k}_h + 1$, for $h \in A$, being the largest amount it can get raised). □

To state our more general sufficient condition for canonical eigenvalue distribution, let Γ denote any of the components of the complement of $R(f)$. Write $i(\Gamma)$ for $i(f - \lambda)$ where $\lambda \in \Gamma$. (This number is independent of the λ chosen.) If $i(\Gamma) \neq 0$ let $A(\Gamma)$, $p(\Gamma)$ denote the set A and integer p, respectively, in the statement of Lemma 5.3 with $m = |i(\Gamma)|$ and denote by $G_h(\lambda)$ $(h = 1, \ldots, H)$ the quantities (4.15) associated with the function $f - \lambda$. Clearly the functions $G_h(\lambda)$ are analytic for $\lambda \in \Gamma$.

THEOREM 5.4. *The eigenvalues of* $T_n(f)$ *are canonically distributed if for each* Γ *with* $i(\Gamma) \neq 0$ *the functions* $G_h(\lambda)$ $(h \in A(\Gamma))$ *have the following property: If*

$$\sum_B c_B \prod_{h \in B} G_h(\lambda) \equiv 0, \tag{5.1}$$

where c_B *are constants and the sum is taken over all subsets* B *of* $A(\Gamma)$ *satisfying* $|B| = p(\Gamma)$, *then* $c_B = 0$ *for some* B.

PROOF. We shall show that (1.3) holds in measure on each Γ. Of course we may assume $i(\Gamma) \neq 0$. Applying Lemma 5.3, and using the notation introduced above, we see that the sum in the statement of theorem 4.3 associated with the function $f - \lambda$ can be rewritten as

$$\prod_{h=1}^{H} G_h(\lambda)^{\bar{k}_h} z_h^{\bar{k}_h n} \cdot \sum_{B} c_{\{\bar{k}_h + \chi_B(h)\}} \prod_{h \in B} G_h(\lambda) z_h^n$$

where the sum here is taken over all B as in the statement of the proposition and where $\chi_B(h)$ equals 1 or 0 depending on whether h does or does not belong to B. Since the $G_h(\lambda)$ are nonzero, it follows from Theorem (4.2) and Lemma 5.1 that it suffices to show that

$$|\sum_{B} c_B \prod_{h \in B} G_h(\lambda) z_h^n + o(f)|^{1/n} \to 1$$

in measure. Here we have written c_B for the nonzero constants $c_{\{\bar{k}_h + \chi_B(h)\}}$. For this it suffices to show that for each $\varepsilon > 0$,

$$\text{meas}\{\lambda : |\sum_{B} c_B \prod_{h \in B} G_h(\lambda) z_h^n| \geq \varepsilon\} \to 0. \tag{5.2}$$

And for this, it suffices to show that every sequence of n's has a subsequence for which this relation holds. By choosing an appropriate subsequence we can assure that each $\{z_h^n\}$ converges, say to a point $\zeta_h \in \mathbb{T}$. But then, uniformly on compact subsets of Γ,

$$\sum c_B \prod_{h \in B} G_h(\lambda) z_h^n \to \sum c_B \prod_{h \in B} G_h(\lambda) \zeta_h,$$

and the function on the right, by assumption, can only have a discrete set of zeros. Thus (5.2) holds for this subsequence. □

The assumption of the theorem is equivalent to the assumption that at least one of the products $\prod_{h \in B} G_h(\lambda)$ is not a linear combination of the others. We believe this always to be true, but we cannot prove it. So we shall make an ad hoc assumption on f (more on the curve it describes, in a sense) which is a sufficient condition for the assumption to be satisfied.

Given a component Γ of the complement of R(f), by a *regular singular value* for Γ we a mean a point $w \in \partial\Gamma$ with the following properties: $w = f(z)$ for a unique point z, which is a singularity of f; if $z = z_h$ then the quantities $f^{(\alpha_h)}(z_h\pm)$ are both nonzero and

$$f^{(\alpha_h)}(z_h+) \neq \pm f^{(\alpha_h)}(z_h-). \tag{5.3}$$

PROPOSITION 5.5. *Assume* f *is piecewise analytic and each* Γ *for which* $i(\Gamma) \neq 0$ *has at least* $p(\Gamma)$ *regular singular values* $w_h = f(z_h)$ *with* $h \in A(\Gamma)$. *Then the eigenvalues of* $T_n(f)$ *are canonically distributed.*

As an example take the case where f is one-one. Then if Γ is the inside of $R(f)$ the set $A(\Gamma)$ consists of the indices corresponding to the singularities of minimal order, and the requirement is that one of these correspond to a regular singular value. For a lemniscate-like curve we require one of these on each loop. If, to give another example, f describes a limaçon with a double loop, and there are two singularities having orders α and $\alpha + 2$ then for the region between the outer and inner loop $m = 1$ and $p = 0$ so we have uniqueness and no condition due to this region is required, but for the region inside the inner loop $m = 1$, $p = 1$ and the set $A(\Gamma)$ consists of both indices so we require that this region have at least one regular singular value.

The analyticity assumption is just a convenience for the proof and can be dropped. We shall not pursue this point. We prove the proposition by showing that the assumption implies the assumption of Proposition 5.4, and we show this by determining the behavior of the various $G_h(\lambda)$ for λ near a particular regular singular value. We begin with a preliminary lemma.

LEMMA 5.6. *Suppose* $\phi(\theta)$ *is analytic in a neighborhood of* $\theta = 0$. Then *for any* $\alpha = 1, 2, \cdots$ *and any* $\eta > 0$ *the function*

$$\Phi(\lambda) = \int_0^{\eta} \frac{\phi(t^{1/\alpha})\, t^{\frac{1}{\alpha} - 1}}{t - \lambda}\, dt \tag{5.4}$$

has near $\lambda = 0$ $(\lambda \notin \mathbf{R}^+)$ *the form*

$$\lambda^{\frac{1}{\alpha} - 1} \phi_1(\lambda^{1/\alpha}) + \log \lambda\, \phi_2(\lambda) \tag{5.5}$$

where ϕ_1 and ϕ_2 are *analytic in a neighborhood of* $\lambda = 0$ *and*

(i) *the order of (vanishing of)* ϕ_1 *at* $\lambda = 0$ *is at least the minimum of* $\alpha - 1$ *and the order* ϕ,

(ii) $\phi_2(0) = \phi^{(\alpha-1)}(0)/(\alpha - 1)!$

PROOF. Clearly Φ is analytic in the plane cut along $[0, \eta]$ and for the limits from above and below the cut we have

$$\Phi(\lambda - i0) - \Phi(\lambda + i0) = \frac{2\pi i}{\alpha}\, \phi(\lambda^{1/\alpha})\, \lambda^{\frac{1}{\alpha}-1}. \tag{5.6}$$

Now one solution to the Hilbert problem

$$\Phi(\lambda - i0) - \Phi(\lambda + i0) = \lambda^q \qquad (\lambda > 0)$$

is

$$(e^{2\pi q} - 1)^{-1}\lambda^q \quad \text{if} \quad q \notin \mathbf{Z}, \qquad (2\pi i)^{-1}\lambda^q \log\lambda \quad \text{if} \quad q \in \mathbf{Z}, \tag{5.7}$$

where here we take $0 < \arg\lambda < 2\pi$. Let us write, for θ and λ sufficiently small, $\phi(\lambda) = \sum_{k=0}^{\infty} d_k\, \theta^k$, and

$$\phi_1(\lambda) = \frac{2\pi i}{\alpha} \sum_{k+l \notin \alpha\mathbf{Z}} \frac{d_k}{(e^{2\pi i(k+l)/\alpha} - 1)}\, \lambda^{\frac{k+1}{\alpha}-1}$$

$$\phi_2(\lambda) = \alpha^{-1} \sum_{q=0}^{\infty} d_{\alpha(q+1)-1}\, \lambda^q .$$

Then it follows from (5.6) and (5.7) that

$$\phi_3(\lambda) := \Phi(\lambda) - \lambda^{\frac{1}{\alpha}-1}\, \phi_1(\lambda^{1/\alpha}) - \log\lambda\, \phi_2(\lambda)$$

is single-valued and analytic in a deleted neighborhood of 0. It is easy to check that as $\lambda \to 0$ $(\lambda \notin \mathbf{R}^+)$

$$\Phi(\lambda) = \begin{cases} O(\log|\lambda|^{-1}) & \text{if } \alpha = 1 \\ O(|\lambda|^{\frac{1}{\alpha}-1}) & \text{if } \alpha > 1 \end{cases}$$

and so ϕ_3 is actually analytic at $\lambda = 0$. This $\phi_3(\lambda)$ can be incorporated into $\lambda^{\frac{1}{\alpha}-1}\, \phi_1(\lambda^{1/\alpha})$ and so Φ has the stated form. Properties (i) and (ii) are easily verified. If different branches of $\lambda^{1/\alpha}$, $\log\lambda$ had been chosen the definitions of the ϕ_i would have been slightly different, but (i) and (ii) would still hold. □

REMARK. If in the statement of the lemma we integrate along the segment $[0, \eta e^{i\delta}]$ instead of $[0, \eta]$ the same conclusion clearly holds, except that now $\lambda \notin \mathbf{R}^+ e^{i\delta}$.

We now consider the behavior of the various $G_h(\lambda)$ near a regular singular value w for a component Γ. For definiteness we assume $w = f(z_1)$ and write α instead of α_1.

LEMMA 5.7. *For* $\lambda \in \Gamma$, λ *near* w, *each* $G_h(\lambda)$ *with* $h \neq 1$ *is of the form*

$$\exp\{\phi_1((\lambda - w)^{1/\alpha}) + (\lambda - w)\log(\lambda - w)\,\phi_2(\lambda - w)\} \tag{5.8}$$

while $G_1(\lambda)$ *is of the form* $(\lambda - w)^\beta$ *times* (5.8) *where*

$$\beta = \frac{i}{\pi\alpha}\,\mathrm{sgn}\, i(\Gamma)\log\frac{f^{(\alpha)}(z_1+)}{(-1)^\alpha f^{(\alpha)}(z_1-)} - 1.$$

Here the functions ϕ_1 *and* ϕ_2 *are analytic and single-valued in a neighborhood of* 0 *and the last logarithm is specified by choosing an appropriate argument.*

PROOF. Let us assume for simplicity that $w = 0$ and $z_1 = 1$. Once again we shall write $f(\theta)$ instead of $f(e^{i\theta})$, so the singularity occurs at $\theta = 0$ and, near $\theta = 0$,

$$f(\theta) = \frac{1}{\alpha!} f^{(\alpha)}(0\pm)\,\theta^\alpha + \cdots$$

where the dots represent functions analytic and of order $\alpha + 1$ at $\theta = 0$. The curves $z = f(\theta)$, $z = f(-\theta)$, with θ positive and small, either coincide or are disjoint. (This follows from the analyticity condition.) In the former case the part of Γ in a deleted neighborhood of zero consists of the complement of this curve in the neighborhood. In the latter case the curve divides the neighborhood into two parts, one of which is Γ. In either case Γ is on one side or the other of the curve $z = f(\theta)$, $z \in (-\eta, \eta)$, with η taken sufficiently small. For definiteness let us assume Γ is to the left of the curve as z runs from $-\eta$ to η.

Recall that $G_h(\lambda)$ is obtained by replacing f in (4.15) by $f - \lambda$. We consider first the exponent in (4.15)—in fact its derivative with respect to λ, which equals

$$\frac{i}{2\pi}\,\mathrm{sgn}\, i(\Gamma)\int_{-\pi}^{\pi}\frac{\cot\frac{1}{2}(\theta - \theta_h)}{f(\theta) - \lambda}\,d\theta.$$

Here, as before, we have written $z_h = e^{i\theta_h}$. By the first property of a regular singular value, f assumes the value 0 only at $\theta = 0$. Thus for any $\eta > 0$ the above is equal to a constant times

$$\int_{-\eta}^{\eta} \frac{\cot \frac{1}{2}(\theta - \theta_h)}{f(\theta) - \lambda} \, d\theta \tag{5.9}$$

plus a function analytic at $\lambda = 0$. Consider first

$$\int_{0}^{\eta} \frac{\cot \frac{1}{2}(\theta - \theta_h)}{f(\theta) - \lambda} \, d\theta \, .$$

Let us write $c^+ = f^{(\alpha)}(0+)/\alpha!$ so that near $\theta = 0$ we have $f(\theta) \sim c^+\theta^\alpha$. If we make the substitution $f(\theta) = c^+t$ then θ is an analytic function of $t^{1/\alpha}$ and the integral takes the form

$$\int_{0}^{f(\eta)/c^+} \frac{\phi(t^{1/\alpha}) \, t^{\frac{1}{\alpha} - 1}}{c^+ t - \lambda} \, dt \tag{5.10}$$

where ϕ is analytic at 0 and

$$\phi(0) = -\frac{i}{2\pi\alpha} \operatorname{sgn} i(\Gamma) \cot \tfrac{1}{2}\theta_h.$$

Now the integral is taken over a curve joining 0 to $f(\eta)/c^+$ (approximately equal to η^α) which is tangent to the positive real axis at 0. The point λ/c^+ lies above this curve (i.e., by our assumption, to the left of it as it is traversed away from 0). Let δ be a positive number (which may be taken to be small if η is small). Cauchy's theorem tells us that we can replace the path of integration in the above integral by the line segment from 0 to $\eta^\alpha e^{-i\delta}$ followed by the line segment from $\eta^\alpha e^{-i\delta}$ to $f(\eta)/c^+$; for λ/c^+ lies outside the corresponding closed contour. The integral over the second line segment is analytic at $\lambda = 0$ since the segment itself does not contain zero. It follows from (the remark following) Lemma 5.6 that the integral over $[0, \eta^\alpha e^{-i\delta}]$ has the form (5.5) with λ replaced by λ/c^+. A similar argument applies to the integral over $[-\eta, 0]$ in (5.9). If we integrate with respect to λ and exponentiate we see that the exponential factor in (4.15) has the form (5.8). The other factor, however, is a constant times $(f(\theta_h) - \lambda)^{-1}$ and so is analytic at $\lambda = 0$. Thus the first statement of the lemma is established.

For the second, observe that the derivative of the exponent in (4.15) is now equal to

$$\frac{i}{2\pi} \operatorname{sgn} i(\Gamma) \int_{-\pi}^{\pi} \frac{\cot \frac{1}{2}\theta}{f(\theta) - \lambda} \, d\theta. \tag{5.11}$$

This in turn equals, for sufficiently small η,

$$\frac{i}{2\pi} \operatorname{sgn} i(\Gamma) \int_{-\eta}^{\eta} \frac{\cot \frac{1}{2}\theta}{f(\theta) - \lambda} \, d\theta$$

plus a function analytic at $\lambda = 0$. We write this as

$$\frac{i}{2\pi}\,\mathrm{sgn}\;i(\Gamma)\left[\int_{-\eta}^{\eta}\frac{\cot\frac{1}{2}\theta - 2\theta^{-1}}{f(\theta)-\lambda}\,d\theta + \int_{-\eta}^{\eta}\frac{2\theta^{-1}}{f(\theta)-\lambda}\,d\theta\right]$$

$$= \mathrm{I} + \mathrm{II},$$

say. Repeating the argument of the first half of this proof, we find that I is of the form (5.5). We can write

$$\mathrm{II} = \frac{i}{\pi}\,\mathrm{sgn}\;i(\Gamma)\int_{-\eta}^{\eta}\left[\frac{1}{f(\theta)-\lambda}+\frac{1}{\lambda}\right]\frac{d\theta}{\theta}$$

$$= \frac{i}{\pi\lambda}\,\mathrm{sgn}\;i(\Gamma)\int_{-\eta}^{\eta}\frac{f(\theta)/\theta}{f(\theta)-\lambda}\,d\theta\,. \tag{5.12}$$

If we now make the substitution $f(\theta) = c^{+}t$ for $\theta \in [0, \eta]$ the part of (5.12) corresponding to this interval becomes, in analogy to (5.4),

$$\lambda^{-1}\int_{0}^{f(\eta)/c^{+}}\frac{\phi(t^{1/\alpha})\,t^{\frac{1}{\alpha}-1}}{t-\lambda/c^{+}}\,dt \tag{5.13}$$

where now $\phi(s)$ is analytic at $s = 0$ and vanishes there to order $\alpha - 1$. Near $s = 0$

$$\phi(s) = \frac{i}{\pi\alpha}\,\mathrm{sgn}\;i(\Gamma)\;s^{\alpha-1} + \cdots\,. \tag{5.14}$$

Proceeding as before and using (i) and (ii) of Lemma 5.6 we see that (5.13) has the form

$$\lambda^{-1}\phi_1(\lambda^{1/\alpha}) + \lambda^{1}\log\lambda\,\phi_2(\lambda) \tag{5.15}$$

with, as usual $\phi_1(\lambda)$ and $\phi_2(\lambda)$ analytic at 0, and

$$\phi_2(0) = \frac{1}{\pi\alpha}\,\mathrm{sgn}\;i(\Gamma).$$

Similarly the contribution of the interval $[0, \eta]$ to (5.12) is of the form

$$-\lambda^{-1}\int_{0}^{f(-\eta)/c^{-}}\frac{\phi(t^{1/\alpha})\,t^{\frac{1}{\alpha}-1}}{t-\lambda/c^{-}}\,dt$$

where ϕ has the same form (5.14) is before and $c^{-} = (-1)^{\alpha} f^{(\alpha)}(0-)$. Hence this also has the form (5.15) but now (because of the minus sign in front of the integral) $\phi_2(0)$ is minus what it was before. Hence adding (and changing notation) shows that (4.12) has the form

$$\lambda^{-1}\phi_1(\lambda^{1/\alpha}) + \log\lambda\, \phi_2(\lambda).$$

Let us evaluate $\phi_1(0)$. This is equal to

$$\frac{i}{\pi} \operatorname{sgn} i(\Gamma) \lim_{\lambda\to 0} \int_{-\eta}^{\eta} \frac{f(\theta)/\theta}{f(\theta)-\lambda}\, d\theta,$$

a limit which must be independent of η since it is the coefficient of λ^{-1} in the asymptotic expansion of (5.11). We have

$$\int_{-\eta}^{\eta} \frac{f(\theta)/\theta}{f(\theta)-\lambda}\, d\theta = \frac{1}{\alpha}\int_{-\eta}^{\eta} \frac{f'(\theta)}{f(\theta)-\lambda}\, d\theta + \int_{-\eta}^{\eta} \frac{f'(\theta)}{f(\theta)-\lambda}\left[\frac{f(\theta)}{\theta f'(\theta)} - \frac{1}{\alpha}\right] d\theta. \tag{5.16}$$

The first term on the right side equals

$$\frac{1}{\alpha}\log\frac{f(\eta)-\lambda}{f(-\eta)-\lambda}$$

where the logarithm is determined by taking a continuously varying $\arg(f(\theta)-\lambda)$, $\theta \in [-\eta,\eta]$. The expression in brackets in the second integral on the right side vanishes to order 1 at $\theta = 0$ and so we can take the limit as $\lambda \to 0$ under the integral sign. Hence the limit of (5.16) equals

$$\frac{1}{\alpha}\log\frac{f(\eta)}{f(-\eta)} + \int_{-\eta}^{\eta} \frac{f'(\theta)}{f(\theta)}\left[\frac{f(\theta)}{\theta f'(\theta)} - \frac{1}{\alpha}\right] d\theta.$$

But since this limit is independent of η it is also equal to

$$\frac{1}{\alpha}\lim_{\eta\to 0}\frac{f(\eta)}{f(-\eta)} = \frac{1}{\alpha}\log\frac{c^+}{c^-}.$$

Putting all these things together shows that (5.11) is of the form

$$\lambda^{-1}\phi_1(\lambda^{1/\alpha}) + \log\lambda\, \phi_2(\lambda)$$

where ϕ_1 and ϕ_2 are analytic at $\lambda = 0$ and

$$\phi_1(0) = \frac{i}{\pi\alpha} \operatorname{sgn} i(\Gamma) \log\frac{c^+}{c^-}.$$

This is the derivative with respect to λ of the exponent in (4.15) for $G_1(\lambda)$. Integrating, and noting that the second factor in (4.15) is in this case a nonzero constant times λ^{-1}, we deduce that $G_1(\lambda)$ has the form given in the statement of the lemma. □

PROOF of Proposition 5.5. We show that the hypothesis of Proposition 5.4 is satisfied. Suppose $i(\Gamma) \neq 0$, write p for $p(\Gamma)$ and assume $w_i = f(z_i)$ $(i = 1, \ldots, p)$ are regular singular values for Γ. Assume (5.1) holds and no c_B is zero. We can rewrite this as

$$G_1 P(G_2, \ldots, G_H) + Q(G_2, \ldots, G_H) \equiv 0 \tag{5.17}$$

where P and Q are polynomials of degree $p-1$ and p respectively and the coefficient of $G_2 \cdots G_p$ in P is nonzero. With the notation of Lemma 5.7 we see, from that lemma, that $P(G_2, \ldots, G_H)$ and $Q(G_2, \ldots, G_H)$ have in the neighborhood of $z = w$ series expansions in which only terms $(\lambda - w)^{\frac{\kappa}{\alpha}} \log^q(\lambda - w)$ $(\kappa = 0, 1 \cdots)$ appear, while in the expansion for G_1 only terms $(\lambda - w)^{\frac{\kappa}{\alpha}+\beta} \log^q(\lambda - w)$ appear and the term $(\lambda - w)^\beta$ does appear. In order for (5.17) to hold without $P(G_2, \ldots, G_H)$ being identically zero β must be α^{-1} times an integer, in other words, we must have

$$f^{(\alpha)}(z_1+)/(-1)^\alpha f^{(\alpha)}(z_1-) = \pm 1.$$

But this violates (5.3) and so $P(G_2, \ldots, G_H)$ must be identically zero. We now repeat the argument using G_2 instead of G_1, and so on, until we reach the absurdity $G_p(\lambda) \equiv 0$. □

6. THE CASE OF A SIMPLE SMOOTH CURVE

In this section we prove a couple of results of a different kind in case f is one-one, $f \in C^1$, and $f' \neq 0$. (The latter conditions can be relaxed.) We shall assume throughout that these conditions hold. We also denote by C the range of f thought of as an oriented curve, Γ the interior of C.

THEOREM 6.1. *Suppose there exists a compact subset* K *of* Γ *such that*

$$\lim_{n \to \infty} n^{-1} |\{i : \lambda_{i,n} \in K\}| = 1. \tag{6.1}$$

Then f *extends analytically to an annulus* $r < |z| < 1$ (resp. $1 < |z| < r$) *if* $i(\Gamma)$ *equals* 1 (resp. -1). *Conversely if* f *does extend analytically as described then there is a compact subset* K *of* Γ *such that for* n *sufficiently large all eigenvalues* $\lambda_{i,n}$ *belong to* K.

PROOF. Since the limit relations (1.3) holds for all λ outside C any weak limit μ of the measures $\mu_{n,f}$ must satisfy

$$\int \log|w-\lambda|\, d\mu(w) = \log G(|f-\lambda|) = \int \log |w-\lambda|\, d\mu_f(w) \tag{6.2}$$

for all λ outside C. If μ is supported on a compact subset K of Γ, which is true if (6.1) holds, then the integral on the left, and so also the integral on the right, extends harmonically into the complement of K. It follows that

$$\int (w-\lambda)^{-1}\, d\mu_f(w) = \frac{1}{2\pi} \int_C \frac{f'(f^{-1}(w))^{-1}}{w-\lambda}\, dw$$

extends analytically to the complement of K. Now it is well known that if we define the piecewise analytic function F by

$$F(\lambda) = \frac{1}{2\pi i} \int_C \frac{f'(f^{-1}(w))^{-1}}{w-\lambda}\, dw \qquad \lambda \notin C$$

then the limits $F_+(\lambda)$, $F_-(\lambda)$ of F as $\lambda \to C$ from inside and outside C respectively satisfy

$$F_+(\lambda) - F_-(\lambda) = f'(f^{-1}(\lambda))^{-1}, \qquad \lambda \in C.$$

(This holds if C is described counterclockwise. Otherwise the right side is replaced by its negative.) Since $F_-(\lambda)$ extends analytically to the complement of K we see that

$$f'(f^{-1}(\lambda))^{-1} = \frac{d}{d\lambda} f^{-1}(\lambda)$$

extends analytically to the region between C and K. Hence so does $f^{-1}(\lambda)$. This extension is one-one in a neighborhood of C and its inverse, restricted to this neighborhood, is an analytic extension of f to an annulus $r < |z| < 1$ or $1 < |z| < r$; which it is depends on whether $i(\Gamma)$ equals $+1$ or -1. (This last follows from the fact than an analytic mapping is orientation-preserving.) Thus the first statement of the theorem is established.

For the converse we begin with the fact that for any compact subset of the invertible elements of $C(\mathbb{T})$, for each of which $i(f) = 0$, there exists an n_0 such that $n > n_0$ implies that $T_n(f)$ is invertible for all f in the set. (See, e.g., [W2, Th. 5.1].) Using this we deduce for our f that for any neighborhood U of $\Gamma \cup C$ there exists an n_0 such that $T_n(f-\lambda)$ is invertible for $n > n_0$, $\lambda \notin U$. Thus all $\lambda_{i,n}$ lie in U for large enough n. Suppose now, for definiteness, that f extends to an analytic function in an annulus $r < |z| < 1$ and write $f_r(z) = f(rz)$ for $|z| = 1$. Then as explained in the introduction the $\lambda_{i,n}$ are also the eigenvalues of $T_n(f_r)$. Now

C_r, the range of f_r, is a curve lying inside C (again by consideration of orientation). Hence, by what we just observed, if Γ_r denotes the inside of C_r and U_r is any neighborhood of $\Gamma_r \cup C_r$ all $\lambda_{i,n}$ belong to U_r for sufficiently large n. But since $C_r \subset \Gamma$ we can choose such a U_r so that $\overline{U}_r \subset \Gamma$, and $\overline{U}_r$ is our desired set K. □

THEOREM 6.2. *If* μ *is a weak limit of the measure* $\mu_{n,f}$ *whose support is a subset of* C *then* $\mu = \mu_f$.

PROOF. Our limiting measure μ must satisfy (6.2). It follows that $\nu = \mu - \mu_f$ is a real-valued measure supported on C, with $\nu(C) = 0$, such that

$$\int \log |w - \lambda| \, d\nu(w) = 0$$

for all λ outside C. We deduce, first, that

$$\int (w - \lambda)^{-1} \, d\nu(w) = 0$$

for all such λ, and then that ν annihilates all functions which are analytic outside C (including ∞) and continuous up to C. If we apply a conformal mapping of the exterior of C to the exterior of the unit circle, ν is transformed into a measure which annihilates all powers z^{-n} $(n = 0, 1, 2 \cdots)$. Since this transformed measure is real-valued it must be zero. Therefore $\nu = 0$ and $\mu = \mu_f$. □

COROLLARY 6.2. *Suppose that for every neighborhood* V *of* C *we have*

$$\lim_{n\to\infty} n^{-1} |\{i : \lambda_{i,n} \in V\}| = 1.$$

Then the eigenvalue of $T_n(f)$ *are canonically distributed.*

PROOF. The limit of any weakly convergent subsequence of $\{\mu_{n,f}\}$ is supported in C and so must equal μ_f. □

REFERENCES

[B] M. Brelot. Élements de la théorie classique du potential, Centre Docum. Univ., Paris 1969.

[B-S1] A. Böttcher and B. Silbermann. Invertibility and asymptotics of Toeplitz matrices. Akademie-Verlag, Berlin 1983.

[B-S2] _______. Toeplitz operators and determinants generated by symbols with one Fisher-Hartwig singularity. Math. Nachr. 127 (1986) 95-124.

[D] K. M. Day. Measures associated with Toeplitz matrices generated by the Laurent expansion of rational functions. Trans. Amer. Math. Soc. 209 (1975) 175-183.

[H] I. I. Hirschman, Jr. The spectra of certain Toeplitz matrices. Ill. J. Math. 11 (1967) 145-159.

[P-S] G. Polya and G. Szegö. Aufgaben und Lehrsätze aus der Analysis, v. 1. Springer-Verlag, Berlin, 1964.

[S-S] F. Schmidt and F. Spitzer. The Toeplitz matrices of an arbitrary Laurent polynomial. Math. Scand. 8 (1960) 15-38.

[S] G. Szegö. Beiträge zur Theorie der Toeplitzchen Formen I. Math. Zeit. 6 (1920) 167-202.

[W1] H. Widom. Toeplitz determinants with singular generating functions. Amer. J. Math. 95 (1973) 333-383.

[W2] _______. Asymptotic behavior of block Toeplitz matrices and determinants II. Adv. in Math. 21 (1976) 1-29.

Department of Mathematics
University of California
Santa Cruz, CA 95064 U.S.A.

Operator Theory:
Advances and Applications, Vol. 48

ON SOME CLASSES OF HYPONORMAL TUPLES OF COMMUTING OPERATORS

Daoxing Xia*

Dedicated to the memory of Professor Ernst Hellinger

Some theorems on hyponormal or cohyponormal n-tuples of commuting operators are given. Using the analytic model, a trace formula for the subnormal n-tuple of operators is established.

1. Let $\mathcal{H}$ be a Hilbert space, and $\mathcal{L}(\mathcal{H})$ be the algebra of all linear bounded operators on $\mathcal{H}$. An operator $H \in \mathcal{L}(\mathcal{H})$ is said to be hyponormal if $[H^*, H] \geq 0$, where $[A, B]$ is the commotator of A and B. An operator $C \in \mathcal{L}(\mathcal{H})$ is said to be cohyponormal if $[C^*, C] \leq 0$. It is obvious that C is cohyponormal iff C^* is hyponormal operator. Recently some authors [1], [14] tried to generalize the concept of hyponormal operators to the n-tuple case. In [1], Athavale introduced a sort of hyponormality for n-tuple of operators. Although, the author still does not know whether this concept is the most natural generalization of the single operator case or not, in the present paper we adopt this hyponormality and study some classes of hyponormal n-tuple of commuting operators.

A tuple $\mathbb{H} = (H_1, \cdots, H_n)$, $H_j \in \mathcal{L}(\mathcal{H})$ is said to be hyponormal [1], if

$$\sum_{i,j} ([H_i^*, H_j]x_i, x_j) \geq 0, \tag{1}$$

for all $x_j \in \mathcal{H}$. The tuple $\mathbb{H}$ is said to be commuting, if $[H_i, H_j] = 0$ for $i, j = 1, 2, \cdots n$. Similar to the single operator case, a tuple $\mathbb{A} = (A_1, \cdots, A_n)$, $A_j \in \mathcal{L}(\mathcal{H})$ is said to be cohyponormal if

$$\sum_{i,j} ([A_i^*, A_j]x_i, x_j) \leq 0$$

for all $x_j \in \mathcal{H}$. An n-tuple $\mathbb{A} = (A_1, \cdots, A_n)$ is said to be t-hyponormal if $\mathbb{A}^* = (A_1^*, \cdots, A_n^*)$ is cohyponormal. For $n = 1$, t-hyponormal is equivalent to hyponormal. The question is whether t-hyponormal is equivalent to hyponormal in the case of $n > 1$. It may not be well-known to some mathematician that for the n-tuple of commuting operators t-hyponormal is not equivalent to hyponormal if $n \geq 2$. The author gives an elementary example to show that fact in §2 of the

* Supported in part by NSF grant DMS-8901506.

present paper. We also prove that suppose $\mathbb{H} = (H_1, \cdots, H_n)$ is a commuting irreducible n-tuple of operators and satisfies the condition that (i) there is no subspace $\mathcal{H}_0$ reducing H_1 such that $H_1|_{\mathcal{H}_0}$ is a linear combination of a unilateral shift with the identity and (ii)

$$\operatorname{rank}\,[H_1^*, H_1] = 1,$$

then $\mathbb{H}$ is hyponormal iff $\mathbb{H}$ is t-hyponormal (Theorem 4 in §3).

The second problem arises from R. Curto's work [6], [7]. He proves that if $\lambda = (\lambda_1, \cdots, \lambda_n) \in \mathbb{C}^n$ is not in the Taylor spectrum $\operatorname{sp}(\mathbb{A})$ of a commuting n-tuple $\mathbb{A} = (A_1, \cdots, A_n)$, then

$$\sum_{j=1}^{n}(A_j - \lambda_j I)(A_j^* - \overline{\lambda}_j I) \quad \text{and} \quad \sum_{j=1}^{n}(A_j^* - \overline{\lambda}_j I)(A_j - \lambda_j I)$$

are invertible. He also shows that this condition is not sufficient for $\lambda \notin \operatorname{sp}(\mathbb{A})$. A natural probelm is to find some classes of n-tuples of communting operators in which that condition is also sufficient. For the hyponormal or t-hyponormal n-tuple of operators the invertibity of $\sum_{j=1}^{n}(A_j - \lambda_j I)(A_j^* - \overline{\lambda}_j I)$ implies the invertibity of $\sum_{j=1}^{n}(A_j^* - \overline{\lambda}_j I)(A_j - \lambda_j I)$. In §4, we prove that if $\mathbb{A} = (A_1, \cdots, A_n)$ is a t-hyponormal n-tuple of commuting operators then the condition that

$$\sum_{j=1}^{n}(A_j - \lambda_j I)(A_j^* - \overline{\lambda}_j I)$$

is invertible is a necessary and sufficient condition for $\lambda = (\lambda_1, \cdots, \lambda_n) \notin \operatorname{sp}(\mathbb{A})$ (see Theorem 5). It is not known that if this condition is also sufficient for hyponormal commuting n-tuple of operators for $n > 1$.

An important class of hyponormal n-tuples of commuting operators is the class of all subnormal n-tuples of operators (cf. [1], [5], [9], [10]). In [17], the author generalized Morrel's Theorem (cf. [11]) to the subnormal n-tuple case. Using Morrel's theorem, Theorem 9 in [17] is equivalent to the following

THEOREM. *Let $(S_1, \cdots, S_n)$ be an irreducible subnormal n-tuple of operators on a Hilbert space $\mathcal{H}$. If rank $[S_1^*, S_1] = 1$, and $[S_j^*, S_j]\mathcal{H} \subset [S_1^*, S_1]\mathcal{H}$ then there are complex numbers α_j, $\beta_j \neq 0$, $j = 2, \cdots, n$ such that*

$$S_j = \alpha_j S_1 + \beta_j I, \qquad j = 2, \cdots, n.$$

In §2 and §3 we prove the following. Let $\mathbb{H} = (H_1, \cdots, H_n)$ be an irreducible commuting n-tuple of operators satisfying $\operatorname{rank}[H_1^*, H_1] = 1$. If $\mathbb{H}$ satisfies that either (i) $\mathbb{H}$ is hyponormal and $\operatorname{rank}[H_j^*, H_j] \leq 1$ for $j = 2, \cdots, n$, and there is no subspace $\mathcal{H}_0$ reducing H_1 such that $H_1|_{\mathcal{H}_0}$ is a linear combination of a unilateral shift with multiplicity one and the identiy, (ii) $\mathbb{H}$ is hyponormal and $[H_j^*, H_j]\mathcal{H} \subset [H_1^*, H_1]\mathcal{H}$, or (iii) $\mathbb{H}$ is cohyponormal or t-hyponormal, then there are $\alpha_j, \beta_j \in \mathbb{C}$ such that

$$H_j = \alpha_j H_1 + \beta_j I, \qquad j = 2, \cdots, n.$$

This is a generalization of above theorem. By the way, the analytic model of a hyponormal operator with rank one self-commutator has been given in [12].

For the subnormal n-tuple of operators, in [17] the author established an analytic model. Using this analytic model, in §5 of this paper, the author establishes a trace formula for the subnormal n-tuple of operators which has some connection with Carey and Pincus theory of local index and principal current (cf. [2], [3], [4]).

2. The following example shows that for the n-tuple of commuting operators, the t-hyponormality is not equivalent to the hyponormality if $n \geq 2$.

Example 1. Let $\mathcal{H} = H^2(\mathbb{T})$ be the Hardy space. Define operators

$$(H_1 f)(\varsigma) = \varsigma f(\varsigma)$$

and

$$(H_2 f)(\varsigma) = M(\varsigma) f(\varsigma)$$

for $f \in \mathcal{H}$, where $M \in H^\infty(\mathbb{T})$. Then H_2^* is the Toeplitz operator $f \mapsto P(\overline{M} f)$, where P is the orthogonal projection from $L^2(\mathbb{T})$ to $H^2(\mathbb{T})$. It is easy to calculate that

$$\begin{aligned}([H_2^*, H_2]f, f) &= (\overline{M} f, \overline{M} f) - (P(\overline{M} f), \overline{M} f) \\ &= \|P_-(\overline{M} f)\|^2,\end{aligned}$$

for $f \in \mathcal{H}$, where $P_- = I - P$. On the otherhand,

$$([H_1, H_1]g, g) = |g(0)|^2 = |P_-((\bar{\cdot})g)|^2$$

and

$$\begin{aligned}([H_2^*, H_1]f, g) &= (\overline{M} f, (\bar{\cdot})g) - (P(\overline{M} f), (\bar{\cdot})g) \\ &= (P_-(\overline{M} f), P_-((\bar{\cdot})g)).\end{aligned}$$

Therefore

$$|([H_2^*, H_1]f, g)|^2 \leq ([H_1^*, H_1]g, g)([H_2^*, H_2]f, f).$$

Hence (H_1, H_2) is hyponormal. But if $M(\varsigma)$ is not linear function of ς, $|\varsigma| < 1$, then there are funcitons $f, g \in H^2(\mathbb{T})$ such that

$$|([H_2^*, H_1]g, f)|^2 \not\leq ([H_1^*, H_1]g, g)([H_2^*, H_2]f, f).$$

Thus (H_1, H_2) is t-hyponormal iff there are constants c and b such that

$$M(\varsigma) \equiv c\varsigma + b.$$

This example is also very useful for Theorem 1.

A tuple $\mathbb{H} = (H_1, \cdots, H_n)$ of operators on $\mathcal{H}$ is said to be irreducible, if the only subspaces reducing $H_1 \cdots, H_n$ are $\{0\}$ and $\mathcal{H}$.

THEOREM 1. *Let* $\mathbb{H} = (H_1, \cdots, H_n)$ *be a commuting hyponormal irreducible* n*-tuple of operators satisfying the condition that* $\operatorname{rank}[H_1^*, H_1] = 1$ *and*

$$\operatorname{rank}[H_j^*, H_j] \leq 1, \qquad j = 2, \cdots, n.$$

Then either (i) there are numbers α_k *and* β_k *such that*

$$H_k = \alpha_k H_1 + \beta_k I, \qquad k = 2, 3, \cdots, n.$$

or (ii) H_1 *is a linear combination of a unilateral shift with multiplicity one and identity and there are numbers* α_k, β_k *and* γ_k *such that*

$$H_k = (\alpha_k H_1 + \beta_k I)(\gamma_k H_1 + I)^{-1}, \qquad k = 2, 3, \cdots, n.$$

PROOF. (1) First let us study the relation between H_1 and H_2.

There exists a vector $e_1 \in \mathcal{H}$, $e_1 \neq 0$ such that

$$[H_1^*, H_1]x = (x, e_1)e_1, \quad \text{for } x \in \mathcal{H}. \tag{2}$$

From (1), it follows that

$$|([H_2^*, H_1]x_2, x_1)|^2 \leq ([H_1^*, H_1]x_1, x_1)([H_2^*, H_2]x_2, x_2).$$

Therefore there is a vector e_2 such that

$$[H_2^*, H_1]x = (x, e_2)e_1 \tag{3}$$

and

$$[H_2^*, H_2]x = c(x, e_2)e_2 \tag{4}$$

for $x \in \mathcal{H}$, $c \geq 1$.

(ii) Let $\mathcal{H}_1$ be the smallest subspace containing e_1 and reducing H_1. We have to prove that $\mathcal{H}_1$ is invariant with respect to H_2^*. It is obvious that

$$\begin{aligned}
\|e_1\|^2 H_2^* e_1 &= H_2^*[H_1^*, H_1]e_1 \\
&= H_1^*[H_2^*, H_1]e_1 + [H_1^*, H_1]H_2^* e_1 - [H_2^*, H_1]H_1^* e_1 \\
&= (e_1, e_2)H_1^* e_1 + (H_2^* e_1, e_1)e_1 - (H_1^* e_1, e_2)e_1.
\end{aligned}$$

Hence $H_2^* e_1 \in \mathcal{H}_1$. Also we have

$$H_2^* H_1^{*m} H_1^n e_1 = H_1^{*m}[H_2^*, H_1^n]e_1 + H_1^{*m} H_1^n H_2^* e_1$$
$$= \sum_{j=0}^{n-1} H_1^{*m} H_1^{n-j-1}[H_2^*, H_1]H_1^j e_1 + H_1^{*m} H_1^n H_2^* e_1.$$

Therefore $H_2^* H_1^{*m} H_1^n e_1 \in \mathcal{H}_1$ for $m, n = 0, 1, 2, \cdots$. Thus $\mathcal{H}_1$ is invariant with respect to H_2^*.

(iii) According to the decomposition $\mathcal{H} = \mathcal{H}_1 \oplus \mathcal{H}_1^\perp$, H_j may be written as matrices

$$H_j = \begin{pmatrix} A_j & 0 \\ B_j & C_j \end{pmatrix}, \qquad j = 1, 2,$$

where $A_j^* = H_j^*|_{\mathcal{H}_1}$, $B_1 = 0$, and C_1 is normal. The relation $[H_1, H_2] = 0$ is equivalent to

$$[A_1, A_2] = 0, \tag{5}$$
$$[C_1, C_2] = 0 \tag{6}$$

and

$$B_2 A_1 = C_1 B_2. \tag{7}$$

Let e_2^1 be the projection of e_2 on $\mathcal{H}_1$ and $e_2^2 = e_2 - e_2^1$. Then (3) is equivalent to

$$[A_2^*, A_1]x = (x, e_2^1)e_1, \qquad \text{for } x \in \mathcal{H}_1, \tag{8}$$
$$(B_2^* C_1 - A_1 B_2^*)x = (x, e_2^2)e_1 \qquad \text{for } x \perp \mathcal{H}_1, \tag{9}$$

and

$$[C_2^*, C_1] = 0. \tag{10}$$

(iv) Let us prove that $e_2^2 = 0$. Suppose on contrary that $e_2^2 \neq 0$.

LEMMA 1. *If $e_2^2 \neq 0$, then $(zI - A_1)^{-1} e_1$ is an eigenvector of A_1^* for $z \in \rho(H_1)$.*

PROOF. From (9), we have

$$(B_2^*(zI - C_1)^{-1} - (zI - A_1)^{-1} B_2^*)x = ((zI - C_1)^{-1}x, e_2^2)(zI - A_1)^{-1} e_1, \tag{11}$$

for $z \in \rho(H_1)$. From (7) and (11), we obtain that

$$(B_2^*(zI - C_1)^{-1} C_1^* - (zI - A_1)^{-1} A_1^* B_2^*)x = ((zI - C_1)^{-1} C_1^* x, e_2^2)(zI - A_1)^{-1} e_1 \tag{12}$$

and

$$[B_2^* C_1^* (zI - C_1)^{-1} - A_1^* (zI - A_1)^{-1} B_2^*] x = ((zI - C_1)^{-1} x, e_2^2) A_1^* (zI - A_1)^{-1} e_1, \tag{13}$$

for $z \in \rho(H_1)$. On the otherhand, it is obvious that

$$[A_1^*, A_1] x = (x, e_1) e_1. \tag{14}$$

Therefore

$$[A_1^*, (zI - A_1)^{-1}] x = ((zI - A_1)^{-1} x, e_1)(zI - A_1)^{-1} e_1. \tag{15}$$

From (12), (13), (15) and $[C_1^*, C_1] = 0$ we have

$$\begin{aligned} &((zI - A_1)^{-1} B_2^* x, e_1)(zI - A_1)^{-1} e_1 \\ &= ((zI - C_1)^{-1} C_1^* x, e_2^2)(zI - A_1)^{-1} e_1 - ((zI - C_1)^{-1} x, e_2^2) A_1^* (zI - A_1)^{-1} e_1. \end{aligned} \tag{16}$$

Let

$$x = (zI - C_1) e_2^2$$

in (16). Then (16) shows that $(zI - A_1)^{-1} e_1$ is an eigenvector of A_1^*, for $z \in \rho(H_1)$, which proves Lemma 1.

COROLLARY 1. *If $e_2^2 \neq 0$, then $\mathcal{H}_0 = \mathcal{H}_1$, where $\mathcal{H}_0$ is the closure of span $\{(zI - H_1)^{-1} e_1 : z \in \rho(H_1)\}$.*

PROOF. It is easy to show that $(zI - H_1)^{-1} e_1 = (zI - A_1)^{-1} e_1$.

By Lemma 1, $\mathcal{H}_0$ is invariant with respect to H_1^*. It is obvious that $e_1 \in \mathcal{H}_0$, $\mathcal{H}_0 \subset \mathcal{H}_1$ and $\mathcal{H}_0$ is invariant with respect to H_1. Therefore $\mathcal{H}_0 = \mathcal{H}_1$ which proves Corollary 1.

Now let us continue to prove theorem 1. By Lemma 1, if $e_2^2 \neq 0$, then let $q(z)$ be the eigenvalue of H_1^* corresponding to the eigen vector $(zI - H_1)^{-1} e_1$, for $z \in \rho(H_1)$. As in [12], denote

$$S(z, w) = ((zI - H_1)^{-1} e_1, (wI - H_1)^{-1} e_1), \qquad z, w \in \rho(H).$$

Then

$$\begin{aligned} S(z, w) &= ((\overline{w} I - H_1^*)^{-1} (zI - H_1)^{-1} e_1, e_1) \\ &= (\overline{w} - q(z))^{-1} ((zI - H_1)^{-1} e_1, e_1). \end{aligned} \tag{17}$$

However, $S(z, w) = \overline{S(w, z)}$, therefore

$$S(z, w) = (z - \overline{q(w)})^{-1} (e_1, (wI - H_1)^{-1} e_1). \tag{18}$$

From $H_1^*(zI - H_1)^{-1}e_1 = q(z)(zI - H_1)^{-1}e_1$, it is easy to see that

$$q(\infty) = \lim_{z \longrightarrow \infty} q(z)$$

exists and is finite. Besides,

$$H_1^* e_1 = q(\infty) e_1.$$

Let H_1 be replaced by $H_1 - \overline{q(\infty)}$. Therefore we may assume that $q(\infty) = 0$. Multiplying (17) and (18) through by $\overline{w}$ and letting $w \longrightarrow \infty$, we get

$$((zI - H_1)^{-1}e_1, e_1) = z^{-1}\|e_1\|^2.$$

Using the formula

$$[H_1^*, (zI - H_1)^{-1}]e_1 = ((zI - H_1)^{-1}e_1, e_1)(zI - H_1)^{-1}e_1,$$

we obtain that

$$q(z) = \|e_1\|^2 z^{-1}.$$

Therefore

$$S(z, w) = \|e_1\|^2(\overline{w}z - \|e_1\|^2)^{-1}.$$

By multiplying a positive number to H_1, we may assume that $\|e_1\| = 1$ and A_1 is a unilateral shift with multiplicity one. Thus we assume that $\mathcal{H}_1 = H^2(\mathbb{T})$, the Hardy space, and

$$(A_1 f)(z) = z f(z), \qquad \text{for } f \in H^2(\mathbb{T}). \tag{19}$$

(v) Now, let us determine the form of the normal operator C_1. By (9), we have

$$C_1^* B_2 f = B_2((f(\cdot) - f(0))(\cdot)^{-1}) + f(0)e_2^2. \tag{20}$$

Hence $C_1^* B_2 1 = e_2^2$ and $C_1^* B_2(\cdot) = B_2 1$. Therefore

$$C_1 e_2^2 = C_1 C_1^* B_2 1 = C_1^* B_2(\cdot) = B_2 1$$

and $C_1 C_1^* e_2^2 = e_2^2$. Thus for polynomials $p(\cdot)$ and $q(\cdot)$, we have

$$C_1(B_2 p(\cdot) + q(C_1^*)e_2^2) = B_2((\cdot)p(\cdot) + q(0)) + q_1(C_1^*)e_2^2, \tag{21}$$

by (7), where $q_1(\varsigma) = (q(\varsigma) - q(0))\varsigma^{-1}$. Similarly, we have

$$C_1^*(B_2 p(\cdot) + q(C_1^*)e_2^2) = B_2(p(\cdot) - p(0))(\cdot)^{-1} + (p(0) + q(C_1^*)C_1^*)e_2^2. \tag{22}$$

Therefore, by (21) and (22) we get

$$C_1C_1^*(B_2p(\cdot)+q(C_1^*)e_2^2)=B_2p(\cdot)+q(C_1^*)e_2^2. \tag{23}$$

Let $\mathcal{H}_2$ be the closure of the set $\{B_2p+q(C_1^*)e_2^2 : p,q \text{ are polynormals }\}$. From (21), (22) and (23), it is easy to see that $\mathcal{H}_2$ reduces C_1 and $C_1\mid_{\mathcal{H}_2}=U$ is unitary. Let the spectral resolution of U be

$$U=\int e^{i\theta}dE(e^{i\theta}),$$

where $E(\cdot)$ is a spectral measure on $\mathbb{T}$. Denote

$$\rho(\cdot)=\|E(\cdot)B_21\|^2$$

By (7), it is easy to calculate that

$$\int |f(e^{i\theta})|^2d\rho(e^{i\theta})=\|f(U)B_21\|^2=\|B_2f(\cdot)\|^2$$
$$\leq \|B_2\|^2\frac{1}{2\pi}\int |f(e^{i\theta})|^2d\theta.$$

Therefore $\rho(\cdot)$ is absolutely continuous. Let $F(e^{i\theta})=d\rho(e^{i\theta})/d\theta$. From (20), (21) and (22), it is easy to see that B_21 is the cyclic vector for $\{U,U^*\}$ in $\mathcal{H}_2$. So we may assume that $\mathcal{H}_2$ is the Hilbert space of all Baire functions $f(\cdot)$ on $\mathbb{T}$ satisfying

$$(f,f)=\frac{1}{2\pi}\int |f(e^{i\theta})|^2F(e^{i\theta})d\theta<+\infty,$$
$$B_2f(e^{i\theta})=f(e^{i\theta}),\qquad \text{for } f\in\mathcal{H}_1 \tag{24}$$

and

$$(Uf)(e^{i\theta})=e^{i\theta}f(e^{i\theta}),\qquad \text{for } f\in\mathcal{H}_2.$$

Denote

$$M(\cdot)=A_21.$$

By (8), we may prove that

$$e_2^1(\varsigma)=([A_1^*,A_2]1)(\varsigma)=(M(\varsigma)-M(0))\varsigma^{-1}. \tag{25}$$

It is easy to see that $M(\cdot)\in H^\infty(\mathbb{T})$ and

$$(A_2f)(\varsigma)=M(\varsigma)f(\varsigma),\qquad \text{for } f\in H^2(\mathbb{T}), \tag{26}$$

since $[A_1, A_2] = 0$, and $([A_1^*, A_2]f)(\varsigma) = f(0)(M(\varsigma) - M(0))\varsigma^{-1}$, by (8). By the way, A_2 is the same operator H_2 defined in the example 1. Besides, A_2^* is the Toeplitz operator

$$A_2^* f = P(\overline{M} f), \qquad f \in H^2(\mathbb{T}). \tag{27}$$

By (24) we have

$$(B_2 f, B_2 h) = \frac{1}{2\pi} \int f(e^{i\theta}) \overline{h(e^{i\theta})} F(e^{i\theta}) d\theta, \qquad \text{for } f, h \in H^2(\mathbb{T}).$$

Therefore

$$B_2^* B_2 f = P(F f). \tag{28}$$

From (4), it is easy to see that

$$([A_2^*, A_2] + B_2^* B_2) f = c(f, e_2^1) e_2^1. \tag{29}$$

From (25)-(29), we conclude that

$$(P((F + |M|^2) f) - M P(\overline{M} f))(\varsigma) = c(f, M_1(\cdot)) M_1(\varsigma), \tag{30}$$

for $f \in H^2(\mathbb{T})$, where $M_1(\varsigma) = (M(\varsigma) - M(0))\varsigma^{-1}$. Put $f(\varsigma) = (z - \varsigma)^{-1}, |z| > 1$ in (30). Notice that

$$P(\overline{M}(z - (\cdot))^{-1})(\varsigma) = \overline{M(\overline{z}^{-1})}(z - \varsigma)^{-1},$$
$$((z - (\cdot))^{-1}, (M(\cdot) - M(0))(\cdot)^{-1}) = \overline{M(\overline{z}^{-1})} - \overline{M(0)}$$

and

$$P(h(z - (\cdot))^{-1})(\varsigma) = (P(h)(\varsigma) + P_-(h)(z))(z - \varsigma)^{-1}$$

where $h = F + |M|^2$, and $P_- = I - P$. Therefore (30) implies that

$$\begin{aligned} &P(F + |M|^2)(\varsigma) + P_-(F + |M|^2)(z) \\ &= M(\varsigma)\overline{M(\overline{z}^{-1})} + c(z - \varsigma)\varsigma^{-1}(M(\varsigma) - M(0))(\overline{M(\overline{z}^{-1})} - \overline{M(o)}). \end{aligned} \tag{31}$$

for $|\varsigma| < 1$ and $|z| > 1$. For $\xi \in \mathbb{T}$, letting $\varsigma \longrightarrow \xi$ and $z \to \xi$ in (31), we obtain that

$$F(\xi) = 0 \qquad \text{for a.e. } \xi \in \mathbb{T}.$$

Thus $B_2 1 = 0$, and hence $e_2^2 = 0$. This is a contradiction.

(vi) Now, we only have to consider the case $e_2^2 = 0$. Under this condition, we have

$$B_2^* C_1 - A_1 B_2^* = 0, \tag{32}$$

by (9). From (7) and (32), it is easy to see that

$$B_2[A_1^*, A_1] = [C_1^*, C_1]B_2 = 0.$$

Thus $B_2 e_1 = 0$. Besides

$$B_2 A_1^{*m} A_1^n e_1 = C_1^{*m} C_1^n B_2 e_1 = 0.$$

Therefore $B_2 = 0$ and $\mathcal{H}_1$ reduces H_2. Similarly, $\mathcal{H}_1$ also reduces $H_3, \cdots, H_n$. Therefore $\mathcal{H}_1 = \mathcal{H}$, otherwise $\mathbb{H}$ is reducible. Hence H_1 is pure on $\mathcal{H}$.

(vii) By the identity $[A,[B,C]] + [B,[C,A]] + [C,[A,B]] = 0$, we have that

$$[H_2, [H_1^*, H_1]] = [H_1, [H_1^*, H_2]]. \tag{33}$$

therefore

$$H_2[H_1^*, H_1]e_1 - [H_1^*, H_1]H_2 e_1 = H_1[H_1^*, H_2]e_1 - [H_1^*, H_2]H_1 e_1. \tag{34}$$

Using (34), (2) and (3), we obtain that

$$(H_2 - k_2 I)e_1 = (H_1 - k_1 I)e_2, \tag{35}$$

where

$$k_j = (H_j e_1, e_1)\|e_1\|^{-1}, \qquad j = 1, 2.$$

By (2), (3) and (35), it is easy to show that

$$(H_1 - k_1 I)[H_1^*, H_2 - k_2 I] = (H_2 - k_2 I)[H_1^*, H_1 - k_1 I]. \tag{36}$$

By the identity $[A, BC] = B[A, C] + [A, B]C$, it is easy to see that (36) is equivalent to

$$[H_1^*, (H_1 - k_1 I)](H_2 - k_2 I) = [H_1^*, (H_2 - k_2 I)](H_1 - k_1 I).$$

Therefore

$$((H_2 - k_2 I)x, e_1)e_1 = (H_1 - k_1 I)x, e_1)e_2. \tag{37}$$

If $(H_1^* - \overline{k}_1 I)e_1 = 0$, then from (37), we have

$$H_2^* e_1 = \overline{k}_2 e_1,$$

For simplicity, in this case we may assume that $k_1 = k_2$ and $\|e_1\| = 1$. Then

$$T(z, w) = ((\overline{w} I - H_1^*)^{-1} e_1, (\overline{z} I - H_1^*)^{-1} e_1) = z^{-1}\overline{w}^{-1}.$$

By the commutator property (e.g. cf. [12]), we have

$$S(z,w) = T(z,w)(1 - T(z,w))^{-1} = (z\overline{w} - 1)^{-1}.$$

Thus we may assume that $\mathcal{H}_1 = H^2(\mathbb{T})$ and that H_1 and H_2 are the operators in the example 1. Now, we have to prove theorem 1 for the case of the example 1. It is easy to see that

$$e_2(\varsigma) = (M(\varsigma) - M(0))\varsigma^{-1}.$$

Thus $H_2^* e_1 = 0$ implies that $P(\overline{M}) = 0$, i.e. $M(0) = 0$. Let $M_1(\varsigma) = M(\varsigma)\varsigma^{-1}$. Then (4) becomes

$$P(|M|^2 f) - MP(\overline{M}f) = c(f, M_1)M_1 \tag{38}$$

Let $f = 1$, in (38), then

$$P(|M_1|^2) = c\overline{M_1(0)}M_1.$$

However $P_-(|M_1|^2)(z) = \overline{P(|M_1|^2)(\frac{1}{\overline{z}})} - \overline{P(|M_1|^2)(0)}$. Therefore

$$P_-(|M_1|^2)(z) = cM_1(0)(\overline{M_1(\overline{z}^{-1})} - \overline{M_1(0)}).$$

Let $f(\varsigma) = (z - \varsigma)^{-1}, |z| > 1$ in (38). Then we get (31) with $F \equiv 0$, i.e.

$$\begin{aligned} &cM_1(0)(\overline{M_1(\overline{z}^{-1})} - \overline{M_1(0)}) + c\overline{M_1(0)}M_1(\varsigma) \\ &= M_1(\varsigma)\overline{M_1(\overline{z}^{-1})}((1-c)\varsigma z^{-1} + c) \end{aligned}$$

If $(c-1)M_1(\cdot) \not\equiv 0$, then

$$M_1(\varsigma) = bM_1(0)(\varsigma + b)^{-1}$$

where b is a constant satisfying

$$b = c(\overline{M_1(0)} - \overline{M_1(\overline{z}^{-1})})z((c-1)\overline{M_1(\overline{z}^{-1})})^{-1}$$

for $|z| > 1$. It is easy to see that $|b|^2 = c(c-1)^{-1} > 1$. Thus

$$H_2 = bM_1(0)(H_1 + bI)^{-1}.$$

(viii) If $(H_1^* - \overline{k}_1 I)e_1 \neq 0$ then (37) implies that there is a $\alpha \in \mathbb{C}$ such that

$$e_2 = \alpha e_1. \tag{39}$$

From (35) there is a number $\beta \in \mathbb{C}$ such that

$$H_2 e_1 = (\alpha H_1 + \beta I)e_1. \tag{40}$$

Now we have to prove that

$$H_2 = \alpha H_1 + \beta I. \tag{41}$$

From (2), (3) and (39), we get $[H_1^*, H_2] = [H_1^*, \alpha H_1 + \beta I]$. Therefore, by (40),

$$\begin{aligned}
&H_2(\overline{\lambda}I - H_1^*)^{-1}(\mu I - H_1)^{-1}e_1 \\
&= (\overline{\lambda}I - H_1^*)^{-1}(\mu I - H_1)^{-1}H_2 e_1 + (\overline{\lambda}I - H_1^*)^{-1}[H_2, H_1^*](\overline{\lambda}I - H_1^*)^{-1}(\mu I - H_1)^{-1}e_1 \\
&= (\overline{\lambda}I - H_1^*)^{-1}(\mu I - H_1)^{-1}(\alpha H_1 + \beta I)e_1 \\
&\quad + (\overline{\lambda}I - H_1^*)^{-1}[\alpha H_1 + \beta I, H_1^*](\overline{\lambda}I - H_1^*)^{-1}(\mu I - H_1)^{-1}e_1 \\
&= (\alpha H_1 + \beta I)(\overline{\lambda}I - H_1^*)^{-1}(\mu I - H_1)^{-1}e_1,
\end{aligned}$$

for $\lambda, \mu \in \rho(H_1)$, which proves (41). Similarly we may study the relation between H_1 and H_i, $i > 2$. Theorem 1 is proved.

Remark. Example 1 also shows that there is a commuting hyponormal irreducible pair (H_1, H_2) of operators satisfying the condition that $\operatorname{rank}[H_1^*, H_1] = 1$ and $\operatorname{rank}[H_2^*, H_2] > 1$. Besides, there is a commuting hyponormal irreducible pair (H_1, H_2) of operators such that the case (ii) of theorem 1 does occur with $\gamma_2 \neq 0$.

From the proof of Theorem 1, it is easy to see the following

COROLLARY 2. *Let $(H_1, \cdots, H_n)$ be an irreducible hyponormal n-tuple of commuting operators. If $[H_1^*, H_1] = 1$ and $[H_j^*, H_j]\mathcal{H} \subset [H_1^*, H_1]\mathcal{H}, j = 2, \cdots, n$, then there are complex numbers α_j, β_j such that $H_1 = \alpha_1 H_1 + \beta_j I, j = 2, 3, \cdots, n$.*

THEOREM 2. *Let $n > 1$ and $\mathbb{H} = (H_1, \cdots, H_n)$ be a commuting hyponormal irreducible n-tuple of operators satisfying the condition that* $\operatorname{rank}[H_1^*, H_1] = 1$. *Then either there is a subspace $\mathcal{H}_0$ which reduces H_1 such that $H_1|_{\mathcal{H}_0}$ is a linear combination of an uniletral shift with multiplicity one and the identity, or there are numbers α_k and β_k such that*

$$H_k = \alpha_k H_1 + \beta_k I, \qquad k = 2, 3, \cdots, n.$$

3. Let $\mathbb{H} = (H_1, \cdots, H_n)$ be a tuple of operators on $\mathcal{H}$. It is obvious that $\mathbb{H}$ is t-hyponormal if and only if

$$\sum_{i,j}([H_j^*, H_i]x_i, x_j) \geq 0, \tag{42}$$

for all $x_j \in \mathcal{H}$.

THEOREM 3. *Let* $\mathsf{H} = (H_1, \cdots, H_n)$ *be a commuting irreducible n-tuple of operators. If* H *is either cohyponormal or t-hyponormal, and* H *satisfies the condition that*

$$\operatorname{rank}[H_1^*, H_1] = 1,$$

then there are numbers α_k *and* β_k *such that*

$$H_k = \alpha_k H_1 + \beta_k I, \qquad k = 2, \cdots, n.$$

PROOF. We only consider the case that H is t-hyponormal. The proof of this theorem is similar to, different from but much simpler then that of theorem 1. It is obvious that there are $e_j \in \mathcal{H},\ e_j \neq 0$ such that (2) still holds good but (3) becomes

$$[H_2^*, H_1]x = (x, e_1)e_2. \tag{43}$$

Let $\mathcal{H}_1$ be the smallest subspace containing e_1 and reducing H_1. By the same method in the proof of Theorem 1 we may prove that $H_2 e_1 \in \mathcal{H}_1$ and then $\mathcal{H}_1$ is invariant with respect to H_2. According to the decomposition $\mathcal{H} = \mathcal{H}_1 \oplus \mathcal{H}_1^{\perp}$, we have

$$H_j = \begin{pmatrix} A_j & B_j \\ 0 & C_j \end{pmatrix}, \ j = 1, 2. \tag{44}$$

where $B_1 = 0$ and C_1 is normal. The relation $[H_1, H_2] = 0$ is equivalent to (5), (6) and

$$A_1 B_2 = B_2 C_1. \tag{45}$$

Let e_2^1 be the projection of e_2 on $\mathcal{H}_1$ and $e_2^2 = e_2 - e_2^1$. Then (43) is equivalent to (10),

$$[A_1^*, A_2]x = (x, e_2^1)e_1 \tag{46}$$

and

$$(A_1^* B_2 - B_2 C_1^*)x = (x, e_2^2)e_1. \tag{47}$$

It is obvious that (47) implies

$$(\overline{w}I - A_1^*)^{-1} B_2 x - B_2(\overline{w}I - C_1^*)^{-1} x = ((\overline{w}I - C_1^*)^{-1} x, e_2^2)(\overline{w}I - A_1^*)^{-1} e_1, \tag{48}$$

for $w \in \rho(H_1)$. From (45) and (48), we get

$$\begin{aligned}&[(\overline{w}I - A_1^*)^{-1}(zI - A_1)^{-1} B_2 - B_2(\overline{w}I - C_1^*)^{-1}(zI - C_1)^{-1}]x \\ &\quad = ((\overline{w}I - C_1^*)^{-1}(zI - C_1)^{-1} x, e_2^2)(\overline{w}I - A_1^*)^{-1} e_1\end{aligned} \tag{49}$$

and

$$[(zI - A_1)^{-1}(\overline{w}I - A_1^*)^{-1}B_2 - B_2(zI - C_1)^{-1}(\overline{w}I - C_1^*)^{-1}]x = ((\overline{w}I - C_1^*)^{-1}x, e_2^2)(zI - A_1)^{-1}(\overline{w}I - A_1^*)^{-1}e_1. \quad (50)$$

It is similar to (15) that

$$[(\overline{w}I - A_1^*)^{-1}, (zI - A_1)^{-1}]B_2 x = ((\overline{w}I - A_1^*)^{-1}(zI - A_1)^{-1}B_2 x, e_1)(zI - A_1)^{-1}(\overline{w}I - A_1^*)^{-1}e_1. \quad (51)$$

Subtracting (50) from (49) and using (51) and $[C_1^*, C_1] = 0$, we get

$$((wI - A_1^*)^{-1}(zI - A_1)^{-1}B_2 x, e_1)(zI - A_1)^{-1}(\overline{w}I - A_1^*)^{-1}e_1 = ((\overline{w}I - C_1^*)^{-1}(zI - C_1)^{-1}x, e_2^2)(\overline{w}I - A_1^*)^{-1}e_1 - ((\overline{w}I - C_1^*)^{-1}x, e_2^2)(zI - A_1)^{-1}(\overline{w}I - A_1^*)^{-1}e_1. \quad (52)$$

If $e_2^2 \neq 0$, then letting $x = (zI - C_1)(\overline{w}I - C_1^*)e_2^2$ for $z, w \in \rho(H_1)$ in (52) we see that $(\overline{w}I - A_1^*)^{-1}e_1$ is an eigenvector for A_1 corresponding to $w \in \rho(H_1)$. Let

$$\xi = (\overline{w}I - H_1^*)^{-1}e_1 = (\overline{w}I - A_1^*)^{-1}e_1$$

where $w \in \rho(H_1)$. Then $\xi \neq 0$, $\xi \in \mathcal{H}_1$ and

$$H_1\xi = A_1\xi = \alpha\xi$$

where α is the eigenvalue corresponding to ξ. Thus

$$H_1^*\xi = \overline{\alpha}\xi,$$

since H_1 is hyponormal. Therefore

$$0 = [H_1^*, H_1]\xi = (\xi, e_1)e_1,$$

i.e. $\xi \perp e_1$. Similarly, we have

$$(\xi, H_1^{*m}H_1^n e_1) = \alpha^m\overline{\alpha}^n(\xi, e_1) = 0.$$

i.e. $\xi \perp \mathcal{H}_1$. This is a contradiction. Therefore $e_2^2 = 0$. Then $\mathcal{H}_1$ reduces H_2 and similarly $H_3, \cdots, H_n$. Hence $\mathcal{H}_1 = \mathcal{H}$.

Now, we have to show that e_2 is a multiple of e_1. By (43), it is easy to calculate that

$$[H_1^*H_1, H_2]x = [H_1^*, H_2]H_1x = (H_1x, e_2)e_1 \quad (53)$$

and

$$[H_1H_1^*, H_2]x = H_1[H_1^*, H_2]x = (x, e_2)H_1e_1. \tag{54}$$

On the otherhand, by (2), we have

$$\begin{aligned}[[H_1^*, H_1], H_2]x &= [H_1^*, H_1]H_2x - H_2[H_1^*, H_1]x \\ &= (H_2x, e_1)e_1 - (x, e_1)H_2e_1. \end{aligned} \tag{55}$$

From (53), (54), and (55), we may conclude that

$$H_2e_1 = \alpha(x)H_1e_1 + \beta(x)e_1$$

where

$$\alpha(x) = (x, e_2)(x, e_1)^{-1}$$

and

$$\beta(x) = ((H_2x, e_1) - (H_1x, e_2))(x, e_1)^{-1}.$$

It is obvious that H_1e_1 and e_1 are linearly independent. Therefore $\alpha(x)$ and $\beta(x)$ must be independent of x. Thus there are numbers α and β such that $e_2 = \overline{\alpha}e_1$ and (40) is satisfied. Theorem is proved.

THEOREM 4. *Let $\mathbb{H} = (H_1, \cdots, H_n)$ be a commuting irreducible n-tuple of operators satisfying the condition that (i) there is no subspace $\mathcal{H}_0$ reducing H_1 such that $H_1|_{\mathcal{H}_0}$ is a linear combination of a unilateral shift with multiplicity one and the identity and (ii)*

$$\text{rank}[H_1^*, H_1] = 1.$$

Then $\mathbb{H}$ is hyponormal iff $\mathbb{H}$ is t-hyponormal.

4. In this section, we are studying the Taylor spectrum [13] of cohyponormal or t-hyponormal operators. First, let us review some basic technique (cf. [13], [6], [7]) in the theory of Taylor spectrum.

Let Λ be the exterior algebra of n generators $e_1, \cdots, e_n$ with identity e_0 over the complex field $\mathbb{C}$ equipped with an inner product such that

$$\{e_{i_1} \cdots e_{i_k} : 1 \le i_1 < \cdots < i_k \le n\} \cup \{e_0\}$$

is an othonormal basis. Let E_i, $i = 1, 2, \cdots, n$ be the creation operators

$$E_i\xi = e_i\xi, \qquad \xi \in \Lambda.$$

Then $E_iE_j + E_jE_i = 0$, and

$$E_i^*E_j + E_jE_i^* = \delta_{ij}I. \tag{56}$$

For $1 \leq k \leq n$, let $\mathbb{N}_k$ be the set of k-tuple of integers $\alpha = (i_1, \cdots, i_k)$ satisfying

$$1 \leq i_1 < \cdots < i_k \leq n,$$

and $\mathbb{N}_0 = \{0\}$. Let $e_\alpha = e_{i_1} \cdots e_{i_k}$ for $\alpha = (i_1, \cdots, i_k) \in \mathbb{N}_k$. Besides, let Λ_k be the subspace of Λ spanned by $\{e_\alpha : \alpha \in \mathbb{N}_k\}$.

Notice that for $\alpha \in \mathbb{N}_k, \ell, m = 1, 2, \cdots, n$, we have

$$(e_\ell e_\alpha, e_m e_\beta) = 0, 1 \text{ or } -1,$$

if $\begin{pmatrix} \ell & \alpha \\ m & \beta \end{pmatrix}$ is not a permutation, is an even permutation or an odd permutation respectively. If $\ell \neq m$ then

$$(E_\ell^* e_\alpha, E_m^* e_\beta) = -(e_m e_\alpha, e_\ell e_\beta)$$

by (56). If $\ell \in \alpha$, then $(E_\ell^* e_\alpha, E_\ell^* e_\beta) = 0$ or 1 for $\alpha \neq \beta$ or $\alpha = \beta$ respectively.

Let $\mathbb{H} = (H_1, \cdots, H_n)$ be a commuting n-tuple of operators on $\mathcal{H}$. Denote $\tilde{\mathcal{H}}_k = \mathcal{H} \otimes \Lambda_k$ and $\tilde{H} = \mathcal{H} \otimes \Lambda$. For $k = 0, 1, 2, \cdots, n-1$, let D_k be the operator in $\mathcal{L}(\tilde{\mathcal{H}}_k \longrightarrow \tilde{\mathcal{H}}_{k+1})$ defined by

$$D_k x \otimes \xi = \xi = \sum_{j=1}^{n} H_j x \otimes e_j \xi \qquad \text{for } x \otimes \xi \in \tilde{\mathcal{H}}_k.$$

Denote

$$\hat{\mathbb{H}} = \begin{pmatrix} D_0 & D_1^* & 0 & \cdots & 0 \\ 0 & D_2 & D_3^* & \cdots & 0 \\ 0 & 0 & D_4 & \cdots & 0 \\ \cdot & \cdot & \cdot & \cdots & \cdot \\ 0 & 0 & 0 & \cdots & D_{n-1}^* \end{pmatrix},$$

for even n and

$$\hat{\mathbb{H}} = \begin{pmatrix} D_0 & D_1^* & 0 & \cdots & 0 \\ 0 & D_2 & D_3^* & \cdots & 0 \\ 0 & 0 & D_4 & \cdots & 0 \\ \cdot & \cdot & \cdot & \cdots & \cdot \\ 0 & 0 & 0 & \cdots & D_{n-1} \end{pmatrix},$$

for odd n.

By Curto's Theorem [7], $0 = (0, \cdots, 0)$ is not in the Taylor's spectrum $\mathrm{sp}(\mathbb{H})$ iff $\hat{\mathbb{H}}$ is invertible.

THEOREM 5. *Let $\mathbb{H} = (H_1, H_2, \cdots, H_n)$ be a commuting t-hyponormal n-tuple of operators on $\mathcal{H}$. Then $(\lambda_1, \cdots, \lambda_n)$ is in the Taylor spectrum $sp(\mathbb{H})$, iff*

$$\sum_{j=1}^{n}(H_j - \lambda_j I)(H_j^* - \overline{\lambda}_j I) \tag{57}$$

is not invertible in $\mathcal{L}(\mathcal{H})$.

PROOF. (i) Without loss of generality, we may assume that $\lambda_j = 0, j = 1, 2, \cdots, n$. By a theorem of Curto [7], if $(0, \cdots, 0)$ is not in the $\mathrm{sp}(\mathbb{H})$, then

$$\sum_{j=1}^{n} H_j H_j^* \tag{58}$$

is invertible in $\mathcal{L}(\mathcal{H})$ which is equivalent to the condition that there is a positive number K such that

$$\sum_{j=1}^{n} \|H_j^* x\|^2 \geq K^2 \|x\|^2, \qquad \text{for } x \in \mathcal{H}. \tag{59}$$

Hence we only have to prove that

$$\|\hat{\mathbb{H}} x\| \geq K\|x\|, \qquad x \in \tilde{\mathcal{H}} \tag{60}$$

and

$$\|\hat{\mathbb{H}}^* x\| \geq K\|x\|, \qquad x \in \tilde{\mathcal{H}} \tag{61}$$

under the condition (59).

(ii) In order to give the estimate of $\|\hat{\mathbb{H}} x\|$ or $\|\hat{\mathbb{H}}^* x\|$ from below, we have to calculate

$$\begin{aligned} S_j &= \|D_j \sum_{\alpha \in \mathbb{N}_j} x_\alpha \otimes e_\alpha + D_{j+1}^* \sum_{\alpha \in \mathbb{N}_{j+2}} x_\alpha \otimes e_\alpha\|^2 \\ &= S_{j1} + S_{j2} + S_{j3}, \end{aligned}$$

for $j = -1, 1, \cdots, n-1$, where

$$S_{j1} = \|\sum_{m=1}^{n} \sum_{\alpha \in \mathbb{N}_j} H_m x_\alpha \otimes e_m e_\alpha\|^2,$$

$$S_{j2} = \|\sum_{m=1}^{n} \sum_{\alpha \in \mathbb{N}_{j+2}} H_m^* x_\alpha \otimes E_m^* e_\alpha\|^2,$$

and

$$S_{j3} = 2\mathcal{R} \sum_{\ell,m=1}^{n} \sum_{\alpha \in \mathbb{N}_j, \beta \in \mathbb{N}_{j+2}} (H_\ell x_\alpha \otimes e_\ell e_\alpha, H_m^* x_\beta \otimes E_m^* e_\beta).$$

Besides, $S_{(-1)1}, S_{(-1)3}, S_{(n-1)2}$ and $S_{(n-1)3}$ are zeros. It is easy to see that

$$\|\hat{\mathbb{H}}x\|^2 = \sum_{k=0}^{[(n-1)/2]} S_{2k}.$$

where $x = \sum_{k=0}^{[n/2]} \sum_{\alpha \in \mathbb{N}_{2k}} x_\alpha \otimes e_\alpha$. It is obvious that

$$S_{j3} = 2\mathcal{R} \sum (H_m H_\ell x_\alpha, x_\beta)(e_m e_\ell e_\alpha, e_\beta) = 0,$$

since $H_m H_\ell = H_\ell H_m$ and $e_m e_\ell = -e_\ell e_m$. On the otherhand,

$$\begin{aligned} S_{j1} &= \sum_{\ell \notin \alpha} \sum_{m \notin \beta} (H_\ell x_\alpha, H_m x_\beta)(e_\ell e_\alpha, e_m e_\beta) \\ &= S_j^{(1)} + S_j^{(2)}, \end{aligned}$$

where

$$S_j^{(1)} = \sum_{\alpha \in \mathbb{N}_j} \sum_{\ell \notin \alpha} \|H_\ell x_\alpha\|^2,$$

$S_0^{(2)} = 0$, and

$$S_j^{(2)} = \sum_{\alpha,\beta \in \mathbb{N}_j} \sum_{\ell \notin \alpha, m \notin \beta, \ell \neq m} (H_m^* H_\ell x_\alpha, x_\beta)(e_\ell e_\alpha, e_m e_\beta),$$

for $j \geq 1$. Similarly, we have

$$\begin{aligned} S_{j2} &= \sum_{\ell \in \alpha \in \mathbb{N}_{j+2}} \sum_{m \in \beta \in \mathbb{N}_{j+2}} (H_\ell^* x_\alpha, H_m^* x_\beta)(E_\ell^* e_\alpha, E_m^* e_\beta) \\ &= S_j^{(3)} + S_j^{(4)}, \end{aligned}$$

where

$$S_j^{(3)} = \sum_{\ell \in \alpha \in \mathbb{N}_{j+2}} \|H_\ell^* x_\alpha\|^2$$

and

$$S_j^{(4)} = - \sum_{\alpha,\beta \in \mathbb{N}_{j+2}} \sum_{\ell \neq m, \ell \notin \beta, m \notin \alpha} (H_m H_\ell^* x_\alpha, x_\beta)(e_m e_\alpha, e_\ell e_\beta).$$

Thus

$$S^{(2)}_{j+2} + S^{(4)}_{j} = \sum_{\alpha,\beta\in \mathsf{N}_{j+2}} \sum_{\ell\neq m, \ell\notin\alpha, m\notin\beta} ([H^*_m, H_\ell]x_\alpha, x_\beta)(e_\ell e_\alpha, e_m e_\beta)$$

and

$$S^{(1)}_{j+2} + S^{(3)}_{j} = \sum_{\ell=1}^{n} \sum_{\alpha\in\mathsf{N}_{j+2}} \|H^*_\ell x_\alpha\|^2 + \sum_{\ell\notin\alpha\in\mathsf{N}_{j+2}} ([H^*_\ell, H_\ell]x_\alpha, x_\alpha).$$

Hence

$$S^{(1)}_{j+2} + S^{(2)}_{j+2} + S^{(3)}_{j} + S^{(4)}_{j} = \sum_{\ell=1}^{n} \sum_{\alpha\in\mathsf{N}_{j+2}} \|H^*_\ell x_\alpha\|^2 + S^{(0)}_{j+2},$$

for $j = 0, 2, \cdots, 2([(n-1)/2] - 1)$, where

$$S^{(0)}_{j} = \sum_{\alpha,\beta\in\mathsf{N}_j} \sum_{\ell\notin\alpha, m\notin\beta} ([H^*_m, H_\ell]x_\alpha, x_\beta)(e_\ell e_\alpha, e_m e_\beta).$$

Therefore

$$\|\hat{\mathsf{H}}\|^2 \geq K^2\|x\|^2 + \sum_{k=1}^{[(n-1)/2]} S^{(0)}_{2k},$$

by (59), since

$$S_{01} = \sum_{m=1}^{n} \|H_m x_0\|^2 \geq \sum_{m=1}^{n} \|H^*_m x_0\|^2$$

and

$$S_{(n-2)2} = \sum_{m=1}^{n} \|H^*_m x_{12\cdots n}\|^2,$$

for even n.

(iii) In order to prove (60), we only have to prove that

$$S^{(0)}_{2k} \geq 0,\ k = 1, 2, \cdots, [(n-1)/2]. \tag{62}$$

Assume $1 \leq j \leq n-1$. Let $\gamma = (\ell_1, \cdots, \ell_{j+1}) \in \mathsf{N}_{j+1}$, and $\gamma_p = (\ell_1, \cdots, \ell_{p-1}, \ell_{p+1}, \cdots, \ell_{j+1}) \in \mathsf{N}_j$. Then

$$\sum_{i,k\in\gamma} ([H^*_{\ell_k}, H_{\ell_i}]x_{\gamma_i}, x_{\gamma_k})(e_{\ell_i} e_{\gamma_i}, e_{\ell_k} e_{\gamma_k})$$
$$= \sum_{i,k\in\gamma} ([H^*_{\ell_k}, H_{\ell_i}]\hat{x}_{\gamma_i}, \hat{x}_{\gamma_k}) \geq 0.$$

by (42), where $\hat{x}_{\gamma_i} = (e_\gamma, e_{\ell_i} e_{\gamma_i})$. Thus (62) is proved.

(iv) It is easy to see that

$$\|\hat{\mathsf{H}}^* x\|^2 = \sum_{k=0}^{[n/2]} S_{2k-1},$$

where $x = \sum_{k=0}^{[(n-1)/2]} \sum_{\alpha \in \mathsf{N}_{2k+1}} x_\alpha \otimes e_\alpha$. Using the same method in the step (ii) and (iii) we may prove

$$\|\hat{\mathsf{H}}^* x\|^2 = \sum_{k=0}^{[(n-1)/2]} \left(\sum_{\ell=1}^{n} \sum_{\alpha \in \mathsf{N}_{2k+1}} \|H_\ell^* x_\alpha\|^2 + S_{2k+1}^{(0)} \right)$$

where $S_n^{(0)} = 0$ and hence (61). Theorem is proved.

THEOREM 6. *Let $\mathsf{H} = (H_1, \cdots, H_n)$ be a commuting cohyponormal n-tuple of operators on $\mathcal{H}$. Then $(\lambda_1, \cdots, \lambda_n)$ is in the Taylor spectrum $sp(\mathsf{H})$, iff*

$$\sum_{j=1}^{n} (H_j^* - \overline{\lambda}_j I)(H_j - \lambda_j I)$$

is not invertible in $\mathcal{L}(\mathcal{H})$.

5. In this section, a trace formula for the subnormal n-tuple of operators is given. A commuting n-tuple of bounded operators $\mathsf{S} = (S_1, \cdots, S_n)$ on a separable Hilbert space $\mathcal{H}$ is subnormal if there is a commuting n-tuple of normal operators $\mathsf{N} = (N_1, \cdots, N_n)$ on a Hilbert space $\mathcal{K}$ containing $\mathcal{K}$ as a subspace satisfying $N_j \mathcal{H} \subset \mathcal{H}$ and

$$S_j = N_j \mid_{\mathcal{H}}, \qquad j = 1, 2, \cdots, n.$$

This N is said to be a normal extension of S. Assume S is a subnormal n-tuple of operators with normal extension N. Let $M = \bigvee_{i,j} [S_i^*, S_j]\mathcal{H}$, then M is an invariant subspace of S_j^*, $j = 1, 2, \cdots, n$. Let C_{kj} be the operators in $\mathcal{L}(M)$ defined by

$$C_{kj} x = [S_k^*, S_j] x, \quad x \in M.$$

There exist operators $\Lambda_j \in \mathcal{L}(M)$ such that

$$\Lambda_j^* x = S_j^* x, \qquad x \in M.$$

Let $E(\cdot)$ be the spectral measure of the normal extension $\mathbb{N}$ on the Taylor spectrum sp $(\mathbb{N})$. Define a positive semi-definite $\mathcal{L}(M)$-valued measure

$$e(\cdot) = P_M E(\cdot) \mid_M$$

on sp$(\mathbb{N})$, where P_M is the projection from $\mathcal{K}$ to M. In this section, we use the analytic model [17] for $\mathbb{S}$ and notions in [17] without further explanation. The trace ideal of the operator algebra $\mathcal{L}(\mathcal{H})$ is denoted by $\mathcal{L}^1(\mathcal{H})$.

LEMMA 2. *Let $\mathbb{S} = (S_1, \cdots, S_n)$ be a subnormal n-tuple of operators. If $[S_i^*, S_i] \in \mathcal{L}^1(\mathcal{H}), i = 1, 2, \cdots, n$ then $[\prod_{i=1}^n (\overline{w}_i I - S_i^*)^{-1}, \prod_{j=1}^n (z_j I - S_j)^{-1}] \in \mathcal{L}^1(\mathcal{H})$ and*

$$tr([\prod_{i=1}^n (\overline{w}_i I - S_i^*)^{-1}, \prod_{j=1}^n (z_j I - S_j)^{-1}] \prod_{k=1}^n (\overline{v}_k I - S_k^*)^{-1})$$
$$= \sum_{i=1}^n tr\left(\int \frac{e(du)(\overline{u}_i I - \Lambda_i^*)}{(\overline{w}_i - \overline{u}_i)\prod_{j=1}^n [(\overline{w}_j - \overline{u}_j)(\overline{v}_j - \overline{u}_j)(z_j - u_j)]}\right). \tag{63}$$

for $z_i, v_i, w_i \in \rho(S_i), i = 1, 2, \cdots, n$.

PROOF. For simplicity, let $A_i = \overline{w}_i I - S_i^*, B_i = z_i I - S_i$ and $C_i = \overline{v}_i I - S_i^*$. Then

$$[A_i, B_j] = C_{ij} P_M.$$

It is easy to calculate that

$$\left[A_i^{-1}, \prod_{j=1}^n B_j^{-1}\right] = \sum_{j=1}^n A_i^{-1} \left(\prod_{p=j}^n B_p^{-1}\right) C_{ij} P_M \left(\prod_{q=1}^j B_q^{-1}\right) A_i^{-1}.$$

Therefore

$$\left[\prod_{j=1}^n A_j^{-1}, \prod_{k=1}^n B_k^{-1}\right]$$
$$= \sum_{i,j=1}^n \prod_{u=1}^i A_u^{-1} \prod_{p=j}^n B_p^{-1} C_{ij} P_M \prod_{q=1}^j B_q^{-1} \prod_{t=i}^n A_i^{-1},$$

and it belongs to $\mathcal{L}^1(\mathcal{H})$. Thus

$$\operatorname{tr}\left(\left[\prod_{i=1}^{n} A_i^{-1}, \prod_{j=1}^{n} B_j^{-1}\right] \prod_{k=1}^{n} C_k^{-1}\right)$$
$$= \sum_{i,j=1}^{n} tr\left(\prod_{k=1}^{n} C_k^{-1} \prod_{\ell=1}^{n} A_i^{-1} \cdot A_i^{-1} \prod_{p=j}^{n} B_p^{-1} C_{ij} P_M \prod_{q=1}^{j} B_q^{-1}\right). \quad (64)$$

It is obvious that $P_M B_j^{-1} = (z_j I - \Lambda_j)^{-1} P_M$ and

$$P_M \prod_{k=1}^{n} C_k^{-1} \cdot \prod_{\ell=1}^{n} A_\ell^{-1} \cdot A_i^{-1} \cdot \prod_{p=j}^{n} B_p^{-1} a$$
$$= \int \frac{e(du)a}{(\overline{w}_i - \overline{u}_i)\prod_{k=1}^{n}[(\overline{w}_k - \overline{u}_k)(\overline{v}_k - \overline{u}_k)] \cdot \prod_{p=j}^{n}(z_p - u_p)}.$$

Therefore (64) equals

$$\sum_{j,j=1}^{n} \operatorname{tr}\left(\int \frac{e(du)C_{ij}\prod_{q=1}^{j}(z_q I - \Lambda_q)^{-1}}{(\overline{w}_i - \overline{u}_i)\prod_{k=1}^{n}[(\overline{w}_\ell - \overline{u}_\ell)(\overline{v}_\ell - \overline{u}_\ell)]\prod_{p=j}^{n}(z_p - u_p)}\right) \quad (65)$$

By (13) of [17], $e(du)C_{ij} = e(du)(\overline{u}_i I - \Lambda_i^*)(u_j I - \Lambda_j)$. Therefore (65) equals

$$\sum_{i=1}^{n} \operatorname{tr}\left(\int \frac{e(du)(\overline{u}_i I - \Lambda_i^*)}{(\overline{w}_i - \overline{u}_i)\prod_{\ell=1}^{n}[(\overline{w}_\ell - \overline{u}_\ell)(\overline{v}_\ell - \overline{u}_\ell)]}.\right.$$
$$\left.\left.\sum_{j=1}^{n}\left(\frac{\prod_{q=1}^{j-1}(z_q I - \Lambda_q)^{-1}}{\prod_{p=j}^{n}(z_j - u_j)} - \frac{\prod_{q=1}^{j}(z_q I - \Lambda_q)^{-1}}{\prod_{p=j+1}^{n}(z_p - u_p)}\right)\right)\right)$$

$$= \sum_{i=1}^{n} \operatorname{tr}\left(\int \frac{e(du)(\overline{u}_i I - \Lambda_i^*)}{(\overline{w}_i - \overline{u}_i)\prod_{j=1}^{n}[(\overline{w}_j - \overline{u}_j)(\overline{v}_j - \overline{u}_j)(z_j - u_j)]}\right.$$
$$\left. - \int \frac{e(du)(\overline{u}_i I - \Lambda_i^*)\prod_{q=1}^{n}(z_q I - \Lambda_q)^{-1}}{(\overline{w}_i - \overline{u}_i)\prod_{j=1}^{n}[(\overline{w}_j - \overline{u}_j)(\overline{v}_j - \overline{u}_j)]}\right)$$

which equals the right-hand side of (63), by [17]. Lemma 2 is proved.

LEMMA 3. *If* $[S_i^*, S_i]^{\frac{1}{2}} \in \mathcal{L}^1(\mathcal{H})$, *then for every bounded Baire function* $f(\cdot)$ *on* $sp(\mathsf{N})$,

$$\int f(u)e(du)(\overline{u}_i I - \Lambda_i^*) \in \mathcal{L}^1(M), \tag{66}$$

and there is a complex measure $\nu_i(\cdot)$ *with finite total variation such that*

$$\mathrm{tr}(\int f(u)e(du)(\overline{u}_i I - \Lambda_i^*)) = \int f(u)\nu_i(du). \tag{67}$$

This measure $\nu_i(du)$ is formally denoted by $\mathrm{tr}(e(du)(\overline{u}_i I - \Lambda_i^*))$. Similarly, we may define the complex measure $\mathrm{tr}((u_i I - \Lambda_i)e(du))$.

PROOF. Notice that

$$\begin{aligned} &\int |(e(du)(\overline{u}_i I - \Lambda_i^*)a, b)| \\ &\leq \left(\int \|\sqrt{e(du)}(\overline{u}_i I - \Lambda_i^*)a\|^2 \int \|\sqrt{e(du)}b\|^2\right)^{\frac{1}{2}} \\ &= \|b\| \; \|C_{ii}^{\frac{1}{2}} a\| \end{aligned} \tag{68}$$

for $a, b \in M$. From (68), it follows (66). Define a complex set function

$$\nu_i(F) = \mathrm{tr}\left(\int_F e(du)(\overline{u}_i I - \Lambda_i^*)\right), \qquad F \in \mathcal{B},$$

on $\mathrm{sp}(\mathsf{N})$, where $\mathcal{B}$ is the Borel field of all Borel sets in $\mathrm{sp}(\mathsf{N})$. Then it is easy to see that

$$\sum_{j=1}^{\infty} |\nu_i(F_j)| = \mathrm{tr}\left(\int f(u)e(du)(\overline{u}_i I - \Lambda_i^*)\right), \tag{69}$$

where $\{F_j\}$ is any sequence of multurally disjoint Borel sets and

$$f(u) = \sum_j 1_{F_j}(u)\mathrm{sign}\ (\nu_i(F_j)).$$

From (68) and (69), it is easy to see that

$$\sum_j |\nu_i(F_j)| \leq \mathrm{tr}(C_{ii}^{1/2}),$$

which proves the lemma.

It is unknown whether the condition $[S_i^*, S_i]^{1/2} \in \mathcal{L}^1(\mathcal{H})$ in Lemma 3 may be replaced by a weaker condition $[S_i^*, S_i] \in \mathcal{L}^1(\mathcal{H})$ or not.

Let $R(\mathbb{S})$ be the algebra of the functions of variables $(u_1, \cdots, u_n)$ generated by $(\overline{u}_i - \overline{z})^{-1}$ and $(u_i - z)^{-1}$, $z \in \rho(S_i), i = 1, 2, \cdots, n$. For $f = f(\overline{u}_1, \cdots, \overline{u}_n; u_1, \cdots u_n) \in \mathrm{R}(\mathbb{S})$, let

$$f(\mathbb{S}^*, \mathbb{S}) = f(S_1^*, \cdots, S_n^*; S_1,, \cdots, S_n).$$

It is obvious that, in general, this operator $f(\mathbb{S})$ depends on the ordering of product.

THEOREM 7. *Let $\mathbb{S}$ be a subnormal n-tuple of operators satisfying $[S_i^*, S_i]^{1/2} \in \mathcal{L}^1(\mathcal{H})$, $i = 1, \cdots, n$ and $\mathbb{N}$ be its normal extension. Then there are complex measures with finite total variations*

$$\nu_j(du) = itr((u_j I - \Lambda_j)e(du)), \quad j = 1, \cdots, n,$$

on $\mathrm{sp}(\mathbb{N})$ *such that for $f, h \in R(\mathbb{S})$, and any ordering of the product of $f(\mathbb{S}^*, \mathbb{S})$ and $h(\mathbb{S}^*, \mathbb{S})$,*

$$\mathrm{tr}(i[f(S^*, \mathbb{S}), h(\mathbb{S}^*, \mathbb{S})]) = \int_{\mathrm{sp}(\mathbb{N})} f(\overline{u}, u) d_\nu h(\overline{u}, u), \tag{70}$$

where

$$d_\nu h(\overline{u}, u) = \sum_{j=1}^{n} \left(\frac{\partial h(\overline{u}, u)}{\partial u_j} \nu_j(du) + \frac{\partial h(\overline{u}, u)}{\partial \overline{u}_j} \overline{\nu_j(du)} \right).$$

PROOF. From Lemma 3 and 4, it is easy to see that for $f(\overline{u}), h(u), q(\overline{u}) \in R(\mathbb{S})$. we have

$$\begin{aligned} &\mathrm{tr}(i[f(\mathbb{S}^*), h(\mathbb{S})]q(S^*)) \\ &= -\sum_{j=1}^{n} \int_{\mathrm{sp}(\mathbb{N})} q(\overline{u}) h(u) \frac{\partial f(\overline{u})}{\partial \overline{u}_j} \overline{\nu_j(du)} \end{aligned} \tag{71}$$

Take the conjugate of both sides of (71), it is easy to verify that for $r(u) \in R(\mathbb{S})$ we have

$$\mathrm{tr}(i[f(\mathbb{S}^*), h(\mathbb{S})]r(\mathbb{S})) = \sum_{j=1}^{n} \int_{\mathrm{sp}(\mathbb{N})} r(u) f(\overline{u}) \frac{\partial h(u)}{\partial u_j} \nu_j(du). \tag{72}$$

Therefore, for $f_1(\overline{u}), h_1(\overline{u}), f_2(u), h_2(u) \in R(S)$, we have

$$\begin{aligned}
&\operatorname{tr}([f_1(S^*)f_2(S), h_1(S^*)h_2(S)]) \\
&= \operatorname{tr}(h_1(S^*)[f_1(S^*), h_2(S)]f_2(S)) + \operatorname{tr}(f_1(S^*)[f_2(S), h_1(S^*)]h_2(S)) \\
&= \operatorname{tr}([h_1(S^*)f_1(S^*), h_2(S)]f_2(S) - [h_1(S^*), h_2(S)]f_1(S^*)f_2(S)) \\
&+ \operatorname{tr}(f_1(S^*)[f_2(S)h_2(S), h_1(S^*)] - f_1(S^*)f_2(S)[h_2(S), h_1(S^*)]) \\
&= \operatorname{tr}([f_1(S^*)h_1(S^*), h_2(S)]f_2(S)) + \operatorname{tr}([f_2(S)h_2(S), h_1(S^*)]f_1(S^*)). \qquad (73)
\end{aligned}$$

Since $\operatorname{tr}([h_1(S^*), h_2(S)]f_1(S^*)f_2(S)) + \operatorname{tr}(f_1(S^*)f_2(S)[h_2(S), h_1(S^*)]) = 0$. From (71), (72) and (73), it is easy to see that (70) holds good for $f(\overline{u}, u) = f_1(\overline{u})f_2(u)$ and $h(\overline{u}, u) = h_1(\overline{u})h_2(u)$. Then the general case may be proved by linear operation.

The author wishes to express his appreciation to Professor L. de Branges for his invitation to visit Purdue University 1988 and his hospitalities. The author is also grateful for the helpful discussions he had with Professor de Branges.

References

[1] Athavale, A., On joint hyponormality of operators. Proc. Amer. Math. Soc. **103**(1988), 417-423.

[2] Carey, R. W., Some remarks on principal currents and index theory for single and several commuting operators, Surveys of Some Recent Results in Operator Theory (edited by J.B. Conway & B. B. Morrel), Vol. I, 133-154, Longman Sci. Tech. N.Y. 1988.

[3] Carey, R. W. and Pincus, J.D., The concept of local index, to appear in Proc. of A.M.S. Summer Inst., New Hampshire; 1988.

[4] Carey, R. W. and Pincus, J.D., Principal currents, Integ. Equat. and Oper. Thy. **8**(1985), 615-640.

[5] Conway, J.B., Towards a functional calculus for subnormal tuples: The minimal normal extension and approximation in several complex variables, to appear in Proc. of A.M.S. Summer Inst., New Hampshire, 1988.

[6] Curto, R.E., Freedholm and invertible n-tuples of operators; The deformation problem, Trans. Amer. Math. Soc. **266**(1981) 129-159.

[7] Curto, R.E., Applications of several complex variables to multiparameter spectral theory, Surveys of Some Recent Results in Operator Theory (edited by J.B. Conway & B. B. Morrel) Vol. II, 25-90, 1988, Longman Sci. Tech. N.Y.

[8] Curto, R., Muhly, P., and Xia, J., Hyponormal pairs of commuting operators (preprint).

[9] Curto, R. E., Joint hyponormality: a bridge between hyponormality and subnormality: to appear in Proc. of A.M.S. Summer Inst., New Hampshire, 1988.

[10] Douglas, R.G., Paulsen, V. and Yan K., Operator theory and algebraic geometry, Bull. Amer. Math. Soc. **20**(1989),67-71.

[11] Morrel, B. B., A decomposition of some operators, Indiana Univ. Math. **23**(1973), 495-511.

[12] Pincus, J.D., Xia, D. and Xia, J., The analytic model of a hyponormal operator with rank one self-commutator, Integ. Equa. and Oper. Thy. **7**(1984), 516-534.

[13] Taylor, J.L., A joint spectrum for several commuting operators, J. Func. Anal. **6**(1970), 172-191.

[14] Xia, D. On the semi-hyponormal n-tuple of Operators, Integ. Equa. and Oper. Thy. **6**(1983), 879-898.

[15] Xia, D. The analytic model of a subnormal operator, Integ. Equa. and Oper. Thy. **10**(1987), 255-289.

[16] Xia, D. Analytic theory of subnormal operators, Integ. Equa. and Oper. Thy. **10**(1987), 880-90.

[17] Xia, D. Analytic theory of subnormal n-tuple of operators, to appear in Proc. of A.M.S. Summer Inst. New Hempshire, 1988.

Department of Mathematics
Vanderbilt University
Nashville, TN 37221
U.S.A.